COSMOLOGY

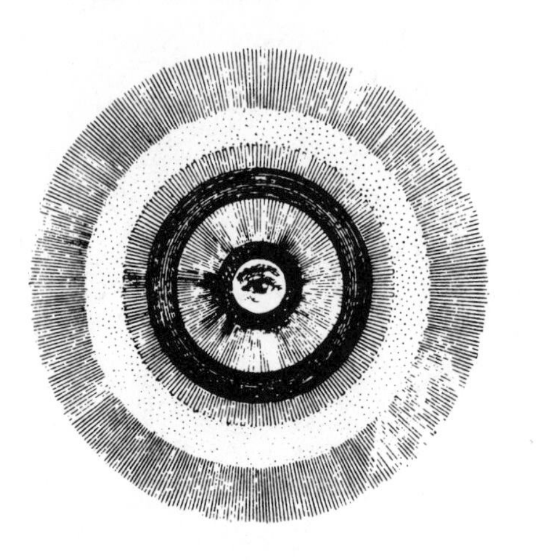

COSMOLOGY

THE SCIENCE OF THE UNIVERSE

EDWARD R. HARRISON
University of Massachusetts at Amherst

CAMBRIDGE UNIVERSITY PRESS
CAMBRIDGE
LONDON NEW YORK NEW ROCHELLE
MELBOURNE SYDNEY

Published by the Press Syndicate of the University of Cambridge
The Pitt Building, Trumpington Street, Cambridge CB2 1RP
32 East 57th Street, New York, NY 10022, USA
296 Beaconsfield Parade, Middle Park, Melbourne 3206, Australia

First published 1981

Printed in the United States of America
Typeset by Science Press, Ephrata, Pennsylvania
Printed and bound by The Book Press, Brattleboro, Vermont

Library of Congress Cataloging in Publication Data
Harrison, Edward Robert.
Cosmology, the science of the universe.
Includes bibliographies and index.
1. Cosmology. I. Title.
QB981.H32 523.1 80–18703
ISBN 0 521 22981 2 hard covers

CONTENTS

PREFACE

Cosmology, the science of the universe, attracts and fascinates most people. In one sense, it is the science of the large-scale structure of the universe: of the realm of extragalactic nebulae, of distant and receding horizons, of the interplay of cosmic forces, and of the dynamic curvature of universal space and time. In another sense, both ancient and modern, cosmology is the grand science that seeks forever to assemble all knowledge into a world picture. Most sciences tear things apart into smaller and smaller constituents, for the purpose of examining the world in progressively greater detail, whereas cosmology is the one science that is devoted to putting the pieces together into a "mighty frame." In yet another sense, cosmology is the history of mankind's quest for understanding of the universe, a quest that began long ago in the Age of Magic and the subsequent Age of Mythology. We cannot study cosmology in the broadest terms unless we pay heed to the pageantry of world pictures that have shaped the history of the human race. This enables us, more specifically, to trace the rise of the scientific method and the way it has played an increasing role in our understanding of the universe. The major theme of this book is the attainment of an elementary understanding of the physical universe of the twentieth century.

Let us be frank and admit that cosmology will rarely provide one with the coinage of the realm. Yet no persons living in the twentieth century can claim to be educated if they are unaware of the modern vision of the physical universe and the history of the magnificent concepts that it embodies. Cosmology forces us, willy-nilly, to examine our deepest and most cherished beliefs. It creates an awareness of ancient and vestigial paradigms, unconsciously accepted, that direct our individual lives and control the destiny of society. If you migrate to California, emigrate to a new land, join in a revolution, go to war, or do almost anything else, it is because you are held in the grip of ancient and modern cosmic beliefs. It is to cosmology that we must turn to learn about these cosmic beliefs that control our lives.

This book endeavors to be elementary and the needed background is minimal at the college level. A good high school education is a sufficient preparation for a large fraction of the material. My main aim has been to present cosmology in a way that will appeal to a wide audience, and at a level that is understandable to a college student who is not necessarily majoring in a natural science. Actually, cosmology is the "natural" science for students who are not majoring in a science. I have avoided mathematics as much as possible, and am inclined to believe that an understanding of the basic principles of science can be gained without the aid of mathematics. Nevertheless, the treatment has not been trivialized to the point where mathematics is avoided at all costs. In any event, the approach is sufficiently varied to meet the diverse needs of both those who enjoy and those who do not enjoy mathematical expressions and demonstrations.

There are several good texts on cosmology and general relativity that in most cases are too technical for the college student. Numerous, less specialized, texts now exist that generally are too brief and of insufficient scope for a one-semester course. These elementary texts and the many articles now available are fine for supplementary reading. At the end of each chapter, I have placed suggestions for further reading. These suggestions will enable the reader to explore alternative treatments, sometimes in greater depth and detail, of the subjects discussed in the chapter. Also at the end of each chapter is a list of sources, which contains references of a more extensive nature. The references selected are usually readable and not too technical; a few are somewhat more technical but are included for their historical interest.

All chapters have a final section entitled "Reflections." This section serves several purposes: It raises questions, presents topics for discussion and debate, and promotes thought and meditation. The fascination of cosmology is that it impels us to ask searching questions, to read widely and think deeply. It is not the sort of subject that lends itself readily to simple yes and no answers. Usually, on every issue, there are conflicting arguments to be investigated and weighed, to be rejected, accepted, or modified, according to the reader's personal tastes and beliefs. From the lecturer's viewpoint this is the drawback of cosmology as a pedagogic subject, for there are too few questions of the kind $2 + 2 = ?$ with straightforward answers. From the reader's or student's viewpoint this is the attraction of cosmology, for it challenges the mind as no other subject, shapes one's way of thinking about the world in general, and leaves impressions and ideas that last a lifetime. It is hopeless for one person – the teacher – to claim to be an authority on such a profound and universal subject. No such authority exists on Earth. The best way to teach cosmology is to let the students take control of the class as much as possible. The best way to learn about cosmology is to read, think, talk, and write.

This book has evolved from class notes used for teaching cosmology in the Five College Astronomy Department of Amherst College, Hampshire College, Mount Holyoke College, Smith College, and the University of Massachusetts. A preliminary draft has been used by Tom Dennis of Mount Holyoke College for first- and second-year science majors and by John Lathrop of Williams College for upper-division nonscience majors. I have used revised and extended notes at Amherst College for the past two years in a one-semester cosmology course. Those attending this course have ranged from first-year students to seniors; the former had some knowledge of astronomy, and most of the latter were majoring in a nonscience field. The ideal number of students in a cosmology course is about 20, but each year the number exceeded 30. Comments made by these students were of great help in preparing the present book.

The method I adopted was as follows. Each week I demanded a paper. This was not too arduous for the student skilled in writing, because the paper had to be a succinct statement, not longer than three handwritten pages. Each paper was graded according to its degree of analytical and imaginative thought. Even poetry was acceptable. Each paper was on any topic of the student's choice within an assigned subject, and usually I suggested a variety of topics, none of which, however, was mandatory. The depth displayed by these papers was a source of surprise, and frequently I found myself being educated by the students. We used various texts and articles, approaching 30 altogether, placed on reserve in the library. This was not thoroughly successful: What was needed, but not available, was a general text that covered sufficient material for such a course – a text of broad scope that might hold the attention of students of different backgrounds, and that in addition would provide the material needed for discussions and the preparation of papers. Because no such text existed, I have written this book.

I am indebted to many people for their comments and helpful criticisms. The comments by Tom R. Dennis and John D. Lathrop were extremely valuable. In addition, I am particularly grateful to Thomas T. Arny, Leroy F. Cook, Norman C. Ford, Robert V. Krotkov, Nicholas Z. Scoville, Eugene Tademaru, and David Van Blerkom of the University of Massachusetts for reading parts of the text and making critical comments. I am also indebted to Wayne A. Christianson of the University of North Carolina, Andrew G. Fraknoi of the Astronomical Society of the Pacific, and Kenneth R. Lang of Tufts University for their suggestions and remarks. I also thank Joan C. Centrella for the use of material from her honors thesis in Chapter 19. I gratefully acknowledge the advice and help of Kyle Wallace of Cambridge University Press, New York, and Janis R. Bolster for their advice and help, and their careful editing and production of this book. My thanks also go to Teresa Gryzybowski and Nellie C. Bristol for typing the manuscript.

I am exceedingly grateful to June Z. Harrison for her continual help and to J. Peter Harrison for his occasional observations. Most of all, I acknowledge a great debt to my wife Photeni for her unflagging encouragement and help. To her, I dedicate this book.

Edward R. Harrison

Amherst, Mass.
November 1980

INTRODUCTION

With equal passion I have sought knowledge. I have wished to understand the hearts of men, I have wished to know why the stars shine. And I have tried to apprehend the Pythagorean power by which number holds sway above the flux. A little of this, but not much, I have achieved.
— Bertrand Russell (1872–1970), *Autobiography,* Prologue

A brief summary of the contents of this book serves as an introduction to the scope of cosmology. In outline only, chapter by chapter, the subjects covered are as follows.

WHAT IS COSMOLOGY? (CHAPTER 1)

The history of cosmology shows us that in every age devout people believe that they have at last discovered the true nature of the Universe, whereas in each case they have devised a world picture – merely a universe – that is like a mask fitted on the face of the still unknown Universe. In this book we use the word *universe* to denote a "model of the Universe" and avoid making pretentious claims to a true knowledge of the Universe.

Cosmology is probably as old as mankind and is the art of creating world pictures. We discuss how world pictures, or universes, have evolved with time. In the Age of Magic, tens of thousands of years ago, the world was explained in terms of the activity of vibrant and ambient spirit folk. In the Age of Mythology, beginning ten or more thousand years ago, human beings were of central importance in an enlarged universe governed by remote gods and goddesses. Now, in the modern Scientific Age, having abandoned the anthropomorphic and anthropocentric viewpoints, we have devised a physical universe. The old universes had many things to offer that no longer fit naturally into the physical world picture. This is a cause of general concern and prompts each of us to think deeply. The last section considers how cosmology affects society and governs our everyday actions and outlook.

STARS AND GALAXIES (CHAPTERS 2 AND 3)

These chapters discuss briefly stars and galaxies and touch on miscellaneous topics of general interest. The treatment is oriented toward cosmology and can be supplemented by reference to one or more of the many introductory astronomy texts now available. Many readers may wish to skip these chapters and proceed directly to the following two chapters, which explore the important subjects of location and containment.

LOCATION AND THE COSMIC CENTER (CHAPTER 4)

The principles of location and containment, which at first sight seem deceptively simple, serve to guide us among the pitfalls that have trapped earlier cosmologists, and still trap students. Generally speaking, location is concerned with the cosmic center and containment is concerned with the cosmic edge, and both subjects bring us into contact with major developments in the history of science.

Location deals with the geocentric and heliocentric universes, and with the rise of the twentieth-century isotropic universe in which all directions are observed to be alike. The observations that reveal an isotropic universe, when coupled with the location principle, lead us to the conclusion that the universe is probably homogeneous. Homogeneity of the universe, meaning that on the average all places are alike in space, is the essence of the cosmological principle. The perfect cosmological principle – all places are alike in space and time – encompasses the original Newtonian universe and the expanding steady state universe.

CONTAINMENT AND THE COSMIC EDGE (CHAPTER 5)

Containment is the subject that deals with the contents of the physical universe. It stresses the important fact that space and time are physically real and are therefore contained within the physical universe. This fact has surprising consequences. The containment principle emphasizes that we are dealing with the physical universe, not the Universe, and this has implications for the social and life sciences. We encounter various topics, such as cosmogony (origin of the components of the universe), cosmogenesis (origin of the universe itself), and the anthropic principle. Because the fundamental constants of nature are "finely tuned," living creatures are able to exist in the universe, and hence the universe is self-aware. The anthropic principle asserts that the universe is the way it is because we are here, whereas the theistic principle asserts that we are here in a finely tuned universe because it was designed by a Creator.

A word of warning comes not amiss while we are on the subject of containment. Cosmology is incomplete in the fundamental sense that we do not know how to put ourselves, as cosmologists, into our world pictures. We can place our bodies and bioelectronic brains in the physical universe, but we cannot insert our minds (whatever that means) into the universe that is conceived by our minds. When we try, we encounter the absurdity of an infinite regression: A cosmologist conceives a universe, which contains the cosmologist conceiving the universe, and so on, indefinitely. Painters, for the same reason, always leave themselves out of the pictures they are painting. This problem is commented on briefly and is referred to as the *containment riddle.*

SPACE AND TIME (CHAPTER 6)

Space and time are considered in the pre-Newtonian and Newtonian universes and in the spacetime continuum of special relativity. Special relativity, contrary to what most students have been taught to think, is easy to understand, and a grasp of the essential ideas does not require tedious algebra. Space travel close to the speed of light provides interesting applications. The "twin paradox" is puzzling only when the most elementary aspects of relativity theory are not understood.

In dealing with the nature of time, we attempt to show that our understanding of time is a hodgepodge of primitive and sophisticated ideas. The time that is used in special relativity is not quite the same as that used in most other sciences, which is not the same as that referred to in everyday speech, which in turn is not quite the same as the time that we experience. Conflict and contrast abound whenever we try to discuss the nature of time. Without hope of significant success we try to clarify the issues involved in this perplexing subject.

CURVED SPACE (CHAPTER 7)

The rise of non-Euclidean geometry in the nineteenth century presents us with one of the most fascinating topics in the history of science and mathematics. Most people are puzzled by curved space, even those who live in curved spaces. Much of our attention is focused on the three homogeneous and isotropic spaces that are of basic importance in modern cosmology.

GENERAL RELATIVITY (CHAPTER 8)

Special relativity and curved space lead us to the rise of general relativity and the labors of Einstein. The first stepping-stone is the principle of equivalence. This is established by means of experiments in imaginary laboratories that move freely in space near to and far from stars. The second stepping-stone is the realization that this dynamic state of affairs is analogous in many ways to the properties of curved space. In a flight of inspiration we are catapulted to general relativity and the Einstein master equation. Various ingenious tests of general relativity are still proceeding and great progress has been made.

We consider the origin and nature of the strange ideas embodied in *Mach's principle,* so named by Einstein. The bootstrap theory, periodically revived, asserts that all things are immanent within each other, and the nature of any one thing is determined by the universe as a whole. So far, science has failed to make sense of the bootstrap theory. Mach's principle is actually a rudimentary bootstrap idea and may be on the wrong track. It claims that the inertia of a body is determined by the rest of the matter distributed in the universe. To this day many people still dislike the idea of an "unclothed" space that exists in its own right; and following the ancients, Bishop Berkeley, and Ernst Mach, they hold that space cannot exist in a physically real sense unless it is decently "clothed" with a distribution of matter. Berkeley's ideas, revamped by Mach, played a historical role in the formulation of general relativity. But this idea – the materialization of space – championed at first by Einstein, was later dropped when Einstein attempted the converse – the geometrization of matter.

BLACK HOLES (CHAPTER 9)

Although black holes were anticipated long ago on the basis of Newtonian theory, the proper theory for their study is general relativity. Of all bodies of the same mass, a black hole has the strongest gravitational force at its surface. In the general relativity picture this means that it is wrapped in its own highly curved spacetime. A black hole is in a state of free-fall collapse; but owing to the extreme distortion of spacetime, an external observer sees it in a frozen, motionless state from which nothing, not even light, can escape. We consider various topics of interest, such as nonrotating and rotating black holes, the energy liberated by the accretion of matter, miniholes and superholes, the temperature of black holes, and the breakdown of certain cherished laws of conservation.

EXPANSION (CHAPTER 10)

The expansion of the physical universe is one of the greatest discoveries of this century. We study the expansion of the universe by performing several imaginary experiments with ERSU – the "Expanding Rubber Sheet Universe." To aid us in our investigations we use the two kinds of observer introduced in Chapter 4: first, an ordinary stay-at-home "observer" who looks out at distant things in much the same way as we do; second, a cosmic gadabout "explorer" who can rush around the universe in no time at all. The explorer in this case is really us looking down on ERSU. Our experiments shed light on many topics, such as homogeneous expansion, cosmic time, the recession of the galaxies, and the velocity–distance expansion law (known also as the Hubble law). The experiments stress that the galaxies are not hurling away through space but are actually at rest in space that is expanding.

This is why the very distant galaxies can recede from us faster than the speed of light. The recession velocity is therefore quite unlike the ordinary sort of velocity with which we are familiar. We introduce comoving coordinates, coordinate distances, the scaling factor, the Hubble term, and the deceleration term, and we show how universes can be classified by the way the scaling factor changes with time.

REDSHIFTS (CHAPTER 11)

Lightrays received from distant galaxies are redshifted because of the expansion of the universe. This cosmic redshift, as distinct from the Doppler and gravitational redshifts, is produced by the stretching of wavelengths as radiation propagates through expanding space. Space expands; therefore wavelengths are stretched; and the cosmic redshift is as simple as that. There are a few redshift curiosities: Some galaxies, it is claimed, have discordant redshifts that are not of cosmic origin, and some quasars have several different redshifts in the absorption lines of their spectra. The oddest curiosity of all, however, is the insistence in popular literature that cosmic redshifts are of Doppler origin. The Doppler explanation assumes that the galaxies are rushing away through space and that special relativity can be used to explain the universe. This is a mischievous view and leads to endless confusion for those who try to understand modern cosmology. Furthermore, it restores the anachronism of a cosmic edge where, in this case, the recession equals the speed of light. Our treatment emphasizes two concepts: The recession is the result of the expansion of space (and the galaxies are stationary in expanding space), and the cosmic redshift is the result of the stretching of wavelengths as radiation moves through expanding space. It now becomes apparent why we have previously insisted that space and time are physically real (this is the essence of general relativity) and are contained within the universe; the universe is not expanding within space but contains expanding space.

DARKNESS AT NIGHT (CHAPTER 12)

The dark night sky paradox, often referred to as *Olbers' paradox,* was discovered nearly four hundred years ago by Kepler. Why the sky at night is dark, and not ablaze with light from countless stars, has perplexed many scientists and has played a prominent role in the history of cosmology. Only in recent years have we begun to understand why the universe, with all its stars and galaxies brightly shining, is an abode of darkness. In the last two decades it has been said many times that the night sky is dark because of the expansion of the universe. This is not true, because even if our universe were static, the sky at night would still be dark. The correct answer is more simple and was anticipated over a hundred years ago by Edgar Allan Poe. The sky at night is dark because the stars are widely separated from each other and they shine for too short a time to fill space with their bright radiation.

THE UNIVERSE IN A NUTSHELL (CHAPTER 13)

The notion that all places in the universe are alike has far-reaching consequences. Distant regions are in the same state as the local region at the same time. We can therefore discover much about the universe by studying only a sample region. This is the idea of "the universe in a nutshell." We imagine that a part of the universe is enclosed within a cosmic box that has perfectly reflecting walls and expands with the universe. What happens inside the cosmic box is exactly the same as what happens elsewhere in the universe. The cosmic box is relatively small and expands very slowly. Hence, as in the laboratory, we are able to use Euclidean space and the ordinary laws of physics to study the behavior of various kinds of things. The cosmic box is a useful tool that helps us to tackle problems that would otherwise be quite difficult. We then become ambitious and tackle such subjects as the entropy of the universe and the problem of where all the energy goes.

NEWTONIAN COSMOLOGY (CHAPTER 14)

The gravity paradox, or war of cosmic forces, was resolved by Newton, who assumed that the universe is homogeneous. Nowadays we realize that gravity cannot travel with infinite speed and therefore the resolution of the so-called gravity paradox is much the same as the resolution of the dark night sky paradox. The dynamics of the universe, or how gravity and the lambda force determine expansion, are discussed with the aid of simple Newtonian ideas. Newtonian theory has the advantage that it is easy to understand, and we try to explain why in cosmology, under certain circumstances, it yields the same results as general relativity.

THE MANY UNIVERSES (CHAPTER 15)

We now take a look at various "mighty frames," such as the Einstein, de Sitter, Friedmann, and Friedmann–Lemaître universes. They can be classified according to whether they are static, bang, whimper, or oscillating. Other schemes of classification are also presented. In this great gallery of universes the lambda force, invented by Einstein, is important. This chapter has some mathematics, but the student who freezes at the sight of a mathematical symbol can skip the mathematics and still get a great deal of information.

THEORIES OF THE UNIVERSE (CHAPTER 16)

There are many scientific theories of the universe, a number of which are quite interesting. From the cosmic bestiary we select universes in compression, tension, and convulsion. Then we introduce the "dream machine" of the scalar–tensor theory. By adjusting the control knobs on the dream machine, the cosmologist converts a physical universe into any one of an infinite number of physically different universes. Many people, slightly out of date in their knowledge of cosmology, are under the impression that life offers only a choice between the big bang and the steady state universes. They do not realize how varied are the options now available in modern supermarket cosmology. The steady state universe (now out of the running) is actually only one of a wide array of universes manufactured by the dream machine. Every man, woman, and child can have his or her own custom-built universe. In this welter of universes we grope our way, guided by observations and the anthropic principle, and for esthetic reasons we naturally favor the simplest and most pleasing.

THE COSMIC NUMBERS (CHAPTER 17)

This chapter deals with the intriguing cosmic numbers that measure the small- and large-scale properties of the universe. These natural numbers exhibit striking coincidences that so far have not been entirely explained. We discuss them with the aid of the *cluster hypothesis* and Dirac's *large-number hypothesis,* and look at them from the point of view of the anthropic principle. This subject began long ago when Archimedes calculated the number of grains of sand the universe could contain.

THE EARLY UNIVERSE (CHAPTER 18)

The discovery in 1965 of the cosmic radiation is now regarded as strong evidence in support of a big bang universe. Here we take a look at the early universe, or the big bang – not journeying in space to the big bang but remaining where we are and traveling back in time billions of years. If the universe is infinite in size, then so also is the big bang. As we journey back in time the density and temperature of the universe rise to higher and higher values. When the universe is 1 million years old, it is flooded with brilliant light: We stand at the threshold of the radiation era from which descends directly the cosmic radiation, cooled by expansion,

Powers of ten

1,000,000,000,000 =	10^{12}	= trillion
1,000,000,000 =	10^{9}	= billion
1,000,000 =	10^{6}	= million
1000 =	10^{3}	= thousand
100 =	10^{2}	= hundred
10 =	10^{1}	= ten
1 =	10^{0}	= one
1/10 =	10^{-1}	= tenth
1/100 =	10^{-2}	= hundredth
1/1000 =	10^{-3}	= thousandth
1/1,000,000 =	10^{-6}	= millionth
1/1,000,000,000 =	10^{-9}	= billionth
1/1,000,000,000,000 =	10^{-12}	= trillionth

Note: In the European system, 10^{9} = milliard, 10^{12} = billion, 10^{18} = trillion.

that nowadays is observed. We push on into the big bang, and when the universe is 1 second old, and the temperature is 10 billion degrees and the density is 100,000 times that of water, we quit the radiation era and enter the bizarre world of the lepton era. Hordes of electrons, muons, and their antiparticles struggle to survive, and from the battlefields of the lepton era flee ghostly neutrinos condemned forever to wander unseen through the universe. It is a tribute to modern physics that we can trace the history of the universe back to an age of 1/10,000 of a second, to the doorway of the mysterious hadron era. Undaunted, using only wild speculations, we continue our journey in a world of warring matter and antimatter, and enter the quark era. Ultimately, our thrust back to the frontier of time grinds to a halt before an impenetrable barrier, where the universe is 1 billion-trillionth of a jiffy old – a jiffy is a billion-trillionth of a second – and the density is 1 followed by 94 zeros times the density of water. At the Planck barrier lies chaos, where quantum fluctuations of space and time are of cosmic magnitude, and space and time are inextricably entangled together in a turbulent foam. Here lie, perhaps, the guarded secrets that foretell the design and architecture of the universe, to be discovered not in our age but by theoreticians of the future.

We return from the early universe, and with our time machine pointing to the future we journey to the end of time. There are two possible scenarios: Either the universe eventually collapses and terminates in a big bang, or it continues to expand forever and dies with a long drawn-out whimper. In the first scenario the galaxies are eventually crushed together. In a holocaust that rises to a crescendo, where dissolving stars zip about at speeds close to that of light, the brilliance of the radiation era returns and the universe reverts to ultimate chaos. In the second scenario the universe continues to expand and the galaxies become dead and lifeless after hundreds of billions of years. In the vast stretches of time that lie ahead, star systems and galaxies dissolve or are engulfed by black holes, and particles slowly decay and become radiation. After unimaginable eons of time all black holes evaporate away into radiation. The universe then contains no matter, only feeble radiation that forever grows feebler.

HORIZONS IN THE UNIVERSE (CHAPTER 19)

How far can we see in the universe? The answer depends on what we are looking at: either events or worldlines. In this chapter we discuss the event and particle (or worldline) horizons, beginning with the static Newtonian universe to illustrate their nature. The horizon riddle is briefly discussed and we then turn to such matters as the observable universe and the Hubble sphere. The event and particle horizons in expanding universes are then discussed in detail.

LIFE IN THE UNIVERSE (CHAPTER 20)

In this last chapter we consider past and present theories of the origin of life and discuss evolution and natural selection. The nature of intelligence is not an easy subject, although of vital importance in cosmology, and among the topics mentioned we consider how humans might have acquired their large brains. As cosmologists, in our finely tuned universe, we feel impelled to believe that life must exist elsewhere out there in the multitudes of galaxies. But is there intelligent life, technologically advanced, elsewhere in our Galaxy? The pursuit of the answer to this question leads to rather startling conclusions that are of great importance to the future of mankind. Natural selection leads to *galactic selection* in the case of technological civilizations, and the implications of a biogalactic law such as galactic selection are examined.

BIBLIOGRAPHY

Elementary books on cosmology divide into two groups: those of 1965 and earlier, prior to the discovery of the cosmic background radiation, and those written subsequently.

SELECTED BOOKS OF 1965 AND EARLIER

Bondi, H. *Cosmology*. Cambridge University Press, Cambridge, 1960. A slightly technical but excellent discussion.

Bondi, H., Bonnor, W. B., Lyttleton, R. A., and Whitrow, G. J. *Rival Theories of Cosmology: A Symposium and Discussion of Modern Theories of the Universe*. Oxford University Press, London, 1960.

Bonnor, W. B. *The Mystery of the Expanding Universe*. Eyre and Spottiswoode, London, 1964.

Boschke, F. L. *Creation Still Goes On*. McGraw-Hill, New York, 1964.

Coleman, J. A. *Modern Theories of the Universe*. Signet Science Library Books, New York, 1963. A well-written book that is easy to understand.

Couderc, P. *The Wider Universe*. Arrow Books, New York, 1960.

Eddington, A. S. *The Expanding Universe*. Cambridge University Press, Cambridge, 1933. Reprint. University of Michigan Press, Ann Arbor Paperback, Ann Arbor, 1958. A classic book of masterly exposition.

Gamow, G. *The Creation of the Universe*. Viking Press, New York, 1952. The first book to discuss the possibility of a hot big bang.

Gardner, M. *The Ambidextrous Universe*. Basic Books, New York, 1964.

Hoyle, F. *The Nature of the Universe*. Blackwell, Oxford, 1950. Rev. ed. Penguin Books, London, 1960. The term *big bang* is first used in this book.

Hubble, E. *The Realm of the Nebulae*. Yale University Press, New Haven, Conn., 1936. Reprint. Dover Publications, New York, 1958.

Hubble, E. *The Observational Approach to Cosmology*. Oxford University Press, Clarendon Press, Oxford, 1937.

Jeans, J. *The Mysterious Universe*. Cambridge University Press, Cambridge, 1930. Reprint. Macmillan, London, 1937.

Lemaître, G. *The Primeval Atom*. Van Nostrand, New York, 1951.

McVittie, G. C. *Fact and Theory in Cosmology*. Eyre and Spottiswoode, London, 1961.

Messel, H., and Butler, S. T., eds. *The Universe and Its Origin*. St. Martin's Press, New York, 1964.

Sciama, D. W. *The Unity of the Universe*. Faber and Faber, London, 1959.

Singh, J. *Great Ideas and Theories of Modern Cosmology*. Constable, London, 1961. Reprint. Dover Publications, New York, 1970.

Sitter, W. de. *Kosmos*. Harvard University Press, Cambridge, Mass., 1932.

The Universe. A Scientific American Book, Simon and Schuster, New York, 1957. A collection of *Scientific American* articles by outstanding authors.

Whitrow, G. J. *The Structure and Evolution of the Universe: An Introduction to Cosmology*. Hutchinson, London, 1959.

Whittaker, E. *From Euclid to Eddington: A Study of Conceptions of the External World*. Cambridge University Press, Cambridge, 1948. Reprint. Dover Publications, New York, 1958.

ELEMENTARY BOOKS PUBLISHED AFTER 1965

Alfvén, H. *Worlds–Antiworlds: Antimatter in Cosmology.* W. H. Freeman, San Francisco, 1966.

Asimov, I. *The Collapsing Universe.* Walker, New York, 1977.

Bergmann, P. G. *The Riddle of Gravitation: From Newton to Einstein to Today's Exciting Theories.* Charles Scribner's Sons, New York, 1968.

Calder, N. *Violent Universe: An Eyewitness Account of the New Astronomy.* Viking Press, New York, 1969.

Charon, J. *Cosmology: Theories of the Universe.* McGraw-Hill, New York, 1969.

Davies, P. C. W. *Space and Time in the Modern Universe.* Cambridge University Press, Cambridge, 1977.

Davies, P. C. W. *The Runaway Universe.* J. M. Dent, London, 1978.

Dickson, F. P. *The Bowl of Night: The Physical Universe and Scientific Thought.* M.I.T. Press, Cambridge, Mass., 1968.

Ferris, T. *The Red Limit.* William Morrow, New York, 1977.

Gardner, M. *The Relativity Explosion.* Random House, New York, 1976.

Geroch, R. *General Relativity from A to B.* University of Chicago Press, Chicago, 1978.

Gingerich, O., ed. *Cosmology + 1.* W. H. Freeman, San Francisco, 1977.

Hoyle, F. *Ten Faces of the Universe.* W. H. Freeman, San Francisco, 1977.

Kaufmann, W. J. *Relativity and Cosmology.* Harper and Row, New York, 1977.

Kilmister, C. W. *The Nature of the Universe.* Thames and Hudson, London, 1971.

Laurie, J., ed. *Cosmology Now.* British Broadcasting Corp., London, 1973. Series of radio talks by eminent cosmologists.

Merleau-Ponty, J., and Morando, B. *The Rebirth of Cosmology.* Alfred A. Knopf, New York, 1976.

Motz, L. *The Universe: Its Beginning and End.* Charles Scribner's Sons, New York, 1975.

Saslaw, W., and Jacobs, K., eds. *The Emerging Universe: Essays on Contemporary Astronomy.* University Press of Virginia, Charlottesville, 1972.

Sciama, D. W. *Modern Cosmology.* Cambridge University Press, Cambridge, 1971.

Shatzman, E. *The Structure of the Universe.* McGraw-Hill, New York, 1968.

Shipman, H. *Black Holes, Quasars and the Universe.* Houghton Mifflin, Boston, 1976.

Silk, J. *The Big Bang.* W. H. Freeman, San Francisco, 1980.

Weinberg, S. *The First Three Minutes: A Modern View of the Origin of the Universe.* Basic Books, New York, 1977.

BOOKS OF GENERAL INTEREST

Blacker, C., and Loewe, M., eds. *Ancient Cosmologies.* George Allen and Unwin, London, 1975.

Eddington, A. S. *Space, Time, and Gravitation.* Cambridge University Press, Cambridge, 1920. Reprint. Harper Torchbooks, New York, 1959. An outline of the theory of general relativity that is still informative.

Faculty of Aligari Museum University. *The Changing Concept of the Universe: Extracts from Lucretius to Hoyle.* Asia Publishing House, London, 1963.

Haber, F. C. *The Age of the World, Moses to Darwin.* Johns Hopkins Press, Baltimore, 1959.

Koyré, A. *From the Closed World to the Infinite Universe.* Johns Hopkins Press, Baltimore, 1957. Reprint. Harper Torchbooks, New York, 1958.

Lanczos, C. *Albert Einstein and the Cosmic World Order.* Interscience Publishers, New York, 1965.

Lewis, C. S. *The Discarded Image.* Cambridge University Press, Cambridge, 1967. Lewis laments the loss of the medieval universe that gave meaning to mankind's place in the universe.

Lovejoy, A. O. *The Great Chain of Being: A Study of the History of an Idea.* Harvard University Press, Cambridge, Mass., 1936.

Munitz, M. K., ed. *Theories of the Universe: From Babylonian Myth to Modern Science.* Free Press, Glencoe, Ill., 1957.

North, J. D. *The Measure of the Universe: A History of Modern Cosmology.* Oxford University Press, Clarendon Press, Oxford, 1965.

Shapley, H. *Of Stars and Men: the Human Response to an Expanding Universe.* Beacon Press, Boston, 1958.

Shklovskii, I. S., and Sagan, C. *Intelligent Life in the Universe.* Holden-Day, New York, 1966.

Whitrow, G. J. *What Is Time?* Thames and Hudson, London, 1972.

MORE TECHNICAL
TREATMENTS OF COSMOLOGY

Berry, M. *Principles of Cosmology and Gravitation.* Cambridge University Press, Cambridge, 1976. A brief treatment at the undergraduate level.

Hawking, S. W., and Ellis, G. F. R. *The Large Scale Structure of Space-Time.* Cambridge University Press, Cambridge, 1973. An advanced graduate-level text.

Heckmann, O. *Cosmological Theories* (in German). Springer-Verlag, Berlin, 1968. Advanced and extremely good.

Landau, L., and Lifshitz, E. *The Classical Theory of Fields.* 3rd ed. Pergamon, Oxford, 1971. A good, advanced text.

Landsberg, P. T., and Evans, D. A. *Mathematical Cosmology: An Introduction.* Oxford University Press, Clarendon Press, Oxford, 1977. An undergraduate text that stresses Newtonian cosmology.

McVittie, G. C. *General Relativity and Cosmology.* 2nd ed. Chapman and Hall, London, 1965. A good, not too advanced, treatment.

Misner, C. W., Thorne, K. S., and Wheeler, J. A. *Gravitation.* W. H. Freeman, San Francisco, 1973. Comprehensive and advanced, with many readable sections of interest in cosmology.

Narlikar, J. *The Structure of the Universe.* Oxford University Press, London, 1977. A brief undergraduate treatment, with a personal slant.

Peebles, P. J. E. *Physical Cosmology.* Princeton University Press, Princeton, N.J., 1971. An authoritative text at the graduate level with many readable passages.

Reines, F., ed. *Cosmology, Fusion and Other Matters: George Gamow Memorial Volume.* Colorado Associated University Press, Boulder, 1972. Various discussions at the first-year graduate level.

Rindler, W. *Essential Relativity: Special, General, and Cosmological.* Van Nostrand Reinhold, New York, 1969. A good treatment for advanced undergraduate students.

Rowan-Robertson, M. *Cosmology,* Oxford University Press, Clarendon Press, Oxford, 1977. A brief outline at the undergraduate level, with an epilogue of "twenty controversies in cosmology."

Sandage, A., Sandage, M., and Kristran, J. *Galaxies and the Universe.* University of Chicago Press, Chicago, 1975. A useful compendium of review articles.

Tolman, R. C. *Relativity Thermodynamics and Cosmology.* Oxford University Press, Clarendon Press, Oxford, 1934. Still a very useful treatment.

Weinberg, S. *Gravitation and Cosmology: Principles and Applications of the General Theory of Relativity.* John Wiley, New York, 1972. A good graduate-level text with many readable sections on cosmology and gravitational theory.

Wald, R. M. *Space, Time, and Gravity: The Theory of the Big Bang and Black Holes.* University of Chicago Press, Chicago, 1977. A brief treatment at the advanced undergraduate level.

SOURCES

Heninger, S. K. *The Cosmographical Glass: Renaissance Diagrams of the Universe.* Huntington Library, San Marino, Calif., 1977.

Russell, B. *The Autobiography of Bertrand Russell.* Vol. 1. Little, Brown, Boston, 1967.

WHAT IS COSMOLOGY?

He has ventured far beyond the flaming ramparts of the world and in mind and spirit traversed the boundless universe.
— Lucretius (99–55 B.C.), *The Nature of the Universe*

THE UNIVERSE

I don't pretend to understand the universe – it's a great deal bigger than I am.
— Attributed to William Allingham (1828–89)

From the outset we must decide whether to use the word *Universe* or the word *universe*. This is not so trivial a matter as it might at first seem. There is only one Earth – unique in all its detail – and similarly we believe that there exists only one Universe. The *Universe* is the *Uni*ty that embraces the di*verse*. *Cosmos* means essentially the same thing and is the harmonious whole that is reality.

But what is the Universe – does anyone know? It seems that the Universe has many faces and means different things to people of different outlooks. To religious people the Universe is a realm of the spirit and is a theistic creation; to the artists, poets, and singers it is a Universe of exquisite forms pierced by sensitive perceptions; to the philosophers it is a Universe of analytic and synthetic structures disciplined by logic; and to the scientists it is a Universe of intricate structures, elucidated by theory that is disciplined by observation and experiment. Each sees a different world picture that is like a mask fitted on the face of the unknown Universe (see Figure 1.1).

If the word *Universe* is to be used, then we must be specific, and always speak of the "models of the Universe." Each of these models, whether religious, artistic, philosophical, or scientific, is a world picture that represents one of many possible interpretations. Thus the Pythagorean model, the Atomist model, the Aristotelian model, and so on are examples taken from the history of science. Because it is not always clear what the model refers to, we must be more precise and say the Pythagorean model of the Universe, the Atomist model of the Universe, the Aristotelian model of the Universe, and so on. Inevitably, the models then receive abbreviated titles: the Pythagorean Universe, the Atomist Universe; and the word *Universe* comes to mean nothing more than "model of the Universe." This can sometimes be confusing and the cause of misunderstanding.

The grandiose word *Universe* has a further disadvantage. When used alone, without a specification of the model we have in mind, it conveys the impression that the Universe is a known entity. We find ourselves, in company with a host of others in the past and even the present, speaking of the Universe as if it had at last been discovered and determined. As a consequence, we mistake the mask for the face, the model for the Universe.

Because we do not know, and in our

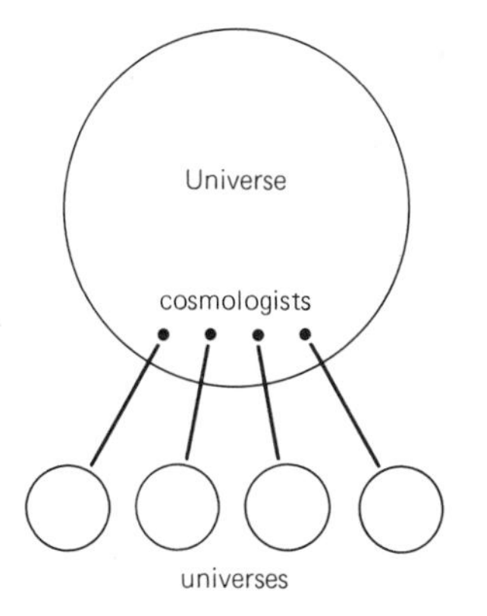

Figure 1.1. The Universe contains us, the cosmologists, who construct the many universes.

Figure 1.2. The universe, one and all-inclusive, *by Filippo Picinelli, 1694. In* The Cosmographical Glass: Renaissance Diagrams of the Universe *(1977), S. K. Heninger writes, "We might conjecture that the artist, not bound by the constraint of cosmological dogma, felt free to engage in cosmological speculations of his own sort. He assumed a licence to create his own universe. The worlds of Hieronymous Bosch, of Leon Battista Alberti, and of John Milton, to name a few examples, are the result." (Courtesy of the Henry E. Huntington Library, San Marino, Calif.)*

wildest dreams cannot imagine, the true nature of the Universe, we may avoid referring to it directly by using the more modest word *universe.* A universe is simply a model of the Universe (see Figure 1.2). Hence we can say the Pythagorean universe, the Atomist universe, the Aristotelian universe, and so on, and each universe is a mask, a world picture, a model that is first invented, then modified in the course of time as knowledge advances, and finally discarded.

The word *universe,* which we shall adopt, has the further advantage that it can be used freely and loosely without any need to remind ourselves continually that the Universe itself is still mysterious and unknown. When the word *universe* is used alone, as in phrases such as "the vastness of the universe," it denotes our modern universe as disclosed by science.

COSMOLOGY

We search the sky, the Earth, and within ourselves, and wonder about the universe: What is it all about? Why did it all begin? How will it all end? And what is the meaning of life? Each of us echoes the words of Erwin Schrödinger – "I know not whence I came nor whither I go nor who I am" – and seeks the answers. The search is hopeless and doomed to go astray unless we familiarize ourselves with the modern picture of the universe.

Cosmology is the study of universes. In the broadest sense it is a joint enterprise by science, philosophy, religion, and the arts that seeks to gain understanding of what unifies and is fundamental. As a science, it is the study of the large-scale structure of the universe; it draws on the knowledge of other sciences and branches of learning, such as physics and astronomy, and assembles a world picture.

In our everyday life we deal with ordinary objects, such as plants and flowerpots, and to understand these things of sensible size we explore the small-scale and large-scale realms of the universe. We delve into the microscopic realm of cells, atoms, and subatomic particles, and we reach out into the macroscopic realm of planets, stars, and galaxies.

Within the last century science has advanced rapidly and the number of sciences has grown enormously. Each science focuses on a domain of the universe and tends in the course of time to fragment into closely related new sciences of greater specialization. Originally, the characteristics of living and nonliving things defined the differences between the broad domains of biology and physics. Each of these basic sciences, as it advanced, has spawned new sciences, which in turn have spawned newer and more specialized sciences. Physics – once known as natural philosophy – has grown and split into subatomic particle physics, nuclear physics, atomic physics, chemical physics, solid-state physics, and so on, and each has its own theoreticians, experimentalists, and technicians. Biology – once the subject of naturalists of broad interests – has grown and split into molecular biology, biochemistry, biophysics, genetics, and so forth, with associated sciences such as botany, zoology, entomology, ecology, and paleontology, among others. And astronomy – once the subject in which almost everybody had equal knowledge – has mushroomed into planetary sciences, stellar structure, stellar atmospheres, interstellar media, galactic astronomy, extragalactic astronomy, and the various fields of radioastronomy, optical astronomy, X-ray astronomy, and so on.

It is evident that the sciences divide the universe in order that each science can rule a small domain of knowledge. Science, as a whole, tears things apart into smaller and smaller constituents; it pays close attention to detail and discloses new domains. A person who studies in depth a branch of science becomes a specialist, engrossed in a maze of knowledge, who knows much about a small domain of the universe and is comparatively ignorant of all the rest.

The physics of atoms and subatomic particles lies at one extreme, in the microscopic realm of the universe, and is the science devoted to the study of the universe on its smallest scale. Cosmology lies at the other extreme, in the macroscopic realm, and is the science concerned with the study of the universe on its largest scale. Cosmology is the one science in which specialization is rather difficult. Its main aim is to assemble the cosmic jigsaw puzzle, not to study in detail any particular jigsaw piece. While all other scientists are pulling the universe apart into more detailed bits and pieces, the cosmologists endeavor to put the pieces together in order to see the picture on the jigsaw puzzle. The cosmologists, unlike other scientists, take the broad view: They are like the impressionistic painters who stand well back from their canvas in order not to see too much distracting detail.

Cosmology is not astronomy. It is not an inventory of the contents of the universe; it is not just a "whole universe catalogue" of descriptive data. It is the study of cosmic things such as space and time, of the primary cosmic constituents, and of the expansion of the universe. The things of importance are those that endure over long periods of time and are distributed over large regions of space. The origin and evolution of stars and galaxies, even the origin of life and intelligence, are important cosmic subjects. Subatomic particles, the role they played during the earliest moments of the universe, their subsequent combination into atoms and molecules, which in turn form the complexity of the living cell and the world around us, are also of great interest. Perhaps most important of all are the physical properties of space and time that form the framework of the expanding universe and are the cradle of all science.

At each turn, the issues of cosmology cause one to pause and reflect. Many subjects of importance are still obscure and not understood, such as the way in which human beings acquired speech and large brains, the ability to construct mental images, and the power to think quantitatively. Whatever it is that determines the way human beings think must also determine the structure and design they perceive in the universe. Human beings are an essential part of cosmology and represent the Universe perceiving itself and thinking about itself.

Who are the cosmologists? Professional cosmologists are relatively few in number; they are well versed in mathematics, physics, and astronomy, and they study the large-scale structure of the physical universe. Everybody else is occasionally an amateur cosmologist. When somebody stands back from the study of a specialized area of the sciences or humanities and reflects on things in general, and tries to see the forest and not just the trees, tries to see the whole picture and not just the dabs of paint, the whole tapestry and not the threads, that person becomes a cosmologist.

THE ANTHROPOMORPHIC UNIVERSE

THE AGE OF MAGIC. The origins of cosmology are lost in the mists of time. Cosmology is the oldest of all intellectual enterprises; it preceded science and arose when primitive mankind made its first faltering attempts to understand and explain the world that it perceived. Tens, and perhaps even hundreds, of thousands of years ago, mankind explained the world as an activity of spirits who were motivated by impulses and emotions similar to those of human beings.

It was the Age of Magic, of benign and demonic spirits incarnate in animal and plant form, of the living earth, water, wind, and fire. Everything that happened was explained, readily and easily, by the motives and actions of ambient spirits. It was an anthropomorphic world – fashioned in the image of mankind – into which men and women projected their own emotions and thoughts as the guiding forces. It was an exhilarating and frightening world, of the kind children read about in fairy tales. From this "golden age" of magic comes our primeval fear of the menace of darkness and the rage of storms, and our enchantment with the wizardry of sunrise, sunset, and rainbows. Through interbreeding mankind remained one species throughout the prehistoric Age of Magic, and cultures (languages, myths, social codes, laws, technologies) diffused everywhere and had much in common. Then, at the dawn of history, almost 10,000 years ago, the early city-states attained more abstract concepts of the universe.

THE ANTHROPOCENTRIC UNIVERSE

THE AGE OF MYTHOLOGY. The long Age of Magic gave way to the Age of Mythology. The spirits that had once been everywhere, activating everything, retreated and became remote gods and goddesses who personified abstractions of thought and language. James Frazer, in *The Golden Bough,* speculated on how magic among primitive people evolved into myths: "But with the growth of knowledge man learns to realize more clearly the vastness of nature and his own littleness and feebleness in the presence of it. The recognition of his helplessness does not, however, carry with it a corresponding belief in the impotence of those supernatural beings with which his imagination peoples the universe. On the contrary, it enhances his conception of their power . . . If then he feels himself to be so frail and slight, how vast and powerful must he deem the beings who control the gigantic machinery of nature! . . . Thus in the acuter minds magic is gradually superseded by religion, which explains the succession of natural phenomena as regulated by will, passion, or caprice of the spiritual beings like man in kind, though vastly superior to him in power."

Mythology is prescientific cosmology; it is the weaving of myths into world pictures that illustrate the earliest known attempts to explain the universe with systematic thought. The Sumerian, Assyro-Babylonian, Minoan, Greek, Judaic, Chinese, Norse, Celtic, and Maya mythologies, to name a few, are of historical interest because they portray mankind's early views of the universe. The creation myths, often difficult to interpret, are of particular interest (see Figure 1.3).

The oldest known myths are those of Mesopotamia. In the beginning, we are told, "when Heaven above and Earth below had not been named," there existed the primal Apsu – the encircling watery abyss – and also the tumultuous blind force of Tiamat. From the inchoate mingling of Apsu and Tiamat arose ultimately dynasties of gods and goddesses whose history of intrigue and warfare, including catastrophes such as the Deluge, serves as a pattern for the study of mythology.

Before the creation, according to the myths of ancient Egypt, there dwelt in Nun,

Figure 1.3. The Ancient of Days *by William Blake (1757–1827): "When he set a compass upon the face of the depths" (Proverbs 8:27).*

the primal oceanic abyss, "a spirit, still formless, that bore within it the sum of all existence," called Atum, whose name signifies "to be complete." Then Atum, in the manifestation of Atum-Ra, created the gods and goddesses, the living creatures, and all the worlds they inhabit. Atum-Ra became personified as Ra the Sun god, and thereafter the gods and goddesses abounded in profusion: No less than 740 are listed in the tomb of Thutmosis III, who lived in the fifteenth century B.C. (see Figure 1.4).

With the Indian myths we are confronted by an imaginative riot of Vedic, Hindu, Buddhist, and Jainic deities. In the early myths of the Rig Veda it is said: "Who verily knows and can declare whence came this creation? He, the first origin of this creation, whose eye in highest heaven controls this world, whether he did or did not form it all, he verily knows it, or perhaps he knows it not." Later, in the Hindu law of Manu, it is said: "All was darkness, without form, beyond reason and perception, as if wholly asleep. Then the self-existent Lord became manifest, making all discernible with his power, unfolding the universe in the form of its elements, and scattering the shades of darkness." Here we see the primal, undifferentiated state develop into elements, and these elements in their most subtle form combine to create living creatures, and then later assume the grosser states of the nonliving world.

The Chinese created more practical myths. Heaven was conceived as a well-organized bureaucracy, in which the gods and goddesses devoted their time to compiling registers, making reports, and issuing directives. Later, in the Confucian scriptures, we find the five elements of ether, fire, air, water, and earth (common to several myths after the rise of Greek science), each possessing its own degree of subtlety and

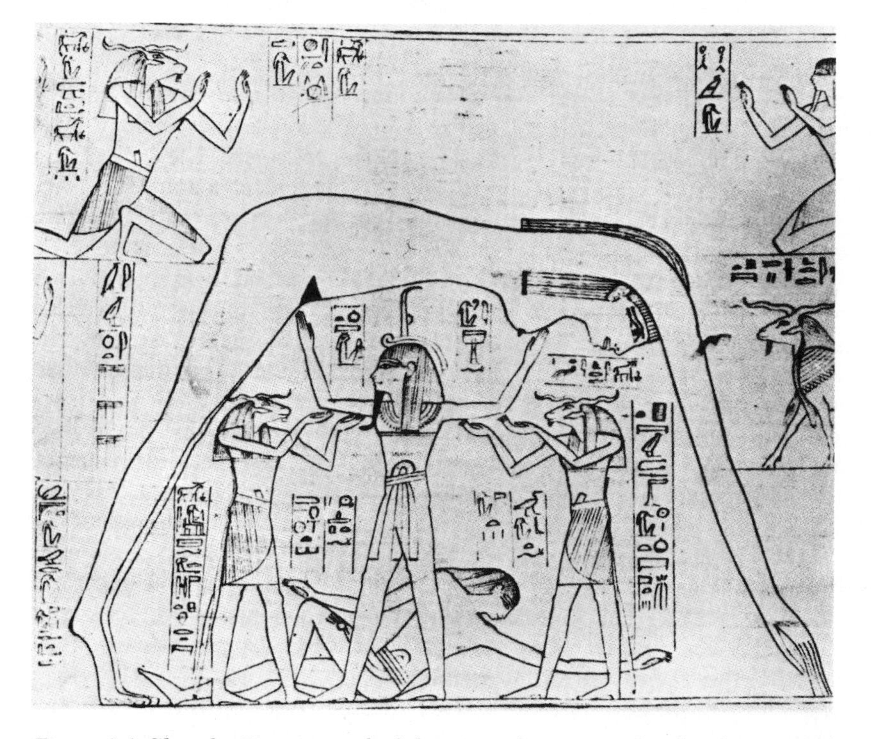

Figure 1.4. Shu, the Egyptian god of the atmosphere, raises his daughter Nut, the sky goddess, above the recumbent body of his son Geb, the Earth god. (Courtesy of the Trustees of the British Museum.)

Figure 1.5. The yin and the yang.

each having correspondence with one of the five notes of harmony, the five flavors, and the five colors. The masculine qualities of light, heat, and dryness, which resemble the Sun, were called the yang; whereas the feminine qualities of shade, coolness, and moisture, which resemble the Moon, were called the yin (see Figure 1.5). The convoluted forces of yang and yin generated order, sense, and the way of all things.

In the beginning, according to Greek myths, there existed four primal beings. First was Chaos (the infinite abyss), then Gaea (the Earth), Tartarus (the lower world), and Eros (the spirit of love). Hosts of gods and goddesses were generated by the four primal beings, by their matings with each other, and by each alone. Literally hundreds of pages are needed to construct the genealogical charts of the various deities. A significant early event was the begetting of Uranus the sky god by Gaea; and from the mating of Uranus with Gaea herself there arose the Titans, who were the first rulers. The ancient Greeks in the sixth century B.C. took the extraordinary and improbable step of laying down the foundations of scientific cosmology. We shall consider the origins of scientific cosmology in Chapters 4 and 5, and here we merely remark that in the whole Age of Mythology the most unlikely miracle was the rise of Greek science.

At the dawn of time, according to the Germanic and Norse myths, there was an earthless and skyless abyss. To the north was a world of cloud and shadow and to the south was a world of smoke and fire. From the north flowed glacial waters, and from the south came bitter waters, which filled the abyss and formed the Earth. A feature of interest is the *Götterdämmerung* – the Twilight of the Gods – which illustrates the eschatology (any doctrine concerning the final state) common to many myths. From the beginning the universe is doomed and the gods are destined to die. The end is foreshadowed by baleful omens, oath breaking, and warfare among men and gods. In the final carnage of Doomsday the Sun becomes swollen and blood red, and the Earth, frozen in the grip of paralyzing winter, sinks back into the abyss of nothingness. Out of the cosmic wreckage, goes the tale, emerges a new universe equipped with fresh gods.

MANKIND IS OF CENTRAL IMPORTANCE. The gods of mythology, no matter how powerful and remote they became, continued to serve and protect mankind, and men and women remained secure and of central importance in an anthropocentric universe. The universe was assembled about a center and human beings were located prominently at the center.

The anthropocentric conception was carried over into the Greek cosmology of an Earth-centered universe. The universe of Aristotle in the fourth century B.C. was geocentric (or Earth centered). The spherical Earth was located at the center of the universe, and the Moon, Sun, planets, and stars were fixed to translucent heavenly spheres that revolved about the Earth. The innermost regions of heaven – the sublunary sphere between the Earth and the Moon – contained earthly and tangible things in an everchanging state, and the outer regions of heaven – the celestial spheres – contained ethereal and intangible things in an unchanging state. Subsequent elaborations of this system, which brought it into closer agreement with astronomical observations, culminated in the Ptolemaic system of about A.D. 140.

The Middle Ages (fifth to fifteenth centuries) were not so terribly dark as we sometimes think. The medieval universe that arose in the thirteenth century and lasted until the sixteenth century was perhaps the most satisfying form of cosmology ever known in history. Christians, Jews, and Moslems were blessed with a cosmic scheme in which they had utmost importance in a finite and bounded Aristotelian universe that revolved about the Earth. By the Arab and European standards of those times it was a rational and well-organized universe that everybody could understand; it gave location and prominence to mankind's place in the firmament, it provided a secure foundation for religion, and it gave meaning and purpose to human life on Earth. Never before or since has cosmology served in so vivid a manner the everyday needs of ordinary people; it was simultaneously their religion, their philosophy, and their science.

THE COPERNICAN REVOLUTION. The transition from the finite geocentric universe to the infinite and centerless universe is known as the Copernican Revolution. In the sixteenth century, Copernicus crystallized trends of astronomical thought that had been anticipated in Greek science 2000 years previously and proposed the heliocentric (or Sun-centered) universe. The heliocentric universe was then transformed into the infinite, centerless Newtonian universe. This revolution in outlook occupied the sixteenth and seventeenth centuries. As a result of the Copernican Revolution, the universe of planets and stars ceased to be anthropocentric, and the foundations were laid for modern cosmology.

But the biological universe, which everybody regarded as far more important than the physical universe of stars, remained firmly anthropocentric. The biological universe was the "great chain of being," a chain of countless links that descended from human beings through all the lower forms of life to inanimate matter, and ascended from human beings through hierarchies of angelic beings to the divine Creator of the universe. Mankind was the unique link connecting the angelic and brute worlds and was the pivot of the living and spiritual world. Even in an infinitely large physical universe, deprived first of the Earth and then of the Sun as its center, it was still possible to cling to old ideas that portrayed human beings as having central importance in the cosmic scheme. The gods were always mysterious, and after the Copernican Revolution, they merely became more mysterious and more remote.

THE DARWINIAN REVOLUTION. Then, in the middle of the nineteenth century, came the most dreadful of all revolutions – the Darwinian Revolution. Human beings, hitherto the central figures in the cosmic drama, were dethroned and became akin to the beasts of the field. The gods who had protected and comforted mankind were thrown out of the physical universe.

The anthropomorphic and anthropocentric universes were wrong in almost every detail, and the medieval universe and the great chain of being have now gone. Science was at last the victor, dispersing the myths of the past. We applaud the Renaissance (fifteenth to seventeenth centuries), with its great revival of art and learning, we applaud the Age of Enlightenment (seventeenth to eighteenth centuries), with its conviction of the power of human reason, and we applaud the rise of the Age of Science (seventeenth to twentieth centuries); yet we tend to forget the growing confusion of ordinary men and women in a universe that century by century became progressively more bleak and senseless. With the overthrow of the old universes – anthropomorphic and anthropocentric – mankind was cast aimlessly adrift in a meaningless universe.

THE ANTHROPOMETRIC UNIVERSE

"Man is the measure of all things."
— Protagoras (fifth century B.C.)

We believe that the universe is not anthropomorphic; it is not made in the image of human beings, and it is not a magic realm, seething with the activity of spirits.

We believe that the universe is not anthropocentric; human beings are not at its center and are not the reason for its existence, and the world is not controlled by a pantheon of gods and goddesses.

Instead, human beings are the measure of the universe, and the universe is therefore *anthropometric*. Let us see if we can understand what this means.

We have minds, or as some would say, we have brains. For our purpose it is not necessary to inquire into the nature of the mind and attempt to probe its mysteries. It does not matter whether we regard the mind as a nonphysical psychic entity, or as a physical brain throbbing with bioelectronic activity. We have minds into which information pours, and from this information we construct within our minds the Aristotelian, Copernican, Newtonian, and other universes. We observe plants and flowerpots and other things and devise great theories to explain them, and these theories reside not in the things observed but in our minds. At each step in the history of cosmology there is a different universe, and each universe is a mask that mankind devises by theorizing about the things observed and experienced. Every mask, or every universe, is therefore anthropometric because it is devised by mankind and consists of ideas invented by mankind.

For those who feel lost in the vast and apparently meaningless modern universe there is comfort in the realization that all universes are anthropometric. The medieval universe was made and measured by men and women, although the Medievalists would have vehemently rejected the thought. The modern universe, with its bioelectronic brains pondering over it, is also man-made. Like the medieval universe it may fade away in due course. The universes of the future will almost certainly differ from our modern version; nevertheless, they will all be anthropometric because "man is the measure of all things" entertained by man. The Universe itself, of course, is not man-made, but we have no absolute knowledge of what it actually is.

COSMOLOGY AND SOCIETY

GODS AND SOCIETIES. Cosmology and society are intimately related. Societies create universes; not only do these universes reflect their societies, but each universe controls the history and destiny of its society.

This intimate relationship is most obvious in primitive cosmology, where mythology and society mirror each other, and the ways of gods and goddesses are the ways of men and women. Mankind creates myths and then is controlled by myths. Cruel people create cruel gods and their myths provide permanent sanction for cruel behavior; gentle people create peaceful gods and their myths foster peaceful behavior. The interplay between cosmology and society still exists today, but often in ways more subtle and less easily recognized.

The most powerful and influential ideas in any society are those that relate to the universe: They shape history, inspire civilizations, foment wars, create empires, and establish political systems. One such idea is the principle of plenitude, which can be traced back to Plato and has been influential since the fifteenth century.

THE PRINCIPLE OF PLENITUDE. The principle of plenitude originated from the anthropocentric belief that the universe is created for mankind by an intelligible Creator. In its simplest form it states that a beneficent Creator has given to mankind, for its own use, an Earth of unlimited bounty. The more formal argument goes as follows. The divine being is without limitation, because limitation of any kind implies imperfection, and imperfection in the divine being is contrary to belief. The divine being is unlimited and has therefore unlimited potential, and the unlimited potential is made manifest in the unlimited actuality of the created universe. Hence the Earth and the other parts of the universe necessarily display every possible form of reality in unlimited and inexhaustible profusion. This is the principle of plenitude that we have learned at our mothers' knees. When people protest that they were never taught this principle, don't believe them; the very language that they first learned to lisp is saturated with the doctrine of plenitude.

In the Renaissance, telescopes disclosed the richness of the heavens, microscopes disclosed a teeming world of microorganic life, and the great voyages by mariners opened up dazzling vistas of a bountiful Earth. An apparently unlimited abundance of every conceivable thing provided sufficient proof of the principle of plenitude. Europeans developed the principle, were guided by it, and have since exported it to the rest of the world.

POLITICAL IDEOLOGIES. Modern political ideologies everywhere have been shaped by the principle of plenitude. The principle guaranteed endless untapped wealth, and free enterprise flourished as never before. To offset depletion and population growth it was necessary only to push further east and west to the glittering prizes of unravished lands. "The real price of anything is the toil and trouble of acquiring it," said Adam Smith. Go east! – the streets are paved in gold. Go west! – beyond the sunset lie lands of unharvested wealth. Husbandry of finite resources was not part of the philosophy. Consciously and unconsciously, people believed that everything existed in unlimited abundance, and when anything showed signs of limitation, they were taken by surprise.

The inevitable question followed, and has since echoed around the world: Why should inequality of personal wealth exist in a world of unlimited abundance? The answer came in the message from Karl Marx, and in *The Communist Manifesto* we read that the less wealthy "have nothing to lose but their chains. They have a world to win." The principle of plenitude that now lies deep within our cultural heritage, and has been disseminated in various forms throughout the world, is unfortunately nothing but a myth.

THE INDUSTRIAL REVOLUTION. Old ideas of cosmological breadth still dominate our everyday thoughts, and many of these ideas are totally unsuitable in the modern world. We are, it seems, locked into the misguiding logic of obsolete universes that are destroying us. We live in an age of crises – unchecked population growth, nuclear weapons, pollution, depletion – and are mesmerized by prophecies of doom.

In 1776 the engineering firm of Boulton and Watt began to sell steam engines that, unlike previous steam devices, were powerful, quick-acting, and easily adapted for driving machinery of various kinds. This event more than any other ushered in the Industrial Revolution that has transformed our way of life. Many people say that the ills of today are the direct outcome of the Industrial Revolution. But our problems were not caused by the Industrial Revolution, nor are they caused by modern technology, and actually the steam engine and the transistor are quite blameless. It is not the tools and technological devices that are to blame, but the ideas that guide the hand.

To make our point clear, let us take an extreme case and imagine that space travelers encounter a planet that has been devastated by unbridled technology and as a consequence is lifeless. The space travelers, in their investigations, find that they cannot automatically assume that the technology once used was the cause of the devastation. They therefore look more carefully, searching for evidence that will indicate the nature of the beliefs of the vanished inhabitants. Finally, when they come to write their reports, they will probably draw the conclusion that they stand within the ruins wrought by an ancient and misguided cosmology, a cosmology of common beliefs that had originated before the rise of technology – founded on principles such as that of plenitude – that in their saner moments the inhabitants had rejected, and that yet had relentlessly driven them to their doom.

Ancient universes in vestigial and disguised forms are interwoven in our religions, ethics, and politics, and they constrain our perception of the grandeur of reality and the pageantry of life to the level of more primitive societies. With the aid of modern cosmology we must search and find a new meaningful unity of mankind and the universe.

REFLECTIONS

1 *Is cosmology yet another science? Or is it, by virtue of its antiquity and breadth of scope, the mother of all sciences, a sort of superscience?*

* *We study an extraordinarily complex Universe, and because nobody knows quite what it is, there are two schools of thought. The first holds that cosmology is the study of many universes, each of which is a model of the Universe. The second school holds that the Universe is studied with the aid of many cosmologies, each of which is a field of inquiry peculiar to a model of the Universe. Thus we have either cosmology and many universes or the Universe and many cosmologies. The first approach refers continually to the "universe," and the second refers continually to the "Universe." In this book we have adopted the practice of the first school, since it has the advantage that we avoid using the word* Universe – *except occasionally to make a point clear – and do not foster the unfortunate impression that the Universe is a known entity.*

* *"Possibly the world of external facts is much more fertile and plastic than we have ventured to suppose; it may be that all these cosmologies and many more analyses and classifications are genuine ways of arranging what nature offers to our understanding, and that the main condition determining our selection between them is something in us rather than something in the external world" (Edwin Burtt,* The Metaphysical Foundations of Modern Physical Science, *1932).*

* *"Natural science does not simply describe and explain nature; it is part of the interplay between nature and ourselves; it describes nature as exposed to our method of questioning" (Werner Heisenberg,* Physics and Philosophy, *1958).*

* *In* The Discarded Image *(1967), C. S. Lewis writes: "The great masters do not take any Model quite so seriously as the rest of us. They know that it is, after all, only a model, possibly replaceable." Later he continues: "It is not impossible that our own Model will die a violent death, ruthlessly smashed by an unprovoked assault of new facts – unprovoked as the nova of 1572. But I think it is more likely to change when, and because, far-reaching changes in the mental temper of our descendents demand that it should. The new Model will not be set up without evidence, but the evidence will turn up when the inner need for it becomes sufficiently great. It will be true evidence. But nature gives most of her evidence in answer to the questions we ask her."*

2 *Religion is not an easily defined subject. A possible interpretation from the cosmological viewpoint is as follows. Religion is an amalgamation of mystical experience and evocative theory; the experience is intensely personal, and the theory, or theology, rationalizes this experience on the basis of faith. Faith consists of beliefs whose truth is not demonstrable by experiments.*

Modern cosmology studies the universe and theology studies religion. In the ancient world this distinction did not exist, and mythology served as both cosmology and theology. The separation of religion into experience and theory indicates that Frazer may be wrong when he traces the roots of religion back to mythology. Religion is probably as old as mankind. This error of confusing religious experience with religious theory seems quite common. When people of different religions insist on retaining their mythological theories, they unwittingly make the mistake of confusing experience with theory, and think that without primitive cosmology they cannot have religion. It is not always realized that scientific rejection of mythological theory does not bring science into conflict with religious experience. After all, the twentieth-century theory of light as quanta of energy has not robbed us of our sensation of color, even though color is not a scientific term.

3 *Astrology, according to which the affairs of human beings are influenced by the heavenly bodies, consists basically of mythological beliefs. Millions of people in*

America read the astrological columns of the daily newspapers; they find astrology interesting and entertaining, for it is anthropocentric and relates human beings and the universe in ways that science cannot. Some people take it seriously, and then, by modern standards, it becomes slightly ridiculous. But I suspect that most people find it entertaining because it appeals to vestigial elements in our cultural heritage. Bart Bok, Lawrence Jerome, and 192 other leading scientists, in Objections to Astrology *(1975), vent their concern: "Scientists in a variety of fields have become concerned about the increased acceptance of astrology in many parts of the world ... It should be apparent that those individuals who continue to have faith in astrology do so in spite of the fact that there is no verified scientific basis for their beliefs, and indeed that there is strong evidence to the contrary." Why is astrology still popular? Is it possible that many people find themselves in a meaningless universe from which their earlier religions, even their philosophies, have retreated? What can be done about this pathetic situation, in which multitudes of people seek comfort from an interest in astrology that science is determined to destroy? Sunday schools did not arrest the flight from religion; will therefore more introductory science courses arrest the flight from the scientific universe?*

* *"Nature is a dull affair, soundless, scentless, colourless; merely the hurrying of material, endlessly, meaninglessly" (Alfred Whitehead,* Science and the Modern World, *1925).*

4 *In* The Great Chain of Being *(1936), by Arthur Lovejoy, we read: "Next to the word 'nature,' the 'Great Chain of Being' was the sacred phrase of the eighteenth century, playing a part somewhat analogous to that of the blessed word 'evolution' in the late nineteenth." It gave rise to the notion of "missing links" in the great chain long before Darwin. The great chain of being, explains Lovejoy, was intimately related to the principle of plenitude. "Not so very long ago," he writes, "the world seemed almost infinite in its ability to provide for man's needs – and limitless as a receptacle for man's waste products. Those with an inclination to escape from worn-out farms or the clutter of urban life could always move out into a fresh, unspoiled environment. There were virgin forests, rich lodes waiting to be discovered, frontiers to push back, and large blank regions marked* unexplored *on the map ... It has, so far as I know, never been distinguished by an appropriate name; and for want of this, its identity in varying contexts and in different phrasings seems often to have escaped recognition by historians. I shall call it the principle of plenitude."*

* *Adam Smith said, "The real price of anything is the toil and trouble of acquiring it." But in all undertakings with nature we should first read carefully the small print in the contract. This might disclose that the real price is to be paid by those who inherit the depletion and despoliation that follows.*

5 *Garrett Hardin, in "The tragedy of the commons" (1968), discusses how individuals strive to maximize their share of a common resource. When herdsmen graze their beasts on common land, each strives to increase the size of his herd. Disease and tribal warfare maintain a state of stability by limiting the numbers of people and beasts below the capacity of the land. Then comes a more orderly and civilized way of life, with diminished war and disease, which places an increased burden on the commons. A herdsman now thinks, "If I increase my herd, the loss owing to overgrazing will be shared by all, and my gain will exceed my loss." All herdsmen reach this conclusion, and therein lies the tragedy. "Each person," states Hardin, "is locked into a system that compels him to increase his herd without limit – in a world that is limited ... Ruin is the destination to which all men rush." Problems created by cosmological myths do not have technological solutions, Hardin explains. "A technical solution may be defined as one that requires a change only in the techniques of the natural sciences,* *demanding little or nothing in the way of change in human values or ideas of morality." The "concern here is with that important concept of a class of human problems which can be called 'no technical solution problems' ... My thesis is that the 'population problem,' as conventionally conceived, is a member of this class ... It is fair to say that most people who anguish over the population problem ... think that farming the seas or developing new strains of wheat will solve the problem – technically."*

Do you think that colonizing space will technologically solve the population problem? Sebastian von Hoerner, in "Population explosion and interstellar expansion" (1975), shows that this could solve the problem, with the present birthrate, for at most only 500 years. The human space bubble would expand faster and faster and in 500 years would expand at the speed of light. Thereafter, each colonized planet would become progressively more crowded and would face the same problem that we now face on Earth. To what extent is the West, with its technology, medicine, and ideas of plenitude, responsible for the rapid growth in the human population?

6 *Articulate cosmological concepts have always had great influence, for good or evil. Consider "Thou shalt not suffer a witch to live." It is estimated that in the Renaissance and the Age of Enlightenment about half a million people confessed sorcery and witchcraft under torture and were burned to death. It was believed that heretics would burn forever in hell, and the temporary anguish of fire on Earth was justified if they could be saved from eternal fire. Here is an instance of the maxim stated in the text: Cruel people create cruel gods and their myths provide permanent sanction for cruel behavior. Can you think of other examples?*

7 *Compare the following and find additional examples of cosmic despair and hope:*

"That man is the product of causes which had no prevision of the end they were achieving; that his origin, his growth, his hopes and fears, his loves and his beliefs, are but the outcome of accidental collocations of atoms; that no fire, no heroism, no intensity of thought or feeling, can preserve a life beyond the grave; that all the labors of the ages, all the devotion, all the inspiration, all the noonday brightness of human genius, are destined to extinction in the vast death of the solar system; and the whole temple of Man's achievement must inevitably be buried beneath the debris of a universe in ruins – all these things, if not quite beyond dispute, are yet so nearly certain, that no philosophy which rejects them can hope to stand. Only within the scaffolding of these truths, only on the firm foundation of unyielding despair, can the soul's habitation be safely built" (Bertrand Russell, A Free Man's Worship, *1923).*

"The same thrill, the same awe and mystery, come again and again when we look at any problem deeply enough. With more knowledge comes deeper, more wonderful mystery, luring one on to penetrate deeper still. Never concerned that the answer may prove disappointing, but with pleasure and confidence we turn over each new stone to find unimagined strangeness leading on to more wonderful questions and mysteries – certainly a grand adventure!" (Richard Feynman, "The value of science," 1958).

FURTHER READING

Blacker, C., and Loewe, M., eds. *Ancient Cosmologies.* George Allen and Unwin, London, 1975.

Hamilton, E. *Mythology: Timeless Tales of Gods and Heroes.* Little, Brown, Boston, 1942.

King, I. R. "Man in the universe." *Mercury,* November–December 1976.

Lewis, C. S. *The Discarded Image.* Cambridge University Press, Cambridge, 1967.

Munitz, M. K., ed. *Theories of the Universe: From Babylonian Myth to Modern Science.* Free Press, Glencoe, Ill., 1957.

SOURCES

Arons, A. G., and Bork, A. M., ed. *Science and Ideas.* Prentice-Hall, Englewood Cliffs, N.J., 1964.

Bok, B. J., Jerome, L. E., and Kurtz, P. "Objections to astrology." *Humanist 35,* 4 (October 1975). Reprint. Prometheus Books, Buffalo, N.Y., 1975.

Burtt, E. *The Metaphysical Foundations of Modern Physical Science.* 1924. Rev. ed. Humanities Press, New York, 1932. Reprint. Doubleday, Garden City, N.Y., 1954.

Butterfield, H. *The Origins of Modern Science, 1300–1800.* Bell & Sons, London, 1957. Rev. ed. Free Press, New York, 1965.

Campbell, J. *The Masks of God: Primitive Mythology.* Viking Press, New York, 1959.

Campbell, J. *The Mythic Image.* Princeton University Press, Princeton, N.J., 1974.

Childe, V. G. *What Happened in History.* Penguin Books, London, 1942.

Conant, J. B. *Modern Science and Modern Man.* Columbia University Press, New York, 1952.

Cornford, P. M. *Plato's Cosmology.* Routledge and Kegan Paul, London, 1937.

Dampier, W. C. *A Shorter History of Science.* Cambridge University Press, Cambridge, 1944. Reprint. World Publishing Co., New York, 1957.

Dreyer, J. L. E. *A History of Astronomy from Thales to Kepler.* Dover Publications, New York, 1953.

Feynman, R. "The Value of Science," in *Frontiers in Science: A Survey,* ed. E. Hutchings. Basic Books, New York, 1958.

Frankfort, H., Frankfort, H. A., Wilson, J. A., and Jacobsen, T. *Before Philosophy.* Penguin Books, London, 1949. First published as *The Intellectual Adventure of Ancient Man.* University of Chicago Press, Chicago, 1946.

Frazer, J. G. *The Golden Bough: A Study in Magic and Religion.* Abridged ed. Macmillan, London, 1922.

Gillispie, C. C. *The Edge of Objectivity.* Princeton University Press, Princeton, N.J., 1960.

Hardin, G. "The tragedy of the commons." *Science 162,* 1243 (1968).

Heisenberg, W. *Physics and Philosophy.* Harper & Row, New York, 1958.

Heninger, S. K. *The Cosmographical Glass: Renaissance Diagrams of the Universe.* Huntington Library, San Marino, Calif., 1977.

Hoerner S. von. "Population explosion and interstellar expansion." *Journal of the British Interplanetary Society 28,* 691 (1975).

Kruglak, H., and O'Bryan, M. "Astrology in the astronomy classroom." *Mercury,* November–December 1977.

Leach, M. *The Beginning: Creation Myths around the World.* Funk and Wagnalls, New York, 1956.

Lewis, C. S. *The Abolition of Man.* Macmillan, New York, 1947.

Lewis, C. S. *The Discarded Image.* Cambridge University Press, Cambridge, 1967.

Lovejoy, A. O. *The Great Chain of Being: A Study of the History of an Idea.* Harvard University Press, Cambridge, Mass., 1936.

Malinowski, B. *Magic, Science and Religion.* Free Press, New York, 1948.

Neugebauer, O. *The Exact Sciences in Antiquity.* Brown University Press, Providence, R.I., 1957.

Russell, B. *A Free Man's Worship.* Mosher, Portland, Maine, 1923.

Schrödinger, E. *Science and Humanism: Physics in Our Time.* Cambridge University Press, Cambridge, 1951.

Schrödinger, E. *Science, Theory and Man.* Dover Publications, New York, 1957.

Solla Price, D. J. de. *Science since Babylon.* Yale University Press, New Haven, Conn., 1961.

Spencer, C., and Schroeer, D. "Critical reading list for teachers of physics-and-society courses." *American Journal of Physics 44,* 139 (February 1976).

Trevor-Roper, H. R. *The European Witch-Craze.* Harper & Row, New York, 1969.

Weisskopf, V. F. *Knowledge and Wonder.* Doubleday, New York, 1963.

Whitehead, A. N. *Science and the Modern World.* Macmillan, London, 1925.

Zeilik, M. "Astrology in introductory astronomy courses for nonscience specialists." *American Journal of Physics 41,* 961 (August 1973).

2 STARS

"The stars," she whispers, "blindly run:
A web is wov'n across the sky;
From our waste places comes a cry,
And murmers from the dying sun."
— Tennyson (1809–92), *In Memoriam*

THE DISTANT STARS

LIGHT TRAVEL TIME. We look out from Earth and see the Sun, the planets, and the stars at great distances (see Figure 2.1). The Sun, our nearest star, is at a distance of 150 million kilometers or 93 million miles. Kilometers and miles, suitable units for measuring distances on the Earth's surface, are evidently much too small for the measurement of astronomical distances (see Table 2.1).

Almost all information from outer space comes to us in the form of light and other kinds of radiation that travel at a speed of 300,000 kilometers per second (see Table 2.2). Light from the Sun therefore takes 500 seconds to reach the Earth, and we see the Sun as it was 500 seconds ago. We say that the Sun is at a distance of 500 light seconds. The time taken by light to travel from a distant object is called the *light travel time*. Light travel time is an attractive way of measuring large distances and has the advantage that we immediately know how far we look back into the past when we speak of a distant object (see Table 2.3). Thus a star 10 light years away (almost 100 trillion kilometers) is seen now as it was 10 years ago. Always, when looking out in space, we look back in time.

Light takes approximately 10 hours to travel across the Solar System – the diameter of Pluto's orbit. Yet our system of circling planets, so large by ordinary terrestrial standards, is dwarfed by the great distances to even the nearest stars. The nearby stars are several light years away, and an amazing variety of billions of stars extends to distances of tens of thousands of light years. Beyond these stars lie the galaxies, at millions of light years' distance, which are themselves vast systems of billions of stars.

Even astronomers, accustomed to large distances, marvel at the lavishness of space in the design of the universe. But the universe is rather sparing in its use of time, for who can marvel at only a few years of light travel time? This cosmic display of generosity with space and economy with time is actually the result of the peculiar units that we on Earth use for the measurement of space and time. By using light travel time as a way of measuring distances we set aside our terrestrial bias of favoring small units of distance and large units of time. One second of light travel time, which seems very small, is equivalent to a distance of 300,000 kilometers, which seems very large.

The greatest distances on the Earth's surface are only 1/20 of a light second. We live for 3 score and 10 years and in that time the Sun travels a distance of only 20 light days through our galaxy. From a cosmic viewpoint we are thus confined to a small

Figure 2.1. Star cluster in the constellation Cancer. (Mount Wilson and Las Campanas Observatories, Mount Wilson Observatory photograph.)

Table 2.1. *Distances and sizes*

Distance to Sun	1 astronomical unit $= 1.5 \times 10^{13}$ centimeters
Sun's radius	7×10^{10} centimeters
Earth's radius	6370 kilometers
Fingerbreadth	1 centimeter approximately
Flea	1 millimeter approximately
Wavelength of yellow light	6×10^{-5} centimeters
Size of hydrogen atom	10^{-8} centimeters

1 kilometer = 10^3 meters = 0.62 miles
1 meter = 10^2 centimeters
1 centimeter = 0.4 inches
1 millimeter = 10^{-1} centimeters
1 angstrom = 10^{-8} centimeters

Table 2.2. *Velocities*

Light (denoted by c)	300,000 kilometers a second
Sun in Galaxy	300 kilometers a second
Earth in Solar System	30 kilometers a second
Escape velocity from Earth	11 kilometers a second
1 mile per hour = 44.7 centimeters a second	
32 kilometers an hour = 20 miles an hour	

region of space but endure for a comparatively long period of time. This may explain why we are more impressed with the vastness of space than with the vastness of time.

DISTANCES TO STARS. Determination of the distances to stars is not easy. Throughout the history of astronomy until recent times our ideas about the size of the universe have been wrong because of incorrect estimates of the distances of the stars. There are various ways of determining astronomical distances, and at this stage we shall mention two that are simple and important.

The first method uses the parallax effect for stars within a few hundred light years (see Figure 2.2). A candle flame at a distance of 3 kilometers is as bright in its appearance as a faint star seen with the unaided eye. When looked at alternately with the left and right eyes, it is seen to lie in slightly different directions. This is the parallax effect. The effect is quite obvious when one holds up a finger at arm's length and notices that the finger moves to and fro when seen alternately with the two eyes. The nearby stars, because of parallax, are also seen in slightly different directions when viewed from different positions on the Earth's orbit about the Sun. A star at a distance of 1 light year, when viewed from opposite sides of the Earth's orbit at six-month intervals, is seen in slightly different directions, and the difference is about the same as when we look with alternate eyes at the candle flame at 3 kilometers distance. The maximum distance of stars that can be determined by the parallax effect is approximately 300 light years.

The second method is used for stars too distant to have detectable parallax. Distant stars are selected for their similarity to stars of known distance within 300 light years. By

Table 2.3. *Time*

Hour	3600 seconds
Day	86,400 seconds
Year	3.2×10^7 seconds
Decade	10 years
Century	100 years
Millennium	1000 years
Age of Earth	4.6 billion years
Age of universe	10 to 20 billion years

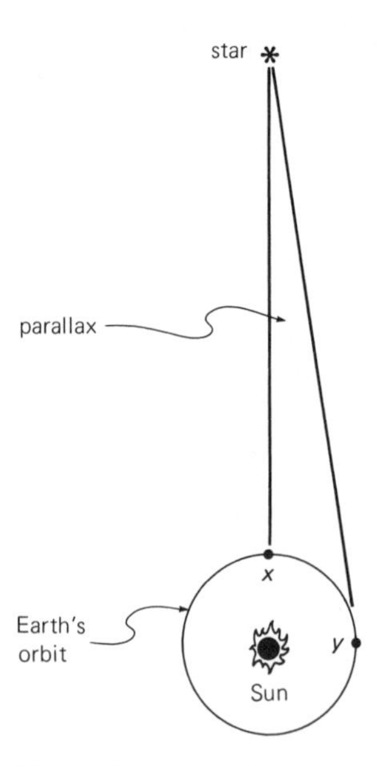

Figure 2.2. At positions x and y of the Earth's orbit about the Sun an observer sees a star in slightly different directions. The small angle between these directions is the parallax.

comparing their apparent brightnesses it is possible to determine the distances of the further stars from the known distances of the nearby stars. Every star has an intrinsic or absolute brightness, and also an apparent brightness (as seen from Earth) that depends on the intrinsic brightness and the distance. As an illustration, suppose we find by parallax measurements that a certain star is 100 light years away. A second star is then observed, of much less apparent brightness, which is believed to have the same intrinsic brightness; we measure its apparent brightness and find it has 1/100 of the apparent brightness of the first star. Apparent brightness decreases as the square of distance, and we therefore know that the fainter star is at a distance of 1000 light years.

A FOREST OF STARS

MULTIPLE STARS. Let us leave the Earth and roam about looking at different stars. First we notice that most stars are much feebler sources of light than the Sun. The second thing we notice is that the majority of stars are grouped together in small families of two, three, or more members. Double stars – known as binary systems – are common: Almost half of all stars about us are members of binary systems. Usually, double stars are separated from each other by many astronomical units (an astronomical unit is the distance from the Sun to the Earth), and they revolve about each other with periods of several years. Some double stars, however, are much closer together and have periods as short as a few hours; these stars exchange matter, have eruptive outbursts, and evolve in strange ways.

The average separating distance between neighboring stars (ignoring binary systems) is about 4 light years. This distance is 30 million times the diameter of the Sun, or 300,000 astronomical units – evidently the space between stars is immense compared with the size of stars and the Solar System.

COLOR AND BRIGHTNESS. What are stars? They are luminous globes of gas that pour out radiation into space. Among their properties of immediate interest are their color and brightness. The color of a star is determined by the temperature of its surface, and

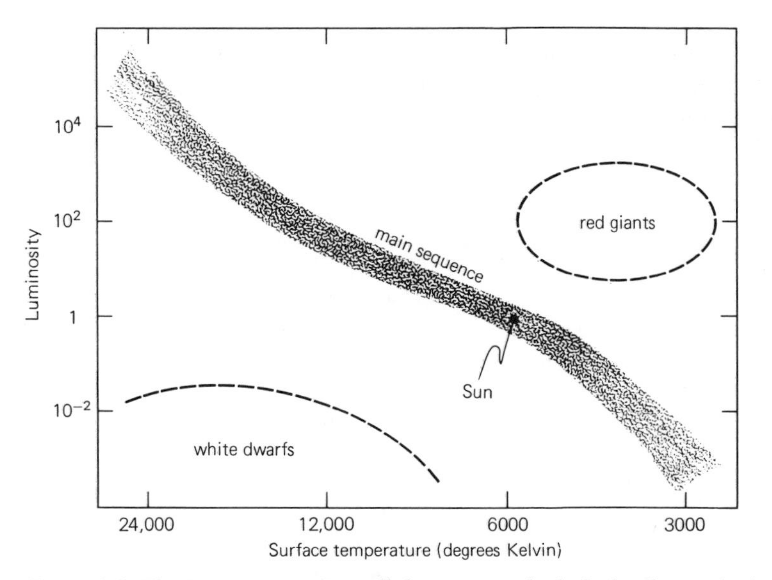

Figure 2.3. The Hertzsprung–Russell diagram in which the brightness (or luminosity) and color (or surface temperature) of the stars are plotted. The luminosity is expressed in units of the Sun's luminosity.

Table 2.4. *Temperature*

Sun's center	2×10^7 degrees Kelvin
Sun's surface	6000 degrees Kelvin
Filament of electric light bulb	2000 degrees Kelvin
Boiling point of water	373 degrees Kelvin = 100 degrees centigrade = 212 degrees Fahrenheit
Freezing point of water	273 degrees Kelvin = 0 degrees centigrade = 32 degrees Fahrenheit
Absolute zero	0 degrees Kelvin = −273 degrees centigrade

the brightness is determined by the amount of light radiated from its entire surface. A brightness–color diagram for stars is known as the Hertzsprung–Russell diagram (after its originators), or more briefly as the H–R diagram (Figure 2.3). Each point in the H–R diagram indicates the brightness and color of a particular star.

The Sun is yellow-white and has a surface temperature of 5800 degrees Kelvin (see Table 2.4); many stars, also yellow-white, are similar in size to the Sun. Other stars are red and large: These are the red giants with surface temperatures of around 3000 degrees. Yet others are white and small, and are the white dwarfs with surface temperatures of approximately 10,000 degrees.

Brightness is the rate at which luminous energy is emitted; it depends only on the surface temperature and the surface area of the star. A red giant has a low surface temperature but a very large surface area and has therefore great brightness. A white dwarf has a high surface temperature but a very small surface area and has therefore small brightness. The low temperature and great brightness make a red giant star similar to a coal fire, whereas a white dwarf star, with high temperature but small brightness, can be compared to a small flashlight bulb.

Often we are interested in the total amount of radiation emitted by a star, some of which is not detectable by the eye. Instead of brightness for the total radiation (visible and invisible) we frequently use the word *luminosity,* usually stated in terms of the luminosity of the Sun (see Tables 2.5 and 2.6). If brightness is expressed as luminosity and color as surface temperature, the brightness–color diagram becomes a luminosity–temperature diagram.

MAIN SEQUENCE. When stars are plotted as points in the H–R diagram a remarkable thing is noticed: The points are not distributed randomly, but tend to concentrate in

Table 2.5. *The 10 nearest stars*

Star	Distance (light years)	Luminosity (solar luminosites)
Sun	1.6×10^{-5}	1
Alpha Centauri	4.3	1.5
Barnard's Star	6.0	5×10^{-4}
Wolf 359	7.6	1.6×10^{-5}
Lalande 21185	8.1	5×10^{-3}
Sirius	8.6	23
Luyten 726	8.9	1×10^{-4}
Ross 154	9.4	4×10^{-4}
Ross 248	10.3	1×10^{-4}
Epsilon Eridani	10.7	0.3

Table 2.6. *The 10 brightest stars seen from Earth (in order of apparent brightness)*

Star	Distance (light years)	Luminosity (solar luminosities)
Sun	1.6×10^{-5}	1
Sirius	8.6	23
Canopus	98	1.5×10^{3}
Alpha Centauri	4.3	1.5
Arcturus	36	114
Vega	26	54
Capella	45	150
Rigel	900	6×10^{4}
Procyon	11	7.2

certain regions. Most stars, including the Sun, lie in a band called the main sequence that runs diagonally across the diagram. In the main sequence, red stars of low temperature have low luminosity, and blue stars of high temperature have high luminosity. All main-sequence stars of the normal kind maintain their luminosity by converting hydrogen into helium. Stars have different masses: The less massive are at the lower end of the main sequence and their rate of converting, or "burning," hydrogen into helium is slow; the more massive are at the upper end of the main sequence and their rate of burning hydrogen is rapid.

Star masses increase steadily up the main sequence from bottom right to top left (see Table 2.7). The majority of stars have masses within the range 1/10 to 10 times the Sun's mass, and normally their surface temperatures range from 2500 to 25,000 degrees and their luminosities from 1/1000 to 10,000 times the Sun's luminosity. Over this entire range, the sizes of stars on the main sequence do not change greatly (the most bright are about 25 times the radius of the least bright), but the luminosity varies enormously. An unknown number of stars have masses less than 1/10 of the Sun's mass; these dull stars, known as red dwarfs, are often the unseen companions of visible stars. A few extreme stars have masses as great as 60 times that of the Sun, and these brilliant stars are 10 million times brighter than the Sun. If a star of this kind were as close as Alpha Centauri, it would shed as much light on Earth as the full Moon.

Table 2.7. *Masses*

Sun	2×10^{33} grams
Earth	6×10^{27} grams
Water in thimble	1 gram approximately
Flea	1 milligram approximately
Hydrogen atom	1.7×10^{-24} grams

10^3 kilograms = 1 ton (metric)
1 kilogram = 10^3 grams = 2.2 pounds
1 milligram = 10^{-3} grams

Above the main sequence, in the H–R diagram, is the region where red giants are found. These stars are distended globes of gas – sometimes larger than the Earth's orbit about the Sun – that are cool and luminous. Although their surface temperatures are low, their surface areas are very large and they are usually exceedingly luminous, radiating hundreds and sometimes thousands of times more energy each second than the Sun. Because red giants have exhausted their central supplies of hydrogen, their prodigal expenditure of energy might at first seem strange. They have quit the main sequence and their central regions are contracting in quest of further sources of nuclear energy.

Below the main sequence are found the

white dwarfs. These stars are approximately the size of Earth; they are dense and hot and are not very luminous because of their very small surface area. White dwarfs are stars that have come to the end of their evolution and are slowly cooling. There are large numbers of these dying stars (about 10 percent of the nearby stars are white dwarfs), but they are difficult to see, owing to their low luminosities. Not all stars terminate their careers as white dwarfs; many become neutron stars, and others quite possibly become black holes, as we shall shortly see.

VARIABLE STARS. The majority of stars shine with almost constant brightness, but a minority vary periodically in brightness and are known as *variable stars.* About a quarter of all variable stars are eclipsing binary systems, in which the apparent brightness varies simply because the orbiting companion stars happen to pass periodically in front of each other. But most of the regularly varying stars are pulsating variables, which rhythmically expand and contract, pulsating in size and brightness.

An important class of pulsating variables are the luminous yellow giants, found above the main sequence and known as cepheids. They are between 100 and 10,000 times as bright as the Sun, and are so named because the first star of this kind discovered was Delta Cephei – a faint naked-eye star in the constellation of Cepheus. Over 700 cepheids are known in our Galaxy; most have periods of between 3 and 50 days, and some vary in brightness by as much as a factor of 5. Polaris, the Pole Star, is a cepheid that changes in brightness by 10 percent with a period of 4 days. Cepheids are stars more massive than the Sun that have evolved beyond the red giant state. They have discovered that by resorting to oscillation they can release the radiation dammed up inside more easily.

Cepheids are extremely important because they can be used as distance indicators. The period of oscillation of a cepheid is related in a known way to the average luminosity: the greater the luminosity, the longer the period. (This period-luminosity relation was discovered in 1912 by the American astronomer Henrietta Leavitt of the Harvard College Observatory.) After measuring the period of oscillation, an astronomer uses the period–luminosity relation to find the luminosity (the intrinsic brightness). Then, by comparing the intrinsic brightness with the observed apparent brightness, the astronomer is able to find the distance to the cepheid. Distance determinations are always tricky, and this is why the relatively rare cepheids are important stars.

INSIDE THE STARS

GLOBES OF HOT GAS. The stars are globes of hot gas that radiate energy away into space. This energy, emitted from the surface, is generated in the deep interior and diffuses slowly outward to the surface. Heat always flows from hot to cool regions and this means that the center of a star is much hotter than its surface. The central temperatures of stars are in fact enormous; in the Sun, for example, the central temperature is about 15 million degrees. The central temperature of normal stars on the main sequence is roughly proportional to stellar mass and hence increases as we go up the main sequence to more massive stars.

Stars are self-gravitating; this means that they are held together by their own gravitational force. Gravity pulls inward and is counterbalanced by a force that pushes outward. The outward-thrusting force is the pressure caused by the hot gas in the interior. If there were no pressure inside the Sun, the Sun would collapse in 1 hour only, and become a black hole.

The balance of pressure (actually the pressure gradient) and gravity is easy to understand. Consider an imaginary shell inside a star consisting of gas contained between two spherical surfaces, as shown in Figure 2.4. On the inside surface of the shell is a pressure pushing outward and on the outside surface of the shell is a pressure

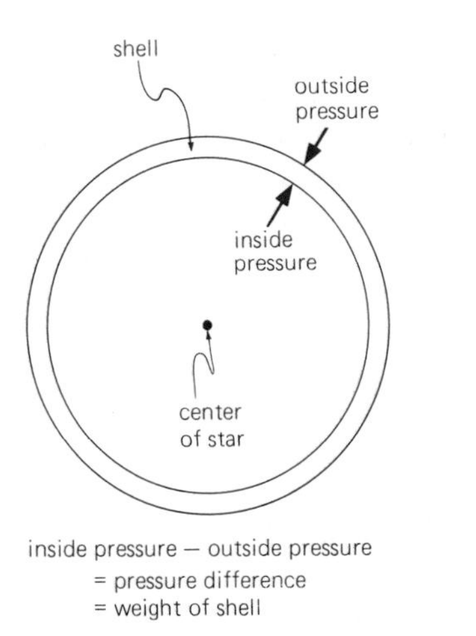

***Figure 2.4.** A thin shell of matter inside a star. The inside pressure is greater than the outside pressure, and the difference equals the weight of the shell.*

pushing inward. The outside pressure can be thought of as the weight of overlying matter, and the inside pressure supports the weight of the shell and the weight of the overlying matter. The difference of the two pressures supports the weight of the shell, and hence

pressure difference = weight of shell

This expression is known as the hydrostatic equation. The star consists of a large number of such imaginary concentric shells, and as we proceed inward, the pressure rises each time a shell is crossed. The pressure progressively increases and attains its maximum value at the center of the star. The central pressures of stars are tremendous; in the center of the Sun, for example, the central pressure is 100 billion atmospheres – equivalent to a weight on Earth of 100 million tons resting on an area equal to that of a dime.

WHY THE TEMPERATURE IS HIGH. The average density of the Sun is 1.4 grams per cubic centimeter (or 1.4 times the density of water) and the central density is approximately 150 grams per cubic centimeter. Nothing in the ordinary solid or liquid states can exist at this kind of density and support such crushing pressures. The only possible form of matter in the Sun and similar stars is a gas that is dense and extremely hot.

A gas at a temperature of millions of degrees is unlike any ordinary gas with which we are familiar. The atoms move at high speeds – hundreds of kilometers per second – and their frequent and energetic collisions with each other strip away their attached electrons. As a result, all atoms are fully ionized (all electrons removed), and the gas consists of negative electrons and positive atomic nuclei, all moving freely as independent particles. The radiation within this hot and dense gas is intense X-rays, and not the gentle beneficent light emitted from the comparatively cool surface. Each ray of this intense radiation in the deep interior travels on the average a distance of 1/10,000 of a centimeter before it is captured or deflected by particles of the gas. Pressure in a gas is proportional to density multiplied by temperature, and the high temperatures in stars are the result of the large pressures needed to withstand gravity.

It is now clear why stars are luminous: Inside they are extremely hot because of the high pressure needed to withstand gravity, and the radiation produced by the high-temperature gas is slowly escaping to the surface. Nuclear reactions replace the energy that is lost and maintain stars in a luminous state for a long period of time, but it is because of their great masses that stars are so hot, and not because of nuclear energy.

CONVECTION AND SOUND WAVES. Radiation deep inside a star is continually scattered, absorbed, and emitted by the particles of gas, and its outward flow to the surface is greatly impeded (see Figure 2.5). This impedance to the flow of radiation is called *opacity*. When the opacity is high, as often happens, the gas dams up the radiation and energy is then transported by convection. This means that the gas is stirred into motion and ascending and descending currents of gas carry heat toward the surface. The outer layers (the envelope) of the Sun, for instance, have high opacity

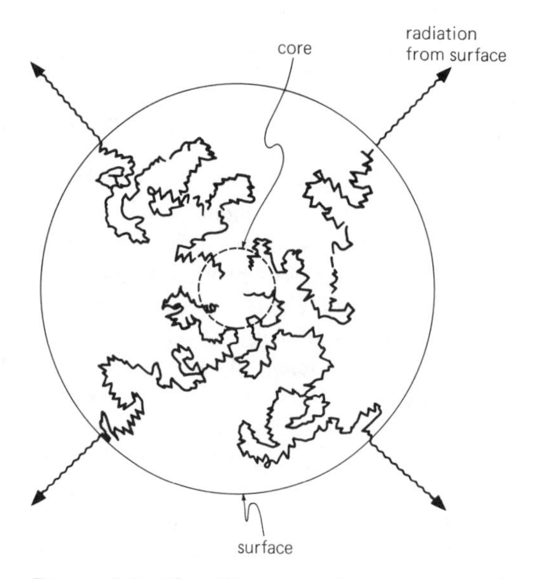

Figure 2.5. This illustrates the way rays of radiation diffuse from the center to the surface of a star. In the case of the Sun the time taken by radiation to diffuse to the surface is about 10 million years. If energy production ceased abruptly in the center of the Sun, we would not know that anything serious happened until 10 million years later.

through which radiation cannot easily diffuse; consequently the envelope of the sun is thrown into a state of convection, like water in a boiling kettle, and in this way heat is transported to the surface. Main-sequence stars of smaller mass than the Sun have deeper convective envelopes; and main-sequence stars of more than twice the Sun's mass do not have convective envelopes, but have instead convective cores. In these more massive stars the nuclear energy is released in a small central region, and the cores are in a convective state because the radiative transport of energy is slow.

The song of a star outrivals the song of the humpback whale. Its interior is a symphony of sound, reverberating with rumbling groans, resounding to a thunder of drums, and quivering with high-pitched wailing shrieks. Nobody has told a star what size and shape it must be, or with what brightness and temperature it must shine, or how it must find the energy that is continuously lost from the surface. Sound waves travel through the star in about 1 hour, and by ceaseless adjustments, with each part sending out signals to all other parts, the star seeks each moment to rediscover its natural equilibrium state. It heaves and readjusts with various slow modes of vibration, and at the other extreme, 60 octaves higher, is the hiss of high-speed particles jostling each other and producing waves that travel only short distances. Not content with this orchestration of sound, the star also acts as an immense loudspeaker. The density decreases steadily from the center to the surface, and each wave, as it travels outward, grows in amplitude like a whiplash. An amplified torrent of sound reaches the surface, passes through, and is dissipated in the outer atmosphere of the star. In the case of the Sun, with its noisy convective envelope, this ceaseless dumping of acoustic energy maintains the corona – the upper atmosphere – at a temperature of a million or so degrees. Because of its very low density the corona is unable to radiate away all the energy it receives, and it therefore does the only thing possible: It expands and carries away the energy. The outer atmosphere of the Sun is like a giant jet engine; it sucks in gas from the Sun and the gas, heated by acoustic energy, then blasts away at high velocity. This is the outward-streaming *solar wind* that carries away each second 100 million tons of gas. Other stars also have stellar winds, generated by their internal acoustic tumult. These winds are sometimes much stronger than the solar wind – so much so that some stars are literally disappearing, blowing themselves away on a time scale of only millions of years.

NUCLEAR ENERGY

ATOMIC NUCLEI. Stars are immense nuclear reactors that generate nuclear power. (Sunlight falling on the Earth originated as nuclear energy deep inside the Sun.) We must therefore take a short detour to try to understand how nuclear energy is generated in stars.

Atoms combine to form molecules, and molecules are held together by electrical forces that are the result of atoms sharing or exchanging their outermost electrons. These molecular electric forces are not very strong compared with nuclear forces, and the assembly and rearrangement of atoms within molecules releases relatively weak chemical energy. Because most of the energy used by mankind comes from burning wood, coal, and oil, it is of a chemical kind.

Each atom has a small positively charged nucleus surrounded by a comparatively large cloud composed of negative electrons. The nucleus consists of particles known as nucleons that are either protons or neutrons. Protons are positively charged, but neutrons have no electric charge. These nucleons are held together by relatively strong nuclear forces. The addition, subtraction, and rearrangement of nucleons in the nuclei of atoms releases or absorbs nuclear energy, and this energy is generally millions of times greater than ordinary chemical energy. Ancient alchemists sought for ways of transmuting the elements, and we have now realized their dream. In practice it is done mainly in nuclear reactors, not for the purpose of producing gold from baser metals, but primarily for research and the generation of electrical power.

Imagine that we have a supply of free nucleons that we combine in various ways to produce the atomic nuclei of the elements. Each time a nucleus is constructed out of free nucleons, no matter what kind of nucleus, energy is released. This energy is released because nucleons attract each other with strong short-range nuclear forces. The total amount of energy that is released is called the binding energy of the nucleus. All objects have binding energy of one kind or another: As an illustration, a stone is bound to the Earth by gravity, and its binding energy is that energy released when the stone falls in from outer space and strikes the Earth's surface. In this case the attractive force is gravity. In the case of a molecule the attractive force between atoms is electrical. In the case of a nucleus the attractive force between nucleons is something with which we are normally not familiar and is called the strong force, or *strong interaction.* In order to tear a thing apart into its components we must expend an amount of energy equal to its binding energy; when it is assembled, the binding energy is released.

It is more convenient to think of the binding energy per nucleon, which is the total binding energy of a nucleus divided by its number of nucleons. Figure 2.6 shows the binding energy per nucleon for elements of

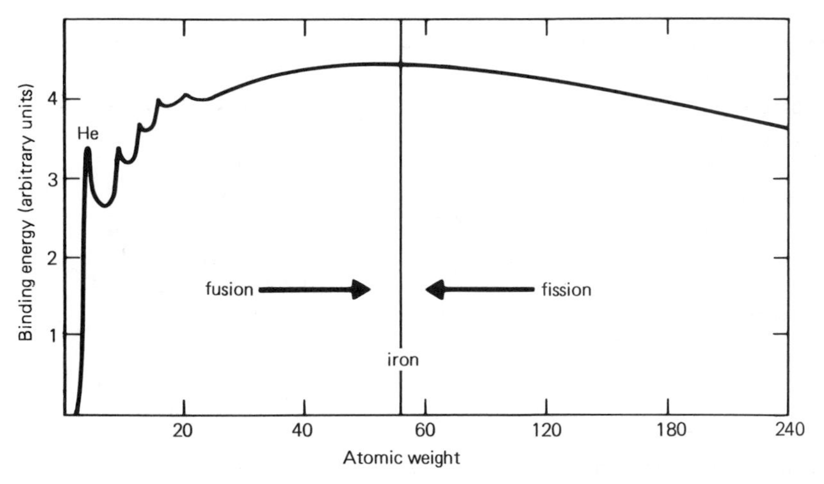

Figure 2.6. The binding energy curve for different nuclei. The maximum binding energy per nucleon occurs in the region of the iron nucleus.

different atomic weight (the atomic weight is roughly the number of nucleons). It rises rapidly at first for the light nuclei, then slowly increases and reaches a maximum at iron (having 56 nucleons), and thereafter diminishes for the heavier nuclei of larger atomic weight. Thus, if we start from scratch with 224 free nucleons, more energy is released by making four iron nuclei than from making one radium nucleus having an atomic weight of 224.

FUSION AND FISSION. Usually we cannot start from scratch with free nucleons. Protons, the nuclei of hydrogen atoms, are easy to find, but neutrons are scarce because in their free state they decay. We must therefore take existing nuclei and either put them together (this is called *fusion*) or break them up (and this is called *fission*). The aim in the nuclear energy game is to increase binding energy and the prize is the energy that is released. To increase nuclear binding energy we must move toward the iron peak shown in Figure 2.6. If the approach is from the left, then energy is obtained by the fusion of light nuclei into heavier nuclei; and if the approach is from the right, then energy is obtained by the fission of heavy nuclei into lighter nuclei. Stars obtain their energy from the fusion of light nuclei, and we at present on Earth derive energy in nuclear reactors by fission of the heavy nuclei of uranium and plutonium.

In all main-sequence stars hydrogen nuclei, or protons, combine to form the nuclei of helium atoms. Four protons are fused together to produce one helium nucleus that weighs almost 1 percent less than the four original free protons. The reason for this loss in weight is that energy has mass, and the released energy carries away a small fraction of the total mass. Energy in every form has mass; a kettle of water, for example, when heated to boiling point weighs a billionth of a gram more than when cold, because heat is a form of energy and therefore has mass. The law that relates energy and mass is

$$\text{energy} = \text{mass} \times c^2$$

where c stands for the speed of light. One ton of matter, if annihilated entirely, would supply the energy needs of the human race for one year. The Sun consumes and radiates away its mass at a rate of 4 million tons per second (See Table 2.8).

BARRIER PENETRATION. Stars on the main sequence produce their energy by burning hydrogen slowly into helium. The energy is released in their central regions, where the temperature and density are highest, and is

Table 2.8. *Energies*

Total energy (mass $\times\ c^2$) in 1 gram	10^{14} joules
Chemical energy in 1 barrel of oil	10^{10} joules
Energy needed to raise 1 gram of water 1 degree Kelvin	4.2 joules
Energy needed to lift flea 1 centimeter	1 erg approximately
1 joule = 10^7 ergs	
Flashlight power	0.2 watts[a]
Power of electrical power station	10^9 watts
Power in sunlight incident on Earth	10^{17} watts
Luminosity of Sun	4×10^{33} ergs per second $= 4 \times 10^{26}$ watts

[a]Power is the rate at which energy is used; a common unit of power is the watt, equal to 1 joule per second.

then transported slowly to the surface. This poses questions for us: Why is the energy released so slowly? Why do stars not explode with an instantaneous and immense release of nuclear energy? There is an important impediment in the generation of nuclear energy by fusion that is easily understood. The positively charged protons deep within a star rush around at high speed and continually approach each other, but because of their electrical repulsion, they rarely come sufficiently close to engage in nuclear reactions. The electrical repulsion therefore acts as a barrier that prevents protons from coming close together.

When a proton approaches another proton it must in effect climb a hill, as shown in Figure 2.7. It moves up the hill, gets only so far, and then comes down again and moves off in a new direction. Protons have different speeds; some are fast, others slow, and the average speed is typically 500 kilometers per second. But this speed is much too small to enable a proton to get anywhere near the top of the hill. To reach the top, a speed of 10,000 kilometers per second is necessary, and in the entire Sun there is not a single proton with this speed.

So far we have spoken of protons as if they were bodies just like stones. In a simpleminded way we can visualize an atom as a miniature Solar System, with the electrons moving in orbits about the nucleus like planets moving around the Sun. Such a picture, however, is a crude mechanistic model that fails to reveal the beauty and intricacy of the atom. All subatomic particles, such as electrons and protons, behave like waves distributed in space that vibrate in various patterns. It is the pulsating electron waves, waltzing about the nucleus, that determine the size and properties of each type of atom.

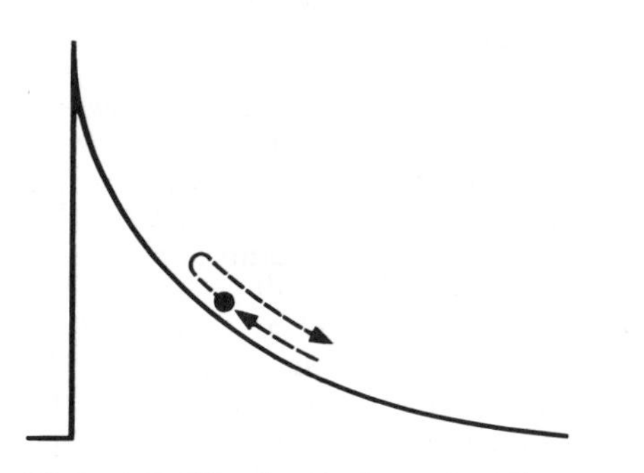

Figure 2.7. The electrical repulsion between two protons is like a hill. Two protons approach each other, and each, in effect, climbs a hill. But, because they lack sufficient energy to reach the top, they climb only part of the way and then roll back and go off in new directions.

Protons, the same as electrons, behave like waves, although their wavelengths normally are much smaller than those of electrons. All elementary particles exhibit corpuscular and wavelike properties and behave in ways quite different from those of familiar bodies such as stones. A stone thrown against a wall will rebound, whereas a wave might penetrate through the wall and emerge on the other side with diminished intensity. Owing to its wavelike nature a particle such as a proton may also penetrate through the wall. This takes us into the world of quantum mechanics, of corpuscle–wave duality, where a particle is a wave distributed in space, and the chance of finding the particle in its corpuscular state at any place is proportional to the square of the wave amplitude at that place. Where the amplitude is largest is where there is the best chance of finding the particle as a discrete entity. A particle such as an electron or a proton is spread over space in the form of waves, and the moment it is detected – because it does something – the waves everywhere collapse into a small region and the particle assumes a corpuscular state. Of course, we cannot have only a bit of a particle, and in the corpuscular state it must always be the whole particle or nothing. Consider what happens when a proton encounters a barrier such as the electrical repulsion barrier between two protons. As a wave, it is partly reflected by the barrier and is also partly transmitted through the barrier. Suppose the amplitude of the wave after penetration is only 1/100 of the original incident amplitude. The chance of finding the particle at a place is proportional to the square of the amplitude, and therefore the chance of finding it on the other side of the barrier is 1/10,000. It is impossible to have only 1/10,000 of a particle in its corpuscular state on one side of the barrier and 9999/10,000 of a particle on the other side. We therefore say that the chance of penetration is 1/10,000, and of every 10,000 particles striking the barrier, on the average 9999 are reflected and 1 is transmitted.

THE IMPEDIMENTS. When protons in a star collide with one another, there is at each collision a small chance of a wavelike penetration of their electrical repulsion barriers (see Figure 2.8). Even in the center of a star, where the temperature is high and protons move fast, the chance of penetration is still extremely small. Each proton makes head-on collisions with other protons about a trillion times a second, and because of the repulsion impediment, about once every second it penetrates a repulsion barrier and comes face to face with another proton.

We now come to the second obstruction to nuclear reactions between protons. When protons meet face to face after penetration they take a long time to react together, and before they have made up their minds that they like each other, they have separated and gone their different ways. Once in 10 billion years, on the average, a proton in the center of the Sun meets another proton face to face, they react together violently, and energy is released as they transform into a new particle called the deuteron. The deuteron is the nucleus of the heavy hydrogen atom (known as deuterium). The deuteron now quickly picks up another proton – for the second impediment does not exist in this case – and with the release of more energy becomes a helium-3 nucleus that contains two protons and one neutron. The helium-3 nuclei then quickly combine with the release of yet more energy and become the helium-4 nuclei of ordinary helium atoms (see Figure 2.9).

The nuclear reactions just described are known as the *proton chain* in which four protons, step by step, become one helium nucleus. As we have seen, there are two main obstructions: Positively charged pro-

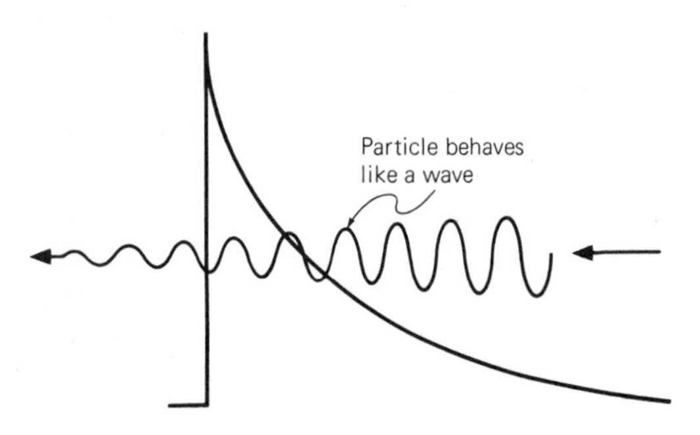

Figure 2.8. A wave penetrating through a barrier illustrates the wavelike penetration of a particle when it approaches another particle.

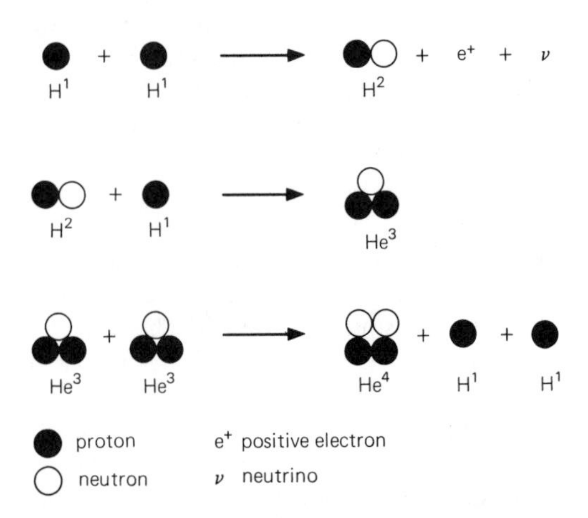

Figure 2.9. The fusion of four protons to give one helium nucleus. First, two protons combine to form a deuteron (H^2), consisting of a proton and a neutron, and a positive electron and a neutrino are created by the reaction. This is a slow reaction, taking about 10 billion years for each proton in the center of the Sun. Next, the deuteron quickly picks up an additional proton and becomes a helium-3 (He^3) nucleus. Two helium-3 nuclei then combine to produce one helium-4 (He^4) nucleus and two protons. This conversion of four protons into one helium nucleus is the proton chain of reactions. There are competing reactions (not shown), in which a helium-3 nucleus combines with a helium-4 nucleus and yields a beryllium nucleus, which becomes, either by radioactive decay or by proton capture, two helium nuclei.

tons have difficulty in penetrating their repulsion barriers; and protons are slow to engage in a reaction even after penetration. The latter delay occurs because a *weak interaction* is involved, creating a positron (a positive electron that is the antiparticle of the common negative electron) and also another particle called the neutrino that moves at the speed of light.

There is an alternative way of converting hydrogen into helium known as the *carbon cycle*. The carbon cycle works as follows: A carbon-12 nucleus combines with a proton and is transformed into nitrogen-13, which then decays and becomes carbon-13; the carbon-13 nucleus combines with a second proton and is transformed into nitrogen-14; the nitrogen-14 nucleus combines with a third proton and is transformed into oxygen-15, which then decays and becomes nitrogen-15; the nitrogen-15 nucleus finally combines with a fourth proton and produces a carbon-12 nucleus and a helium nucleus. At each step of the cycle energy is released and the final outcome of the carbon cycle is that four protons are transformed into one helium nucleus. The carbon-12 nucleus itself acts as a catalyst and is not consumed in the process. The electrical repulsion barriers of carbon-12, carbon-13, nitrogen-14, and nitrogen-15 are more difficult to penetrate than the barriers encountered in the proton–proton chain, and the barrier penetration difficulty is therefore greater in the carbon cycle. But the reluctance to engage in a nuclear reaction after penetration is very much less in the carbon cycle because no weak interaction is involved in the brief encounters. In the carbon cycle there is thus a trade-off in impediments, in which the first is increased and the second is decreased. In lower main-sequence stars, including the Sun, the proton chain dominates, and in upper main-sequence stars the carbon cycle dominates. The upper main-sequence stars have higher central temperatures, and protons are therefore able to penetrate more easily the stronger repulsion barriers of carbon and nitrogen nuclei.

BIRTH OF STARS

DARK CLOUDS. James Jeans, a British physicist, proposed early in this century a general theory of "fragmentation" (see Figure 2.10). In the beginning, he said, the universe was filled with chaotic gas, and astronomical systems were formed in succession by a process of fragmentation "of nebulae [galaxies] out of chaos, of stars out of nebulae, of planets out of stars, and of satellites out of planets." It is still thought that the universe has in some way fragmented into separate galaxies, and that the galaxies have fragmented into stars, but we no longer think that stars fragment into planets, and we doubt that planets have fragmented into satellites such as the Moon.

Most stars in our Galaxy were formed long ago. Yet many are young, and stars are still being born at a rate of a few each year in the large and dark clouds of gas in interstellar space. These clouds consist of hydrogen and helium and contain a small amount – 1 or 2 percent by mass – of heavier elements. Most young stars are still close to their birthplaces and are surrounded by tattered remnants of the clouds from which

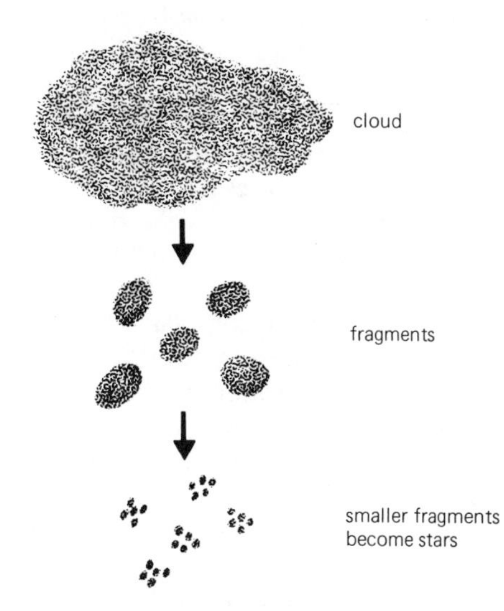

Figure 2.10. The sequential fragmentation of an interstellar gas cloud (or part of a cloud) into smaller and smaller fragments.

Figure 2.11. These majestic stars of the Pleiades were born 60 million years ago and are still festooned with the remnants of the gas cloud from which they were born. (Mount Wilson and Las Campanas Observatories, Mount Wilson Observatory photograph.)

they were born. Sometimes hundreds of young stars are seen clustered together, as in the Pleiades; in such cases it is apparent that the stars were born at the same time from the same dark cloud (see Figure 2.11).

ORIGIN OF THE SOLAR SYSTEM. The Solar System consists of the Sun, Earth, and other planets, and was born almost 5 billion years ago when the universe was nearly half its present age. In the beginning there was a globe of gas and dust, dark and cool, perhaps twice as massive as the Sun. It was a blob, a denser part, of a large cloud where other stars also were forming. The globe consisted mainly of hydrogen gas, with about 25 percent by mass of helium; all the other elements (such as carbon, nitrogen, and oxygen) amounted at most to only 2 percent.

In a somewhat speculative vein, let us change to the present tense, as if we were there watching the sequence of events (Figure 2.12 illustrates the process). The globe of gas rotates, and at first is more or less spherical. But as it contracts, it rotates faster, and then slowly flattens and becomes oblate. The central region (the core) supports the globe and hence contracts faster than the rest; as a result, the core gets denser and spins faster than the outer regions of the globe. Eventually, further contraction of the core becomes slow and difficult because of centrifugal force. But other forces, equally important, are now at work; these are the viscous forces that act

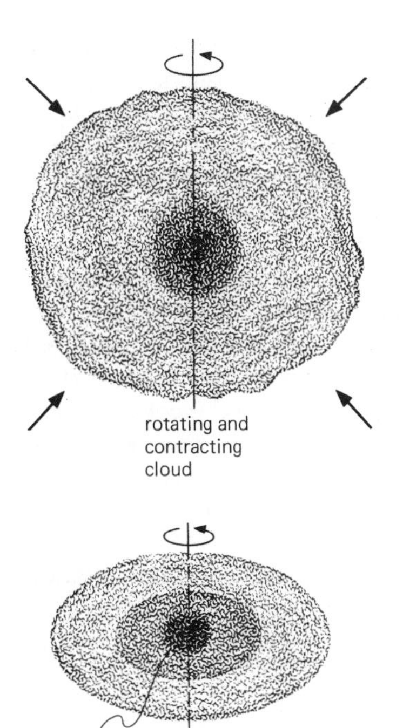

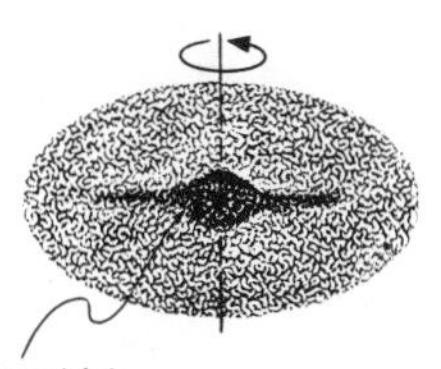

Figure 2.12. A speculative representation of the formation of the Solar System. A rotating globe of gas slowly contracts; as it gets smaller it spins faster, and as a consequence the outer regions flatten. The central region meanwhile becomes dense and hot and eventually becomes the newborn Sun. The planets are formed in an encircling disk of meteoroidal material.

like friction. Because the rotation velocity increases toward the center, the globe in effect consists of layers of gas that rub against each other. This rubbing acts as a braking mechanism that slows down the spinning core and transfers its rotation to the outer parts of the globe. With the aid of viscous forces, the core is able to continue its slow contraction to higher density.

Outside the central region, or core, the gas remains moderately cool. A fraction of the elements heavier than hydrogen and helium consolidate into dust grains about a thousandth of a millimeter in diameter. These grains of dust repeatedly collide with one another and begin to stick together. They coagulate and form small pieces of meteoroidal rock and ice. Perhaps more than 50 percent of the meteoroidal material is ices of frozen water, ammonia, and methane, and the rest consists of such elements as metals and their oxides and silicates. When the meteoroids have grown and become pebble-sized, they move more or less freely through the gas and begin to behave like tiny planets. But friction resulting from motion through the gas causes their orbits slowly to become circular and to concentrate into a flat disk. Within this rotating disk the meteoroids ceaselessly jostle each other and either break up into smaller pieces or coalesce to form larger and larger chunks of matter. This is a game of survival of the biggest; the bigger the meteoroid, the more often it eats up the smaller meteoroids, and the less likely it is to be shattered by collisions. The large meteoroids grow into planetesimals, hundreds and thousands of meters in diameter. At the same time there is a continual downpour into the disk of newly formed meteoroids. The planetesimals themselves occasionally collide and either dissolve into far-flung fragments or amalgamate into larger planetesimals. In the planetesimal struggle of survival of the biggest the victors sweep up all they encounter and become the protoplanets. The protoplanets – massive earthlike objects – now attract and begin to retain the gaseous elements of the globe and acquire atmospheres that are rich in hydrogen and helium. These atmospheres grow and the protoplanets become great planetary spheres of gas, not unlike Jupiter and Saturn at present.

Meanwhile, the core of the globe becomes dense and hot and approaches its final state. It is the primordial Sun, blanketed from view by great quantities of swirling gas and dust. It has discovered that by burning hydrogen into helium it has gained access to an immense reservoir of nuclear energy. The primordial Sun adopts a convulsive state, becoming a flaring T Tauri–type star, while seeking to find an internal structure that will match the rate at which nuclear energy is released in the center to the rate at which energy is radiated from the surface. (T Tauri is an irregularly varying star that is surrounded with dense gas and dust, and is believed to be a newborn star approaching the main sequence. Similar stars are referred to as T Tauri–type stars.) As the primordial Sun approaches the main sequence it is in an eruptive state, and from its upheavals issues an intense wind of fast-moving gas that rushes outward and carries away large quantities of matter. The fierce wind and brilliant radiation from the newborn Sun thrust the remnants of the gaseous globe back into interstellar space.

Now commences a final struggle between the planets and the Sun. The planets seek to retain their massive gaseous outer layers, while the bright Sun with its fierce wind seeks to strip away these outer garments. The inner planets – Mercury, Venus, Earth, Mars – lose the struggle and are stripped down to their rocky cores; the outer planets – Jupiter, Saturn, Uranus, Neptune – win because of their greater distances, and retain forever the lighter elements that constitute the major fraction of their total mass (see Figure 2.13).

The Sun at last settles down into a quiescent state. At first the Solar System is cluttered with debris from the birth process.

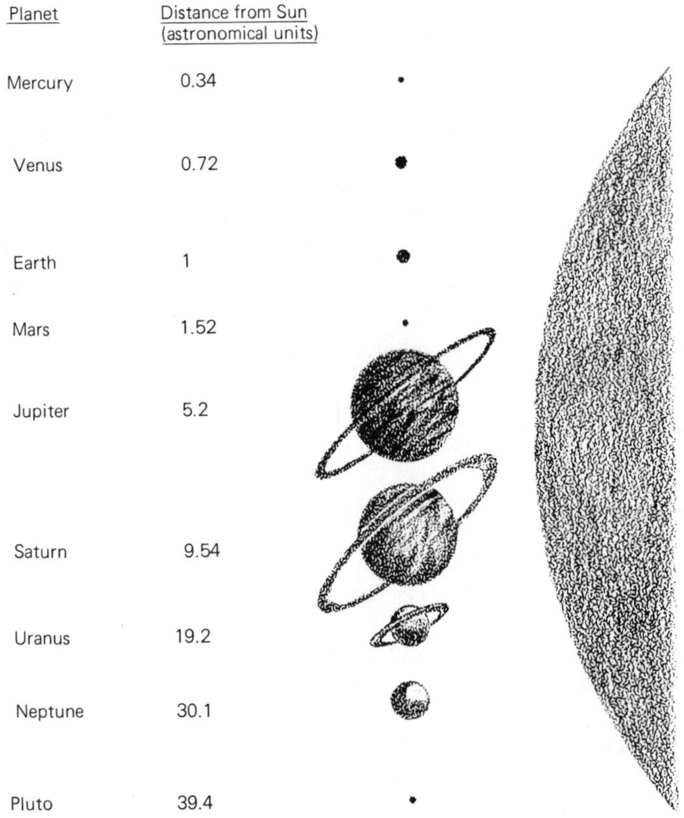

Planet	Distance from Sun (astronomical units)
Mercury	0.34
Venus	0.72
Earth	1
Mars	1.52
Jupiter	5.2
Saturn	9.54
Uranus	19.2
Neptune	30.1
Pluto	39.4

Figure 2.13. The relative sizes of the planets of the Solar System and their distances from the Sun.

Scattered planetesimals encircle the Sun and the planets in great numbers, and hordes of meteoroids drift in interplanetary space. The mopping up of this debris by the Sun and planets is at first rapid, and then slows down after several hundred million years. This is the bombardment era that lasts for roughly half a billion years, and the evidence of this era of intense bombardment is still visible on the surfaces of the Moon and the planets such as Mercury and Mars.

Between the orbits of Mars and Jupiter are concentrated tens of thousands of planetesimals that survive as the asteroids; for reasons unknown they failed to amalgamate into a planet. Other planetesimals also survive as moons distributed about the various planets; the inner planets have 3 of which the Moon is the largest, and the outer planets have at least 29.

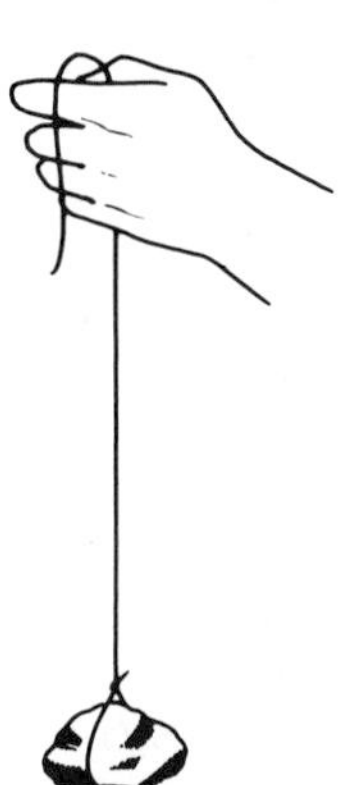

Figure 2.14. If you have difficulty in understanding the release of gravitational energy, try the following experiment. Allow a string, with a weight attached to the end, to slip through the palm of your hand. The heat generated in the palm of the hand comes from the release of gravitational energy.

THE STAR IS DEAD! LONG LIVE THE STAR!

HYDROGEN EXHAUSTION. Advanced stellar evolution is still not fully understood. We know that stars less massive than the Sun remain on the main sequence for longer than 10 billion years and therefore have not evolved appreciably in the lifetime of our Galaxy. But more massive stars evolve much quicker and we think they terminate their careers, according to their masses, as white dwarfs, as neutron stars, or as black holes. Here we consider what happens to these more massive stars.

After a star consumes its central supply of hydrogen, it leaves the main sequence and moves off in the direction of the red giants in the H–R diagram. The core consists almost entirely of helium and is no longer generating nuclear energy. But energy is still radiated from the hot surface into space, and this energy is drained from the core of the star. In response, the core does the only thing possible: It contracts to higher density and temperature, thus releasing gravitational energy (see Figure 2.14). Two things now happen.

First, hydrogen just outside the helium core begins to burn. Surrounding the core there is now a hydrogen-burning shell in which the production of helium steadily increases the mass of the helium core. Second, the energy released by core contraction and the hydrogen-burning shell causes the envelope to expand. As the core contracts, the star swells up, and the envelope becomes partly convective. The star has become a red giant.

Once a star leaves the main sequence it has entered old age and has comparatively little time left to live. It realizes too late that life on the main sequence has been dull and sedentary, and resolves therefore to have a last glorious fling before death. Unfortunately, not much nuclear energy is left. The burning of hydrogen into helium throughout its life has expended most of the available nuclear energy, and to draw on the remaining small fraction requires the prodigious feat of burning helium step by step all the way to nickel and iron. At each step, higher central temperatures and densities are needed; hence the star grows in luminosity, and its diminishing nuclear reserve is consumed at an ever-increasing rate.

RED GIANTS AND WHITE DWARFS. We start by

considering stars less massive than about 2 solar masses. These low-mass stars evolve into white dwarfs. The central region, or core, of such stars continues to contract during the red giant phase until its density and temperature are sufficient to burn helium into carbon. The ignition of helium occurs abruptly, in fact explosively, at a temperature of a 100 or so million degrees, and is referred to as the *helium flash*. Either at this stage or a little later, depending on the mass of the core, the star puffs away its distended envelope and the bright core is all that is left. Further contraction and nuclear burning soon cease and the naked core settles down as a white dwarf. The star consists of an ejected gaseous remnant, called a planetary nebula, and a slowly cooling white dwarf that has a size approximately equal to that of Earth.

The Sun, in about 5 billion years, will become a white dwarf. It will then shine in the sky as a pale light for several billion years while slowly cooling.

SUPERNOVAS. We now consider what happens to a more massive star of the upper main sequence. It squanders its central supply of hydrogen at a rapid rate and lives on the main sequence for only a few tens of millions of years. Then it becomes a monstrous red giant, having a helium core that is surrounded by a hydrogen-burning shell. As the core contracts to higher density and temperature, the helium begins to burn into carbon and oxygen. The star then has a core of carbon and oxygen, surrounded by a helium-burning shell, which is surrounded by a hydrogen-burning shell. Having evolved beyond the red giant stage, the star now becomes even more luminous, passing through spasms of pulsation and ejecting gas into space at high velocity.

To meet the ever-mounting demand for more energy the core continues to contract. When the central temperature exceeds 3 billion degrees, and the density approaches a million grams per cubic centimeter, the carbon and oxygen burn progressively, stage by stage, into neon, magnesium, silicon, phosphorous, sulphur, and so on, to nickel and iron. The nuclear energy released by this multitude of reactions is comparatively small and is soon radiated away.

During these latter stages of advanced evolution an additional loss of energy steadily increases and now becomes serious. Hordes of neutrinos, produced in the core by nuclear reactions and the high-temperature gas, stream out through the star freely and in vast numbers. The neutrino luminosity of the core rises until it exceeds the radiation luminosity from the surface. The only reserve now left is gravitational energy, and to meet the growing loss of energy the core must contract even faster. The central density and temperature soar and energy spills over and drains into the production of elements heavier than iron.

The star is only seconds away from death. The neutrinos no longer easily escape from the star; instead they transport energy to the outer layers, thus heating and igniting nuclear reactions in the mantle of the star. The inward-falling core crushes and breaks up its heavy elements back into helium, and the energy previously acquired by fusion of light elements into heavier elements must be paid back by the release of further gravitational energy. The neutrino wind intensifies, becoming a blast that lifts the exploding mantle and hurls it into space. In the last moments of the imploding core, helium is crushed into protons and neutrons, and all the energy that was radiated for millions of years while the star was on the main sequence must be paid back immediately. The core obtains this energy by catastrophic collapse. The electrons are squeezed into the protons and together they become neutrons; the collapsed core, divested of its mantle, emerges as a neutron star.

The titanic blaze of energy released by the imploding core and the exploding mantle results in a supernova, which for a short time shines as bright as all the stars in our Galaxy together: A supernova at the distance of Alpha Centauri would be as bright as the Sun.

NEUTRON STARS AND BLACK HOLES. A neutron star has a radius of little more than 10 kilometers and a density of nearly 1000 trillion grams per cubic centimeter. A thimbleful of neutron matter would weigh on Earth a billion tons. The neutron star has a magnetic field of 10^{12} gauss – a trillion times stronger than the Earth's magnetic field – and at first rotates rapidly at hundreds of times per second.

Long live the star! From the ashes of the old star a pulsar has been born, a star that ululates across space a pulselike message of matter at its uttermost limits. For millions of years, pulsing more and more slowly, the pulsar radiates away its rotational energy (see Figure 2.15).

Neutron stars have masses less than about three times the Sun's mass, and this is because neutron matter cannot withstand the gravitational pull of greater masses. The imploding cores of massive stars may therefore not always terminate as neutron stars, but continue to collapse and become black holes. These intriguing objects are discussed in Chapter 9; at present it is sufficient to say they have very strong gravitational fields and are enclosed within curved spacetime, and that matter may fall into a black hole but can never escape.

REFLECTIONS

1 *A total of roughly 2000 stars can be seen by the naked eye from the Earth's surface. With a pair of good binoculars this number is increased to about 12,000. The nearest star, Alpha Centauri, consists of two stars too close together to be resolved*

Figure 2.15. The Crab Nebula in the constellation Taurus is a strong source of radio waves and has a total luminosity 100,000 times that of the Sun. This immense output of energy of the nebula comes from the pulsar in its center, which rotates 33 times a second. (Mount Wilson and Las Campanas Observatories, Mount Wilson Observatory photograph.)

by the naked eye. It is actually a triple system: The third star, Proxima Centauri, is faint and slightly closer to us than the other two stars of the system. Sirius, the brightest star seen from the Earth (other than the Sun), has also a faint white dwarf companion.

2 *Stars have different colors and apparent brightnesses. Why? Sketch a color–brightness diagram and show where you would put ordinary sources of light, such as a campfire, a flashlight bulb, a candle flame, a spark, a firefly, and so on.*

3 *Kepler's three laws of planetary motion (see Figure 2.16) are:*

First law: A planet moves about the Sun in an orbit that is an ellipse with the Sun at one focus of the ellipse.

Second law: A straight line joining the planet and Sun sweeps out equal areas within the orbit in equal intervals of time.

Third law: This important law, published in 1619 in the Harmony of the Worlds, *states that the square of the period of the planet is proportional to the cube of the average distance of the planet from the Sun. If P is the orbital period measured in years, and R is the average distance of the planet from the Sun measured in astronomical units, then* $P^2 = R^3$ *for all planets. For Venus:* $P = 0.615$, $R = 0.723$; *for Earth:* $P = 1$, $R = 1$; *for Mars:* $P = 1.881$, $R = 1.524$.

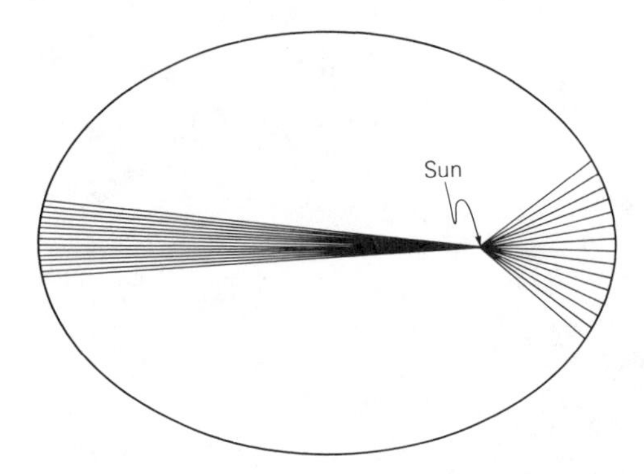

Figure 2.16. Kepler's first two laws of planetary motion: (1) a planet moves about the Sun in an orbit that is elliptical, with the Sun at one focus of the ellipse; (2) a straight line joining the planet and the Sun sweeps out equal areas within the orbit in equal intervals of time.

Kepler's laws were later explained by Newton and shown to be the consequence of the laws of motion and the inverse-square law of gravity. Kepler's laws are quite general and also apply to double stars. Consider two stars, labeled 1 and 2, and let their masses be M_1 *and* M_2 *measured in solar mass units. A star 10 times the mass of the Sun has therefore a mass of 10 units. The period P of revolution of the two stars about each other measured in years and their average distance R apart measured in astronomical units are related by*

$$(M_1 + M_2)P^2 = R^3$$

For example, if $M_1 = 50$, $M_2 = 50$, *and* $R = 1$, *then P is 1/10 of a year.*

4 *William Herschel (1738–1822) was the first to point out that astronomy has much in common with botany. In* Construction of the Heavens *he wrote: "This method of viewing the heavens seems to throw them into a new kind of light. They are now seen to resemble a luxuriant garden, which contains the greatest variety of productions, in different flourishing beds; and one advantage we may at least reap from it is, that we can, as it were, extend the range of our experience to an immense duration. For, to continue the simile I have borrowed from the vegetable kingdom, is it not always the same thing, whether we live successively to witness the germinations, blooming, foliage, fecundity, fading, withering, and corruption of a plant, or whether a vast number of specimens, selected from every stage through which the plant passes in the course of its existence, be brought at once to our view?" Herschel was the greatest astronomer of his time; he discovered the planet Uranus, showed that the motions of double stars are in accord with Kepler's laws, made endless observations, and attempted to determine the position of the Sun within the Milky Way. He believed that the Moon and planets were inhabited and also thought that beneath the hot atmosphere of the Sun was a cool surface that might also be inhabited.*

5 *Matter in bulk exists in four common states: the solid, liquid, gaseous, and* *plasma states. Often, but not always, each exists at a higher temperature than the preceding state. A solid is rigid and consists of atoms, vibrating about fixed points, held together by interatomic electrical forces. Liquids, gases, and plasmas are fluids. When the temperature of a solid is raised, the atoms vibrate more strongly and either singly or in clumps are able to slide around each other, and the solid has become a liquid. At still higher temperatures the vibrating atoms are able to break their bonds and evaporate to form a gas of freely moving particles. At even higher temperatures (more than several thousand degrees) the atoms become ionized and the gas changes into a plasma consisting of free electrons and partially or fully ionized atoms. Most matter in the universe is in the plasma state.*

The number of particles (atoms or molecules) in 1 cubic centimeter of gas at room temperature and atmospheric pressure at sea level is 2.7×10^{19}. They all rush about, hither and thither, separated by comparatively large regions of empty space, and each collides with the others about a billion times a second.

6 *The four forces that rule the universe, in order of increasing strength, are*

gravitational interaction:	*1*
weak interaction:	10^{25}
electromagnetic interaction:	10^{39}
strong interaction:	10^{40}

Gravitational and electromagnetic forces are long range (they fall off slowly in strength as the inverse square of distance) and account for all the rich diversity in the universe from atoms to galaxies. Weak and strong forces are short range (they operate over only very short distances) and account for the diversity of atomic nuclei and the behavior of subatomic particles.

Electrical forces are very much stronger than gravitational forces, as shown by the following illustration. An electron has a negative charge denoted by e, and a proton has a positive charge of e. The attractive force between a proton and an electron is e^2 divided by the square of their separating distance. The gravitational force between the same particles, an electron of mass m_e and a proton of mass m_p, is Gm_em_p divided by the square of their separating distance, where G is the universal constant of gravity. The value of G is found by measurement (no scientist has yet been clever enough to explain the value). The ratio of the two forces between an electron and a proton is

$$\frac{\text{electrical force}}{\text{gravitational force}} = \frac{e^2}{Gm_em_p} = 2 \times 10^{39}$$

and the electrical force is thus shown to be vastly stronger than the gravational force in a hydrogen atom. Gravity is far too weak to hold atoms together. The universe contains equal numbers of positive and negative electric charges, which all tend to cancel each other out in systems much larger than atoms and molecules. In our electrically neutral universe the electromagnetic forces are therefore rarely strong except over short distances. Gravity, on the other hand, is not neutralized and gets progressively stronger as the number of particles in a system increases. Gravitational forces between single particles are the weakest in nature, but in stellar and galactic systems gravity is the dominant force, and on the cosmic scale it is by far the strongest of all forces.

The strong interaction is a short-range force that acts between certain particles and between the nucleons in a nucleus. It is stronger than the electromagnetic interaction and is therefore able to hold together the protons in a nucleus despite their electrical repulsion. If it were not stronger than the electromagnetic force, the universe would consist only of hydrogen and we would not be here discussing the subject. The strong forces operate over a distance of 10^{-13} centimeters (roughly the size of a nucleon) and act rapidly in a time of 10^{-23} seconds (given by the light travel size of a nucleon).

The weak interaction is of even shorter range and acts very much more slowly than the strong force. Various weak interactions between particles produce neutrinos, ghostly particles with no electric charge *that move at the speed of light. The slowness of the weak interaction is responsible for the main impediment in the proton–proton reaction in stars, and is also the reason why neutrinos pass easily through ordinary matter. Nuclear reactions in the Sun produce 10^{38} neutrinos each second, and "like ghosts from an enchanter fleeing," in Shelley's words, they all stream out freely through the Sun into space. Every second 1000 trillion solar neutrinos pass through the human body, even at night, when the Sun is on the other side of the Earth.*

7 *The technological dream of modern science is to find a way on Earth to produce controlled nuclear energy by fusion, as in stars. The aim is to use deuterium – there is plenty of heavy hydrogen in the oceans – instead of ordinary hydrogen, and thus avoid the second obstruction in the proton chain, the difficulty protons have in reacting together. But the first impediment still remains. Although deuterons react together willingly and release plenty of energy, there always exists the problem of penetrating their repulsion barriers. It is widely thought that the best way to achieve fusion is to imitate the stars and create a hot gas at a temperature of tens, and even hundreds, of millions of degrees. This thermonuclear release of energy has already been done in an uncontrolled fashion with the heavy hydrogen bomb, using a fission bomb as a detonator. The Sun also would explode if it were made solely of heavy hydrogen. We desperately need a gentle, controlled way of releasing fusion energy, but nobody yet knows how to maintain a confined gas at a very high temperature for a sufficiently long period of time.*

8 *Suppose that in a glass of water we can radioactively tag each molecule of H_2O. Now assume that we pour the glass of water in the ocean and wait until it is thoroughly mixed with all the water on the Earth's surface. Show that when the glass is dipped back in the ocean it contains roughly 1000 of the original molecules.*

∗ *What would happen to us if all nuclear reactions in the center of the Sun were to cease abruptly at this moment?*

∗ *A 100-watt light bulb radiates about 10^9 ergs of energy each second. How many grams of mass does it radiate in 1 year, and where does this mass come from and go to?*

∗ *The Sun's luminosity (energy radiated per second) is denoted by the symbol $L_\odot$, and $L_\odot = 4 \times 10^{33}$ ergs per second. A watt is equal to 10^7 ergs per second; therefore, the Sun radiates 4×10^{26} watts, or 4×10^{20} megawatts. A large power station generates 1000 megawatts, and the Sun is therefore equivalent to 400,000,000,000,000,000 large power stations. How much mass does the Sun radiate each year?*

∗ *Give examples from everyday life of the various states of matter.*

∗ *The universe is ruled not by the Furies, nor by the Four Horsemen of the Apocalypse, but by the four forces of nature. Discuss these forces.*

∗ *Are you for or against nuclear power stations? Explain why. Should we consider sunlight as nuclear energy?*

∗ *What are the impediments in fusion reactions?*

9 *About 400 supernovas have been observed in other galaxies this century. Each, for a few weeks, may outshine all the rest of the stars in its galaxy. For the last 2000 years we have records of only 14 supernovas in our Galaxy; the last seen was in 1604. Of the famous supernovas, the one in 1054 in the constellation Taurus was recorded by the Chinese and has produced the Crab Nebula; Tycho's star of 1572 was in the constellation Cassiopeia; and Kepler's star of 1604 was in the constellation Serpens. In 1572, Tycho Brahe wrote: "One evening when I was contemplating, as usual, the celestial vault, whose aspect was so familiar to me, I saw, with inexpressible astonishment, near the zenith, in Cassiopeia, a radiant star of extraordinary magnitude."*

During the last 3 to 4 billion years several supernovas have no doubt occurred within 100 light years' distance, each filling the sky with a blaze of light and drenching the Earth with high-energy particles and X-rays. Do you think it possible that mutations caused by these celestial explosions have affected the evolution of life?

FURTHER READING

Readers interested in stars (and galaxies) can consult one of the many elementary astronomy texts that are now available.

SOURCES

Allen, C. W. *Astrophysical Quantities*. Athlone Press, London, 1973. A valuable collection of astronomical and astrophysical data.

Bahcall, J. M. "Neutrinos from the Sun." *Scientific American,* July 1969.

Berendzen, R., and DeVorkin, D. "Educational materials in astronomy and astrophysics: resource letter." *American Journal of Physics 41,* 783 (June 1973).

Bok, B. "The birth of stars." *Scientific American,* August 1972.

Burchfield, J. D. *Lord Kelvin and the Age of the Earth.* Science History Publications, New York, 1975.

Cameron, A. G. W. "Origin and evolution of the Solar System." *Scientific American,* September 1975.

Davies, P. C. W. *The Forces of Nature.* Cambridge University Press, New York, 1979.

Franknoi, A. "The music of the spheres: astronomical sources of musical inspiration." *Mercury,* May–June 1977.

Gursky, H., and Heuvel, E. van den. "X-ray emitting double stars." *Scientific American,* March 1975.

Heel, A. van, and Velzel, C. *What Is Light?* McGraw-Hill, New York, 1968.

Helfand, D. J. "Pulsars: physics laboratories in our Galaxy." *Mercury,* May–June 1977.

Herbig, G. "The youngest stars." *Scientific American,* August 1972.

Hoskin, M. A. *William Herschel and the Construction of the Heavens.* Science History Publications, New York, 1963.

Iben, I. "Stellar evolution: comparison of theory and observation." *Science 155,* 785 (1967).

Jastrow, R., and Cameron, A. G. W., eds. *Origin of the Solar System.* Academic Press, New York, 1963.

Jeans, J. H. *Astronomy and Cosmogony.* Cambridge University Press, Cambridge, 1929. See p. 415.

Ostriker, J. "The nature of pulsars." *Scientific American,* January 1971.

Pacini, F., and Rees, M. "Rotation in high-energy astrophysics." *Scientific American,* February 1973.

Parker, E. "The Sun." *Scientific American,* September 1975.

Reeves, H. "The origin of the Solar System." *Mercury,* March–April 1977.

Richardson, R. S. *The Star Lovers.* Macmillan, New York, 1967. Biographies of astronomers.

Ruderman, M. A. "Solid stars." *Scientific American,* February 1971.

Shapley, H. *Beyond the Observatory.* Charles Scribner's Sons, New York, 1967.

Strom, S. E., and Strom, K. M. "The early evolution of stars." *Sky and Telescope,* May 1973, June 1973.

Wheeler, J. A. "After the supernova, what?" *American Scientist,* January–February 1973.

GALAXIES

The fires that arch this dusty dot –
Yon myriad-worlded ways –
The vast sun-clusters' gathered blaze,
World-isles in lonely skies,
Whole heavens within themselves amaze
Our brief humanities.
— Tennyson (1809–92), *Epilogue*

OUR GALAXY

Those who have lived in deserts and out-of-the-way places, far from the light of cities, can appreciate why the night sky was so impressive to the people of earlier times. On clear moonless nights the vault of heaven swarms with twinkling lights, and the Milky Way – the *via lactea* – is a wraithlike arch stretching across the sky. The Milky Way is part of our Galaxy, an enormous system of clouds of glowing gas and 100 billion stars: Light takes 100,000 years to cross the Galaxy from side to side. The center of the Galaxy lies in the constellation of Sagittarius and is obscured from view by dusty clouds of gas that drift among the stars. Far from the center of the Galaxy is our star the Sun.

THE DISK AND HALO. The Galaxy consists of two basic parts: the *disk* and the *halo* (see Figures 3.1 and 3.2). The Milky Way is actually our panoramic view of the disk, which has a diameter of about 100,000 light years and a thickness of approximately one-twentieth its diameter. The disk is composed of stars and gas and contains perhaps as much as half the total mass of the Galaxy. The gas accounts for one-tenth of all the matter in the disk, and the dust, scattered everywhere, accounts for about 1 percent of the mass of the gas. The disk of stars and gas rotates about the center, or nucleus, of the Galaxy like a giant merry-go-round. Most of the stars we see in the sky belong to the disk and are separated from one another by distances of a few light years. The Sun lies at a distance of 30,000 light years from the nucleus; it moves at 300 kilometers per second and takes 200 million years to travel around the Galaxy in a circular orbit. In its lifetime the Sun has journeyed 25 times around the Galaxy.

The halo is spherical, with a diameter of roughly 200,000 light years, and has its center at the nucleus of the Galaxy. The central region of the halo consists of a vast concentration of stars that forms the *nuclear bulge* of the disk. Elsewhere the halo contains gas of very low density, and also widely separated stars and 120 globular clusters. The globular clusters are compact systems; each contains hundreds of thousands of stars, and each moves in an elliptical orbit about the nucleus of the Galaxy (see Figure 3.3). The halo, or spherical component of the Galaxy, does not rotate with the disk.

TWO POPULATIONS OF STARS. The disk is rich in gas and dust and rotates; the halo has very little gas and dust, and it rotates, if at all, much more slowly. There is yet another distinction, discovered by Walter Baade in

Figure 3.1. A pictoral side view of our Galaxy, showing the disk with its nuclear bulge and the halo containing globular clusters. (With permission from J. S. Plaskett, Popular Astronomy 47, 255, 1939.)

1942: There are two kinds of stars, called populations I and II. The disk contains population I stars and the halo has population II stars.

Population I stars are the kind we have considered so far. They are the disk stars that usually are much younger than the Galaxy and contain heavy elements (heavier than helium) amounting to 1 or 2 percent by mass. The Sun is a population I star. All such stars are born from gas clouds in the disk, and when they die, they eject heavy elements back into space. The reservoir of gas in the disk is slowly diminishing and is also slowly being enriched with heavy elements, and each newborn population I star contains heavy elements made in stars that previously have died. In interstellar space the atoms of heavy elements sometimes stick together, when they collide, and form grains of dust usually less than 1/10,000 of a centimeter in size. These grains of dust ride the atomic winds and collect in dark clouds where new stars are continually born.

Population II stars are found in the nuclear bulge and in the halo with its many globular clusters. They are very old stars, born long ago when the Galaxy was young, with estimated ages of between 8 and 15 billion years. There is very little gas in the halo, and it is certainly not dense enough for the formation of new stars; this is why the halo contains only old stars. In the nuclear bulge we find a mixture of disk and halo stars, that is, a mixture of populations I and II.

The population II stars were formed from gas that consisted of hydrogen and helium and practically no heavy elements. It is therefore unlikely that they have planetary systems. The younger population I stars are

Figure 3.2. The spiral galaxy M31 in Andromeda, which resembles our own Galaxy and is in the Local Group of galaxies at a distance of 2 million light years. (Mount Wilson and Las Campanas Observatories, Mount Wilson Observatory photograph.)

still being born and are formed from gas that is contaminated with the heavy elements made in earlier stars. It is possible that many, if not most, population I stars have planetary systems. These are the stars where we might expect planets supporting other living creatures to exist. Although it is convenient for us to think of two distinctly different populations – the young disk stars rich in heavy elements, and the old halo stars poor in heavy elements – there are naturally many stars of an intermediate kind.

STAR CLUSTERS AND THEIR DISTANCES. Clusters of stars are of two types: open and globular. The open clusters are found in the disk and are loose groups of young popula-

Figure 3.3. The globular star cluster M3 consists of hundreds of thousands of old (population II) stars. This cluster, held together by its gravitational attraction, is one of many such clusters in the halo of the Galaxy. (Association of Universities for Research in Astronomy, Inc., The Kitt Peak National Observatory.)

tion I stars; they usually consist of hundreds of stars and are often associated with clouds of gas, as in the Pleiades Cluster. Globular clusters are spherical in shape, have diameters of about 100 light years, and consist of hundreds of thousands of old population II stars. All clusters of both types are important because they enable the astronomer to estimate distances.

Distances of the nearby stars are found by parallax measurements. The distances of stars much further away can be guessed by comparing their apparent brightnesses with those of the nearby stars of similar kinds. But this latter method is not very reliable, and better results are obtained by comparing clusters of stars. All stars in a single cluster are at approximately the same distance from us and all are approximately the same age. They form a family of stars, and we take their family portrait. We make an H–R diagram by plotting their apparent brightnesses against their surface temperatures. These stars were all born from the same gas cloud and have therefore the same composition of heavy elements. As a result they fit beautifully on a well-defined main sequence. By comparing the various family portraits, or H–R diagrams, of different clusters we can find their relative distances. For instance, if the main sequence in one H–R diagram is four times fainter than the main sequence in a second H–R diagram, we know that the first cluster is at twice the distance of the second cluster. This method works well, particularly when allowance is made for differences in the heavy-element composition of the various clusters.

The nearest open cluster is the Hyades, which outlines the face of the Bull in the constellation Taurus. It consists of at least 200 stars and is at a distance of 140 light years. Although it is sufficiently near for its distance to be determined by parallax measurements, its distance can be determined more accurately by the moving-cluster method: The cluster is moving away from us and therefore appears to be slowly shrinking in size; by measuring its velocity and apparent rate of shrinkage, we can determine its distance. From the known distance of the Hyades cluster it is then possible to find the distances of other open clusters.

A few of the open clusters, now at known distances, contain cepheid stars that obey the period–luminosity law. These pulsating stars are 10,000 times more luminous than the Sun and can be observed at large distances. It is thus possible to take giant strides and use the cepheids as the yardsticks for measuring the distances of nearby galaxies. The Hyades cluster is hence very important, and its distance determines the size of the Galaxy and even the size of the universe. The estimation of astronomical distances is beset by many difficulties, however, not the least of which is the absorption of starlight by dust in interstellar space.

CLOUDS OF GAS AND DUST. The word *nebula* – meaning cloud – was once used in astronomy to denote any fuzzy patch of light. Charles Messier in the eighteenth century catalogued many nebulae, several of which, as we now know, are distant galaxies (such as the Andromeda Nebula). It is the custom nowadays to use *nebulae* only when referring to clouds of interstellar gas (see Figure 3.4), and we distinguish between reflection and emission nebulae. The reflection nebulae reflect light from adjacent stars and are reddish in color; the emission nebulae are hot – heated by adjacent or embedded stars – that emit their own light and are bluish. Many clouds are not luminous, and these dark clouds, often very large, are distributed throughout the disk and obscure from view the more distant stars and galaxies.

For a long time astronomers suspected that starlight is absorbed by matter floating in space between the stars. The existence of obscuring matter was finally settled in 1930 by Robert Trumpler of the Lick Observatory. When we look outward from the disk, away from the Milky Way, we can see distant galaxies; but galaxies cannot be seen in the direction of the Milky Way because we look into the obscuring disk. Trumpler showed that the light from stars, owing to absorption, is reduced by half for every 3000 light years traveled in the disk. Thus a disk star 6000 light years away has its brightness reduced to one-quarter the amount it would have if there were no absorption, and we cannot see far in the plane of the disk because gas and its attendant dust act like a fog. Disk stars are visible out to distances of several thousand light years, but at greater distances, particularly in the direction of the galactic center, they are totally obscured from view. When we look at the Milky Way we see nearby stars but no galaxies, but when we look away from the Milky Way we see distant galaxies and fewer stars.

Radiation at wavelengths greater than the size of dust particles is not easily absorbed or scattered (see Figure 3.5 for wavelengths). This is why yellowish-red light penetrates better than bluish light in foggy weather. The inner region of the Galaxy, obscured from view in visible light, can therefore be observed with long infrared and radio waves. Observations at these longer wavelengths show that the nuclear bulge, extending out to a distance of 10,000 light years from the nucleus, is chaotically complex and contains swirling gas streams and multitudes of stars. Several strong sources of radiation within the nucleus are still not fully understood. Old and young stars are thickly clustered toward the nucleus and are hundreds of times closer together than the stars in our part of the Galaxy. Some astronomers have conjectured that the nucleus may contain massive black holes. These monsters of the deep cannot by themselves radiate, but the gas that falls

Figure 3.4. This dark interstellar cloud – the Horsehead Nebula in the constellation Orion – is seen silhouetted against a background of stars and luminous clouds. (Mount Wilson and Las Campanas Observatories, Mount Wilson Observatory photograph.)

inward is heated and radiates strongly before it is captured.

SPIRAL ARMS. Only in the last fifty years have astronomers begun to understand the structure of the Galaxy. How this knowledge evolved from earlier ideas and how it has affected our understanding of the universe are matters discussed in Chapter 4. Careful observations of the way that stars and gas are distributed have established that the disk of the Galaxy contains spiral arms.

Many external galaxies are disk-shaped and have a spirallike appearance (see Figure 3.6). Their spiral arms are delineated by luminous young stars and bright clouds of gas, and we see the spiral arms clearly because that is where the brightest stars and

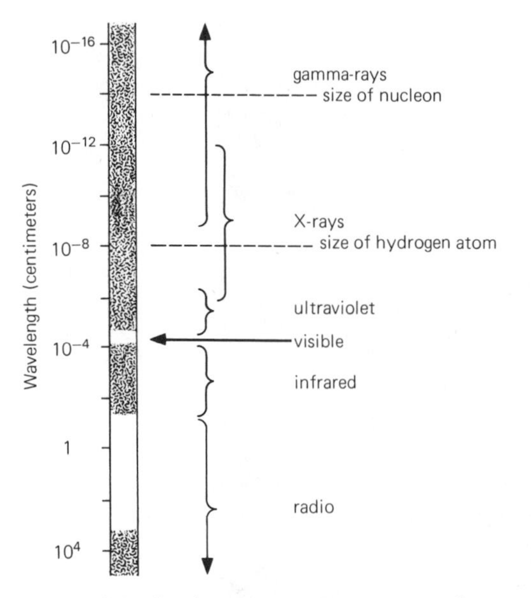

Figure 3.5. The electromagnetic spectrum, showing the gamma-ray, X-ray, ultraviolet, visible, infrared, and radio wave regions. The Earth's atmosphere is transparent to radiation in the unshaded wavelength regions.

gas clouds are found. The other stars in the disk are not arranged into a conspicuous spiral pattern.

At first astronomers thought that spiral arms were composed always of the same stars and that the stars and spiral arms rotated together in the disk. Then it was realized that this explanation led to a dilemma. The inner stars of the disk must revolve about the galactic center more rapidly than the outer stars, in the same way that the inner planets revolve about the Sun more rapidly than the outer planets, and therefore the arms would slowly wind up and progressively form a tighter spiral. Every few hundred million years each spiral arm would gain an additional turn. Yet many spiral galaxies have only one or two turns and are billions of years old. To overcome the "winding dilemma," the Swedish astronomer Bertil Lindblad suggested a density-wave theory, and in more recent years this theory has been developed further by C. C. Lin, an American mathematician, and others.

According to the density-wave theory, the spiral arms are ripples or density waves that move around the disk like sound waves traveling through air. Each spiral arm is actually a spiral-shaped ripple, of increased gas density, that rotates around the disk and preserves its spiral shape. The increased gas density in the spiral ripple triggers star formation (in a way not yet understood), and the spiral arms therefore contain most of the newborn stars. The gas and stars take tens of millions of years to pass through a spiral arm. But the brightest newborn stars have lifetimes at most only a few millions of years, and by the time they move out of a spiral arm they have died. Between the spiral arms there is a deficiency of bright stars and their associated luminous gas clouds. Hence spiral arms are conspicuous because they are regions containing the brightest stars and gas clouds. The older stars in the disk are only slightly perturbed by the passing spiral ripple, and their aggregate light shows almost no spiral appearance.

THE DISTANT GALAXIES

ELLIPTICAL AND SPIRAL GALAXIES. The distant galaxies, once referred to as *nebulae* because they are fuzzy and cloudlike, are separated from each other by millions of light years. They stretch away in countless numbers, seemingly endlessly, and each is a magnificent celestial city of stars moving in the depths of space.

Most galaxies have distinguishing features that enable the astronomer to classify them as either ellipticals or spirals.

The elliptical galaxies have an oval appearance; some are actually spherical, but most are like flattened spheres (see Figure 3.7). They have bright centers, with outer regions that slowly fade away and do not have sharp outside boundaries. An elliptical galaxy contains almost no gas and dust; it has no reserves from which new stars can form and therefore consists mainly of old population II stars. The ellipticals cover a wide range of masses and sizes: from dwarf galaxies not much larger than globular clusters to rare giant galaxies such as M87 (see Figure 3.8) and even rarer supergiant galax-

Figure 3.6. The spiral galaxy M51 of type Sc. (Mount Wilson and Las Companas Observatories, Mount Wilson Observatory photograph.)

ies that are a thousand times more massive than our Galaxy and have diameters of half a million light years. Giant ellipticals, usually type E0 (pronounced E-zero), the most spherical in appearance of the ellipticals, have conspicuously bright nuclei, whereas dwarf ellipticals lack bright nuclei.

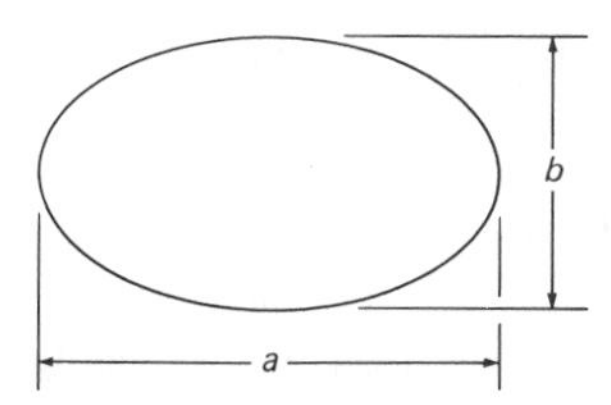

Figure 3.7. The ellipticity of an elliptical galaxy is obtained from the expression 10(a − b)/a, where a is the apparent major diameter and b is the apparent minor diameter. An E5 elliptical galaxy, for example, has a major diameter that is apparently twice its minor diameter.

The majority – more than 60 percent – of all galaxies are ellipticals, and most ellipticals are dwarf galaxies.

Spirals, like our Galaxy, have disks with nuclear bulges and halos not always easily seen because of the brightness of the nuclear bulges and disks. Around the nuclear bulge and within the disk of a spiral galaxy are coiled the spiral arms. The spiral galaxies form two separate sequences: normal spirals denoted by S and barred spirals denoted by SB.

Normal spirals are arranged into a sequence of three types: The Sa have large nuclear bulges and tightly wound arms; the Sb have smaller nuclear bulges and more loosely wound arms; and the Sc have the smallest nuclear bulges and the most loosely wound arms. Our Galaxy and M31 (the Andromeda Nebula) are type Sb.

About a third of all spirals are of the

Figure 3.8. This giant elliptical galaxy, M87, in the constellation Virgo, is a radio galaxy and has a peculiar jet extending from the nucleus. (Association of Universities for Research in Astronomy, Inc., The Kitt Peak National Observatory.)

barred kind. They are classified as SBa, SBb, and SBc in exactly the same way as normal spirals, according to the size of the nuclear bulge and the tightness of the spiral arms. In addition, they have a bright central bar that projects beyond the nuclear bulge and connects with the spiral arms (see Figure 3.9). Why this bar exists is not fully understood.

Spirals are rich in gas and dust that concentrate within the disk, and consequently new stars are continually being formed. The spiral galaxies, unlike the ellipticals, therefore have stars of all ages, with population I stars distributed throughout the disk and population II stars distributed in the halo and nuclear bulge. But they do not have the great range of masses and sizes of the ellipticals: Usually their masses lie between 10 and a few hundred billion solar masses. The majority of the brightest galaxies are spirals; but it should not be forgotten that ellipticals are the most numerous, and the brightest of all galaxies are the rare giant and more rare supergiant ellipticals.

THE TUNING FORK DIAGRAM. Edwin Hubble, of the Mount Wilson Observatory, arranged the various galaxies rather neatly in a diagram that looks like a tuning fork (Figure 3.10). The ellipticals form a sequence in one branch, arranged in order of increasing ellipticity, and the normal and barred spirals form sequences in two parallel branches. This "tuning fork diagram" classifies galaxies according to their appearance; an E0 galaxy, for example, might be a flattened elliptical seen face on. At the junction of the

Figure 3.9. NGC 1300, a barred spiral of type SBb. (Mount Wilson and Las Campanas Observatories, Mount Wilson Observatory photograph.)

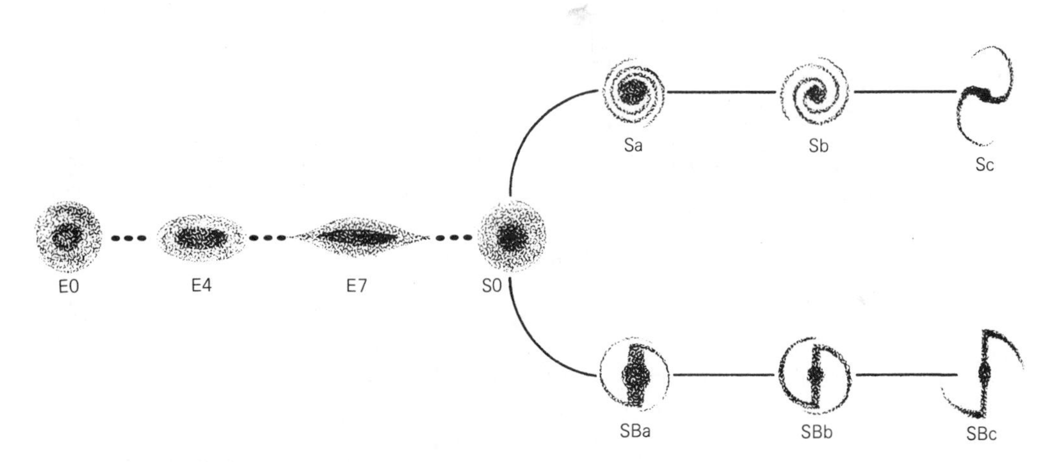

Figure 3.10. The Hubble "tuning fork diagram," showing the ellipticals arranged on the left, in a sequence of increasing apparent ellipticity, and the spirals arranged on the right in two parallel sequences.

three branches, Hubble placed the intriguing S0 (pronounced S-zero) type of galaxy, which combines the properties of ellipticals and spirals. The S0 galaxies are disk-shaped, like spirals, yet lack gas and spiral structure and therefore resemble the flat ellipticals. They in fact look just like spirals that have been swept clean of gas.

The amount of gas in galaxies increases from left to right in the tuning fork diagram, from ellipticals to spirals. In ellipticals the amount of gas is extremely small; it is also small in S0 galaxies, but it progressively increases in the spirals as we go from Sa and SBa to Sc and SBc. The effect of rotation also seems to become more pronounced as we move from the ellipticals to the spirals, and probably spiral disks rotate faster than elliptical galaxies. It was originally thought that galaxies might evolve along the tuning fork diagram from left to right; later it was conjectured that they perhaps evolved from right to left; but astronomers now realize that evolution along the tuning fork diagram is not likely to be significant, and the basic properties of galaxies are probably determined at the time of their birth.

A small percentage of all galaxies have an irregular and chaotic appearance and are classified as irregulars, denoted by Irr. Our nearest galactic neighbors, the Small and Large Magellanic Clouds, are typical examples of irregular galaxies. There are also compact galaxies of high density and unusual brightness that were discovered by the Swiss-American astronomer Fritz Zwicky. Also, some galaxies have distorted forms, such as those displayed in Halton Arp's *Atlas of Peculiar Galaxies*; in some cases the peculiar appearance is possibly the result of interaction with adjacent galaxies, whereas in others the cause of the disruption or contortion is not obvious.

CLUSTERS OF GALAXIES. Galaxies are not uniformly distributed in the universe. They are grouped together in clusters, and the majority of galaxies are members of clusters.

The regular clusters are spherical and have a strong central condensation of galaxies. They look like glorified globular clusters in which stars are replaced by galaxies. These clusters are rich – meaning that they have many members – and contain a thousand or more galaxies that are mostly of the elliptical and S0 kind. Often in their central regions are found supergiant ellipticals that have conceivably grown to their colossal sizes by gobbling up smaller galaxies. The regular clusters typically are 10 million light years in diameter and lack visibly sharp outer boundaries. They contain intergalactic gas (and perhaps intergalactic stars and globular clusters) through which the galaxies rush at velocities of more than 1000 kilometers per second. The galactic wind experienced by the rapidly moving galaxies is sufficiently strong to strip away their interstellar gas. This suggests that the S0 galaxies in the great clusters were perhaps once ordinary spirals and that the gas in their disks has been swept out by galactic winds.

All other clusters of galaxies are of the irregular type. These irregular clusters have various degrees of richness and are far more numerous than the regular clusters. They lack spherical symmetry and strong central condensation, and contain a mixture of all types of galaxies. The great irregular clusters are clumpy and actually look like swarms of small clusters fused together. Irregular clusters range from rich aggregations of more than 1000 members, such as the Virgo Cluster, to groups of 10 or fewer members. The small groups, with a typical size of 3 million light years, are the most numerous. Our Galaxy is a member of the Local Group, which is an irregular cluster of approximately 20 galaxies (and many more, if we count midget systems that look like escaped globular clusters).

Spread out in space, beyond the Local Group, are multitudes of other groups of galaxies. The nearest great cluster is the irregular Virgo Cluster at a distance of about 70 million light years. The nearest regular cluster is the Coma Cluster (Figure 3.11) at a distance of approximately 450 million light years. Beyond about 100 million light years the galaxies appear to

Figure 3.11. Coma Cluster of galaxies. (Association of Universities for Research in Astronomy, Inc., The Kitt Peak National Observatory.)

thin out slightly, and it is now believed that clusters of galaxies are themselves grouped together into superclusters. The Local Supercluster that we occupy has its center somewhere in the vicinity of the Virgo Cluster.

DISTANCES OF GALAXIES. The determination of the distances of galaxies is very important and unfortunately is extremely difficult. It is possible, by measuring the apparent brightness of the most luminous stars, to estimate the distances of nearby galaxies. The bright cepheids are a great help for surveying the nearby extragalactic systems. The distances of galaxies further away are not so readily determined. The apparent brightness of the most luminous stars and globular clusters can be used for estimating distances up to about 80 million light years. The apparent size of highly luminous clouds of gas and the brightness of supernovas take us a little further. Beyond 100 million light years all distances are uncertain by a factor of 2, and perhaps even more. The galaxies themselves must now be used as distance indicators, and it is often assumed that the brightest galaxies in all great clusters are of similar luminosity.

BIRTH OF GALAXIES

The galaxies were born long ago. As far as we know, most if not all contain old population II stars and therefore have ages of 8 or so billion years. The universe is now too old to form new galaxies, because the present intergalactic medium – the gas spread out between galaxies – is too low in density to give birth to them.

A rule-of-thumb average density of matter in a galaxy such as our own is 1 hydrogen atom per cubic centimeter; this would be the density if all stars were dissolved into gas. Very roughly this is a million times greater than the average density of matter in the universe. The universe is expanding, which means that in the past the universe was denser, and the galaxies were crowded closer together. Five billion years ago, when the Solar System was born, the clusters of galaxies were closer together and the average density of the universe was about four times its present value. Very much earlier, when the average density was a million times greater than now, the galaxies did not exist in their present form. Obviously, the galaxies would be crushed together beyond recognition in a universe more dense than galaxies. The galaxies therefore originated in their present form when the expanding universe had an average density of less than 1 hydrogen atom per cubic centimeter. According to cosmological theory, to be laid out in later chapters, this means that the galaxies originated when the universe had an age much greater than 10 million years.

PROTOGALAXIES. Nobody knows why the galaxies were born. Theories on the origin of galaxies start by assuming that the expanding universe had a tendency to break into fragments, but it is not understood why this fragmentation occurred nor why the fragments have galactic masses. In our discussion we shall therefore assume that the universe, for unknown reasons, has separated into fragments.

The fragments are the protogalaxies. They are large globes of gas, composed almost entirely of hydrogen and helium (the helium was made from hydrogen in the big bang), and have masses ranging from millions to hundreds of billions of solar masses. At first, these great globes of cool gas continue to expand with the universe. Then, because of the internal pull of gravity, their expansion slows down, and their separation from one another increases. The universe has in effect been shattered into fragments and the cracks are now widening between the fragments. Eventually, when the universe is hundreds of millions of years old, the globes cease to expand and begin to collapse. A possible picture of how these protogalactic globes of gas form into galaxies is as follows (see Figure 3.12).

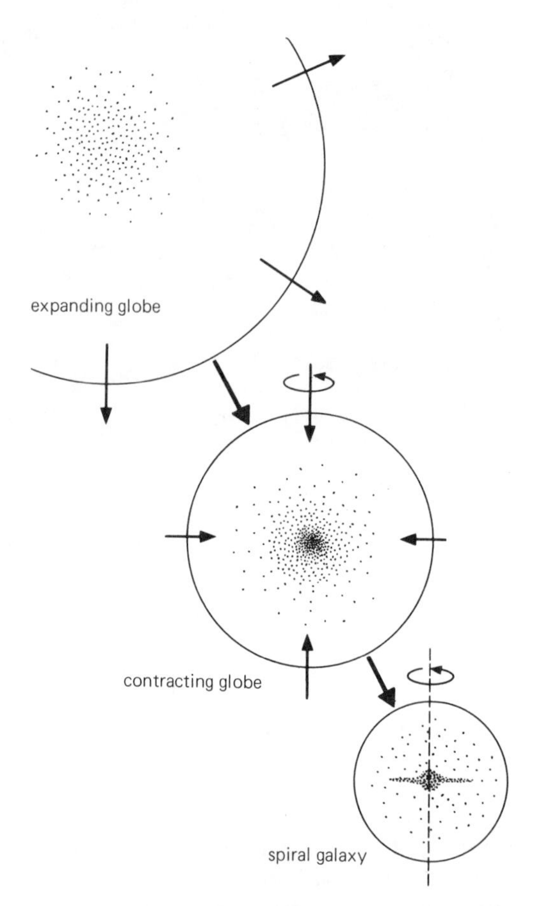

Figure 3.12. This figure illustrates one possible way in which a spiral galaxy might be formed. The expanding universe, already hundreds of millions of years old, has fragmented into large globes of gas. At first, each globe continues to expand; then, after a time, it begins to collapse. Population II stars in large numbers form in the central region. The outer regions of the globe fall inward and create a rotating disk of gas in which population I stars are subsequently formed.

COLLAPSE THEORY. At first each globe of gas has roughly uniform density. But by the time a globe has attained maximum size – and has a diameter of about 500 million light years in the case of a giant galaxy – the density, no longer uniform, has increased in the central region. At about this time the earliest population II stars begin to form in the center. The brightest of these first-generation stars evolve rapidly, and in their final catastrophic moments they erupt as supernovas and eject into space gas that is enriched with heavy elements. Meanwhile the globe of gas has begun to collapse; the collapse lasts hundreds of millions of years, and in some cases perhaps even billions of years. Stars all the while are continually born in the central region, and the very brightest last only millions of years. The collapsing gas of the protogalaxy is therefore steadily enriched with heavy elements.

We must understand that stars, once they are formed, cease to participate in the general collapse of the protogalaxy. Each newborn star initially moves inward, but follows an elliptical orbit, and its average distance from the center thereafter does not change greatly. At this stage we have a picture of a collapsing protogalaxy, as proposed in 1962 by Olin Eggen, Donald Lynden-Bell, and Allan Sandage, in which the central region has become a swarm of population II stars. This swarm of stars is roughly spherical, and in the case of our Galaxy has a diameter of between 100 and 200 million light years. The central region has in fact transformed itself into a galactic halo.

Stars are rarely born in isolation; instead, they usually form in groups. Perhaps most of the stars in the halo are formed in gas clouds and are clumped together in clusters; many of these clusters later dissolve, whereas others, having lost their fastest stars, became compact globular clusters.

Through the halo of stars falls the rest of the collapsing protogalaxy. This is the "outside-in" theory of galaxy formation. The outside of the protogalaxy falls in and becomes the inside of the final galaxy. What happens to the infalling gas determines the kind of galaxy that is produced.

Let us consider a protogalactic globe of gas that is initially in uniform rotation. The outer parts have higher rotational velocity than the inner. The core of the globe is transformed into stars, and because of the core's low rotational velocity, these halo stars form a swarm that is almost spherical in shape. The infalling gas from the outer regions has higher rotational velocity, and as it moves inward, its rotational velocity increases further, just as a ballet dancer rotates faster when she lowers her arms. This swirling gas, as it descends through the halo stars, becomes contaminated with heavy elements and is progressively consumed by the formation of new stars. These new stars are intermediate between the extreme population I and extreme population II stars and have a distribution that is flatter (like a squashed sphere) than that of the first-generation stars. The vast quantity of surviving gas cannot fall all the way into the center because of its relatively rapid rotation. Instead, it settles into a rotating gaseous disk consisting of hydrogen, helium, and most of the heavy elements ejected from stars born before the formation of the disk. Population I stars then begin to form in the gaseous disk. But because of rotation and the presence of magnetic fields, there is now an impediment to star formation, and only a relatively few stars are born each year. The original gigantic globe of gas has collapsed and finally produced a spiral galaxy.

One might well ask where the initial rotation comes from. We do not know for certain; perhaps the protogalactic globes of gas pull each other into rotation with their interacting gravitational forces; or perhaps there are spinning whirls in the cosmic medium long before the birth of galaxies. Whatever theory is preferred, one comes to the conclusion that some protogalaxies have slower rotation than others.

Let us then consider a globe of gas that has very slow rotation. As before, the first-generation stars form in the central region, and the infalling gas of the outer regions is partly consumed by the formation of new stars. But the infalling gas now lacks sufficient rotation to create a large disk of swirling gas. Instead, the gas continues to fall, and is continually consumed by star formation. The gas that ultimately survives converges on the center and settles in the nucleus. The original globe of gas has finally collapsed and produced a large elliptical galaxy. There was never enough rotation to form a large gaseous disk, but there was enough to produce the flattened distribution of stars we see in some ellipticals.

Rotation, however, is not the only thing that determines the final structure of galaxies. The density of gas in protogalaxies is also very important. This is because the rate at which stars are born depends on the density of the gas from which they form. If a protogalaxy has higher density – because it becomes a separated fragment at a slightly earlier epoch in the expanding universe – then stars are born more quickly and the infalling gas is consumed more rapidly. There is no gas left that can make a disk or can fall into the center. Perhaps this is the way most small ellipticals are formed. In these galaxies the ejected gas from stars, which is contaminated with heavy elements, never has a chance to be incorporated in later stars, because all the stars are already born. The ejected gas just lingers around and is later swept out by galactic winds.

FRAGMENTATION AND CLUSTERING. The most numerous of all galaxies are the small systems not much larger than globular clusters. It has been suggested by some cosmologists that the universe first fragments into midget protogalaxies, having masses not much greater than those of globular clusters, which then cluster together to form larger and larger galaxies.

There are actually two ways of trying to understand the large-scale architecture of the universe. The first is the fragmentation scenario, originally proposed by James Jeans, in which the universe first breaks up into large objects, which then fragment successively into smaller and smaller systems, ultimately terminating with stars. The second is the clustering scenario, advocated by David Layzer of Harvard University, in which the universe first breaks up into small objects, which then cluster successively into larger and larger systems, ultimately terminating with clusters and superclusters of galaxies.

Our present picture of galaxy formation tends to draw from both scenarios. The universe first fragments into globes of gas, some of which may cluster and then amalgamate to form large turbulent protogalaxies. The collapsing globes subsequently fragment into clusters of stars, which then fragment into isolated stars. The galaxies also aggregate to form clusters of galaxies of different sizes. It is possible that fragmentation and clustering are of equal importance in the structural design of the universe.

RADIO GALAXIES AND QUASARS

In 1931, Karl Jansky of the Bell Telephone Laboratories detected radio signals from the Milky Way. This exciting discovery received wide publicity, and the signals from Jansky's receiver were relayed and broadcast in a radio program, with the commentator announcing, "I want you to hear for yourself this radio hiss from the depths of the universe."

Grote Reber, a radio engineer, in 1939 detected radio signals from the Milky Way and noticed that the signals were strongest from the galactic center and from constellations such as Cygnus and Cassiopeia. Reber built his radio telescope – a dish reflector 31 feet in diameter – in his backyard at a cost of $1300. (His neighbors thought it was for either collecting rainwater or controlling the weather.) Radioastronomy was Reber's hobby, and for a few years he was the only radioastronomer. After World War II radioastronomy flourished first in England and Australia, and there are now many radio observatories scattered around the world.

RADIO GALAXIES. Hundreds of radio sources were discovered after World War II in the early years of radioastronomy, and many were successfully identified by optical astronomers. Thus Taurus A – the strongest of the observed radio sources in the constellation Taurus – was found to be the Crab Nebula, a chaotic cloud of gas produced by the supernova of 1054. Cygnus A – or 3C 405 – was identified with a disturbed-looking giant galaxy at a distance of 1 billion light years. It was found that the most powerful of the optically identified radio sources are galaxies, and these radio galaxies, such as Cygnus A, emit millions of times more energy in radio waves than our Galaxy. The radio galaxies, when seen on photographic plates, often have bright central regions and occasionally display protruding jets, as does M87, and wispy extensions. The strong radio galaxies radiate more energy in radio waves than in visible light and are usually giant ellipticals that are the brightest members in clusters of galaxies.

The radio waves from these sources are emitted by fast electrons that spiral in magnetic fields. The electrons in radio sources have extremely high energies, and the magnetic fields are much weaker than the Earth's magnetic field. The waves emitted by electrons gyrating in magnetic fields are called synchrotron radiation because this kind of radiation is produced in high-energy accelerators called synchrotrons. Radio galaxies generate, in a way that is not yet understood, hordes of energetic electrons that are dispersed throughout the radio-emitting regions.

EXTENDED AND COMPACT SOURCES. Great strides have been made in radioastronomy, and it is now possible to study in detail the structure of radio sources. The majority of sources have double structure, and radio emission is detected in two extended components that lie on opposite sides of each source (see Figure 3.13). The two extended components of a source are separated from each other by distances of hundreds of thousands, sometimes millions, of light years. Between the components sits the radio galaxy.

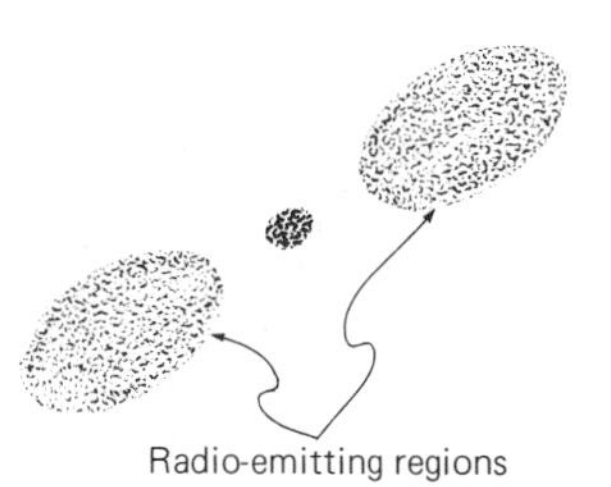

Figure 3.13. An extended two-component radio source. Radio waves are emitted by the extended components.

Often the central region, or nucleus, of a radio galaxy is also a powerful radio source that may contain two radio-emitting components separated by distances of typically only hundreds of light years. There are thus two varieties of radio sources: extended sources and compact sources, and some – for example, Cygnus A – are both extended and compact.

QUASARS. By 1960 the radioastronomers had found and catalogued hundreds of radio galaxies and about 50 had been identified as giant ellipticals by the optical astronomers. It was thought that the remaining unidentified radio galaxies were too faint to be recorded on photographic plates. Then strange things were noticed, and a sequence of events began in 1960 that led to the discovery of the most puzzling objects in the universe – the quasars.

It was first noticed that the radio source 3C 48 is not nebulous and fuzzy like a galaxy but starlike in appearance, and that it emits strong ultraviolet radiation. Similar objects, such as 3C 273, were soon found. It seemed that they were just an unusual kind of star within the Galaxy that happened to emit radio waves. Then in 1963 came startling news. Maarten Schmidt of Mount Wilson Observatory had discovered that emission lines in the spectrum of 3C 273 were shifted by 16 percent to longer wavelengths. In other words, the radio source had a redshift of 0.16, and was receding from us at about 16/100 of the speed of light.

All objects at great distances in the expanding universe are receding from us, and the light we receive from these receding objects is redshifted. This is a subject to be discussed later, and here we need only point out that a redshift of 0.16 corresponds to a distance of roughly 2 billion light years. In 1963 it was realized that the only logical and simple explanation for the redshift of 3C 273 is that it is caused by the expansion of the universe. Many other radio sources were soon optically identified as starlike objects with large redshifts, and it became apparent that these *quasi-stellar radio sources* are remote beacons radiating enormous quantities of energy.

Allan Sandage then discovered that many extragalactic starlike objects are radio quiet. The word *quasar,* coined by the astronomer Hong Yee Chui at the State University of New York at Stony Brook, is a contraction of *quasi-stellar radio source* and is now used to denote all starlike objects of large redshift, whether or not they emit radio waves. The nearest quasar so far discovered is 3C 273, and the most distant have redshifts greater than 3.

The quasars lie at distances of billions of light years, and it is estimated that at least 10 million of them are observable with the largest telescopes. When we look through telescopes to such vast distances we also look back billions of years in time. We look back in time and notice that quasars were once far more numerous in the universe than at present. They were sufficiently plentiful at the time when the Solar System was born that perhaps at least one was near enough to be seen gleaming in the sky like a brilliant jewel. Earlier still, shortly after the birth of galaxies, when the universe was one-tenth its present age, the quasars were thousands of times more plentiful than now.

WHAT ARE QUASARS? Quasars radiate energy at a rate approximately 100 times that from all the stars of a galaxy such as our own, and they squander energy at this extravagant rate for typically a billion years. If matter could be annihilated and converted entirely into energy (equal to mass times the square of the speed of light) and then radiated, some quasars would need the mass of at least 100 million stars.

The light from many quasars fluctuates in brightness, changing in a time sometimes as short as weeks and even days. Generally, a luminous object cannot significantly change its light output in a time less than the time taken by light to travel a distance equal to its size. When any object suddenly increases its light output we always notice a slower increase; this is because we see first the light emitted from the nearest parts of its surface and later the light emitted from parts of its surface that are further away. Hence, when light varies in brightness in 1 week, this means that the source has a size smaller than 1 light week and is therefore smaller than about 10 times the size of the Solar System. Quasars not only are powerful sources of radiation but by astronomical standards are also astonishingly small. Imagine that a large room represents the size of the Galaxy; on this scale the highly luminous quasar is no more than a mere speck of dust floating in the air.

The radio quasars are remarkably similar to the radio galaxies and have either extended or compact radio-emitting regions. In fact, radioastronomers cannot tell the difference, and even optical astronomers cannot always determine the difference between a quasar and the bright nucleus of a radio galaxy. Possibly quasars and the nuclei of radio galaxies are basically the same type of object, manifesting itself in various ways during its evolution. The nuclei of giant ellipticals may become quasars at some time or other, and possibly most quasars pass through an active radio-emitting phase during their lifetime.

DESPERATE ANSWERS. Several years have now passed since the discovery of quasars and we still do not know how such compact objects, perhaps not much larger than the Solar System, are able to generate and dispense immense amounts of energy at ultraviolet, infrared, and radio wavelengths. In the excitement that followed their discovery, many ideas were proposed, such as annihilation of matter and antimatter, dense stellar systems in which stars collide cataclysmically and supernovas are frequent, and even alterations in the known laws of nature.

It is widely thought that the energy radiated comes in some way from the release of gravitational energy. A tentative picture is the following. During the formation of a galaxy there is infalling gas that is not consumed by star formation. In a giant elliptical the contaminated infalling gas accumulates in the nucleus and forms a reservoir of matter having a mass of perhaps one-tenth the total mass of the galaxy. The slow subsequent contraction of this reservoir of matter creates in some way a quasar that shines for the next billion years.

Quite likely the first step is the formation of numerous stars in the nucleus, many of which evolve and become pulsars and some of which even become black holes. Martin Rees of Cambridge University has suggested that perhaps the combined output in waves and energetic particles from hundreds of thousands of pulsars is beamed away from the nucleus in two opposite directions. The energy in these intense beams is then distributed over large regions that radiate the radio waves we detect from quasars and radio galaxies. According to another theory, advocated by Soviet astronomers and by Philip Morrison of the Massachusetts Institute of Technology, the gas in the nucleus contracts and forms one or more supermassive stars that rotate and have strong magnetic fields. These objects, named spinars, are millions of times more massive than the Sun and behave like titanic pulsars capable of radiating intense beams into intergalactic space.

Another theory, which was proposed by Edwin Salpeter of Cornell University and has growing popularity, suggests that the dominant role is played by black holes. There seems no way of escaping the fact that many stars at the end of their evolution must collapse totally and become black holes. The laws of nature as we understand them lead inevitably to this conclusion.

Imagine then a black hole, a few times the mass of the Sun with a diameter of about 10 kilometers, located in a nucleus where the gas is dense and millions of stars are huddled close together. It is a situation that recalls those terrible words "cry 'Havoc!' and let slip the dogs of war." The cry "Havoc!" was a command to massacre without quarter. (In the reign of Richard II of England the cry was forbidden on pain of death.) Gas immediately begins to drain in on the black hole. Incautious stars that come too near either collide with each other or are torn to shreds by strong gravitational forces, and their gaseous remnants in either case add to the headlong rush that spirals in on the black hole. Next time you pull the plug out of a bath and see the water draining away into a dark vortex, think of a voracious black hole. The black hole rapidly grows in mass and even occasionally swallows other black holes. Provided the galactic nucleus is sufficiently rich in gas and stars, a black hole will grow in mass to 100 million times that of the Sun in a few hundred million years, and have a size of 1 light hour. All the while, during its growth, a torrent of energy is unleashed by the gas that drains inward: As the gas approaches the black hole it is compressed and heated to a very high temperature, and in this way the released energy is radiated away. A fraction – a tenth or more – of the mass of the captured gas is converted directly into radiant energy that escapes. Thus a monster black hole that grows to a billion solar masses radiates at least 100 million solar masses of energy. This picture of a quasar accounts for its brilliance and small size and might also even explain why it becomes a radio source. It has been suggested that the hot and therefore electrically conducting gas that swirls inward on the black hole might act as a vast electrical dynamo that generates oppositely directed beams of high-energy particles. These beams of fantastic intensity travel out and energize the radio-emitting regions of radio sources.

According to the black-hole scenario, a quasar dies when most of the matter in a galactic nucleus has been either swallowed or violently ejected. Thereafter the black hole lies dormant, erupting into occasional displays of activity whenever fresh supplies of gas become available. It is conceivable that lurking in the nucleus of our Galaxy are black holes that might be millions of times the mass of the Sun. From radio studies it is known that the nucleus of the Galaxy is in a disturbed state with gas clouds flying outward, and many astronomers believe that an eruption of some kind occurred about 10 million years ago. Perhaps this was the result of a wandering supermassive black hole encountering an accumulation of gas within the nucleus.

Black holes that might exist in the nuclei of spirals are probably smaller than those that develop in the nuclei of giant ellipticals. This is because of the rotational impediment in spirals that distributes gas throughout the disk and prevents it from concentrating in the nucleus. Some spirals, known as Seyfert galaxies, have bright active nuclei that resemble quasars; these nuclei, however, are generally not as bright as quasars, and this may be because they contain smaller concentrations of matter than the nuclei of giant ellipticals.

REFLECTIONS

1 *Imagine the Solar System reduced to the size of an atom 10^{-8} centimeters in diameter. On this scale, how far apart are the stars, how large is the Galaxy, and how far apart are the galaxies? Assume that stars are separated by 5 light years and galaxies by 10 million light years.*

* *Consider how the moving-cluster method can be used to estimate the distance of a flock of birds flying in the sky. Suppose the flock is flying away at a speed of 60 kilometers (40 miles) per hour, and its size is shrinking by 50 percent a minute. Show that the distance in this case is 1 kilometer.*

* *If a spiral and an elliptical are at equal distances, and have equal numbers of stars, why is the spiral the brighter galaxy?*

* *Suppose that each galaxy contains 100 billion stars and that galaxies are separated from each other by 10 million light years. If we can look out to a distance of 10 billion light years, how many galaxies and how many stars are there in the observable universe? Compare the number of stars with the number of grains of sand on all the beaches and deserts of Earth. (Assume that a grain of sand has a volume of 1 cubic millimeter and sand covers the Earth's surface to a depth of 1 centimeter.)*

* *Is it likely that extreme population II stars have planets? Why is it that intelligent life (in a form that we could recognize) is more likely to exist in spirals than in ellipticals?*

* *Take any piece of metal, such as a coin, and ask yourself, When and where was this metal made?*

* *Imagine a universe that failed to fragment into galaxies and stars. What would be the consequences as far as we are concerned?*

2 *We can estimate the mass of the Galaxy with Kepler's third law. Let P be the period of revolution in years and R the radius of the orbit in astronomical units. Then $P^2 = R^3$ for planets moving about the Sun. For objects moving about a body of much larger mass, then, $MP^2 = R^3$, where M is the mass measured in solar masses. The Sun moves around the Galaxy in $P = 2 \times 10^8$ and the Sun's distance from the center (30,000 light years) is $R = 2 \times 10^9$. Therefore*

$$M = (2 \times 10^9)^3/(2 \times 10^8)^2 = 2 \times 10^{11}$$

and the mass of the Galaxy is 200 billion times that of the Sun. This calculation is not exact for two reasons. It supposes that the mass of the Galaxy, inside the Sun's orbit, is distributed as in a sphere. A sphere exerts a gravitational pull as if all its mass were concentrated at the center. The Galaxy is not a sphere, but the error in this case is not very large. Our calculation ignores the mass of the Galaxy outside the Sun's orbit. It is usually assumed that this neglected mass is not large, although some astronomers have suggested there is more mass outside the Sun's orbit than within it.

* *A cluster of galaxies is held together by its own gravity – it is gravitationally bound – and the galaxies have orbits determined by the gravitational forces of all the galaxies in the cluster. The average speed of the galaxies and the total mass of the cluster are related by the virial theorem, which applies to all self-gravitating systems. This theorem states that the total kinetic energy is equal to half the binding energy. This means that the mean square of the speed of the galaxies is equal to twice the square of the escape speed from the surface. Let M stand for the mass and R for the radius of the cluster, and let v be a typical speed of a galaxy. Then the virial theorem says:* $v^2 = GM/R$. *This means that*

$$v^2 = \frac{1}{100} \times \frac{\text{mass in solar masses}}{\text{radius in light years}}$$

where v is in kilometers per second. If a cluster is 1000 times the mass of the Galaxy and has a radius of 2 million light years, then the speed v is 1000 kilometers per second. By determining v from observations it is possible to find the mass of a cluster by means of the virial theorem. This "virial theorem mass" often exceeds what is observed by counting galaxies, and creates the problem of "missing mass." The missing mass in many clusters is still an unsolved problem.

3 *Charles Messier (1730–1817), a French astronomer, was an ardent comet hunter who compiled a famous catalogue of 103 nebulae. He did this, he said, "so that astronomers would not confuse these same nebulae with comets just beginning to shine." The Messier nebulae are prefixed with the letter M; thus M1 is the Crab Nebula, which he described as "whitish light and spreading like a flame," and M31 is the galaxy in Andromeda.*

William Herschel published catalogues in which he listed the positions and descriptions of thousands of nebulae. His son, John Herschel (1792–1871), carried on this work and published in 1864 the General Catalogue – referred to as GC – which was the first systematic survey of the entire sky and contained 5000 nebulae and star clusters. The GC was replaced by the New General Catalogue – NGC – in 1890, and later supplements called the Index Catalogues – IC – were added. Thus M31 is also known as NGC 224.

4 *Edwin Hubble (1889–1953), an American astronomer, is best known for his work on the classification of galaxies, the determination of the distances of galaxies, and the Hubble law of expansion of the universe. He studied Roman and English law at Oxford, but after one year of practice decided to "chuck the law for astronomy." In his classic book,* The Realm of the Nebulae *(1936), Hubble said: "Research men attempt to satisfy their curiosity, and are accustomed to use any reasonable means that may assist them toward the receding goal. One of the few universal characteristics is a healthy skepticism toward unverified speculations. These are regarded as topics for conversation until tests can be devised. Only then do they attain the dignity of subjects for investigation." The book closes with the words: "Thus the explorations of space ends on a note of uncertainty. And necessarily so. We are, by definition, in the very center of the observable region. We know our immediate neighborhood rather intimately. With increasing distance, our knowledge fades, and fades rapidly. Eventually, we reach the dim boundary – the utmost limits of our telescopes. There, we measure shadows, and we search among ghostly errors of measurement for landmarks that are scarcely more substantial."*

5 *"We have found that, as Newton first conjectured, a chaotic mass of gas of approximately uniform density and of very great extent would be dynamically unsta-*

ble: nuclei would tend to form in it, around which the whole of the matter would ultimately condense ... We may conjecture, although it is improbable that we shall ever be able to prove, that the spiral nebulae were formed in this way. Any currents in the primaeval chaotic medium would persist as rotations of the nebulae, and, as these would be rotating with different speeds, they might be expected to shew all the various types of configurations" (James Jeans, Astronomy and Cosmogony, *1929).*

"It is not too much to say that the understanding of why there are these different kinds of galaxy, of how galaxies originate, constitutes the biggest problem in present-day astronomy. The properties of the individual stars that make up the galaxies form the classical study of astrophysics, while the phenomenon of galaxy formation touches on cosmology. In fact, the study of galaxies forms a bridge between conventional astronomy and astrophysics on the one hand, and cosmology on the other" (Fred Hoyle, Galaxies, Nuclei, and Quasars, *1965).*

6 *"To sum up:*

1) The cosmos is a gigantic fly-wheel making 10,000 revolutions a minute.

2) Man is a sick fly taking a dizzy ride on it.

3) Religion is the theory that the wheel was designed and set spinning to give him the ride."

— Henry Mencken (1880–1956), in *Smart Set,* December 1920.

FURTHER READING

Berendzen, R., Hart, R., and Seeley, D. *Man Discovers the Galaxies.* Science History Publications, New York, 1976.

Bok, B., and Bok, P. *The Milky Way.* Harvard University Press, Cambridge, Mass., 1974.

Burbidge, G., and Burbidge, M. "Stellar populations." *Scientific American,* November 1958.

Hodge, P. *Galaxies and Cosmology.* McGraw-Hill, New York, 1966.

Mitton, S. *Exploring the Galaxies.* Charles Scribner's Sons, New York, 1976.

Shipman, H. *Black Holes, Quasars and the Universe.* Houghton Mifflin, Boston, 1976.

Weaver, H. "Steps toward understanding the large-scale structure of the Milky Way." *Mercury,* September–October 1975, November–December 1975, January–February 1976.

Whitney, C. A. *The Discovery of Our Galaxy.* Alfred A. Knopf, New York, 1971.

SOURCES

Allen, R. H. *Star Names: Their Lore and Meaning.* 1899. Reprint. Dover Publications, New York, 1963.

Arp, H. Atlas of Peculiar Galaxies. Astrophysical Journal, Supplement, *14,* 1 (1966).

Baade, W. *Evolution of Stars and Galaxies,* ed. C. Payne-Gaposchkin. Harvard University Press, Cambridge, Mass., 1963.

Burbidge, M., and Burbidge, G. *Quasi-Stellar Objects.* W. H. Freeman, San Francisco, 1967.

Davis, M. "Galaxies and cosmology," in *Frontiers of Astrophysics,* ed. E. Avrett. Harvard University Press, Cambridge, Mass., 1976.

Fowler, W. A., and Stephens, W. E. "Origin of the elements: resource letter." *American Journal of Physics 36,* 1 (April 1968).

Greenstein, J. "Quasi-stellar radio sources." *Scientific American,* December 1963.

Herbig, G. H. "Interstellar smog." *American Scientist,* March–April 1974.

Hey, J. S. *The Evolution of Radio Astronomy.* Neale Watson Academic Publications, New York, 1973.

Hoyle, F. *Galaxies, Nuclei, and Quasars.* Harper and Row, New York, 1965.

Hubble, E. *The Realm of the Nebulae.* Yale University Press, New Haven, Conn. 1936. Reprint. Dover Publications, New York, 1958.

Jeans, J. *Astronomy and Cosmogony.* Cambridge University Press, Cambridge, 1929.

Kellermann, K. "Extragalactic radio sources." *Physics Today,* October 1973.

King, I. "Stellar populations in galaxies." *Publications of the Astronomical Society of the Pacific,* August 1971.

Pannekoek, A. *A History of Astronomy.* Interscience Publishers, New York, 1961.

Payne-Gaposchkin, C. *Introduction to Astronomy.* Methuen, London, 1954.

Rees, M., and Silk, J. "The origin of galaxies." *Scientific American,* June 1970.

Roberts, M. "Hydrogen in galaxies." *Scientific American,* June 1963.

Sandage, A. R. *The Hubble Atlas of Galaxies.* Carnegie Institution of Washington, Washington, D.C., 1961.

Sanders, R., and Wrixon, G. "The center of the Galaxy." *Scientific American,* April 1974.

Schmidt, M. "Quasars and cosmology." *Observatory 91,* 209 (1971).

Schmidt, M., and Bello, F. "The evolution of quasars." *Scientific American,* May 1971.

Shapley, H. *Flights from Chaos: A Survey of Material Systems from Atoms to Galaxies.* McGraw-Hill, New York, 1930.

Shapley, H. *Through Rugged Ways to the Stars.* Charles Scribner's Sons, New York, 1969.

Struve, O., and Zebergs, V. *Astronomy in the 20th Century.* Macmillan, New York, 1962.

Toomre, A., and Toomre, J. "Violent tides between galaxies." *Scientific American,* December 1973.

Verschuur, G. *The Invisible Universe.* Springer-Verlag, New York, 1974.

Verschuur, G., and Kellermann, K., eds. *Galactic and Extragalactic Radio Astronomy.* Springer-Verlag, New York, 1974.

Weymann, R. "Seyfert galaxies." *Scientific American,* January 1969.

Woltjer, L., ed. *Galaxies and the Universe.* Columbia University Press, New York, 1968.

Wright, H. *Explorers of the Universe: A Biography of George Ellery Hale.* E. P. Dutton, New York, 1966.

LOCATION AND THE COSMIC CENTER

The Sun is lost, and the earth, and no man's wit
Can well direct him where to look for it.
And freely men confess that this world's spent,
When in the Planets, and the Firmament
They seek so many new; then see that this
Is crumbled out again to his Atomies.
'Tis all in pieces, all coherence gone;
All just supply, and all Relation.
— John Donne (1572–1631), *The Anatomy of the World*

LOCATION PRINCIPLE

The Greeks developed the "two-sphere" universe, which endured for 2000 years and consisted of a spherical Earth surrounded by a distant spherical surface studded with stars. This geocentric picture was finally discarded in the Copernican Revolution and replaced by the heliocentric picture with the Sun at the cosmic center. But revolutions, once begun, do not readily stop, and by the late seventeenth century the heliocentric picture had also been jettisoned. Out of the turmoil of the revolution emerged an infinite, centerless universe that has since had a checkered history. In the eighteenth century the idea arose of a hierarchical universe of many centers, and the nineteenth century brought the idea of an island universe in which the Sun had central location in the Galaxy. In the twentieth century, as the result of advances made in astronomy and cosmology, we once again have the centerless universe.

As we watch the history of cosmology unfold we see a steady growth in the conviction that mankind does not occupy the center of the universe. The cosmic center was displaced first from the tribe and nation and then from the Earth, the Sun, and finally the Galaxy. Simultaneously, notions concerning God and the universe became increasingly grandiose. Medieval theology developed far-reaching concepts of the nature of God that were subsequently transformed into ideas about the nature of the universe. Concepts of God as unconfined, ubiquitous, and infinite were transformed into scientific ideas of the universe as unconfined, infinite, and having its center everywhere or nowhere. From theology, philosophy, science, astronomy, and our cultural background has emerged a cosmic outlook that is expressed by the principle of location.

The location principle is this: *It is unlikely that we have special location in the universe.* How "unlikely" or improbable is "special location," and what special location means, are matters to be considered in each application of the principle. The principle can, of course, be rejected. In that case evidence must be produced in support of the contrary belief that we enjoy special location, and such evidence nowadays is not easy to find.

The Earth, Sun, Galaxy, and Local Group are all unique objects, and in an astronomical sense we therefore undoubtedly have special location. Yet other creatures, perhaps, living on other planets, encircling other stars, in other galaxies, in other clusters have also special location in precisely the same sense. The cosmologist is concerned not with the uniqueness of local astronomical detail but only with the main outline of the cosmic picture, just as hills and valleys are blurred and smoothed out when we think of the overall shape of the Earth. When all the local irregularity of astronomical detail is ignored, special location means location that is special within the universe as a whole.

The location principle is not an assertion that special location, such as a cosmic center, does not exist somewhere. It merely states that of all the planets, stars, and galaxies of the universe, it is unlikely that the Earth, Sun, and Galaxy are privileged places. It is a revolutionary manifesto proclaiming that man is not king of the cosmic castle, and it demands that we abdicate our throne and become citizens of a large republic. The prospect of dethronement, of forsaken cosmic privilege, was naturally strongly resisted until recent times. Why abdicate, it was asked, when nothing is gained and everything is lost? Physics, astronomy, and biology have remorselessly forced abdication on us, and their advances in the Age of Science have transformed the universe. We can be kings in a medieval universe but not in the vastness and complexity of the modern physical universe.

THE SURVEYORS. As an illustration of the location principle, and of how it works, we imagine that a team of small creatures called Surveyors has been placed on a very large surface. Their mission is to determine the large-scale shape of the surface.

The Surveyors set up their instruments on hilltops and begin to make measurements. After much observation, calculation, and discussion they come to an important conclusion: If the hills and valleys are ignored, or imagined to be smoothed out, the surface is the same in all directions. On finding that the surface is isotropic – the same in all directions – they exclaim, "How fortunate we are to be on this special spot!" The Surveyors have previously agreed that the surface has no edge, for reasons that need not detain us until Chapter 5, and they start theorizing about the shape of the surface. A popular guess is their camp is on the summit of a large hill-shaped surface that falls away in all directions and extends to infinity, as in Figure 4.1.

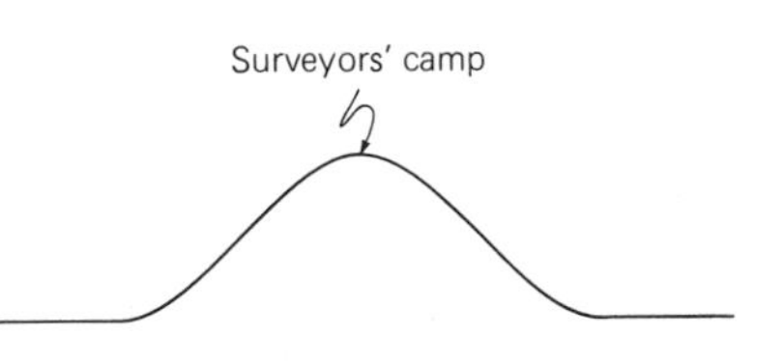

Figure 4.1. A hill-shaped surface that stretches away to infinity. The Surveyors' camp is at the summit of the hill.

After some time one of the more thoughtful Surveyors says, "Isn't it odd that we happen to be just on this special spot?" The Surveyors believe that elsewhere on the surface are other Surveyor teams, and very quickly they realize that the chance of occupying a small region, around which the surface is isotropic, is extremely small. They therefore formulate a location principle that says, "We have no right to assume we have been chosen to occupy a special spot." This means that the other Surveyor teams have perhaps also found that the surface is isotropic about their camps. For if isotropy exists at one place, and is an unlikely privilege, then probably it exists everywhere and is shared by all. The Surveyors conclude that if the surface is isotropic everywhere, and extends to infinity, then it is flat.

From the beginning of the mission some Surveyors have disliked the idea of an infinite surface and have said that they much prefer to live on a surface that is finite in extent. They have declared that the measurements, because of uncertainties, are also consistent with the possibility of an egg-shaped surface, as illustrated in Figure 4.2. Their camp, they have previously pointed out, need not be on the summit of a large hill, but might be situated at the end of a large egg where the surface is also the same in all directions. These Surveyors, now confronted with the location principle and the probability that the surface is isotropic everywhere, come to the conclusion that they live on the surface of a sphere. The

Figure 4.2. A finite surface that is egg-shaped. The Surveyors' camp in this case is at one end of the egg.

Surveyors thereupon divide into two groups: the "open" group, which thinks that the surface is open, infinite, and flat; and the "closed" group, which thinks that the surface is closed, finite, and spherical (see Figure 4.3). All agree that more careful measurements are needed to determine which group is correct.

After further surveying it is discovered that the surface is apparently not quite isotropic. An anisotropy of about 1 percent is detected that is possibly due to an overlooked surface irregularity. The alternative explanation, that it is due to a real center of symmetry at an estimated distance l, is at first a cause of worry. Perhaps the surface is hill-shaped after all and their camp is not quite at the top? Or perhaps the surface is really egg-shaped and their camp is not quite at the end? The Surveyors meet this challenge by arguing as follows. Their camp and also the center of symmetry, if it exists, both occupy a small region of the surface of an approximate area πl^2. Now L is the distance out to which they can survey, and the chance of occupying a small privileged area of πl^2 that contains the center, in a large area of πL^2, is the ratio of these two areas, equal to l^2/L^2. Their measurements indicate that $L = 100l$ (because l/L is roughly equal to the 1 percent anisotropy), and the chance of special location near a center of symmetry is therefore only 0.0001. This means, they argue, that the chance of a center existing at all within their field of view is only 1 in 10,000, and the probability that the surface is the same everywhere (apart from local irregularities) is 99.99 percent. They conclude that the apparent anisotropy is probably the result of an overlooked irregularity in the surface.

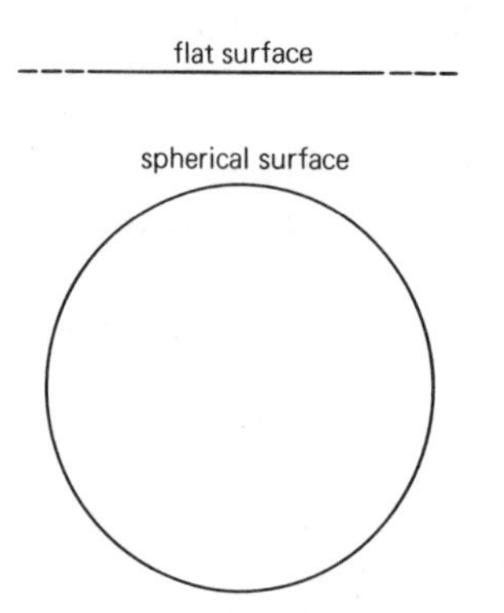

Figure 4.3. After the discovery of the location principle, the "open" group concludes that the surface is actually flat, and the "closed" group concludes that it is spherical.

A few Surveyors are not convinced that the surface everywhere is the same and start to talk about a "hierarchical" surface. Perhaps, they say, we live on a very large hill or in a very large valley, and this hill or valley is perhaps a comparatively small irregularity in an even larger hill or valley, which in turn is also perhaps only a comparatively small irregularity and so on. The rest of the Surveyors at this point, finding that such bewildering thoughts give them headaches, refuse to listen any further to the hierarchists.

Around the campfire one evening a Surveyor tells of a science-fiction story he has read about a violent universe filled with cosmic radiation. Ferocious two-legged creatures in this universe, the shuddering Surveyors are told, find that this cosmic radiation is 99.9 percent isotropic. They believe that the 0.1 percent anisotropy is perhaps because they inhabit a body that whirls through three-dimensional space at high velocity. The storyteller then says that these warlike creatures worship a terrible god called Big Bang who consumed their universe and erupted in fiery cosmic radiation long ago. Many of the Surveyors, overcome with horror, have by this time crept away to their tents. Those who remain are told that in this universe $L = 1000l$, and the probability that it has a center is $l^3/L^3 = 0.000,000,001$. The listening Surveyors are greatly impressed. But they become appalled and also scatter to their beds when told that among the warlike creatures of the violent universe the "closed" group pray for the return of Big Bang.

GEOCENTRIC UNIVERSE

BABYLONIC WIZARDRY. Four thousand years ago the Babylonians divided the sky into the constellations of the Zodiac and began to compile star catalogues and to record the movements of the planets. They invented multiplication tables and established rules of arithmetic, and they were skilled in the arts of computation. By their study of the cyclic variations of the heavens they were able to predict eclipses and prepare calendars that forecast the seasons and the times of full and new Moon. All this was done for the purpose of religious divination and prediction. The Babylonian wizards chartered the heavens, but did not theorize on the nature of celestial regularity and did not invent geometrical models of the heavens. They had no science and their explanations, when offered, were purely mythological.

PYTHAGOREAN HARMONY. The Babylonian scholars were priests and prophets, whereas the Hellenic (or Greek) scholars were philosophers and scientists. They had different types of mind and different ways of thinking. The Babylonic mind excelled in arithmetic manipulation and symbolic representation, and the Hellenic mind excelled in geometric manipulation and the portrayal of things by analogy and symmetry.

Pythagoras, in the sixth century B.C., visualized a universe of geometrical harmony governed by mathematical laws. Spheres, circles, and vortical motions were basic in the design of the universe. "It was said," according to Diogenes (third century), "that Pythagoras was the first to call the heavens cosmos and the Earth a sphere." Pythagoras, and others of the same school (referred to as Pythagoreans), believed that the heavenly bodies were divine, and like the Earth were perfect spheres that moved in perfect circles around a central cosmic fire that lay beyond the reach of mortal eye. In their motions the heavenly bodies emitted harmonious notes, but the celestial symphony or "harmony of the spheres" was beyond the reach of mortal ear.

THE ARISTOTELIAN UNIVERSE. The "two-sphere" universe emerged and became popular at about the time of Plato in the fourth century B.C. It consisted of a spherical Earth at the center that was surrounded by an outer sphere of stars, and the planets moved in undetermined ways between the two spheres (see Figure 4.4). "To all earnest students," Plato proposed the problem, "What are the uniform and ordered movements of the planets?" Eudoxus (about 408–355 B.C.), a student of Plato, provided the first important answer. The universe, he said, could be represented by a spherical Earth surrounded by concentric and rotating spheres. The outermost sphere rotated daily and supported the stars, and the intermediate spheres rotated at different rates about different inclined axes and supported

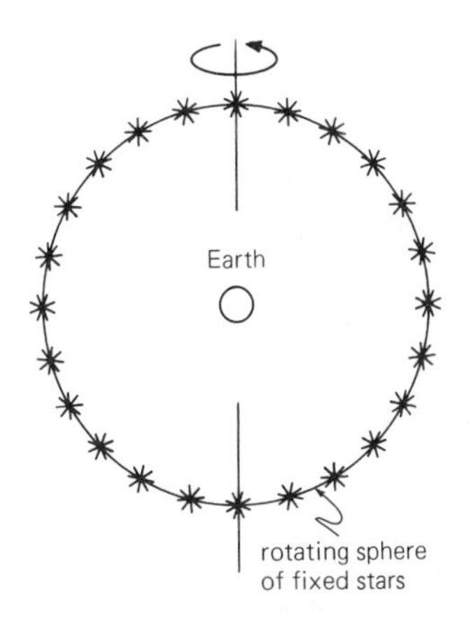

Figure 4.4. The "two-sphere" universe. The central sphere is the Earth, and the outer sphere is studded with stars and rotates daily.

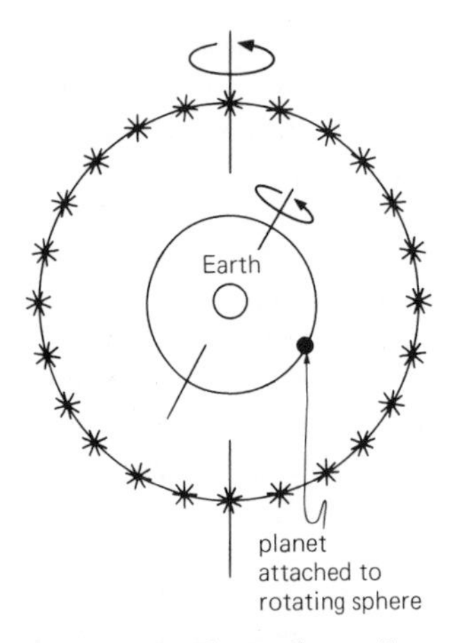

Figure 4.5. The Eudoxus "many-sphere" model. Additional intermediate spheres, rotating about inclined axes, support the planets. (Only one intermediate sphere is shown in this figure.)

the planets (see Figure 4.5). The Eudoxus solution was presented as a hypothetical model, and its geometrical contrivances were intended to be little more than illustrations of celestial motion, much like the mechanical contrivances of a modern planetarium.

Aristotle immediately adopted the model and invested the spheres with spiritual and physical reality (see Figure 4.6). From the geometrical scribbles of Eudoxus it blossomed into a full-blown universe. The planets, starting with the innermost, were the Moon, Mercury, Venus, the Sun, Mars, Jupiter, and Saturn, and were attached to spheres of a translucent substance. Altogether 56 concentric spheres were needed to explain the planetary motions and the rotation of the outermost sphere of stars. The four elements of earth, water, air, and fire belonged to the inner sublunar sphere, and a fifth element called ether (later known as *quintessence*) pervaded the outer celestial regions. The natural motion of the terrestrial elements was up and down, and by this motion they sought to find their proper places according to weight; the natural motion of the etheric element, and of the spheres created from it, was endless revolution around the Earth.

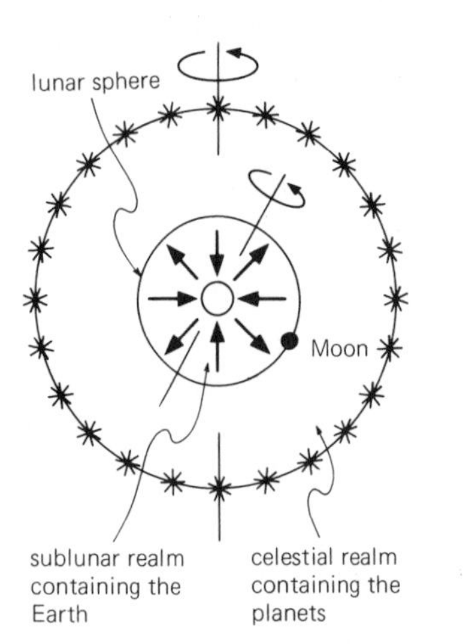

Figure 4.6. The Aristotelian universe (planetary spheres not shown). The many-sphere model is invested with physical and spiritual reality: The physical things occupy the sublunar realm, and the spiritual things occupy the celestial realm.

The whole universe was a cosmic sphere of finite size – necessarily finite, because its outer boundary rotated about the Earth. As Aristotle said, "If the heaven be infinite, and revolve in a circle, it will traverse an infinite distance in a finite time . . . But this we know to be impossible." Straight lines therefore cannot be infinite in length because they would then extend beyond the universe. They are hence incomplete and imperfect, whereas circles are forever complete and perfect. It was thus fitting that the terrestrial elements of perishable form should have only imperfect vertical motion, toward and away from the cosmic center, and this explained why the Earth did not rotate. It was also fitting that the etheric element of the spheres should have only perfect circular motion. "So the unceasing movement of the heavens is clearly understandable," said Aristotle, and "everything ceases to move when it comes to its natural destination, but for the body whose natural path is a circle, every destination is a fresh starting point." Aristotle's cosmos was a steady state universe that had existed unchanged throughout eternity; its perfect motions had no beginning or end and were controlled by intelligent agents that existed in the etheric realm.

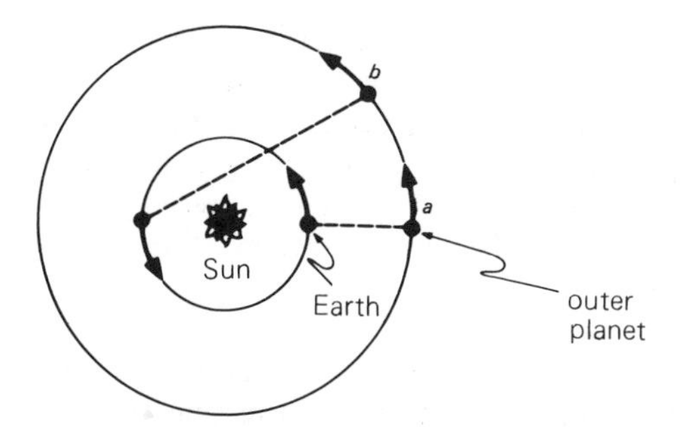

Figure 4.7. This figure shows why the planets, as seen from the Earth, appear to seesaw across the sky. The Earth and an outer planet, such as Mars, are shown revolving about the Sun. The Earth revolves faster than the outer planet: At position a the planet is closer and appears to move backward, but at position b the planet is further away and appears to move forward.

PTOLEMAIC ADDITIONS. The rotating spheres of Eudoxus did not explain planetary retrogression satisfactorily (see Figure 4.7) and failed to explain why the planets have increased brightness during their periods of retrograde motion. One step in the solution of this problem was the idea of eccentric circles: The Earth remained at the center of the universe, but each celestial circle had its center displaced by a certain amount from the Earth. Another and more important

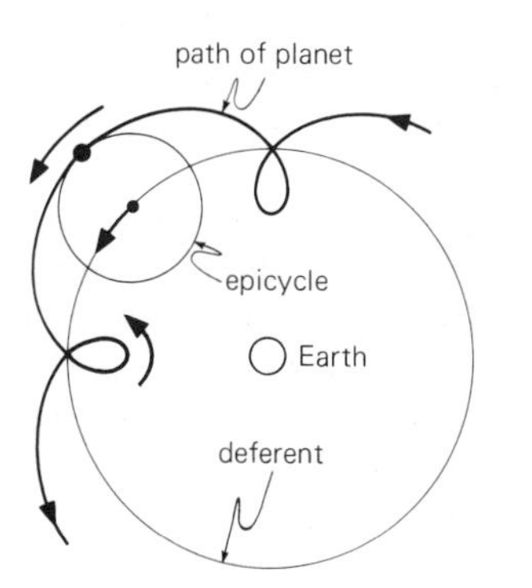

Figure 4.8. The motion of the planets according to the epicyclic theory. A planet revolves around a small circle (an epicycle), which moves around on a larger circle (the deferent) with the Earth at the center of the universe. As a result, the planet is seen to move backward and forward, and the backward (retrograde) motion occurs when the planet is closer to Earth and appears brighter.

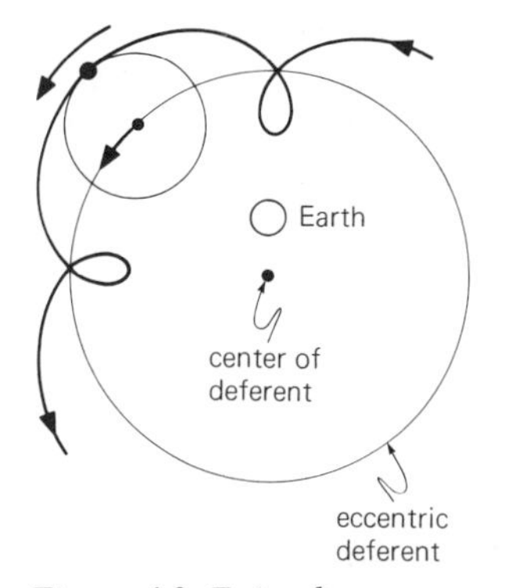

Figure 4.9. Epicycle on an eccentric deferent.

step, taken in the third century B.C., was the introduction of epicycles. Epicycles were additional circular motions, as shown in Figures 4.8 and 4.9, and they explained why the planets appear to move backward (retrogression) and forward across the sky and are brighter during the intervals of backward motion.

Claudius Ptolemy, an astronomer and mathematician of the second century A.D., did for astronomy what Euclid had done for geometry. He brought together the ideas and observations of previous centuries and geometrically mechanized the Aristotelian universe. His main work, *Almagest* (meaning "The Great System" and so named by the Arabs), used not only eccentrics and epicycles but also equants. The equant – or equalizing point – was an off-center point, as shown in Figure 4.10, about which the

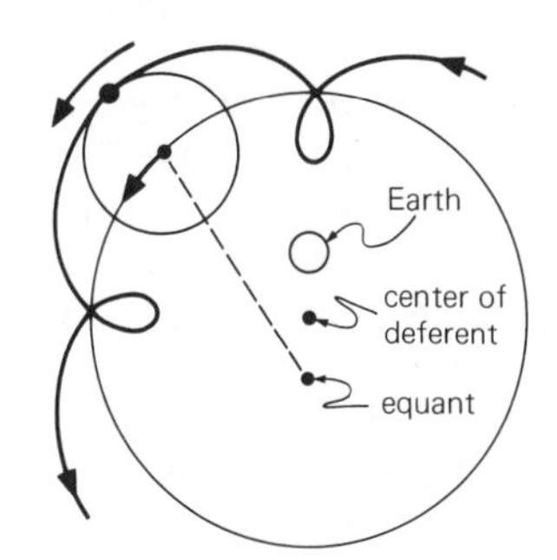

Figure 4.10. The equant introduced by Ptolemy. The equant, or equalizing point, is a noncentral point about which the epicycle moves at a constant angular rate.

epicycle moved at uniform angular rate. With various combinations of eccentrics, epicycles, and equants, and by dint of long laborious calculations, it had at last become possible to account for the motions of the planets in a geocentric universe. The result was a geometric marvel that endured until the Renaissance, when it was overthrown by the efforts of Copernicus, Kepler, and Galileo.

THE DARK AGES. From the sixth century B.C. to the second century A.D., from Thales of Ionia to Ptolemy of Alexandria, Greek science flourished for 700 years. By the end of the second century A.D. the blaze of Greek inspiration had died to a feeble glow, and the Ptolemaic system was among the most conspicuous achievements during the declining years. We must realize that the Aristotelian universe, as elaborated by Ptolemy, did not incorporate many of the developments of Greek science; it rejected the notion of a boundless Atomist universe in which the interplay of atoms creates endless worlds in development and decay; it rejected the Democritean suggestion that the Milky Way is an agglomeration of stars; it rejected the proposal of Heracleides that the Earth rotates daily about its axis; and it rejected utterly the theory of Aristarchus, accepted by Archimedes, that the Earth rotates daily about its axis and revolves annually about the Sun.

With the fall of the Roman Empire in the fifth century darkness descended on Europe, and all intellectual pursuits languished in the hands of barbarians whose languages at first could not encompass the concepts of the Greco-Roman world. Christianity was rationalized by Judaic and Greek scholarship that still survived in Alexandria, and the crumbling Roman Empire provided a communication network for its dissemination. But in the ensuing Dark Ages (the early phase of the Middle Ages) the universe reverted to a mythological polarization of heaven and hell with the Earth in the shape of a flat, rectangular tabernacle surrounded by an abyss of water. Some scholars (such as Boethius and the Venerable Bede) were aware of Greek science through the Latin commentaries of Cicero, Pliny, and others. While European learning was at its lowest ebb, remnants of ancient knowledge survived mainly in Byzantium, Syria, and Persia.

THE RISE OF ISLAM. In the seventh century the Arabs poured out of their deserts and created the great Islamic Empire that extended from the Atlantic to India. Its driving force was the cosmological message of an eternal and universal brotherhood of the faithful. The crafts, arts, and sciences once again flourished; libraries of ancient manuscripts were established; and scholars migrated to Damascus, Baghdad, Cordoba, and other centers of the new civilization. Greek, Egyptian, Persian, Chinese, and Indian literature was gathered, translated first into Syriac and later into Arabic, and then synthesized by industrious scholars. In Europe, by the ninth century, the Earth had regained its rotundity and the universe was once more geocentric. The modern world is indebted to the Islamic Empire for its preservation and transmission of ancient knowledge that ultimately reawakened Europe.

THE MEDIEVAL UNIVERSE. In the eleventh century the Dark Ages were dispelled by the rise of new ideas, such as the conception that it is necessary to understand in order to believe. Schools and then universities arose. In the twelfth and thirteenth centuries, during the decline of the Islamic Empire (hastened in the fourteenth century by the Mongolian invasion), the works of Aristotle, Euclid, Ptolemy, and many others of the ancient world were translated into Latin, first from Arabic and then directly from the original Greek.

Thomas Aquinas in the thirteenth century showed how Christianity could be accommodated within the Aristotelian universe with relatively slight modifications. Human beings retained immortality, but the adopted universe lost its eternity because it had been created by God. Further adaptations soon followed, as portrayed by Dante in *The Divine Comedy;* hell became a nether region within the Earth, purgatory was the sublunar region, and the ethereal regions were found to be ideal for the residence of superlunary hierarchies of angelic beings (see Figure 4.11). By the fourteenth century the medieval universe had attained its peak. It was entirely anthropocentric and was sanctified by religion, sanctioned by philosophy, and rationalized by geocentric science.

HELIOCENTRIC UNIVERSE

THIRTEENTH TO FIFTEENTH CENTURIES. The Copernican Revolution began with the Pythagoreans and ended with the Newton-

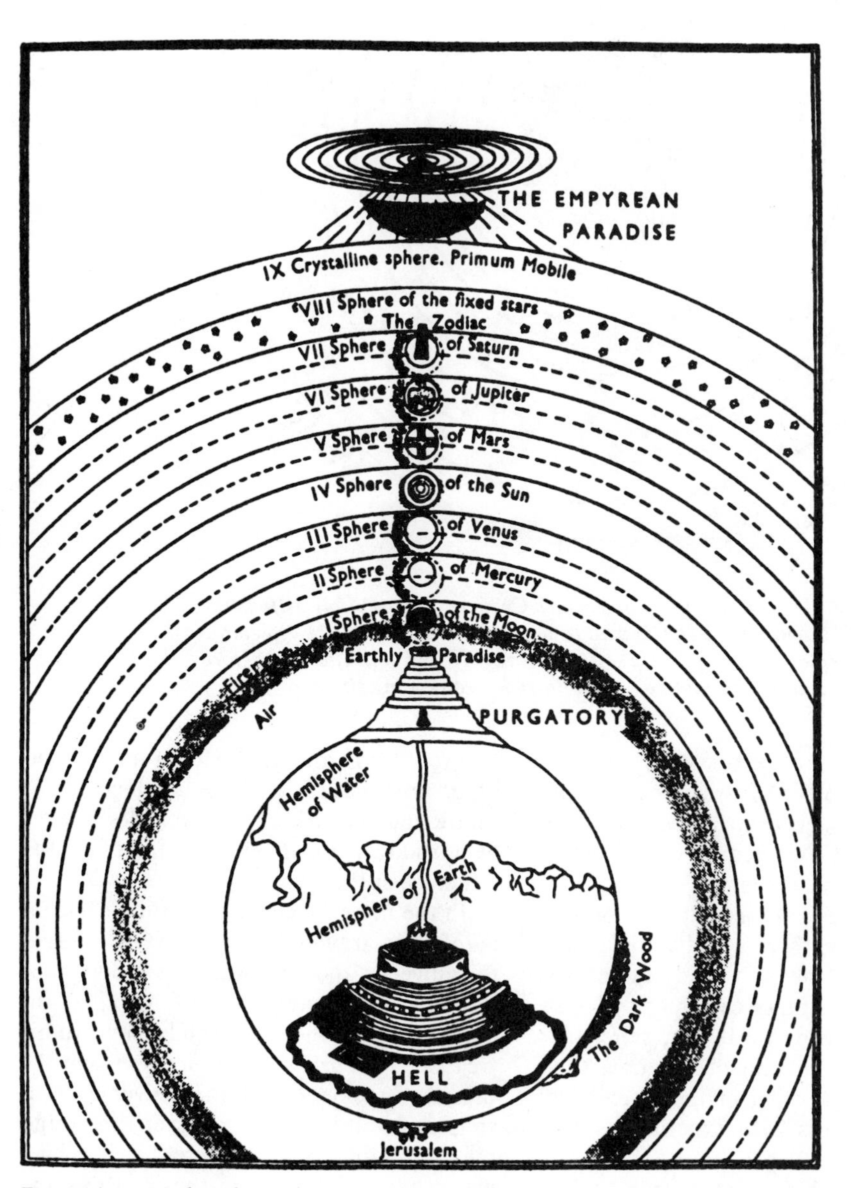

Figure 4.11. Medieval universe at the time of Dante (1265–1321), as presented in The Divine Comedy.

Figure 4.12. The Ptolemaic system from Apian's Cosmographia *(1553). Alfonso X (1221–84), king of Castile and Leon, is said to have remarked, after the Ptolemaic system had been explained to him, "If the Lord Almighty had consulted me before embarking upon Creation, I should have recommended something simpler."*

ians, and its greatest heroes were Aristarchus and Copernicus. It released human beings from their geocentric obsession and paved the way for the ultimate overthrow of the anthropocentric universe.

Already, in reawakened Europe of the thirteenth century, there was dissatisfaction with Aristotle's physics and Ptolemy's astronomy. The Franciscan monk Roger Bacon emphasized that the scientific method consists of making observations (not reading old texts), using mathematics, and checking the results with experiments and further observations.

In the fourteenth century, Bishop Nicholas Oresme emphasized that "motion can be perceived only when one body alters its position relative to another." He refuted the old arguments that claimed the Earth could not rotate and showed that Heracleides' theory of a rotating Earth greatly simplified the structure of the heavens.

In the fifteenth century, Cardinal Nicholas of Cusa argued that because God created the universe, and God is infinite and without location, the universe itself is also unbounded and without edge and center. In his treatise *Of Learned Ignorance* he made the famous statement that the universe "is a sphere of which the center is everywhere and the circumference is nowhere." It is convenient, he said, to regard the Earth as the center of the universe, although there is nothing in reality that compels us to do so.

An actual center does not exist, and he therefore could find no reason why the Earth should not be in motion.

COPERNICUS. In the sixteenth century, Nicolaus Copernicus (1473–1543), a canon of the Catholic Church, demonstrated the feasibility of a heliocentric universe. As a student he had studied the Ptolemaic system and had been impressed by the well-known fact that it violated the Platonic ideal of perfect circular motion. This ideal had been adopted by Aristotle, but when Ptolemy introduced equants that eliminated the uniformity of circular motion, the Ptolemaic system became a contradiction of the basic postulate of perfect circular motion. Copernicus was also aware of the heliocentric theory proposed by Aristarchus in the third century B.C. According to this theory the Earth revolves about the Sun and hence causes the other planets to appear to move backward and forward across the sky. It occurred to Copernicus that in this theory equants might be discarded and the original ideal of perfect circular motion could be restored. Starting from this simpler heliocentric picture, Copernicus devoted the rest of his life to the construction and computation of heliocentric orbits.

Copernicus's great work, *Revolutions of the Celestial Spheres,* rivaling the *Almagest* in its scope, was printed in 1543 shortly before his death. In an earlier work he had written, "All orbs revolve about the Sun, taken as their center point, and therefore the Sun is the center of the universe" (see Figure 4.13). In the *Revolutions of the Celestial Spheres* he wrote, "Why then do we hesitate to allow the Earth the mobility natural to its spherical shape, instead of proposing that the whole universe, whose boundaries are unknown and unknowable, is in rotation?" The arguments he used to justify the Earth's rotation were those proposed earlier by Oresme. From a rotating Earth it is a simple step to a moving Earth: "We therefore assert that the center of the Earth, carrying the Moon's orbit, passes in a great orbit among the other planets in an annual revolution around the Sun; that near

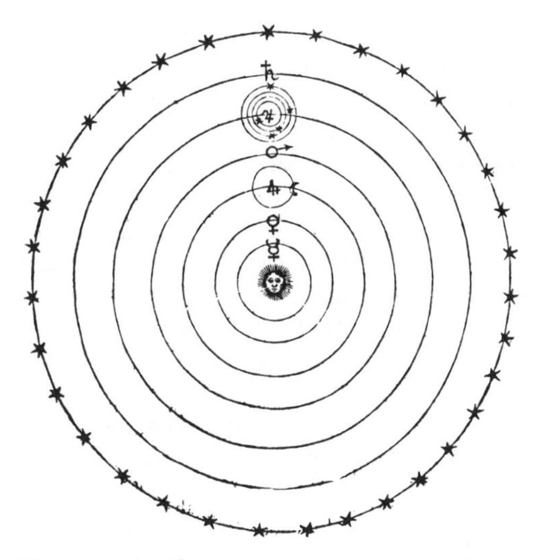

Figure 4.13. The Copernican universe with the Sun at the center. Note that the universe still retains its outer sphere of stars. "But Aristarchus of Samos brought out a book consisting of certain hypotheses in which the premises lead to the conclusion that the universe is many times greater than that now so called. His hypotheses are that the stars and the sun remain motionless, that the earth revolves about the sun in the circumference of a circle, the sun lying in the middle of the orbit" (Archimedes [about 287–212 B.C.], The Sand Reckoner).

the Sun is the center of the universe, and that whereas the Sun is at rest, any apparent motion of the Sun can be better explained by motion of the Earth." The Copernican universe was finite in size; its center was occupied by the Sun, and its outer edge was the sphere of fixed stars.

Alas! the Copernican dream of a simpler universe was not fulfilled and most astronomers at first were not convinced. The more Copernicus labored to bring the heliocentric system into conformity with observations, the larger became the number of circles needed. By sacrificing equants, he required more circles than ever before and was still unable to achieve greater precision than Ptolemy.

DIGGES, BRUNO, AND TYCHO. The astronomer and mathematician Thomas Digges (1543–95) was born in the year in which Copernicus died. In the popular work *A Perfit*

Description of the Caelestiall Orbes, first published in 1576, he expounded on the Copernican system and introduced an important new development. Digges's modification of the heliocentric theory consisted of removing the outer edge and dispersing the fixed stars throughout an unbounded space (see Figure 4.14). Thus only 33 years after the publication of the *Revolutions of the Celestial Spheres* (and 200 years before the Industrial Revolution and the Declaration of Independence), the heliocentric system had its outer edge torn away. The *Perfit Description* passed through many editions in the latter part of the sixteenth century and set in motion ideas that finally toppled the heliocentric universe.

The fiery monk Giordano Bruno (1548–1600) was living in London while Digges's book was the talk of the town. He enthusiastically adopted the idea of an edgeless universe and drew attention to the logical conclusion, previously pointed out by Nicholas of Cusa and others, that the universe is also centerless. He wrote, "In the universe no center and circumference exist, but the center is everywhere." By writing and traveling, Bruno sought to broadcast the

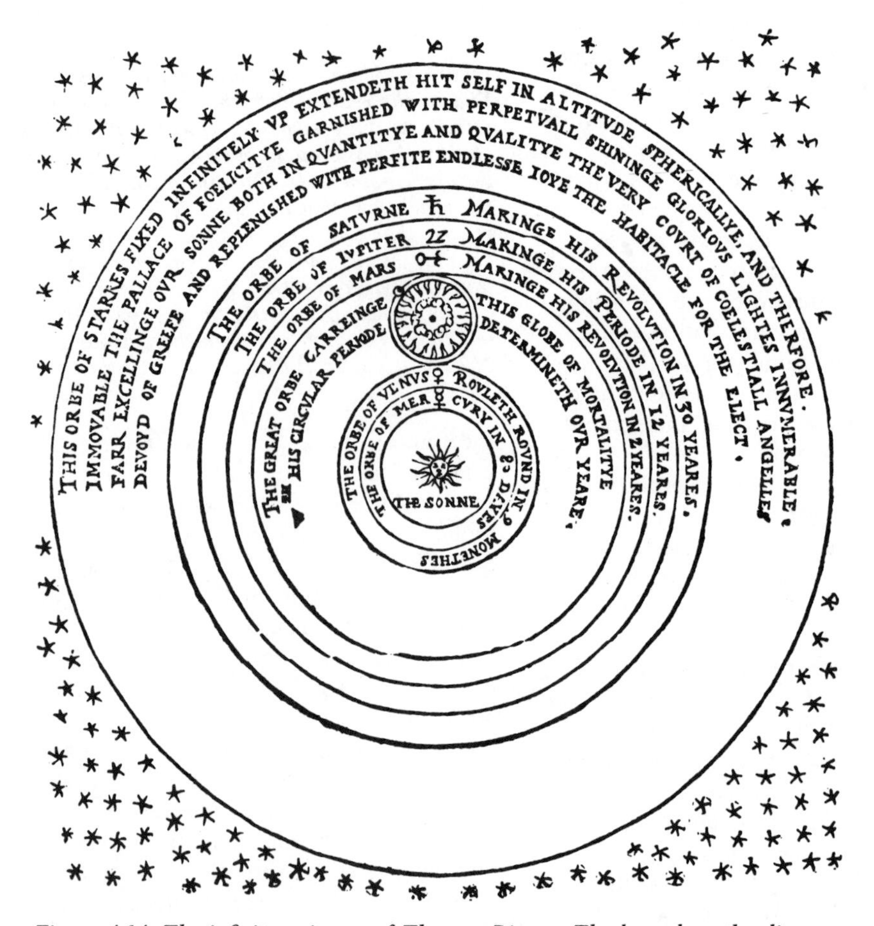

*Figure 4.14. The infinite universe of Thomas Digges. The legend on the diagram reads: "This orbe of starres fixed infinitely up extendeth hit self in altitude sphericallye, and therefore * immovable the pallace of foelicitye garnished with perpetuall shininge glorious lightes innumerable * farr excellinge our sonne both in quantitye and qualitye the very court of coelestiall angelles * devoyd of greefe and replenished with perfite endlesse joye the habitacle for the elect."*

Atomist theory, known through the work of the Roman poet Lucretius, of an infinite universe, uniformly populated with numberless solar systems and teeming with life. Bruno was the revolutionary champion of the Copernican Revolution, and as is true of most revolutionary fanatics, his life was a tragedy and his success questionable. With one blow he attempted to smash geocentricity and heliocentricity and, by ridiculing religious beliefs, to smash also anthropocentricity. He contributed nothing that was not already known, and by premature attacks on cherished beliefs he aroused unnecessary hostility against the idea of a heliocentric universe. He was a martyr: His last seven years were spent in an ecclesiastical prison; he was tortured, but refused to recant; and he was burned at the stake in Rome in 1600. He died not for astronomy, for he was not an astronomer, but for his heretical attacks on orthodox religion.

Tycho Brahe (1546–1601) was a Danish nobleman who made careful and continual observations of the planets, using the utmost precision possible without telescopes. He rejected the Copernican system for a simple and important reason. If the Earth moves in a great circle around the Sun, he said, then the stars should be seen in slightly different directions at different times of the year. But his observations failed to detect stellar parallax. The stars, he argued, have a certain size and are not vanishing small points of light; if they must be banished to distances so great that parallax is unobservable, they become enormously larger than the Sun. Nowadays we know that the eye itself is responsible for the apparent size of stars, and the stars are even further away than Tycho supposed possible.

Tycho constructed a compromise system of his own. The Tychonic system was geocentric: The Sun revolved about the Earth, and all other planets revolved, not about the Earth, but about the Sun (see Figure 4.15).

KEPLER. Johannes Kepler (1571–1630) was an extraordinarily imaginative scientist who overcame ill health and became imperial mathematician of the Holy Roman Empire (an "agglomeration" that, according to Voltaire, though it "was called and still calls itself the Holy Roman Empire was neither holy, nor Roman, nor an empire in any way"). This was a lowly position, earning him only a pittance, whose principal function was to prepare astrological forecasts. Kepler wholeheartedly accepted the finite Copernican universe, with the Sun at its center and the sphere of fixed stars at its outer edge, and vehemently opposed all suggestions of an infinite and centerless universe. The thought of an infinite universe repelled him: In 1606 he wrote, "This very cogitation carries with it I don't know what secret, hidden horror; indeed one finds oneself wandering in this immensity to which are denied limits and center and therefore also all determinate places."

In Kepler's first book, *Mystery of Cosmography,* published in 1596, he sought to unravel the secrets of the cosmos and speculated on the nature of the harmonious and mathematical universe that had been conceived in the mind of the Creator. His cosmic description (i.e., cosmography) made no lasting contribution to the advance of science but provided him with a framework of ideas that directed his later research.

Tycho had been the previous imperial mathematician, and Kepler inherited Tycho's observations of the planets, which were the most precise and detailed in existence. For years Kepler struggled to explain the motion of Mars. He ultimately succeeded and thereby freed astronomy from the paradigm of epicyclic motion. Kepler's three laws of elliptical planetary motion, already mentioned (in Reflection 3 of Chapter 2), provided the astronomical concepts that were essential for the rise of the Newtonian universe.

GALILEO. Galileo Galilei (1564–1642) was born in the year in which Michelangelo died and Shakespeare came into the world, and he died in the year in which Newton was born. His great contribution to astronomy

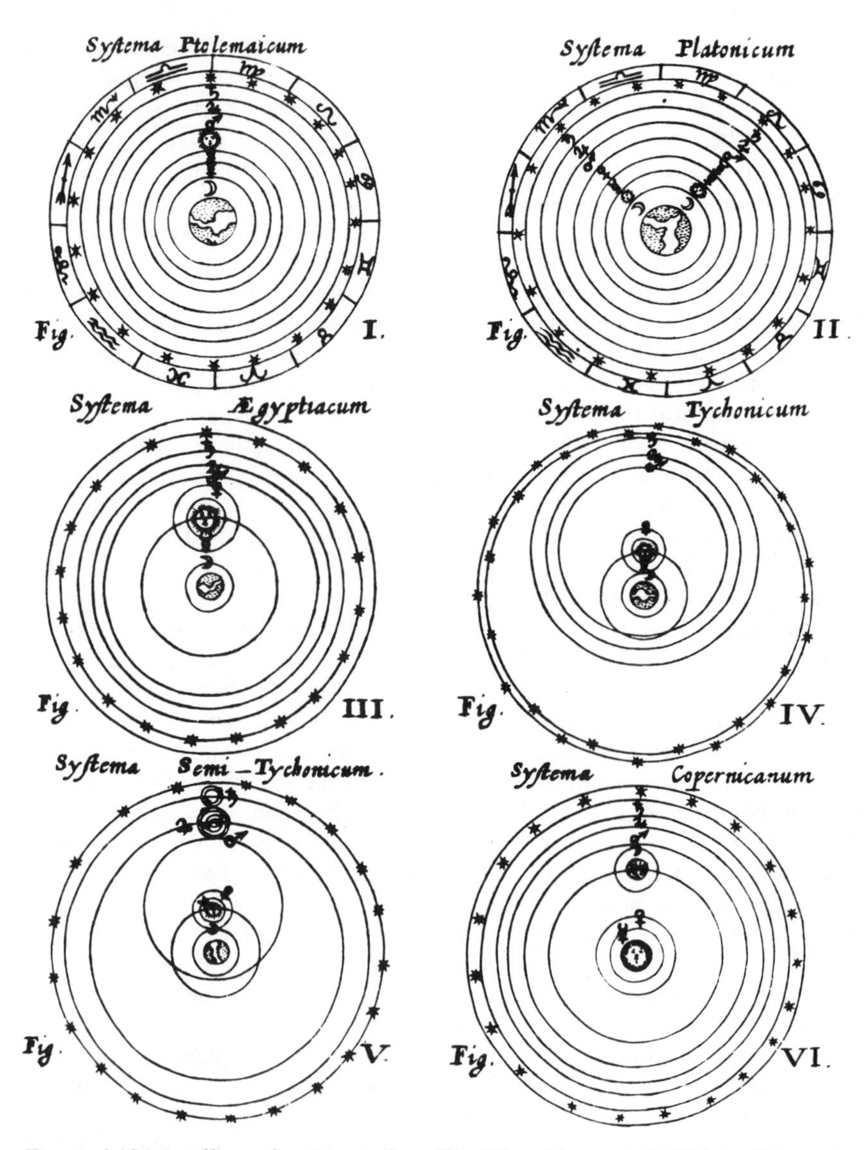

Figure 4.15. A gallery of universes from The New Almagest *(1651) by Giovanni Riccioli. The Ptolemaic system (I), the Tychonic system (IV), and the Copernican system (VI) are discussed in the text. In the Platonic system (II), the Sun is interior to the orbits of Mercury and Venus, whereas in the Ptolemaic system it is exterior to these planetary orbits. In the Egyptian system (III), the inner planets Mercury and Venus revolve about the Sun, which revolves with the outer planets about the Earth. In the Semi-Tychonic system (V), Mercury, Venus, and also Mars revolve about the Sun, which revolves with Jupiter and Saturn about the Earth. (Courtesy of the Henry E. Huntington Library, San Marino, Calif.)*

was the use of telescopes. He believed in the Copernican system but was not impressed with Kepler's theories. His own discoveries – mountains on the Moon, satellites of Jupiter, and numerous hitherto unresolved stars of the Milky Way – were published in 1610 in a book called *The Starry Message*. He found also an answer to Tycho's difficulty.

Although his telescope magnified 30 times, the stars were the same apparent size as when seen with the naked eye, and he therefore concluded that their observable size is misleading.

What Galileo saw through his telescope was not in accord with Ptolemaic teaching. Galileo's ideas also ran counter to the Aristotelian belief that the celestial realm is the abode of spirits. He declared in a forthright way that the sublunar and celestial realms are both physical, and to the observing eye and the critical mind it was obvious that the Earth rotates and also revolves about the Sun. His most hostile opponents were scholars steeped in the orthodox teachings of Aristotle. His famous dialogue of *The Two Great Systems of the World* contrasted the geocentric and heliocentric systems, and by innuendo poured scorn on the physics of Aristotle and the astronomy of Ptolemy. This brought him into conflict with the less moderate elements of the Catholic Church, and under the threat of torture, he recanted and abjured the heliocentric system.

ISLAND UNIVERSES

THE NEWTONIAN UNIVERSE. Leaders of the Protestant Reformation were at first hostile to Copernicus ("this fool," said Luther, "wishes to reverse the entire history of astronomy"), but later they relented, and science fled from the Mediterranean before the mounting intolerance of the Counter Reformation. While Rome continued to cling to geocentrism for a further 200 years, the northwestern countries of Europe discarded it and very soon abandoned even heliocentrism.

Isaac Newton (1642–1727), the most illustrious of all men of science, lived in the company of brilliant scientists and gathered together and made systematic the threads of ideas of great thinkers since the Middle Ages. He developed the theories of motion and gravity, and from his mind emerged the Newtonian universe governed by equations and quantitative laws of nature. It was a clockwork universe that had at last attained the power and perfection to which all science had aspired from the very beginning.

HIERARCHICAL UNIVERSES. The Newtonian universe was infinite and at first was centerless and uniformly populated with stars similar to the Sun. But with improvements in telescopes, astronomy was widening its horizons; and it became increasingly difficult to ignore the fact that stars are not distributed uniformly throughout space.

Thomas Wright (1711–86), of Durham in the north of England, had novel thoughts concerning the Milky Way that were published in his book *An Original Theory of the Universe* of 1750. At first he had supposed that the stars were "promiscuously distributed through the mundane space," but later he realized that they are scattered "in some regular order" (see Figure 4.16). An essential feature of Wright's universe was the existence of a supernatural cosmic center: At "this centre of creation, I would willingly introduce a primitive fountain, perpetually overflowing with divine grace, from whence all the laws of nature have their origin." He imagined two possible constructions of the Milky Way: either a ring-shaped distribution of stars, similar to the rings of Saturn, encircling the center; or a spherical shell of stars, around the center, in which the Milky Way was seen in a plane tangential to the shell (see Figure 4.17). He went even further and speculated on the possibility of many centers of creation. The distant nebulae seen as faint and fuzzy objects are perhaps other creations or "abodes of the blessed," he said, similar to the Milky Way, and "the endless immensity is an unlimited plenum of creations not unlike the known universe" (see Figure 4.18).

Immanuel Kant (1724–1804) read a review of Wright's work and concluded that the theory entailed a lens-shaped Milky Way surrounded by similar systems, such as the elliptical nebulae. In his book *The Theory of the Heavens,* Kant in 1755 elaborated on Wright's ideas and described the

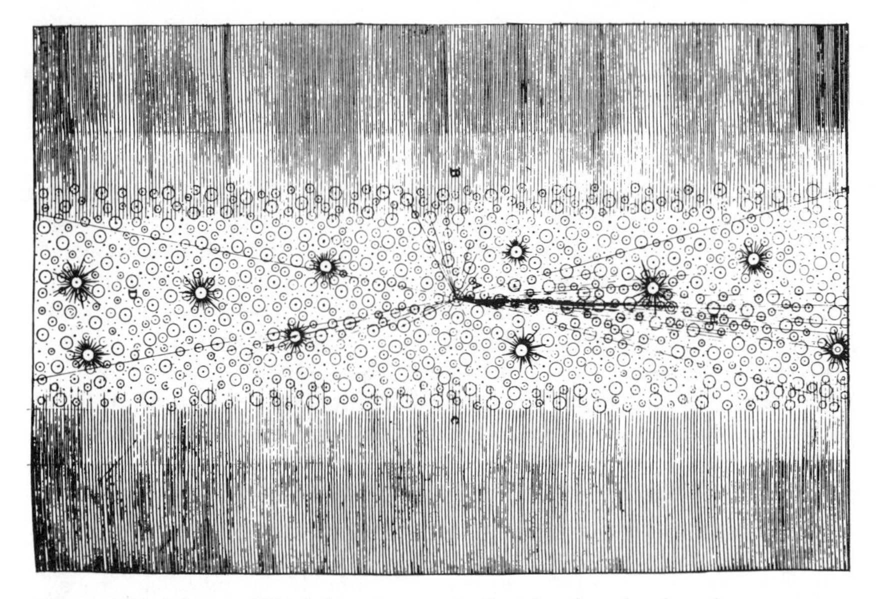

Figure 4.16. Thomas Wright's universe. At first he thought that the stars were uniformly distributed, but the Milky Way caused him to realize that they are distributed within a disk, as shown by this illustration from his book An Original Theory of the Universe.

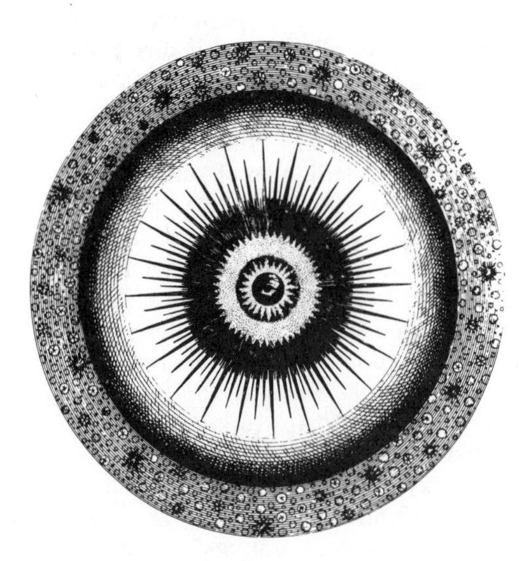

Figure 4.17. Wright considered two possibilities. First, the MilkyWay might be a disk that rotates about a mysterious galactic center, and the universe might be filled with such milky way disks. Second, as shown in this illustration, the stars might be distributed in a shell about the galactic center, so that the Milky Way would be seen in a plane tangential to the shell.

most stupendous picture of the universe that had ever been imagined. According to Kant, the stars of the Milky Way form a disk that rotates about its center, and the distant nebulae are also rotating disklike milky way systems. These numerous milky ways (which we now call galaxies) are themselves clustered together about a common center, and form a vast system of galaxies.

Kant did not stop at this stage, but went on to suppose that there were many such vast systems and that these vast systems were themselves again clustered together to form even vaster systems, and that these vaster systems were also clustered together to form yet vaster systems, and so on, in an awesome universe of infinite space (see Figure 4.19). At each level of the hierarchy existed an array of centers clustered about the centers of the next higher level. At the highest level, of infinite order, was the ultimate center that dominated the structure of the universe. In a universe of many centers, Kant had restored the cosmic center.

Figure 4.18. Wright's "endless immensity" of galaxies, as illustrated in An Original Theory of the Universe.

Johann Lambert (1728–77), a Swiss-German mathematician, entertained similar ideas; the main difference was Lambert's assumption that each center was occupied by a body called a "dark regent." In his *Cosmological Letters* of 1761, Lambert wrote, "The eye, assisted by the telescope, may at length penetrate all the way to the centers of the milky ways, and why not even to the center of the universe?"

NEBULA HYPOTHESIS. Half a century later, Pierre de Laplace (1749–1827) expounded on the famous nebula hypothesis (previously proposed by Kant) concerning the origin of the Solar System. According to this theory, now accepted in modernized form, the Sun and planets condensed from a swirling cloud of gas. The theory further suggested that the Andromeda Nebula and other similar nebulae were perhaps solar systems in the process of formation. The Wright–Kantian interpretation of the faint nebulae was that they were distant milky ways, and the Laplacian interpretation was that they were clouds of gas within a vast Milky Way. These opposing hypotheses ushered in a dramatic era in which astronomers were at loggerheads for more than 100 years. The earlier debate on geocentric versus heliocentric universes (in which most participants were not scientists) had now switched to a higher plane and become a Great Debate among astronomers on a Milky Way universe versus milky ways or "island universes": a single Milky Way universe, containing nebulous clouds, versus an infinity of island universes, each similar to the Milky Way. The term *island universe* (meaning galaxy) was introduced by Friedrich von Humboldt in his book *Kosmos* of 1845.

THE GREAT DEBATE. The leading figure in the opening stages of the Great Debate was William Herschel, with his improved telescopes and years of careful observation. He placed the Sun at the center of the Galaxy and thus initiated the *galactocentric* theory

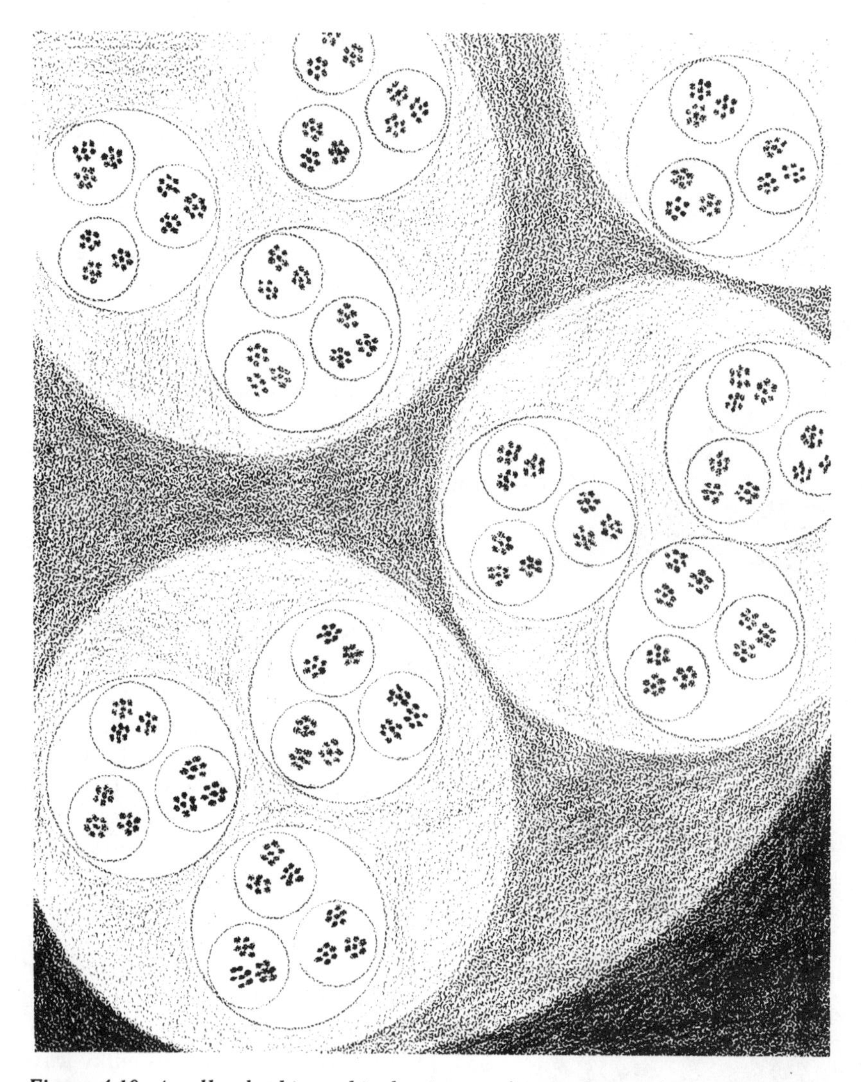

Figure 4.19. A polka-dot hierarchical universe of stars clustered into galaxies and of galaxies clustered into larger systems, which in turn are clustered into yet larger systems, and so on, indefinitely, as conceived by Immanuel Kant and Johann Lambert in the eighteenth century.

(see Figure 4.20). Astronomical advances thereafter were at first slow. Better telescopes had to be developed, photography introduced, and observational techniques improved; distance indicators had to be found and calibrated; nebulae within the Galaxy had to be distinguished from extragalactic nebulae; globular clusters had to be identified as systems of stars lying within and on the outskirts of the Galaxy; and last of all, confusing obscuration caused by interstellar absorption had to be recognized and proper allowance made for it.

By studying the distances and the distribution of globular clusters, Harlow Shapley in 1918 overthrew the Herschel galactocentric universe. From their positions he found that the globular clusters form a spherical

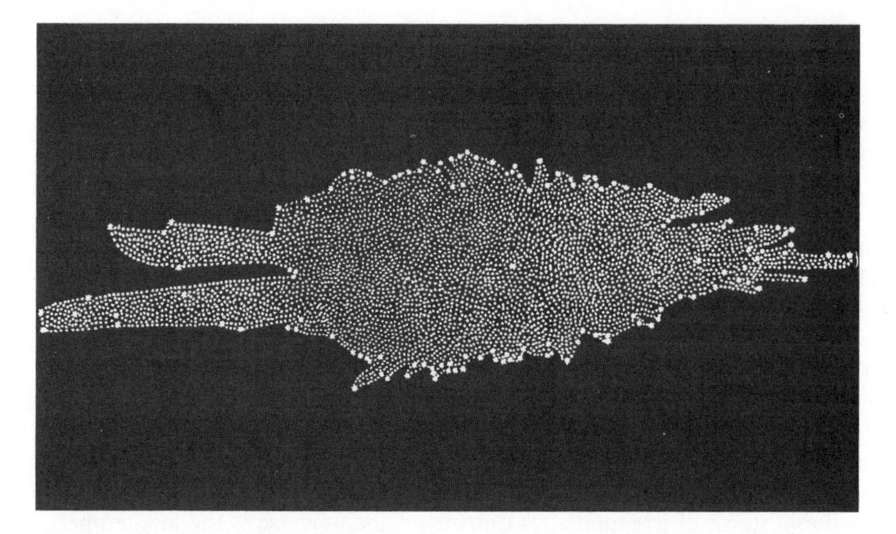

Figure 4.20. The Stellar System, or the Galaxy as we now call it, according to William Herschel in 1785. The Sun is close to the center of Herschel's Stellar System.

swarm, and the center of the swarm is tens of thousands of light years away in the direction of the constellation Sagitarrius. The globular clusters are members of the Galaxy, and Shapley's discovery indicated that the center of the swarm is also the center of the Galaxy. Jan Oort, a Dutch astronomer, confirmed this result by studying the motions of stars and showed that the stars of the Milky Way are all orbiting about the distant center of the Galaxy.

In the 1920s, during the closing stages of the Great Debate, Shapley championed the Milky Way universe theory, and Heber Curtis, another American astronomer, upheld the opposing view of a universe of many galaxies. Because of inadequate allowance for absorption, Shapley, as we now know, overestimated the distances of the globular clusters and made the Galaxy too large, whereas Curtis underestimated the distances of the stars and made the Galaxy too small. The dream of a giant Milky Way universe with small outlying galaxies was championed by Shapley until 1930 and then dropped. The long controversy extending from the eighteenth century to the twentieth, during which astronomy matured into an advanced science, had at last come to an end. In 1936, Edwin Hubble declared in *The Realm of the Nebulae:* "The nebulae are great beacons, scattered through the depths of space . . . Observations give not the slightest hint of a super-system of nebulae. Hence, for purposes of speculation, we may invoke the principle of the Uniformity of Nature, and suppose that any other equal portion of the universe, chosen at random, will exhibit the same general characteristics. As a working hypothesis, serviceable until it leads to contradictions, we may venture the assumption that the realm of the nebulae *is* the universe – that the Observable Region is a fair sample, and that the nature of the universe may be inferred from the observed characteristics of the sample."

A difficulty still remained: Our Galaxy seemed much larger than most other galaxies. This was finally resolved in 1952 by Walter Baade, who distinguished between population I and II stars and their cepheid variables, and thereby doubled the distances of the galaxies; this also doubled the size of the distant galaxies, and the Milky Way no longer seemed disproportionately large.

THE ISOTROPIC UNIVERSE

The universe stretches away . . . just the same in all directions without limit.
— Lucretius (95–55 B.C.), *The Nature of the Universe*

In a state of isotropy things are the same in all directions. Thus from a ship at sea we see the surface stretching away isotropically to the horizon. Everywhere we see local irregularities – the waves – but these are mere details and are small compared with the distance to the horizon. But to the naked eye the universe is not as isotropic as Lucretius implied. We are, it seems, surrounded by large waves, a veritable storm of irregularities.

The Galaxy is not isotropic about us: Stars are concentrated in the rotating disk in which we have noncentral location. All directions beyond the Galaxy also do not appear alike because we look out through the dusty gas clouds of the disk. We see swarms of stars in the Milky Way, but cannot see easily through the Milky Way to the galaxies beyond because of obscuration; away from the Milky Way are fewer stars and less absorbing dust and we see more easily the galaxies beyond. The astronomer looks out at the distant universe through "dirty windows" and must always make allowance for absorption.

Around us, outside the Galaxy, are scattered unevenly the galaxies of the Local Group. Beyond the Local Group are other clusters of galaxies, scattered higgledy-piggledy, which stretch away in their multitudes to the distant horizon. Galaxies are bunched into clusters, and clusters are lumped into superclusters, and the irregularities we perceive extend over vast distances. The Galaxy is like a boat tossed in a stormy ocean. Beyond about 300 million light years the optically observed universe begins to look reasonably isotropic. When allowance is made for obscuration within the Galaxy, what is seen in one direction looks very similar to what is seen in other directions at great distances.

The incoming signals from radio sources are not absorbed by gas and dust in the Galaxy and by the Earth's atmosphere. Radioastronomers find that the very distant and numerous radio sources are distributed about us isotropically.

X-rays are generated in galaxies and in the hot and tenuous gas between galaxies. These X-rays fill the universe and come to us from all directions. They cannot penetrate the Earth's atmosphere and must be detected with instruments mounted on rockets and artificial satellites. The X-ray background has been found to be at least 99 percent isotropic.

Our main source of evidence for the isotropy of the universe is the low-temperature cosmic radiation discovered in 1965 by Arno Penzias and Robert Wilson of the Bell Telephone Laboratories. This cosmic radiation floods the universe and comes, it is now believed, from the big bang. The radiation was once extremely hot, but has been cooled by expansion and has now a temperature of only 3 degrees Kelvin. It has a typical wavelength of 1 millimeter and consists of approximately 400 photons per cubic centimeter. Each of these photons, or waves of energy, has traveled for 10 or more billion years at the speed of light and has traversed the observable universe. Careful measurements show that this 3-degree radiation deviates from isotropy by about 1 part in 1000.

The discovery of the low-temperature cosmic radiation has opened the way to great developments in cosmology. It provides evidence that the universe was once dense and hot and also enables us to understand what happened in the early universe. We now know, for instance, that most of the helium existing at present was produced when the universe was approximately 200 seconds old.

The cosmic radiation has a small "24-hour" anisotropy. This means that its temperature is very slightly higher in one direction than in the opposite direction. As the Earth rotates, the observed temperature of the incoming cosmic radiation rises and falls by about 1 part in 1000, with a cyclic variation of 24 hours. This amount of anisotropy indicates that the Earth (and the Sun) is moving in the universe at a velocity somewhere between 300 and 600 kilometers per second. The cosmic radiation that floods the universe consists of electromagnetic waves moving in all directions. Consider an observer who moves at constant velocity. This observer sees the waves Doppler shifted: The waves coming from the forward direction have slightly shorter wavelengths, and the waves coming from the backward direction have slightly longer wavelengths. The waves from the forward direction have more energy and are received more frequently than the waves from the backward direction, and consequently the temperature is slightly greater in the forward direction. The Earth moves about the Sun at a velocity of 30 kilometers per second, and the Sun revolves about the center of the Galaxy at a velocity of 300 kilometers per second; in addition, the Galaxy moves within the Local Group of galaxies at a velocity of about 100 kilometers per second, and the Local Group itself has a velocity of 200 or more kilometers per second within the Local Supercluster. The cosmic radiation provides us with a cosmic framework relative to which it is possible to measure velocities in an absolute sense.

COSMOLOGICAL PRINCIPLE

ALL PLACES ARE ALIKE. Most treatments of modern cosmology begin with the *cosmological principle.* This principle, given its name by the cosmologist Edward Milne in 1933, is the foundation of modern cosmology. Einstein in 1931 expressed the principle in the words "all places in the universe are alike." When stated in this way it is reminiscent of Rudyard Kipling's *The Cat That Walked by Himself:* "I am the Cat who walks by himself, and all places are alike to me." The cosmological principle asserts that the universe is the same everywhere in space, apart from irregularities of a local nature. This principle at first sight is attractively simple; it is, however, a proposition of utmost generality, so sweeping in scope and far-reaching in implication that it deserves thought and study.

THE "OBSERVER" AND THE "EXPLORER." An ordinary observer can travel only short distances as measured on the cosmic scale and is therefore localized to a small region of space. The best such an observer can do is to look around and see whether the universe is isotropic. This we have done, and have found that all directions in the universe are alike, as seen from our local region of space.

An observer in a cosmic sense is immobile. To escape this constraint we shall invent, for illustrative purposes, a mobile cosmic "explorer" who is able to move infinitely rapidly from place to place. Sex cannot be omitted from cosmology, and by tossing a coin, I have determined that the stay-at-home observer is *he* and the gadabout explorer is *she.*

If the mobile explorer finds in her travels that all places are alike, she will be justified in asserting that the universe is the same everywhere and is hence *homogeneous.* She will declare that the universe is invariant to translations in space. If the immobile observer finds that all directions are alike, he will assert that the universe is *isotropic,* and will declare that the universe at his place is invariant to rotations in space. Hence the gadabout explorer translates from place to place and discovers homogeneity (according to which all places are alike) in space, and the stay-at-home observer rotates and discovers isotropy (according to which all directions are alike) about a point in space.

When the observer asserts that the universe is homogeneous, on the basis of observed isotropy, he in effect postulates the cosmological principle. Lacking essential information, he covers his ignorance with an impressive-sounding principle. Our task is to discuss the difference between isotropy and homogeneity and to show how homogeneity can be deduced from an observed state of isotropy.

HOMOGENEITY AND ISOTROPY. The explorer in her travels in a homogeneous universe sees at every place similar distributions of objects of various kinds – creatures, planets, stars, galaxies, and clusters of galaxies – all steadily evolving and changing with time. Suppose, after waiting a million years, the explorer repeats her instantaneous tour around the universe. Again all places are in similar states, but now everything is slightly older and more evolved. Repeated tours separated by intervals of time reveal a homogeneous universe that remains homogeneous because everywhere things are changing in the same way. From these observations the explorer draws the conclusion that to maintain a state of homogeneity, in which things everywhere change in the same way, it is necessary that the laws of nature be also everywhere the same. The laws, whatever they may be, might conceivably change in time and be different from tour to tour, but during each tour they are everywhere the same.

In a homogeneous universe all places are alike and things everywhere evolve according to the same laws. Now suppose that at one place an observer notices that all directions are alike and the universe is therefore isotropic about that place. In a homogeneous universe all places are alike, and if one place is isotropic, then it follows that all other places are isotropic. A homogeneous universe, isotropic at one place, is therefore isotropic at all places. If the observer perceives that all directions are alike, and is told by the explorer that all places are alike, then the observer knows that the universe is isotropic at all places. The universe, in other words, has no center.

Let us pursue this theme further. A homogeneous universe can be anisotropic (all directions are not alike). Consider a flat and limitless plain of grass. A prevailing wind will cause the grass on the plain to grow tilted in one direction. The plain is still homogeneous – all places are alike – because the tilt is the same everywhere. But the plain is no longer isotropic: In one direction the grass leans away from, and in the opposite direction it leans toward, a person standing anywhere on the plain. This is an example of a "24-hour" anisotropy in which we have to rotate through 360 degrees to observe again the same state. We can imagine a homogeneous universe in which galaxies tend to have their axes of rotation all pointing in the same direction, or in which there is a magnetic field that tends to point in one direction only, and such a universe is not isotropic. When the axes of galactic rotation are randomly oriented, and the magnetic field is tangled in all directions, then the universe becomes isotropic, apart from negligible local irregularities.

An inhomogeneous universe (in which all places are not alike) might be isotropic at one place but cannot be isotropic at all places. From the summit of a hill the surrounding countryside can appear isotropic, but when seen from anywhere else in the neighborhood, the countryside is anisotropic because of the presence of the hill. To sum up: A state of isotropy at one place does not prove homogeneity, and a state of anisotropy does not disprove homogeneity (see Figure 4.21).

HOW THE LOCATION PRINCIPLE WORKS. The mobile explorer perceives a state of homogeneity, and the immobile observer perceives a state of isotropy. Can the stay-at-home observer, by looking carefully, perceive also a state of homogeneity? The answer is no, and for a very simple reason. When we look out into the universe we also look back into the past. We see objects at different distances corresponding to various epochs of the past. We see the universe in progressively different states of evolution and expansion. The universe remains isotropic; yet distant things are different from local things, and we therefore perceive inhomogeneity. How then can we prove that the universe is homogeneous?

We cannot prove by observation that the universe is homogeneous and unfortunately

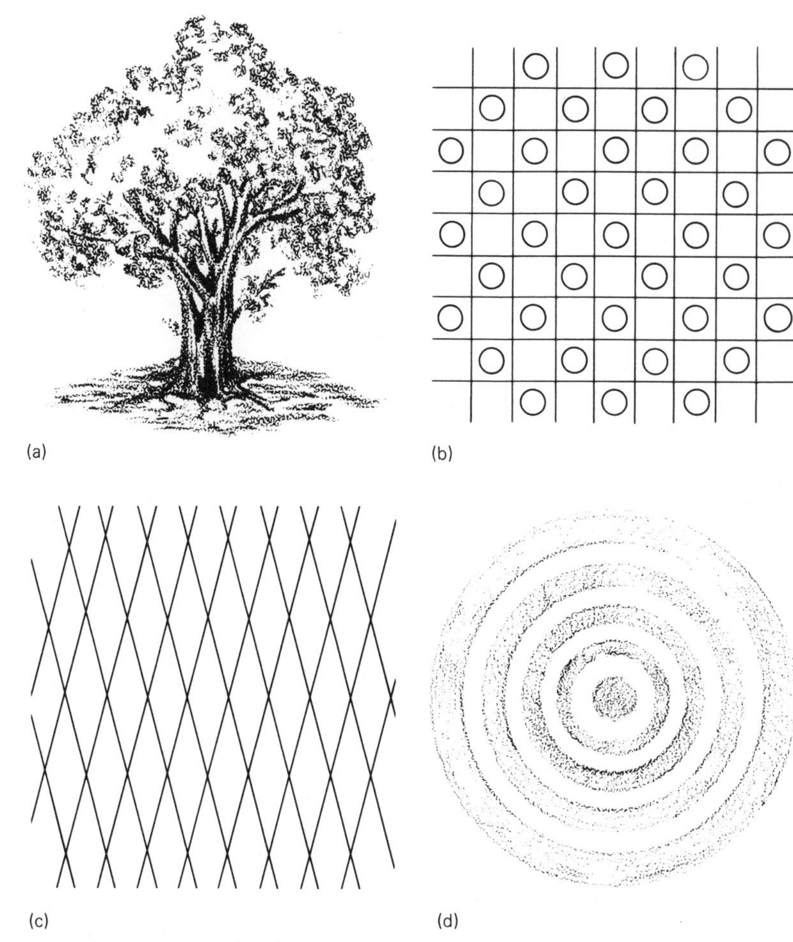

(a) (b) (c) (d)

Figure 4.21. Which of these four illustrations is homogeneous and isotropic? homogeneous and anisotropic? inhomogeneous and isotropic? inhomogeneous and anisotropic?

we do not know a gadabout explorer. Either we must postulate the cosmological principle as an article of faith, or we must become philosophers and use the location argument as it was used by the Surveyors at the beginning of this chapter.

If we use the location principle, then the situation is as simple as this: From observations we have established that the universe is isotropic about us, and if it is inhomogeneous, then we have special location. But special location in a vast universe is improbable, as stated by the location principle, and hence we must deduce that inhomogeneity is also improbable. Observed isotropy and the location principle lead to the conclusion that homogeneity is probable.

The *Copernican principle* is similar to the location principle; it serves the same purpose and states, "We are not at the center of the universe." Copernicus believed that the Sun occupies a central place in the universe, and the Copernican principle, formulated by Hermann Bondi, reflects the Copernican view that we on Earth are not in this privileged position. The Copernican principle, unfortunately, has the disadvantage of appearing to perpetuate the belief that a center, somewhere or other, does exist. It might also be said that this principle asserts too much: We say with certainty not that we do not occupy a central location, but only that there are philosophical reasons for thinking that central location is improbable. Also, the Copernican principle is perhaps incorrectly named, for if historical precedence is of importance, then a more appropriate name is the *Aristarchean principle.*

PERFECT COSMOLOGICAL PRINCIPLE

STATIC AND EXPANDING STEADY STATE UNIVERSES. In 1948, Hermann Bondi and Thomas Gold proposed that the universe was homogeneous in both space and time. Thus "all places are alike in space" became "all places are alike in space and time." This widened homogeneity postulate was called the *perfect cosmological principle,* and the named chosen is not inappropriate, for it recalls the Platonic ideal of a universe whose perfection is unmarred by transience. The perfect cosmological principle meant that the universe was in a steady state and nothing would ever change in appearance. The cosmic explorer, perceiving that all places were alike in each tour, would also notice that nothing had changed between successive tours. Everything would be the same everywhere in space and time, apart from inconsequential irregularities.

The Newtonian universe of the eighteenth century was in a steady state. It was also a static universe. Notice that static means that the universe is neither expanding nor contracting, whereas steady state means that nothing changes in appearance. (A river can be in a steady state, but the water is flowing and not static.) As science advanced, it was realized that everything evolves and nothing remains eternally the same, and the Newtonian universe then became static but evolving (i.e., not in a steady state). In this century we have discovered that the universe is expanding, and for a decade or so the steady state concept was revived in this context.

In a steady state expanding universe the rate of expansion is constant and never changes. Individual things grow old, but new things are born to replace them, and age distributions never change. As in a long-established society with zero population growth, births cancel deaths, thus maintaining a steady state distribution of ages. In an expanding steady state universe matter must be created continuously everywhere to maintain a constant density: new galaxies are formed from the created matter, and old galaxies drift apart and are thinned out by the expansion. The expanding steady state picture, when looked at in detail, is marvelously self-consistent.

The location principle assures the observer that an isotropic universe is also homogeneous in space. But the principle is not of much help in establishing homogeneity in time. This is because time is itself peculiarly asymmetrical, and we cannot perceive the future with the same clarity as the past. If we could see the future, and could see a past–future symmetry, then by means of the location principle we would establish homogeneity in time on the grounds that special location in time is improbable.

There is now overwhelming evidence to show that the universe is not in a steady state. The numbers of radio sources and quasars were greater in the past than they are at present; and the demand of the steady state cosmologists to be shown the ashes of the big bang has been met by the discovery of the low-temperature cosmic radiation. It is not easy to conceive an idea of general cosmological importance, particularly one that can be disproved within a human lifetime. The perfect cosmological principle, although disproved, was a rare achievement that deserves our admiration and a secure place in the history of cosmology.

REFLECTIONS

1 *Discuss the location principle.*

∗ *"The use of the sea and air is common to all; neither can a title to the ocean belong to any people or private persons, forasmuch as neither nature nor public use and custom permit any possessions thereof" (Elizabeth I [1533–1603], to the Spanish ambassador, 1580).*

2 *Pythagoras (ca. 582–497 B.C.), a Greek philosopher born on the Aegean island of Samos, is reputed to have traveled widely in Egypt and other countries. He founded a school in southern Italy and taught that the Earth is round and that all things in the universe are governed by mathematics. The Pythagoreans held that the universe was constructed of numbers. A point was 1, a line 2, a surface 3, and a solid 4; and the number $1 + 2 + 3 + 4 = 10$ was sacred and omnipotent. Points had finite size, and therefore lines and surfaces had finite thickness because they were constructed from points. The concept of geometric atomism, of a universe constructed from points, was one of the paths that led to the atomic theory of matter. Problems were analyzed in terms of triangular numbers, such as*

```
                    .
          .        . .
 .       . .      . . .
. .     . . .    . . . .
3         6         10
```

and square numbers, such as

```
                  . . . .
         . . .    . . . .
. .      . . .    . . . .
. .      . . .    . . . .
 4         9         16
```

Great importance was attached to the mean of two numbers a and b:

$$\text{arithmetic mean} = \tfrac{1}{2}(a + b)$$

$$\text{geometric mean} = \sqrt{(ab)}$$

$$\frac{1}{\text{harmonic mean}} = \frac{1}{2}\left(\frac{1}{a} + \frac{1}{b}\right)$$

The harmonic mean in music was stressed; note also that the 8 corners of a cube are the harmonic mean of its 6 faces and 12 edges.

The Pythagoreans proved (but the Babylonians discovered) the important rule: The square of the diagonal of a right triangle equals the sum of the squares of the sides, thus giving 3,4,5 and 5,12,13 for the magnitudes of the sides and diagonals of right triangles. They also discovered that in general the sides and diagonal are incommensurable and cannot be expressed with integer numbers. A right triangle of equal sides of unit length has a diagonal length of $\sqrt{2} = 1.41421 \ldots$ This irrational result – "devoid of logos" – came as a shock and created a Pythagorean crisis, and to this day numbers such as $\sqrt{2}$, $\sqrt{3}$, and $\sqrt{5}$ are referred to as irrational numbers. The realization that lines are infinitely divisible implied that points, basic to the construction of the universe, had no size in a physical sense.

3 *Socrates (about 470–399 B.C.) lived in Athens and taught that humanistic studies are more important than science. Through his disciple Plato, he changed the course of science and philosophy. He wrote almost nothing that has survived, and his ideas are known mainly through the work of Plato. Socrates believed that the soul was an immortal activity of ideas inhabiting a temporal house of clay. He stressed those questions that commence with* why, *whereas the Greek scientists, with the aid of analogies from the mechanical arts, stressed* how *things worked. If atoms explain how matter is constructed, what of it? – it is surely more important to know why atoms exist, and only pure reason searching within the soul can determine their necessity and purpose. "How fine it would be," said Socrates, "if wisdom were a sort of thing that could flow out of the one of us who is fuller into him who is emptier, by our mere contact with each other, as water will flow through wool from the fuller cup into the emptier." At the age of seventy he was condemned to death by a jury of 500 (280 voted guilty) on a trumped-up charge of misleading the youth of Athens with fallacious and heretical ideas. Plato elaborated on the Socratic philosophy and stressed the difference between appearance and reality, or matter* *and mind. The phenomenal world, he argued, is a shadowy image of the eternal real world, and matter is docile and disorderly, governed by Mind that is the source of coherence, harmony, and orderliness.*

4 *Aristotle (384–322 B.C.), a student of Plato, was the tutor of Alexander the Great. He founded in Athens his own school, the Lyceum, known as the peripatetic school because Aristotle often lectured while walking around with his students. His interests were universal and his lecture notes filled 150 volumes. He was an observer and a thinker, not an experimentalist, and his main field tended to be biology. He elaborated on the great chain of being, supporting the idea of progressive evolution of the various forms of life. When Alexander, his patron, died, he feared accusations of impiety against the gods and fled from Athens, so that Athens could not, in his words, "sin twice against philosophy."*

The universal ideas of the Mind that impose form on disorderly matter, which Plato had made abstract and divine, were invested with physical reality by Aristotle and given an inseparable association with material things. In mythology it had been the gods that ruled; in Plato's universe it was the Mind that ruled; in Aristotle's universe it was the Ideas in their ceaseless interplay with matter that ruled. Aristotle restored the spirits of the Age of Magic, but with a vast difference; the spirits were now the forces and innate properties of matter. The spirits, in their new form, had become the souls of material things, and they still exist, masquerading under the names of forces, masses, momenta, energies, potentials, wave functions, and so forth. Once again humankind lives in an Age of Magic, but the capricious spirits have become the disciplined dancers in a ballet of weaving forces and waves.

5 *Roger Bacon (1220–92) had a prophetic and practical turn of mind and anticipated telescopes, submarines, steamships, automobiles, and flying machines. Thus he wrote: "Machines for navigation can be made without rowers so that the largest ships on rivers or seas will be moved by a single man in charge with greater velocity than if they were full of men. Also cars can be made so that without animals they will move with unbelievable rapidity. . . . Also flying machines can be constructed so that a man sits in the midst of the machine revolving some engine by which artificial wings are made to beat the air like a flying bird." Europe was in the throes of a technological revolution – with the introduction of the stirrup, the heavy plough, water mills, windmills, the magnetic compass, gunpowder, pattern-welded steel, papermaking, and textile manufacture – and Bacon's extrapolations were not so wild or remarkable when we consider the great technological developments then occurring. Many of the technical inventions were made outside Europe (the stirrup, water mills, windmills, the magnetic compass, gunpowder, papermaking). But the Europeans were the first to use them widely and to make technology an indispensable part of society. See* Science in the Middle Ages *(1978), edited by David Lindberg.*

* *William of Ockham (ca. 1280–1349), an Oxford scholar, was opposed to the Platonic theory that ideas are the true realities; he said that ideas are frequently nothing more than empty names and that the true realities are the objects observed. His viewpoint has been summarized in the words "entities must not be needlessly multiplied." This idea, known as Ockham's razor, means that the preferred theory is the one that has the fewest and simplest assumptions, and Ockham used this argument to criticize Thomas Aquinas's theological elaborations of the Aristotelian universe. Ockham's razor is very important in cosmology when we are comparing different universes.*

6 *In his dialogue in* The Two Great Systems of the World, *"Galileo keeps harping on how things happen, whereas his adversaries had a complete theory as to why things happen. Unfortunately the two theories did not bring out the same results. Galileo insists upon 'irreducible and stub-*

born facts,' and Simplicius, his opponent, brings forward reasons, completely satisfactory, at least to himself. It is a great mistake to conceive this historical revolt as an appeal to reason. On the contrary, it was through and through an anti-intellectual movement. It was the return to the contemplation of brute fact; and it was based on a recoil from the inflexible rationality of medieval thought. In making this statement I am merely summarizing what at the time the adherents of the old regime themselves asserted" (Alfred Whitehead, Science and the Modern World, *1925).*

* *Which do you prefer, a "sensible" universe regulated by divine spirits, or an insensible clockwork universe? "In the clockwork universe God frequently appeared to be only the clockmaker, the Being who had shaped the atomic parts, established the laws of their motion, set them to work, and then left them to run themselves" (Thomas Kuhn,* The Copernican Revolution, *1957).*

* *We commonly use the word revolution to denote a social upheaval, but this usage originates from the Copernican Revolution. The following comments apply to the social form of revolution:*

"It is a quality of revolutions not to go by old lines or old laws, but to break up both, and make new ones" (Abraham Lincoln, 1848).

"Revolutions have never lightened the burden of tyranny: they have merely shifted it to other shoulders" (George Bernard Shaw, The Revolutionist's Handbook, *1903).*

7 *Edwin Hubble was able to resolve and study cepheid variable stars in M31 in 1923, and in other nearby galaxies shortly afterward. This was an important step in the history of astronomy and cosmology. Previously it had seemed that the Galaxy was immense in size and that all other galaxies were comparatively small. Hubble's observations of cepheids showed that the extragalactic nebulae are at larger distances than previously supposed, and therefore are larger in size and rank as galaxies in their own right. One puzzle still remained: Why was our Galaxy conspicuously larger than other similar galaxies, such as M31? This was not resolved until Baade discovered the two populations of stars.*

Walter Baade (1893–1960), a German-American astronomer, discovered in 1942 that stars are of two kinds, which he referred to as populations I and II. During the wartime blackout of Los Angeles the observing conditions with the 100-inch telescope at Mount Wilson Observatory were exceptional, and with great care and skill, Baade was able to resolve many of the stars in the nuclear bulge of the spiral galaxy M31. He discovered in this way the difference between the bluish population I stars in the disk and the reddish population II stars in the nuclear bulge. After World War II, with the 200-inch telescope at Palomar, he found that the cepheids of the two stellar populations obey different period-luminosity laws. With the same period of oscillation, population I cepheids are four times as bright as population II cepheids. This was very important, because Shapley and Hubble had assumed that the cepheids in the globular clusters in our Galaxy were the same as the cepheids in the disk of M31. Baade's discovery implied that M31 was not 1 million but 2 million light years distant, and also that all other extragalactic distances had to be increased twofold. This increased the size of the galaxies and our Galaxy was no longer conspicuously large. Baade thus doubled the size of the universe.

8 *Immanuel Kant of Königsberg (which he never left in his whole lifetime) published a comprehensive philosophy in 1781 under the title* Critique of Pure Reason. *A simple and rather naive picture of the role of mind in science is that it works by inductive and deductive processes: Induction consists of drawing theoretical conclusions from observed facts, and deduction consists of deriving further facts from theoretical laws and principles. Kant made this simple picture more complex by arguing in favor of* *a priori ideas; according to this theory, the world of fragmentary sensations is organized into rational and interlocking perceptions by an activity of innate, primitive, a priori ideas. These primitive ideas, issuing from all past experience, precede and accompany each element of new experience, and are essential for making sense of our fleeting and disjointed world of sensations. In Plato's philosophy the ideas that precede experience belong to the universal Mind; in Kant's philosophy the ideas that precede and accompany experience belong to our own minds.*

Kant was a scientist as well as a philosopher, and in his work Theory of the Heavens *he was the first to consider the possibility that the Solar System had formed from a swirling cloud of gas. It is perhaps true to say that the development of new ideas usually depends on a combination of inventive and disciplined thought, and Kant indeed was a daring thinker who had a disciplined mind. In the* Theory of the Heavens *he wrote: "It is natural to assume that these nebulae are systems of numerous suns, which appear crowded because of their distance into a space so limited as to give a pale and uniform light. Their analogy with our own system of stars; their shape, which is just what it should be according to our theory; the faintness of their light which denotes great distance; are in admirable agreement and lead us to consider these elliptical spots as systems of the same order as our own – in a word, to be milky ways similar to the one whose constitution we have explained." He then went on to say: "With what astonishment are we transported when we behold the multitude of worlds and systems that fill the extension of the Milky Way! But how this astonishment is increased, when we become aware of the fact that all these immense orders of star-worlds again form but one of a number whose termination we do not know, and which perhaps, like the former, is a system inconceivably vast – and yet again but one member in a new combination of numbers! We see the first members of a progressive relationship of worlds and systems; and the first part of this infinite progression enables us already to recognize what must be conjectured of the whole. There is here no end but an abyss of a real immensity, in the presence of which all the capability of human conception sinks exhausted."*

9 *Pierre Simon de Laplace (1749–1827), a French mathematician and astronomer, wrote a monumental work on dynamics and gravitational theory entitled* Celestial Mechanics. *Of this work, Napoleon said, "You have written this huge book on the system of the world without once mentioning the author of the universe." Laplace replied, "Sir, I had no need of that hypothesis." Laplace is best known in astronomy for his work on the nebula hypothesis, previously proposed by Kant, according to which the Sun and planets were formed together in a rotating cloud of gas.*

10 *The clustering together of stars, galaxies, and clusters of galaxies constitutes what astronomers called a* hierarchy. *This is, strictly speaking, a misuse of the word. When used in its correct sense, it denotes pecking-order organizations, usually but not always of human membership, in which each member bosses those below and obeys those above. A business or government agency is a hierarchy, and the medieval universe with its various levels of angelic and other beings was also hierarchical. The technically correct word for the structured universes of Kant and Lambert is* multilevel. *A complete multilevel universe has two consequences: The first, recognized by Kant and Lambert, is that the universe retains a primary center and is therefore nonuniform on the largest cosmic scale; the second, recognized by the Irish physicist Fournier d'Albe and the Swedish astronomer Carl Charlier early in this century, is that the total amount of matter is much less than in a uniform universe. This second consequence can be understood in the following way. The mass and volume of each cluster increase as we ascend to higher levels, and in order for clusters to be separated from one another, it is necessary for the volume* *of a cluster to increase faster than its mass. The density of any cluster – its mass divided by its volume – therefore steadily decreases at higher and higher levels. Ultimately, at the highest level, occupied by a single universal cluster, the density may be vanishingly small. Nowadays it is believed that the clumpiness of the observed universe is best represented by a finite hierarchy, or a finite number of levels, consisting of stars, star clusters, galaxies, clusters, and superclusters, and that the universe is probably uniform on scales larger than superclusters.*

11 *Variations on the cosmological principle:*

"God is a circle whose centre is everywhere and whose circumference is nowhere" (Empedocles, fifth century B.C.).

"Whatever spot anyone may occupy, the universe stretches away from him just the same in all directions without limit" (Lucretius, The Nature of the Universe, *first century B.C.).*

"The fabric of the world has its center everywhere and its circumference nowhere" (Nicholas of Cusa, fifteenth century).

"It is evident that all the heavenly bodies, set as if in a destined place, are there formed unto spheres, that they tend to their own centres and that around them there is a confluence of all their parts" (William Gilbert [1544–1603]). Gilbert was influenced by the work of Digges and drew the natural conclusion that the universe has many centers.

"Wee . . . were a consideringe of Kepler's reasons by which he indeavors to overthrow Nolanus [Nicholas of Cusa] and Gilbert's opinions concerninge the immensitie of the spheere of the starres and that opinion particularlie of Nolanus by which he affirmed that the eye being placed in anie part of the universe, the appearance would be still all one as unto us here" (William Lower, in a letter, 1610).

"But if the matter was evenly disposed throughout an infinite space, it could never convene into one mass; but some of it would convene into one mass and some into another, so as to make an infinite number of great masses scattered at great distances from one to another throughout all that infinite space" (Isaac Newton, seventeenth century).

"It has been said that if the spiral nebulae are islands, our own galaxy is a continent. I suppose that my humility has become a middle-class pride, for I dislike the imputation that we belong to the aristocracy of the universe. The earth is a middle-class planet, not a giant like Jupiter, nor yet one of the smaller vermin like the minor planets. The sun is a middling sort of star, not a giant like Capella but well above the lowest classes. So it seems wrong that we should happen to belong to an altogether exceptional galaxy. Frankly I do not believe it; it would be too much of a coincidence. I think that this relation of the Milky Way to the other galaxies is a subject on which more light will be thrown by further observational research, and that ultimately we shall find that there are many galaxies of a size equal to and surpassing our own. Meanwhile the question does not much affect the present discussion. If we are in a privileged position, we shall not presume upon it" (Arthur Eddington, The Expanding Universe, *1933).*

"Not only the laws of nature, but also the events occurring in nature, the world itself, must appear the same to all observers, wherever they may be" (Edward Milne, "World-structure and the expansion of the universe," 1933).

"It is impossible to tell where one is in the universe" (Edmund Whittaker, From Euclid to Eddington, *1958).*

Homogeneity is a cosmic undergarment, and "the frills and furbelows required to express individuality can be readily tacked onto this basic undergarment" (Howard Robertson, "Cosmology," 1957).

"The Earth is not in a central, specially favoured position. This principle has become accepted by all men of science, and it is only a small step from this principle to the statement that the Earth is in a typical position" (Hermann Bondi, Cosmology, *1960).*

"Is it not possible, indeed probable, that our present cosmological ideas on the structure and evolution of the universe as a whole (whatever that may mean) will appear hopelessly premature and primitive to astronomers of the 21st century? Less than 50 years after the birth of what we are pleased to call 'modern cosmology,' when so few empirical facts are passably well established, when so many oversimplified models of the universe are still competing for attention, is it, may we ask, really credible to claim, or even reasonably to hope, that we are presently close to a definitive solution of the cosmological problem?" (Gerard de Vaucouleurs, "The case for a hierarchical cosmology," 1970).

12 *Postulates of impotence are rare, perhaps because of the difficulty of phrasing definitions of what is truly impossible. Whittaker's postulate, "It is impossible to tell where one is in the universe," is an example that illustrates the cosmological principle. Others are: "Perpetual motion is impossible"; "it is impossible to measure precisely and simultaneously the position and momentum of an electron"; "it is impossible for heat to flow from one region to another region of higher temperature." Garrett Hardin, in* Nature and Man's Fate *(1959), has proposed an evolutionary postulate of impotence: "Competition is inescapable"; he argues that a species that eliminates all competition from other species ends by becoming its own competitor. Some postulates of impotence have been shown to be wrong, such as "absolute motion is unmeasurable." We can in principle now determine absolute motion in the universe using the 3-degree cosmic radiation as a frame of reference.*

Can you think of any other postulate of impotence? (For instance, "It is impossible to verify the truth of Whittaker's postulate of impotence"!)

FURTHER READING

Christianson, J. "The celestial palace of Tycho Brahe." *Scientific American,* February 1961.

Geller, M. J. "Large-scale structure in the universe." *American Scientist,* March–April 1978.

Gingerich, O. "Copernicus and Tycho Brahe." *Scientific American,* December 1973.

Hoskin, M. A. "The 'Great Debate': what really happened." *Journal for the History of Astronomy 7,* 169 (1976).

Johnson, F. R. "Thomas Digges and the infinity of the universe," reprinted in *Theories of the Universe: From Babylonian Myth to Modern Science,* ed. M. K. Munitz. Free Press, Glencoe, Ill., 1957.

Karp, W. "Sir Isaac Newton." *Horizon 10,* 16 (Autumn 1968).

Kuhn, T. S. *The Copernican Revolution: Planetary Astronomy in the Development of Western Thought.* Harvard University Press, Cambridge, Mass., 1957.

Lerner, L., and Gosselin, E. "Giordano Bruno." *Scientific American,* April 1973.

Lucretius. *The Nature of the Universe.* Trans. in prose by R. E. Latham. Penguin Books, Harmondsworth, Middlesex, 1951.

Ravetz, J. "The origins of the Copernican Revolution." *Scientific American,* October 1966.

Toulmin, S., and Goodfield, J. *The Fabric of the Heavens: The Development of Astronomy and Dynamics.* Harper and Row, New York, 1961.

SOURCES

Berry, A. *A Short History of Astronomy.* Murray, London, 1896. Reprint. Dover Publications, New York, 1961.

Bondi, H. *Cosmology.* Cambridge University Press, Cambridge, 1960.

Bryant, W. W. *A History of Astronomy.* Methuen, London, 1907.

Butterfield, H. *The Origins of Modern Science.* Free Press, New York, 1957.

Caspar, M. *Kepler.* Abelard-Schuman, New York, 1959.

Cornford, F. M. *Plato's Cosmology.* Routledge and Kegan Paul, London, 1937.

Crombie, A. C. *Augustine to Galileo.* Vol. I, *Science in the Middle Ages.* Vol. II, *Science in the Later Middle Ages and Early Modern Times.* Mercury Books, London, 1961.

Drake, S. *Discoveries and Opinions of Galileo.* Doubleday, Garden City, N.Y., 1957.

Dreyer, J. L. E. *A Shorter History of the Planetary Systems from Thales to Kepler.* Cambridge University Press, Cambridge, 1905. Reprint. Dover Publications, New York, 1953.

Eddington, A. S. *The Expanding Universe.* Cambridge University Press, Cambridge, 1933. Reprint. University of Michigan Press, Ann Arbor Paperback, Ann Arbor, 1958.

Hardin, G. *Nature and Man's Fate.* Rinehart, New York, 1959.

Heninger, S. K. *The Cosmographical Glass: Renaissance Diagrams of the Universe.* Huntington Library, San Marino, Calif., 1977.

Hoskin, M. A. *William Herschel and the Construction of the Heavens.* Science History Publications, New York, 1963.

Hubble, E. *The Realm of the Nebulae.* Yale University Press, New Haven, Conn. 1936. Reprint. Dover Publications, New York, 1958.

Kant, I. *Universal Natural History and Theory of the Heavens.* Trans. by W. Hasties. University of Michigan Press, Ann Arbor, 1969.

Koestler, A. *The Sleepwalkers.* Grosset and Dunlap, New York, 1970. The section on Kepler, entitled "The Watershed," is particularly good and is available as a separate volume: *The Watershed: A Biography of Johannes Kepler.* Doubleday, Garden City, N.Y., 1960.

Koyré, A. *From the Closed World to the Unfinite Universe.* Johns Hopkins Press, Baltimore 1957. Reprint. Harper Torchbooks, New York, 1958.

Kuhn, T. S. *The Copernican Revolution: Planetary Astronomy in the Development of Western Thought.* Harvard University Press, Cambridge, Mass., 1957.

Kuhn, T. S. *The Structure of Scientific Revolutions.* Chicago University Press, Chicago, 1970.

Lindberg, D. C., ed. *Science in the Middle Ages.* University of Chicago Press, Chicago, 1978.

Milne, E. "World-structure and the expansion of the universe." *Zeitschrift für Astrophysik 6,* 1 (1933).

Murchie, G. *Music of the Spheres.* Vol. I, *The Macrocosm: Planets, Stars, Galaxies, Cosmology.* Vol. II, *The Microcosm: Matter, Atoms, Waves, Radiation, Relativity.* Dover Publications, New York, 1967.

Robertson, H. P. "Cosmology." *Encyclopaedia Brittanica,* 1957.

Sambursky, S. *The Physical World of the Greeks.* Routledge and Kegan Paul, London, 1956.

Santillana, G. de. *The Crime of Galileo.* University of Chicago Press, Chicago, 1955.

Sidgwick, J. B. *William Herschel, Explorer of the Heavens.* Faber and Faber, London, 1955.

Vaucouleurs, G. de. "The case for a hierarchical cosmology." *Science 167,* 1203 (1970).

Whitehead, A. N. *Science and the Modern World.* Macmillan, London, 1925.

Whittaker, E. *From Euclid to Eddington: A Study of Conceptions of the External World.* Dover Publications, New York, 1958.

Wright, T. *An Original Theory of the Universe.* Reprint with an introduction by M. Hoskin. American Elsevier, New York, 1971.

CONTAINMENT AND THE COSMIC EDGE

To see a world in a grain of sand,
And heaven in a wild flower:
Hold infinity in the palm of your hand,
And eternity in an hour.
— Blake (1757–1827), *Auguries of Innocence*

CONTAINMENT PRINCIPLE

Much of cosmology in the past has been concerned with the center and edge of the universe (see Figure 5.1), and our attitude nowadays on these matters is summarized in the principles of location and containment. Broadly speaking, the location principle refers to the cosmic center and the containment principle refers directly and indirectly to the cosmic edge, and both principles help us to avoid various pitfalls that trapped earlier cosmologists.

The containment principle states: *The physical universe contains everything that is physical, and nothing else.* It is the battle cry of the physical sciences. To some people it will seem that the principle is so simple and obvious that it is hardly worth stating; to other people it is the declaration of an outrageous philosophy. Before it is condemned as either too simple or too stupid, let us take a look at the implications of the containment principle.

Modern cosmology studies a physical universe that includes all that is physical and excludes all that is nonphysical. The definition of *physical* is deceptive and at first sight often exceeds what common sense deems proper. It includes *all* those things that are observed, that are studied by controlled experiments and are explained by quantitative and predictive theories vulnerable to disproof. Atoms and galaxies, cells and stars, creatures and planets, space and time are all of a physical nature. Particles with their corpuscular–wavelike duality, atoms with their choreography of electron waves, DNA with its genetic coding, forces and waves that reach out and propagate through space, the special relativity properties of spacetime, the dynamic deformation of spacetime in general relativity, and the astronomical universe are all things of a physical nature.

We, as physical creatures, are captives of the physical universe and cannot escape. Space and time are not voids into which the universe has been placed; if they were, we could escape by searching out those places in space and time that are not occupied by the universe. But spacetime, which is the combination of space and time, is not a mere nothingness; it is a continuum that has physical reality in its own right. Space and time, in combination, are active participants in the physical universe, and are therefore contained in and do not extend beyond and exist outside the universe. If you believe in a nonphysical realm, such as heaven, then you must not endow it with space and time; if you do, you will bring it into the physical universe and expose it to the methods of physical inquiry.

The physical nature of spacetime is demonstrated by its dynamic properties. It

Figure 5.1. A nineteenth-century woodcut that supposedly presents the medieval view of the universe. Beyond the sphere of stars lie the celestial machinery and other wonders.

supports the propagation of gravitational waves – ripples in spacetime – that travel at the same speed as light. Because spacetime is physical, it is possible for regions of spacetime to act on and influence one another, and this is the essence of general relativity. Gravitation, which was once a mysterious agent that acted across a vacuum on distant bodies, has become the dynamic and geometric curvature of spacetime. It is possible, such are the bewildering properties of spacetime, to have a universe that contains nothing but a medley of gravitational waves; and the dynamic behavior of this universe is governed by the gravitation, not of matter, but of the energy in the gravitational waves. A black hole need not consist of matter; it may contain only rippling spacetime waves whose total energy has a mass that accounts for the strong gravitational field of the black hole.

Space and time in the Newtonian universe were passive and served as a background, whereas in the modern universe they are welded together and have become active participants. Who can doubt the physical reality of spacetime when it guides the Moon around the Earth and raises the tides, and will tear apart astronauts and their spaceships in the vicinity of neutron stars and black holes?

It is possible to have a physical universe in which space is curved and is finite in extent without an edge. The two-dimensional surface of a balloon is an easily visualized analogy: The surface is finite and yet has no edge. An ant crawling on the surface of a spherical balloon in a straightforward direction comes back to its starting point without encountering an edge. A mobile cosmic explorer in a finite and unbounded universe, traveling in a straight line, also comes back to the starting point from the opposite direction.

Some people will protest that the containment principle leaves out all that they believe to be most valuable. What about our souls, our minds, our ideas and emotions, and all the richness of the inner subjective world: Where do they fit in? The response that must be made is quite simple: "You are confusing universe with Universe; the physicists have made their universe, and if you do not like it, you must make your own." Modern cosmology deals with a physical model of the Universe that is yet another mask on the face of the unknown. But what a fantastic mask it is! All the inventive genius of the greatest thinkers in the history of science has gone into its making. Can one wonder that many people, including even scientists, when confronted with the majesty of the physical universe have mistaken this latest mask for the real face?

THE PHYSICAL UNIVERSE

THE BEGINNING OF SCIENCE. From the time of Thales, the Greek philosopher–scientists sought to account for the complexity of the world by reducing it to an interplay of primary constituents. The proposed primary constituents, or elements, were

Thales (sixth century B.C.) :	water
Anaximenes (sixth century B.C.) :	air
Heraclitus (fifth century B.C.) :	fire
Xenophanes (fifth century B.C.) :	earth
Empedocles (fifth century B.C.) :	earth, water, air, fire

There is an ultimate substance, said Thales, "from which all things come to be, it being conserved." These earlier thinkers did not ask why, but were content to ask how, and used analogies drawn from pottery, metalworking, and other mechanical arts and crafts. They believed in the conservation of elements that combine to form objects of various shapes and compositions. They rejected magic and myth, made careful observations, and were critical in their thinking. Hence they had ideas, made observations, argued by analogies, analyzed into constituents, and used laws of conservation. All this is the essence of the scientific method.

With Pythagoras, Heraclitus, and Parmenides (sixth century B.C.) we see the budding of many ideas and the beginning of a more rigorous use of logic. Pythagoras, who was possibly the first to use the word *philosopher,* believed in the mathematical harmony of the universe. Heraclitus, known as the "weeping philosopher" because of his pessimistic nature, said that everything changes, nothing endures, and the basic element is therefore fire. He conceived a universe of tempestuous flux governed by a conflict of opposing forces, and said that wisdom consists of knowing how things change. Parmenides said that nothing changes, everything endures, and wisdom consists of rejecting sensory illusions. He conceived a universe that was changeless, in which all change was an illusion of the senses. The Heraclitean flux of warring forces and the changeless Parmenidic continuum are both powerful ideas; we see the former in the Newtonian universe of dynamic motions and the latter in the Riemannian spacetime of relativity theory.

THE ATOMIST UNIVERSE. Emphasis by Parmenides on the unity of the One was countered by emphasis on the plurality of the Many. The idea of a changeless continuum containing nothing discrete was opposed by the idea of a void containing discrete entities in motion. Anaxagoras showed the way: He said that the universe was infinite in extent and contained an infinite number of small seeds. These seeds – or atoms – were endowed with properties that affected the senses and were controlled by a universal Mind. The heavens were made of the same substances as those of Earth, and the universe was not ruled by gods. Anaxagoras was accused of impiety and was the first scientist to be tried for heresy; he was defended by powerful friends, was acquitted, and fled from the hostility of Athens.

Leucippus (fifth century B.C.) is credited with the invention of the atomic theory. He was the first to state clearly the principle of causality: All events are the effects of preceding causes, and "nothing happens at random; everything happens out of reason and necessity." Democritus, who was a student of Leucippus, said, "Nothing can be created out of nothing, nor can it be destroyed and returned to nothing." The atomic theory of Leucippus, elaborated by Democritus, was not accepted by Socrates, Plato, and Aristotle, and we are indebted to the later teachings of Epicurus for keeping alive the ideas of the atomic theory. The Atomist universe contained only two things: atoms and the void. The atoms were infinite in number and the void was infinite in extent. All atoms were made of the same substance, but they differed in shape and size; they moved freely through the void, forever colliding with one another, forming new objects by their aggregation, which then decayed back into atoms. The Atomist universe of endless worlds, occupied by living creatures, was in a continual state of flux, and its unlimited variety was the result of the motion and aggregation of atoms.

EPICUREANS AND STOICS. Epicurus (341–270 B.C.) settled in Athens and founded the Epicurean school of philosophy, which was the first school to admit women students. The Epicureans believed that the human being was an evolved and superior form of animal, that the gods existed not in the external world but in ourselves, and that the greatest pleasures in life stemmed from moderate living. Their philosophy, based on Atomist ideas, flourished in the Greco-Roman world among thoughtful people and perished with the rise of Christianity. *The Nature of the Universe,* written by the Roman poet Lucretius in the first century B.C., is in praise of Epicureanism. "Bear this well in mind," wrote Lucretius in his great poem, "that nature is free and uncontrolled by proud masters and runs the universe without the aid of gods. For who . . . can rule the sum total of the measureless? Who can hold in coercive hand the strong reins of the unfathomable? Who can spin all the firmaments alike and foment with the fires of ether all the fruitful earths? Who can be in all places at all times . . . ?" Who indeed! According to the Epicureans it was nature itself. Mythology responded by making the gods more powerful than ever before and by suppressing the atheistic Atomist theory.

Stoicism in the first century B.C. was more popular than Epicureanism, at least among the Romans. The Stoic school of philosophy stressed the importance of duty, justness, tenacity of purpose, and fortitude in the face of adversity. The Stoics believed that fate was the cause of all movement and change, and that Mind, manifesting through the gods, governed the universe. They also believed that the stars were alive and the universe was a living organic whole. Their philosophy was a combination of Pythagorean and Platonic ideas, and they taught that the cosmos was finite and in a state of flux. The Stoics said that beyond the finite cosmos there stretched an infinite void and that the finite cosmos pulsated in size and periodically passed through upheavals and conflagrations of the kind that nowadays we call big bangs.

RISE OF THE MODERN ATOMIC THEORY. Atomism, although an inspired theory, did not enter the mainstream of science until the seventeenth century. Pierre Gassendi, a French philosopher of that century, revived the atomic theory, and it later played a prominent role in the thinking of men like Robert Boyle and Isaac Newton. In his book *Opticks,* Newton wrote: "It seems probable to me that God in the beginning formed matter in solid, massy, hard, impenetrable particles, of such size and figures, and with such other properties, in such proportions to space, as most conduced to the end for which He formed them; even so very hard as never to wear or break in pieces." Newton interwove the atheistic atomic doctrine into the religious doctrine of Creation and God's purpose, and he triumphed by compromise. In prophetic words, he wrote: "There are therefore agents in nature able to make the particles of bodies stick together by very strong attractions. And it is the business of experimental philosophy to find them out."

We now have high-energy particle accelerators for precisely this purpose. Although Newton spoke of "ether waves" that vibrate like air, and are like waves on the surface of water, he believed nonetheless that light is composed of particles. "Are not the rays of light very small bodies emitted from shining bodies?" Robert Hooke and Christiaan Huygens, also of the seventeenth century, advanced various ideas on the wave propagation of light that were developed a century later by Thomas Young. In the nineteenth century, Michael Faraday, a prince among experimental physicists, and James Clerk Maxwell, a prince among theoretical physicists, unified electricity and magnetism with the properties of light and developed the modern theory of electromagnetism. Early in the twentieth century, Max Planck and Albert Einstein showed that light is both wavelike and corpuscular, and the particles of radiation are now called photons.

The modern atomic theory of matter began in 1803 with John Dalton's book *New System of Chemical Philosophy,* which was an immediate success. Dalton introduced the Greek word *atom* into chemistry and was the first scientist to make the atomic theory quantitative. From the known fact that elements combine in definite proportions to form chemical compounds he was able to show that matter is composed of atoms of different weights. Joseph Thomson discovered the electron in 1897; Ernest Rutherford discovered the atomic nucleus in 1911; Niels Bohr in 1913 constructed a mechanical model of the atom that consisted of electrons in orbits about the nucleus; and then later Erwin Schrödinger, Werner Heisenberg, and many other physicists developed the modern wave-mechanical model of the atom using quantum mechanics. "When it comes to atoms," said Niels Bohr, "language can be used only as in poetry."

COSMIC EDGE

COSMIC-EDGE RIDDLE. The early Atomists had championed the idea of an infinite universe without center or edge. But the idea was rejected and it made little impact. The preferred idea was that the universe was finite and had therefore a center and an outer edge. The nature of the outer boundary of the universe, or cosmic edge, has puzzled many of the ablest minds in the history of cosmology. Lucretius wrote: "It is a matter of observation that one thing is limited by another. The hills are demarcated by air, and air by the hills. Land sets bounds to seas, and the sea to every land. But the universe has nothing outside to limit it." To those who believed in a finite universe with an outer edge he proposed the following riddle (see Figure 5.2): "Suppose for a moment that the whole of space were bounded and that someone made his way to its uttermost boundary and threw a flying dart. Do you choose to suppose that the missile, hurled with might and main, would speed along the course on which it was aimed? Or do you think something would block the way and stop it? You must assume one alternative or the other . . . With this argument I will pursue you. Wherever you may place the ultimate limit of things, I will ask you: 'Well, then, what does happen to the dart?' " Lucretius gives the Atomist answer: "Learn, therefore, that the universe is not bounded in any direction."

The cosmic-edge riddle – what happens to a spear when it is hurled across the outer boundary of the universe – can be traced back to the Pythagorean soldier–philosopher Archytas of the fifth century B.C., and was known in the Middle Ages through the writings of Cicero. The poem by Lucretius, after its discovery in 1417 in a monastery, was widely read and influenced many people, including perhaps Nicholas of Cusa and Thomas Digges. Without doubt, Giordano Bruno was greatly impressed. In the dialogue of *The Infinite Universe,* written while he was still living in England, Bruno gave Burchio (an imaginary Aristotelian) this argument: "I think that one must reply to this fellow that if a person would stretch out his hand beyond the convex sphere of heaven, the hand would occupy no position in space, nor any space, and in consequence would not exist."

Figure 5.2. The cosmic-edge riddle: What happens when a spear is thrown across the cosmic edge?

Bruno replied by arguing that space inside and outside the universe must be continuous and the same, and added, in the words of Philotheo (Bruno himself): "Thus, let the surface be what it will, I must always put the question: what is beyond? And if the reply is: nothing, then I call that the void, or emptiness. And such a Void or Emptiness hath no measure nor outer limit, though it hath an inner; and this is harder to imagine than is an infinite or immense universe."

Everything in the finite universe had its position determined in relation to a center and an outer edge, and this appealing picture, in which things have absolute location, is as old as cosmology. But when scrutinized by logical minds this picture encounters the cosmic-edge riddle: How can space terminate when by its very nature it is continuous?

WALL-LIKE COSMIC EDGE. A common conception of the outer edge was that it ended abruptly like a wall; it was a spherical edge with the stars attached or adjacent to the inner surface (see Figure 5.3). Even Kepler believed that the universe was enclosed within a dark cosmic wall, and he was therefore able to explain why the sky at night is dark. Kepler argued that in an infinite universe of stars the sky in every direction would be as bright as the Sun. This *dark night sky paradox,* discussed in Chapter 12, has since played a prominent role in the history of cosmology. We do not know Kepler's answer to the cosmic-edge riddle. The spear either rebounds or passes through the cosmic wall, and according to the Atomists, the first is impossible because space cannot terminate and be bounded by nothing, and the second proves that a cosmic edge does not exist.

ARISTOTELIAN COSMIC EDGE. In the Aristotelian and medieval universes the outer edge was not sharp but gradual, beginning in the sublunar sphere and terminating with the outer sphere of stars. As one moved outward away from Earth the physical realm was progressively transformed into an etheric or

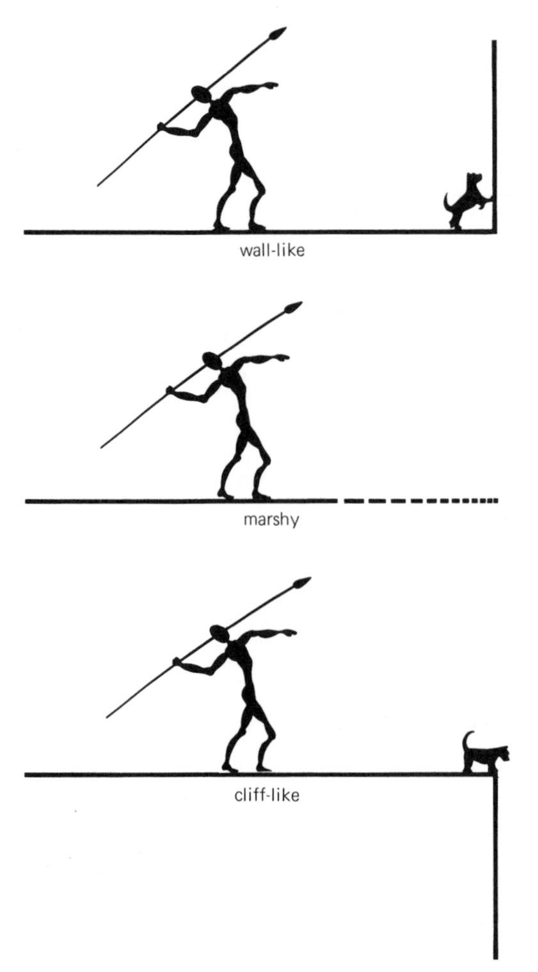

Figure 5.3. Illustrations of the wall-like edge, the gradual Aristotelian edge, and the Stoic clifflike edge.

spiritual realm. In the medieval universe the outer etheric realm was itself surrounded by the empyrean realm occupied by God. To the question What happens to a physical object when it moves away from Earth? there were two answers: first, that the object was made of earthly elements and therefore had to turn around and return to Earth; and second, that if it did not return, its earthly elements were transmuted into the etheric element, and its natural motion was then circular and not up and down.

An abrupt boundary therefore did not exist in either the Aristotelian or medieval universes, and the cosmic-edge riddle was easily evaded. The force of the statement "with this argument I shall pursue you" was completely lost because the pursuer was led into a spiritual outer realm where physical arguments were invalid. This kind of cosmic edge, or Aristotelian edge, was not abrupt like a wall or cliff edge; it was instead similar to a gradual fading of firm ground into a marshland of insubstantial things.

The Aristotelian and medieval rebuttal of the riddle is unacceptable in the twentieth century. Observations show that the physical world does not fade into a nonphysical world at great distances. We know also that space is itself physical and the physical world therefore extends as far as space extends.

STOIC CLIFFLIKE EDGE. The Stoic universe consisted of a finite cosmos surrounded by an infinite void. The Stoic edge was sharp like that of a cliff and divided the universe into two parts: an inner cosmos and an outer empty space that extended to infinity. The answer in this case to the cosmic-edge riddle was quite simple: The act of throwing the spear enlarged the cosmos and extended its outer edge.

The Stoic cosmos, isolated in infinite and empty space, became in the eighteenth century the Milky Way cosmos of William Herschel and survived in modified form until the early decades of this century. It had the attraction that in principle one could travel to a place outside the Milky Way and have a grandstand view of the material content of the universe. Observations have now shown that the material universe extends to vast distances beyond the Milky Way, and there is no sign of an abrupt edge.

When people first take an interest in cosmology, they often have in mind a picture that resembles the Stoic universe. Sad to say, they are sometimes misled by writers who do not stress the importance of containment. In popular literature the universe is described as if space extended miraculously into an extracosmic realm that provided an overview of what is happening. The universe is portrayed in illustrations as if it were a cloud expanding in space, and the impression is thus created that the universe is contained within space and has a center and an edge. This is wrong and violates the containment principle, because space is contained within the universe and does not extend beyond it.

The big bang did not occur somewhere in space, as in the Stoic universe. It occupied the whole of space. An expanding infinite universe remains always infinite in extent, and if space is infinite, the big bang was also infinite in extent. Regardless of where we are located, we only have to travel back in time, staying where we are, to find ourselves immersed in the big bang.

If space is finite, the universe is not like a ball expanding in space. We should try to imagine that the surface of the ball represents our three-dimensional space. The surface is finite in extent, like finite three-dimensional space, and yet has no center and edge. The galaxies are not hurling through space, but sit in space, and the space itself expands in the same way as the surface of an expanding balloon that is slowly inflated. As we shall later show, it is the space between the galaxies that expands, and the galaxies are carried apart by the expansion of space.

Cosmic edges do not exist. We cannot therefore cross an edge and adopt a grandstand view of the universe, either in space or in time. Time, like space, is physical and is contained within the universe, and time cannot therefore extend beyond a timelike cosmic edge. We cannot ask what the universe looks like from outside space, and similarly, we cannot ask what happened in time before the universe began and what will happen after it has ended. Such questions violate the containment principle.

COSMOGONY

ORIGIN OF STRUCTURE IN THE UNIVERSE. *Cosmogony* (the word means the begetting of cosmic progeny) is the subject that deals with the origin of astronomical structures such as planets, stars, and galaxies. It covers also the origin of the elements and even the origin of life.

The constraints set on cosmogony by containment are elementary. Containment requires that all component things have sizes smaller than or equal to the size of the universe. Also, all component things have ages shorter than or equal to the age of the universe. A possible cosmogonic sequence in time is

age of human beings
≤ age of mammals
≤ age of life on Earth
≤ age of Earth
≤ age of Sun
≤ age of Galaxy
≤ age of helium produced in the early universe
≤ age of universe

where the symbol ≤ means "is less than or equal to."

The light elements, deuterium and helium, were formed from hydrogen in the big bang while the universe was still young, dense, and hot. Most other elements are formed in stars and later ejected into space in supernova outbursts. The elements that compose the Earth were produced in stars that died before the Sun was born. Many of the heavy elements are radioactive and are slowly decaying; for example, uranium-238 decays eventually into lead with a half life of 4.5 billion years. By discovering how fast radioactive elements decay, and by measuring their present abundances, it is possible to determine the age of the Earth, the Solar System, and the Galaxy. From these results it is found that the Solar System has an age of approximately 4.6 billion years and the Galaxy has an age somewhere between 10 and 15 billion years. Nucleochronology is the study of the origin and history of the elements and of the various dating techniques that are used.

"Whirl is king," said Aristophanes in the fourth century B.C. Whirlwinds and whirlpools were important in mythology. Vortex theories were also important in early science, and many philosophers thought that

Figure 5.4. "If we don't know how big the whole universe is, then I don't see how we could be sure how big anything in it is either, like the whole thing might not be any bigger than maybe an orange would be if it weren't in the universe, I mean, so I don't think we ought to get too uptight about any of it because it might be really sort of small and unimportant after all, and until we find out that everything isn't just some kind of specks and things, why maybe who needs it?" (With permission from John Milligan. First appeared in Saturday Review, 1971.)

the Earth and stars were formed in a primeval vortex. To this day whirl is king in our theories of star and galaxy formation. René Descartes, the famous French mathematician and philosopher of the seventeenth century, said that all space was filled with matter that swirled around in large and small vortices. By this argument he sought to explain the rotation of the Earth and the revolution of the planets around the Sun. The Cartesian vortex theory was at last abandoned after Newton had shown that planetary orbits are explained by gravity and the laws of motion. The idea of swirling primeval matter was later developed by Kant and Laplace into the solar nebula hypothesis. Nowadays, rotating and contracting gas clouds play an essential role in our cosmogonic theories of star and galaxy formation.

The universe is expanding, and from the rate it expands cosmologists have estimated that it has an age of between 10 and 20 billion years. But after the discovery of expansion it seemed at first, until the middle of this century, that the universe had an age of little more than a billion years. A universe younger than the Earth violated containment and for many years this anomaly was the major problem in cosmology. Various attempts were made to get around the problem, as in the hesitation universe, according to which expansion was slow for a long period in the past, and in the steady state universe, which had an infinite age.

Those cosmologists who favored a big bang universe, such as Georges Lemaître and George Gamow, thought that most elements were made in the big bang. This idea turned out to be wrong but had one great virtue: It started Gamow and his colleagues, Ralph Alpher and Robert Herman, thinking about a hot early universe and led them to a prediction of the low-temperature cosmic radiation almost 20 years before it was discovered. The steady state idea turned out to be wrong but also had one great virtue: The steady state cosmologists could not accept the idea of big bang nucleogenesis and they therefore had to show that most elements are manufactured in stars. The pioneers in the successful theory of stellar nucleogenesis were Al Cameron, Margaret and Geoffery Burbidge, William Fowler, and Fred Hoyle.

Revised estimates of extragalactic distances made by Walter Baade in 1952, and by Alan Sandage and others since, have increased the size and age of the universe, and it is now possible to accommodate the ages of the elements, the Earth, Solar System, stars, and galaxies within a big type of universe.

COSMOGENESIS

Eer time and place were, time and place were not,
When primitive Nothing *something streight begot,*
Then all proceeded from the great united – What.
— John Wilmot (1647–80), "Upon nothing"

Cosmogeny, or cosmogenesis, is the subject that deals with the origin of the universe as a whole. The words *cosmogeny* and *cosmogony* are often confused despite their different derivations, meanings, and pronunciations. How did the universe originate? Scientists usually avoid this question and are inclined to think that cosmogeny is a nonscientific subject. Cosmogeny was once an important branch of mythology, and we shall see that it is still important even in modern cosmology.

According to the Mosaic chronology the universe originated only a few thousand years ago. (The word *Mosaic* pertains to Moses.) Kepler in his book *Mysterious Cosmography* set the date of creation at 3877 B.C., Sunday, April 27, at 11:00 A.M. local Prussian time. In 1658, sixty years later, Archbishop Ussher in *The Annals of the World Deduced from the Origin of Time* pushed the date of creation back to 4004 B.C., Sunday, October 23, at 6:00 A.M. From the time of Kepler until the present day, the estimated age of the universe has increased on the average by a factor of 10 every 50 years. Possibly this growth in estimated age has at last come to an end.

Mark Twain in *Letters from the Earth* said that Earthlings who believe in the Mosaic chronology are unable to explain the light they receive from stars more distant than a few thousand light years. But he underestimated the ingenuity of Earthlings. There is nothing logically wrong with the idea that the universe originated 6000 years ago, or even yesterday, provided we are willing to accept a highly complex initial state. The universe could have been created with light already in transit from the stars.

Eugene Wigner in *Symmetries and Reflections* (1967) writes: "The world is very complicated and it is clearly impossible for the human mind to understand it completely. Man has therefore devised an artifice which permits the complicated nature of the world to be blamed on something which is called accidental and thus permits him to abstract a domain in which simple laws can be found. The complications are called initial conditions, the domain of regularities, the laws of nature." Initial conditions are accidental and not easily explained; the subsequent conditions are explained by means of the known regularities or laws of nature. The initial conditions of star formation, for instance, are the accidental and little-understood irregularities of interstellar gas clouds; the laws of nature enable us to study how stars form from the irregularities and evolve onto the main sequence.

Science advances by progressively simplifying the assumed initial conditions. More and more the "accidental" content of nature is understood and transferred to the "domain of regularities." Thus, by studying interstellar gas clouds, astronomers learn about their structure and evolution and as a result are able to whittle away the accidental content of the initial conditions of star formation. At each step in the advance of science more of the accidental is explained and shown to be the natural outcome of preceding simpler initial conditions. There is a limit to this process, of course, and initial conditions cannot be thrust back to a time preceding the origin of the universe. There will always remain the irreducible initial conditions of the universe. The aim of cosmology is to explain the universe with initial conditions that are as simple as possible.

If the universe was created 6000 years ago, or even yesterday, it was created in a fantastically complex state. Stars were created already in their various stages of evolution; the light that stars had supposedly emitted tens of thousands of years ago was created in the act of traveling toward us; fossils were created in the surface of the Earth with deceptively long ages; and so on. From a scientific viewpoint the universe becomes a giant hoax.

Belief that things are created within the universe is as old as cosmology. Then came science with its insistence that elementary constituents are conserved and only their combinations are created and destroyed. We now have conservation laws of energy, momentum, electric charge, and other things, and whenever a conservation law fails we search for a new and deeper law of conservation.

Some scientists in this century have speculated on the possibility that matter is created in the universe. The steady state theory of Bondi, Gold, and Hoyle is a well-known case. Matter is equivalent to energy and the creation of matter violates the law of conserved energy, but instead of conservation of energy, these cosmologists advocated a different law: conservation of the universe in its present state. In the controversy between the advocates of the big bang and steady state universes it was often said that we have a choice between the instant creation of a big bang or the continuous creation of a steady state, a choice between creation all at once or creation little by little. The assumption is that both kinds, instant creation and continuous creation, are on equal footing. We now show that this assumption is false.

There are two kinds of creation that in this section we shall distinguish by the words *Creation* and *creation.* On the one hand there is Creation, as in cosmogenesis, of the whole universe complete with space and time. On the other hand there is creation of matter within the space and time of a universe already existing. Everything in the big bang universe, including space and time, is Created; in the steady state universe matter only is created within the space and time of a universe already Created. Failure to distinguish between the two is actually a violation of the containment principle. The steady state theory uses creation in the old magical sense that at a certain place and time there is nothing, and at the same place a moment later there is something. But Creation does not have this meaning unless we reject the containment principle and revert to the mistaken idea that time and space exist outside the universe. Contained creation is quite unlike uncontained Creation.

If we cannot say that at one moment there exists nothing and the next moment there exists the universe, what then does Creation or cosmogenesis mean? We do not know. But at least we can make the following important remark. Some cosmologists, such as Eddington, have disliked the idea of a big bang because it seems to have cosmogenic implications, and they have therefore pushed the origin of the universe back into an infinite past. But this kind of trick does not evade the problem of cosmogenesis, for we are considering the Creation of space and time. An infinite span of time has to be Created in the same way as a finite span of time. An infinitely old universe, such as the steady state universe, is Created in the same way, no more and no less, as a big bang universe of finite age. Universes of every kind, with and without big bangs, are confronted with the problem of cosmogenesis.

It is an advantage if a universe has a simple initial state from which all structure can evolve. Its complexity is then implicit in the initial state and can be explained by means of the laws of nature. The steady state universe does not evolve from a simple state and its complexity is a permanent feature throughout eternity. The Creation of a big bang universe is therefore, perhaps, a simpler matter than Creation of a steady state universe, for a steady state universe must be Created complete with all its accidental features, and this is just as difficult as supposing that the universe originated 6000 years ago. Furthermore, it is a Created universe that contains creation, and this is more than was supposed by either Kepler or Ussher.

Cosmogeny is an important branch of cosmology. In some respects it is not exempt from the rules of logic and is therefore capable of aiding us in our study of the universe. The lesson we learn from cosmogeny is this: A universe that evolves from initial conditions, that has all its complexity implicit in a simple initial state, is a preferred universe. We are thus provided with a cosmogenic Ockham's razor.

ANTHROPIC AND THEISTIC PRINCIPLES

AN ENSEMBLE OF UNIVERSES. Hitherto we have spoken of universes in the sense that each is a model of the Universe. In this section we shall now suppose that not one, but many, physical universes exist and that each is self-contained and unaffected by all the rest.

In each physical universe the fundamental constants of nature have different values. By *constants of nature* we mean permanent things, so far unexplained, such as the speed of light c, the gravitational constant G, Planck's constant h, the electric charge e of the proton and the electron, and the masses of the subatomic particles (see Chapter 17). We have hence an array, or ensemble, of universes that covers all values of the constants of nature. Each universe is a workshop in which we can study what happens when the constants are assigned specific values different from those in our own universe.

ANTHROPIC AND THEISTIC PRINCIPLES. We first notice that alterations in the known values of c, h, and e cause huge changes in the structure of atoms and atomic nuclei. Even when the changes are only slight, most atomic nuclei are unstable and cannot exist. The majority of the universes in the ensemble contain little more than hydrogen, and therefore lack earthlike planets and are without elements such as carbon, nitrogen, and oxygen that are essential for organic life.

We also find that slight changes in the values of c, G, h, e, and the masses of subatomic particles cause huge changes in the structure and evolution of stars. The majority of universes will actually not contain any stars at all, and in the few that do, the stars either are nonluminous or are so luminous that their lifetimes are too short for biological evolution.

Life forms that we can recognize as living depend for their complexity on the existence of a variety of elements. Life requires also a habitable environment, such as a planet warmed by a long-lived star, in which it can originate and evolve. Careful examination of the ensemble indicates that these requirements are met only in universes closely similar to our own. Our universe is therefore finely tuned, and we would not exist if the constants of nature had different values. It is even possible that, of the entire ensemble, our universe is the only member that contains life.

What we have said can be interpreted in two entirely different ways. The first is that the ensemble is real and only our universe and perhaps others closely similar contain living creatures. Life exists in one universe at least – and we occupy that universe. Our existence determines the design of our universe (see Figure 5.5). This is the anthropic principle that has been expressed in different ways by Robert Dicke, Brandon

Figure 5.5. "Man is one world, and hath/Another to attend him" (George Herbert [1593–1633], Proverbial Expressions). Hand with a Reflecting Globe, *by M. C. Escher. (Courtesy of the Collection Haags Gemeentemuseum – The Hague.)*

Carter, and John Wheeler, and the name *anthropic principle* comes from Brandon Carter. In *Twelfth Night* the clown said only half the truth when he declared, "That that is is." The full truth is, "I that am am; hence that that is is."

The second interpretation is that a Creator has designed our finely tuned universe specifically for the containment of life. This is the theistic principle, as elaborated in mythology and theology. All other previously imagined universes of the ensemble can now be discarded. They served merely as convenient fictions enabling us to realize that our universe is finely tuned for habitation by life.

Whichever interpretation is chosen is a matter of personal choice. If you ask why the universe is the way it is, or almost any other probing question that begins with *why,* the answer is that we are here and the universe is self-aware (see Figure 5.6). Both principles are subject to limitations. The existence of all other creatures on Earth determines the design of the universe just as much as do human beings. Wild flowers do not exist in other universes. Hence, when one is given a wild flower, one is also given our universe. Creatures with intelligent minds impose no more constraint on the design of the universe than do wild flowers. Both principles (the anthropic principle in particular) also assume that the constants of nature have values that are accidental and forever inexplicable. But science advances

Figure 5.6. The self-aware universe.

by explaining what previously was thought to be accidental and irreducible, and in the future it is possible that the constants of nature will not be regarded as inexplicable within our universe. We shall occasionally refer to the anthropic principle, and the reader may, if it is preferred, substitute the alternative theistic principle.

WHITHER THE LAWS OF NATURE?

THE REALMS OF PHYSICS. The laws of nature are those regularities whereby we perceive harmonious order in the universe. Often they are interpretations of the basic equations in physics that mimic the behavior of the observed world. We believe in them until they fall into discord with observation, and they are then either modified or discarded; no doubt in the future almost all known laws will be different from those of today.

Most branches of science are occupied with the study of objects whose sizes range from those of atoms to those of galaxies. Atoms have sizes typically 10^{-8} centimeters and galaxies have sizes typically 10^{22} centimeters. Galaxies are therefore 10^{30} times larger than atoms, and this entire range of sizes covers 30 orders of magnitude. Most phenomena of interest to science lie in this middle realm of physics. At the atomic end of the scale practically everything is explained with electromagnetic forces, and at the galactic end of the scale practically everything is explained with gravitational forces. By moving up 5 orders of magnitude to a cosmic size, typically 10^{27} centimeters, and down 5 orders of magnitude to a subatomic size, typically 10^{-13} centimeters, we quit the middle realm and enter the lesser-known outer realms of physics.

In the cosmic realm we find that gravity in the guise of curved spacetime continues to be the governing force. Yet at the other extreme, in the subatomic realm, electromagnetism does not continue to rule but is joined by weak and strong forces. The universe at present is thus strangely asymmetrical: On the cosmic scale there is comparative simplicity, and on the subatomic scale there is obscure complexity. But this apparent asymmetry may not be genuine; it is perhaps the result of cosmology lagging behind subatomic physics because of the difficulty of performing experiments of cosmic significance. After all, we must not forget that the universe contains the subatomic particles and should therefore be at least as complex as these particles.

DOCTRINES OF INTERNAL AND EXTERNAL RELATIONS. It is probable that the universe contains hitherto undiscovered physical properties that unify particles and the cosmos. This idea is implicit in the bootstrap principle that has been around for a long time.

The bootstrap principle states that all things are immanent within one another. Anaxagoras had the idea when he said, "In everything there is a portion of everything." The physicist Geoffery Chew has in recent years said, "Nature is as it is because this is the only possible nature consistent with itself." The way things are on the largest scales determines the way things are on the smallest scales, and vice versa. In Gottfried Leibniz's theory that "this is the best of all possible worlds," the microcosm is thought to be an image of the macrocosm (or cosmos). The answer to Victor Hugo's question – "Where the telescope ends, the microscope begins. Which of the two has the grander view?" – might be that in a fundamental sense both views are ultimately identical. Bishop Berkeley in the eighteenth century proposed a bootstrap principle when he argued that the inertia of any body is determined by the distribution and masses of all other bodies in the universe. Ernst Mach proposed a similar idea in the nineteenth century; it is now called Mach's principle and will be discussed later (Chapter 8). Needless to say, nobody yet knows how to make the bootstrap idea work.

It seems that we have organized the physical universe in accordance with the old Atomist relation:

universe = sum of all subatomic particles

The philosopher Alfred Whitehead refers to this kind of arrangement as the doctrine of external relations. Everything works in a push–pull manner, and this way of explaining the universe derives perhaps from the human experience of living in a competitive and not always friendly environment. Questions such as why all electrons have the same electric charge are not easily answered in the Atomist world picture and tend to be suppressed. Questions beginning with *how,* which relate to the way things are pushed about, are answered much more easily.

The bootstrap relation takes the form

universe ≡ subatomic particle

(where the symbol ≡ means "equivalent to"), and each particle in some way represents a facet of the universe and is not just a small part of it. Whitehead refers to this kind of arrangement as the doctrine of internal relations. Why do all electrons have the same electric charge? Because all electrons represent some single aspect of the universe. In a hypothetical world where intelligent creatures are highly sensitive and nonaggressive it is possible that things are explained by means of the doctrine of internal relations. The old Atomist relation tells us how, but not always why; the bootstrap relation tells us why, but so far has failed to tell us how.

HEAVEN IN AN ELECTRON. The basic quantities of the physical world, such as the constants of nature, seem at present to be accidental and inexplicable. But in the bootstrap picture the universe is a self-consistent whole and therefore can contain nothing of a fundamental nature that is purely accidental. According to the bootstrap picture the universe is what it is because it is consistent with itself, and it follows therefore that we are not free to separate out accidental properties and distribute them with various values among different universes, as in the anthropic principle.

If the bootstrap principle were formulated scientifically, it would revolutionize cosmology. Until that happens, if ever, it is interesting to note that the anthropic principle serves as a makeshift or poor-man's bootstrap. It relates organisms and the universe, and although it is only a partial bootstrap, it nonetheless is useful until better ideas are found. At present, when we are given a wild flower, we are also given our universe. Perhaps, in the future, when we are given an electron, we shall also be given our universe. We shall then have not only heaven in a wild flower, but heaven in an electron, or in any other subatomic particle.

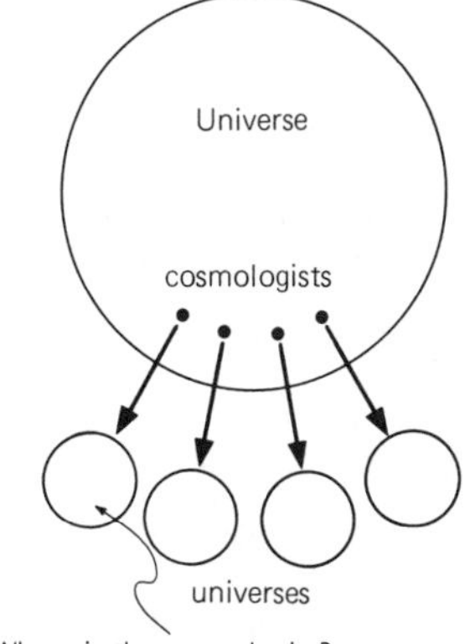

Figure 5.7. The containment riddle.

CONTAINMENT RIDDLE

WHERE IN A UNIVERSE IS THE COSMOLOGIST CONCEIVING THAT UNIVERSE? The physical universe is remarkably successful at explaining how various things work and in enabling us to take control over our environment. But it does not explain the nature of mental or psychic things. A person who persistently asks where life and mind are in the physical world will be passed from science to science like a person with a mysterious illness who is passed from specialist to specialist. The biologists might say that life and mind are the outcome of elaborate systems of billions of organized cells, in which each cell is an organization of billions of atoms. If this fails to satisfy the inquirer, the biologists will probably recommend a visit to the psychologists. The psychologists, not knowing much about the physical universe, cannot help and will recommend seeing the physicists, who, of course, will suggest a visit to the biologists. The inquirer, in sheer desperation, might then think of consulting the cosmologists.

The cosmologists confess that they have not the foggiest notion of an answer. They know instead that universes spring from the imaginative fertility of the human mind, and they cannot help but wonder whether a product of the mind is itself capable of containing the mind. The buck stops with the cosmologists because of the containment riddle: *Where in a universe can be found the cosmologist conceiving that universe?* (see Figure 5.7) All other scientists can evade the riddle by claiming that it does not fall within their province of special knowledge. But the cosmologists cannot and the riddle stares them full in the face. "When we know how to solve the riddle," say the cosmologists to the inquirer, "we might be able to answer your question."

Consider a painter who paints a picture of a studio. The painter (assume it is *he*) stands within the studio and yet does not ordinarily include himself within the picture in the act of painting the picture. An attempt to portray the act of painting leads to the absurdity of an infinite regression of pictures. The picture would contain the painter painting a picture, which contains the painter painting a picture, and so on, indefinitely. A universe is a world picture and the cosmologist is in the same sort of situation as the painter. The cosmologist constructs a world picture that contains his physical body but not his mind that constructs the picture. If his mind is not excluded he also encounters the absurdity of an infinite regression. The universe would contain the cosmologist conceiving a universe, which contains the cosmologist conceiving a universe, and so on, indefinitely. Where then is the cosmologist conceiving the universe? Can image making ever contain the image maker?

The physical universe, consisting of multitudes of facts woven together with ideas, apparently does not contain the thing that has conceived the ideas. Those people who cannot agree with this proposition, and claim that life and mind are no more than a collective dance of atoms, must answer the containment riddle. The people who find that they cannot agree are generally those who confuse the physical universe with the unknown Universe, and who mistake the mask for the face. We do not know the answer to the containment riddle, and that is the only comfort the cosmologists can offer any person who is concerned and seeks to know the meaning of life and mind in the physical universe.

REFLECTIONS

1 *Know then thyself, presume not God to scan;*
The proper study of mankind is man.
Placed on this isthmus of a middle state,
A being darkly wise, and rudely great:
With too much pride for the sceptic side,
With too much weakness for the stoic's pride,
He hangs between; in doubt to act or rest;
In doubt to deem himself a god, or beast;
In doubt his mind or body to prefer;
Born but to die and reasoning but to err;
Alike in ignorance, his reason such,
Whether he thinks too little or too much;
Chaos of thought and passion, all confused;
Still by himself abused, or disabused;
Created half to rise, and half to fall;
Great lord of all things, yet a prey to all;
Sole judge of truth, in endless error hurled;
The glory, jest, and riddle of the world!
— Alexander Pope (1688–1744), *An Essay on Man*

2 *"Democritus of Abdera said that there is no end to the universe, since it was not created by an outside power. Moreover, the universe is boundless. For that which is bounded has an extreme point, and the extreme point is seen against something else" (Epicurus [341–270 B.C.]).*

* *"But multitudinous atoms, swept along in multitudinous courses through infinite time by mutual clashes and their own weight, have come together in every possible way and realized everything that could be formed by their combinations. So it comes about that a voyage of immense duration, in which they have experienced every variety of movement and conjunction, has at length brought together those whose sudden encounter normally forms the starting-point of substantial fabrics – earth and sea and sky and the races of living creatures"* (*Lucretius,* The Nature of the Universe).

* *Zeno of the fifth century B.C. was a Greek philosopher who denied that truth can be attained through perceptions by the senses. He is best known for his paradoxes that seemed to disprove the possibility of motion as perceived by the senses. The paradoxes are all similar and the one most frequently quoted is that of the race between Archilles and the tortoise. Suppose the tortoise has a 100-meter head start and Archilles runs 100 times faster than the tortoise. While Archilles runs 100 meters, the tortoise moves 1 meter, while Archilles runs 1 meter, the tortoise moves 1 centimeter; and so on, in an infinite number of steps, and Archilles therefore never reaches and overtakes the tortoise. The paradox is fallacious because the infinite number of steps occupy only a finite interval of time, and Archilles is therefore able to overtake the tortoise, as observed by the senses.*

Moses Maimonides, a Jewish-Arab scholar of the twelfth century, refers to a resolution of Zeno's paradox that had apparently been proposed by the Atomists. According to this resolution, time is itself divided into indivisible atomic intervals, and motion is therefore not continuous but consists of a series of small jerks (like motion in a movie). Archilles is thus able to overtake the tortoise with jerky movements. Maimonides, like other Arab philosophers, was greatly influenced by Aristotle, and in his work The Guide for the Perplexed *he was scathing in his condemnation of the Atomists. The Atomist theory of time meant, he said, that the universe is created not once, but in each interval of atomic time. It is repeatedly created and destroyed, and in each creation it is complete in all its accidental detail.*

3 *Until the eighteenth century it was widely believed that the Earth was only thousands of years old. Mounting evidence in geology and paleontology (the study of fossils) indicated a much greater age, and doctrines of compromise were developed on the basis of the catastrophic principle, according to which the Earth had been periodically visited by such catastrophes as life-destroying deluges, and natural and supernatural laws had alternated in their control over the Earth.*

James Hutton, a Scottish farmer and physician, advanced in 1785 the theory that nature behaves in an orderly and uniform way, so that the formation and erosion of mountains are continuous processes that have been acting steadily for an indefinitely long period of time. From the evidence he concluded that there is "no vestige of a beginning – no prospect of an end." This was the new uniformitarian principle that was later powerfully argued by Charles Lyell: The landscape is continually modified by natural forces and the surface as a whole remains essentially unchanged. The catastrophists believed in created states periodically established by divine intervention, and the uniformitarians believed that divine creation had established a steady state world controlled by natural laws. The controversy that raged between the two until the middle of the nineteenth century was far more heated than the controversy between the big bang and steady state advocates of this century.

In the second half of the nineteenth century the uniformitarians were attacked by physicists under the leadership of Lord Kelvin. Various calculations (now known to be inapplicable) on tidal effects and terrestrial heat losses showed that the Earth was not as old as the geologists had thought. The calculation of the age of the Sun was the most important of all. Kelvin first assumed that the Sun derived its energy from the infall of meteoroids. He later adopted Helmholtz's idea that the Sun derives its luminous energy from slow contraction, and estimated an age for the Sun of 20 million years. Because the sun is essential for life on Earth, this implied that fossils and rock strata containing fossils could not be older than 20 million years. Insistence by the physicists on this short time span created dismay in the earth and life sciences, and many attempts were made to fit geological history and the evolution of life into the Kelvin chronology. In the early years of this century the physicists redeemed themselves with radioactive dating methods, and were able to show that the age of the Earth is measured in billions, not millions, of years. Ernest Rutherford, in his book Radiation and Emanation *of 1904, said, "The discovery of the radioactive elements, which in their disintegration liberate enormous amounts of energy, thus increases the possible limit of the duration of life on this planet, and allows the time claimed by the geologist and biologist for the purpose of evolution."*

4 *"Behold a universe so immense that I am lost in it. I no longer know where I am. I am just nothing at all. Our world is terrifying in its insignificance!" (Bernard de Fontenelle [1657–1757],* Conversations with a Lady on the Plurality of Worlds*).*

* *"It is impossible to contemplate the spectacle of the starry universe without wondering how it was formed: perhaps we ought to wait, and not look for a solution until we have patiently assembled the elements . . . but if we were so reasonable, if we were curious without impatience, it is probable we would never have created Science and we would always have been content with a trivial existence. Thus the mind has imper-*

iously laid claim to this solution long before it was ripe, even while perceived in only faint glimmers – allowing us to guess a solution rather than wait for it" (Henri Poincaré, French mathematician, 1913).

* *"The type of conjecture that presents itself, somewhat insistently, is that the centers of the nebulae are of the nature of 'singular points,' at which matter is poured into our universe from some other, entirely extraneous, spatial dimension, so that to a denizen in our universe, they appear as points at which matter is being continually created" (James Jeans,* Astronomy and Cosmogony, *1929*).

5 *"An adequate cosmology will only begin to be written when an adequate philosophy of mind has appeared, and such a philosophy of mind must provide full satisfaction both for the motives of the behaviorists who wish to make mind material for experimental manipulation and exact measurement, and for the motives of idealists who wish to see the startling difference between a universe without mind and a universe organized into a living and sensitive unity through mind properly accounted for" (Edwin Burtt,* The Metaphysical Foundations of Modern Physical Science, *1932*).

* *In* Mind and Matter (*1958*), *Erwin Schrödinger writes: "Without being aware of it, we exclude the Subject of Cognizance from the domain of nature that we endeavour to understand. We step with our own person back into the part of an onlooker who does not belong to the world, which by this very process becomes an objective world." He then says, "A moderately satisfying picture of this world has only been reached at the high price of taking ourselves out of the picture, stepping back into the role of a nonconcerned observer." Schrödinger then turns to religion: "Can science vouchsafe information on matters of religion? Can the results of scientific research be of any help in gaining a reasonable and satisfactory attitude towards these burning questions which assail everyone at times? Some of us . . . succeed in shoving them aside for long periods; others, in advanced age, have satisfied themselves that there is no answer and have resigned themselves to giving up looking for one; while others again are haunted throughout their lives by this incongruity of our intellect, haunted also by serious fears raised by time-honoured superstition. I mean mainly the questions concerned with the 'other world,' with 'life after death,' and all that is connected with them."*

6 *Olaf Stapledon, in his imaginative book* The Star Maker (*1937*), *describes how the Star Maker designs and creates a sequence of universes of increasing complexity. In each universe intelligence is interwoven in a different and highly intricate fashion. Some extracts are as follows.*

"In vain my fatigued, my tortured attention strained to follow the increasingly subtle creations which, according to my dream, the Star Maker conceived. Cosmos after cosmos issued from his fervent imagination, each one with a distinctive spirit infinitely diversified, each in its fullest attainment more awakened than the last; but each one less comprehensible to me.

"Sometimes the Star Maker fashioned a cosmos which was without any single, objective, physical nature. Its creatures were wholly without influence on one another; but under the direct stimulation of the Star Maker each creature conceived an illusory but reliable and useful physical world of its own, and peopled it with figments of its imagination. These subjective worlds the mathematical genius of the Star Maker correlated in a manner that was perfectly systematic." This is reminiscent of Leibniz's theory of monads.

"But at the close of his maturity he willed to create as fully as possible, to call forth the full potentiality of his medium, to fashion worlds of increasing subtlety, and of increasing harmonious diversity. As his purpose became clearer, it seemed also to include the will to create universes each of which might contain some unique achieve-

ment of awareness and expression. For the creatures' achievement of perception and of will was seemingly the instrument by which the Star Maker himself, cosmos by cosmos, woke into keener lucidity.

"Sometimes the Star Maker flung off creations which were in effect groups of many linked universes, wholly distinct physical systems of very different kinds, yet related by the fact that the creatures lived their lives successively in universe after universe, assuming in each habitat an indigenous physical form, but bearing with them in their transmigrations faint and easily misinterpreted memories of earlier existences. In another way also, this principle of transmigration was sometimes used. Even creations that were not thus systematically linked might contain creatures that mentally echoed in some vague but haunting manner the experience or temperament of their counterparts in some other cosmos.

"In some creations each being had sensory perception of the whole physical cosmos from many spatial points of view, or even from every possible point of view. In the latter case, of course, the perception of every mind was identical in spatial range, but it varied from mind to mind in respect of penetration or insight. This depended on the mental calibre and disposition of particular minds. Sometimes these beings had not only omnipresent perception but omnipresent volition. They could take action in every region of space, though with varying precision and vigour according to their mental calibre. In a manner they were disembodied spirits, striving over the physical cosmos like chess-players, or like Greek gods over the Trojan Plain." It is interesting that Stapledon does not mention universes in which the creatures are omnipresent in time.

"In one inconceivably complex cosmos, whenever a creature was faced with several possible courses of action, it took them all, thereby creating many distinct temporal dimensions and distinct histories of the cosmos. Since in every evolutionary sequence of the cosmos there were very many creatures and each was constantly faced with many possible courses, and the combination of all their courses were innumerable, an infinity of distinct universes exfoliated from every moment of every temporal sequence in this cosmos.

"At length, so my dream, my myth, declared, the Star Maker created his ultimate and most subtle cosmos, for which all others were but tentative preparations. Of this final creation I can say only that it embraced within its own organic texture the essence of all its predecessors; and far more besides. It was like the last movement of a symphony, which may embrace, by the significance of its themes, the essence of the earlier movements; and far more besides . . . I strained my fainting intelligence to capture something of the form of the ultimate cosmos. With mingled admiration and protest I haltingly glimpsed the final subtleties of world and flesh and spirit, and of the community of those most diverse and individual beings, awakened to full self-knowledge and mutual insight. But as I strove to hear more inwardly into that music of concrete spirits in countless worlds, I caught echoes not merely of joys unspeakable, but of griefs inconsolable."

Stapledon teaches us one important lesson: that when we consider an ensemble of universes, as in the anthropic principle, the imaginative creation of many physical universes should be matched with an equally imaginative creation of many forms of intelligent life. Some people may find Stapledon's metaphysics more appealing than the anthropic principle with its emphasis on the terrestrial form of life. One of Stapledon's ideas on cosmogenesis has already emerged in modern cosmology: Electrons in atoms make transitions between different states, and each transition is to one of a number of permissible states. Hugh Everett in 1957 proposed that at each transition the universe splits so that all permissible states are reached, each in a separate universe. Thus from each atom there exfoliate many new universes each time a transition occurs, and these many universes are identical except that in each the particular atom is in a different state. An object, such as a human being, having at any moment numerous quantum transitions, is continually splitting into numerous universes, in each of which the object is in a slightly different state, and follows a different history.

In Men Like Gods, *H. G. Wells introduced the idea of parallel universes, occupying a "superspace," which normally are cut off from each other, but occasionally make contact. John Wheeler has proposed an intriguing physical theory that also enlarges the notion of containment. According to this theory our universe is a probability wave, of extraordinary complexity, that propagates in superspace, and all possible paths of propogation represent different universes.*

7 *Norman Campbell, in* What Is Science? *(1955), says, "Science deals with judgments concerning which it is possible to obtain universal agreement." If Campbell is correct, and if cosmology is a science similar to other sciences, then it cannot claim complete exemption from this rule. Yet universal agreement is surely impossible concerning the reality of an ensemble of physical universes that can never be verified by direct observation. Cosmology invents universes that are investigated as potential representatives of the Universe. In this sense it is scientific. But when cosmology invents a plurality of coexisting universes, and claims that each is real in its own right, then it ceases to be scientific according to Campbell's dictum. In this sense it becomes metaphysical and the notion of containment becomes vague and once again at the mercy of metaphysics, as in earlier ages. What do you think?*

8 *Kurt Gödel in 1931 showed that mathematical systems are not fully self-contained. In a self-consistent system (free of internal contradictions) it is possible to formulate statements whose truth is undecidable. If the system is enlarged with additional axioms, the previous uncertain statements can be proved to be true; but the enlarged system now contains new undecidable statements, which can only be proved to be true by making the system still larger. One possible conclusion is that the mathemetician cannot be left out of mathematics, just as the cosmologist cannot be left out of cosmology. For a discussion of Gödel's ideas see* Gödel's Proof *(1958) by Ernst Nagel and James Newman, and* What Is the Name of This Book? *(1978) by Raymond Smullyan.*

* *"Mr. Podsnap settled that whatever he put behind him he put out of existence . . . Mr. Podsnap had even acquired a peculiar flourish of his right arm in often clearing the world of its most difficult problems, by sweeping them behind him" (Charles Dickens,* Our Mutual Friend). *Have you ever noticed the "Podsnap flourish" when a scientist is asked, "What is life?" and a humanist, "What is physics?"*

9 *Discuss the principle of containment. Some people (materialists and reductionists), overwhelmed by the grandeur of the physical universe, believe that what is not contained does not exist. What do you think?*

* *Discuss the cosmic-edge riddle.*

* *Contrast the Epicurean and Stoic universes. See* The Physical World of the Greeks *(1956) by S. Sambursky.*

* *What is the difference between cosmogeny and cosmogony?*

* *Why is Creation of a universe different from creation within a universe?*

* *Discuss the anthropic and theistic principles.*

* *Comment on the statement that the universes are the Universe seeking to understand itself.*

* *Consider the statement that everything in the universe is false. This statement is itself part of the universe and cannot therefore be true, for then it contradicts itself; hence it must be false.*

* *Discuss the containment riddle.*
* *"... that fashioning by Nature of a picture of herself, in the mind of man, which we call the progress of Science" (Thomas Huxley [1825–95]).*

FURTHER READING

Dampier, W. C. *A Shorter History of Science.* Cambridge University Press, Cambridge, 1944. Reprint. World Publishing Co., New York, 1957.

Flew, A. *Body, Mind, and Death.* Macmillan, New York, 1979.

Rosenfeld, L. "Niels Bohr's contribution to epistemology." *Physics Today,* October 1963. Rosenfeld shows how quantum theory has led to the containment riddle.

Schramm, D. "The age of the elements." *Scientific American,* January 1974.

Toulmin, S. E. "The evolutionary development of natural science." *American Scientist,* December 1967.

Wald, G. "Fitness in the universe: the choices and necessities," in *Cosmochemical Evolution and the Origin of Life,* eds. J. Oró, S. L. Miller, and C. Ponnamperuma. Reidel Publishing Co., Dordrecht, Holland, 1974.

Wheeler, J. A. "The universe as home for man." *American Scientist,* November-December, 1974.

SOURCES

Allen, R. E. *Greek Philosophy: Thales to Aristotle.* Free Press, New York, 1966.

Burchfield, J. D. *Lord Kelvin and the Age of the Earth.* Science History Publications, New York, 1975.

Burtt, E. A. *The Metaphysical Foundations of Modern Physical Science.* 1924. Rev. ed. Humanities Press, New York, 1932. Reprint. Doubleday, Garden City, N.Y., 1954.

Butterfield, H. *The Origins of Modern Science, 1300–1800.* Bell & Sons, London, 1957. Rev. ed. Free Press, New York, 1965.

Campbell, N. *What Is Science?* Dover Publications, New York, 1955.

Carr, B. J., and Rees, M. J. "The anthropic principle and the structure of the physical world." *Nature 278,* 605 (April 12, 1979).

Carter, B. "Large number coincidences and the anthropic principle in cosmology," in *Confrontation of Cosmological Theories with Observational Data,* ed. M. S. Longair. D. Reidel, Dordrecht, Netherlands, 1974.

Chew, G. F. " 'Bootstrap': a scientific idea?" *Science 161,* 762 (1968).

DeWitt, B. S. "Quantum mechanics and reality." *Physics Today,* September 1970. "Could the solution to the dilemma of indeterminism be a universe in which all possible outcomes of an experiment actually occur?"

Dreyer, J. L. E. *A History of Astronomy from Thales to Kepler.* Dover Publications, New York, 1953.

Eddington, A. S. *New Pathways in Science.* Cambridge University Press, Cambridge, 1935.

Farrington, B. *Greek Science: Its Meaning for Us.* Vol. I, *Thales to Aristotle,* Vol. II, *Theothrastus to Galen.* Penguin Books, Harmondsworth, Middlesex, 1944.

Frank, P. *Modern Science and Its Philosophy.* Harvard University Press, Cambridge, Mass., 1949.

Gardner, M. *Mathematical Magic Show.* Alfred A. Knopf, New York, 1977. See chapter 1, "Nothing"; chapter 2, "More about Nothing"; chapter 19, "Everything."

Grant, E. *Physical Sciences in the Middle Ages.* Cambridge University Press, Cambridge, 1977.

Haber, F. C. *The Age of the World, Moses to Darwin.* Johns Hopkins Press, Baltimore, 1959.

Harrison, E. R. "Cosmological principles II: physical principles." *Comments on Astrophysics and Space Physics 6,* 29 (1974).

Henderson, L. J. *The Fitness of the Environment: An Inquiry into the Biological Significance of the Properties of Matter.* Macmillan, New York, 1913. Reprint with an introduction by G. Wald. Beacon Press, Boston, 1958.

Jeans, J. *Astronomy and cosmogony.* Cambridge University Press, Cambridge, 1929.

Kemble, E. C. *Structure and Development: From Geometric Astronomy to the Mechanical Theory of Heat.* M.I.T. Press, Cambridge, Mass., 1966.

Koyré, A. *Newtonian Studies.* Chapman and Hall, London, 1965.

Lucretius. *The Nature of the Universe.* Trans. in prose by R. E. Latham. Penguin Books, Harmondsworth, Middlesex, 1951.

MacDonald, D. K. C. *Faraday, Maxwell and Kelvin.* Doubleday, Anchor Press, New York, 1964.

Maimonides, M. *The Guide for the Perplexed.*

Trans. by M. Friedlander. Dover Publications, New York, 1956.

Munitz, M. K., ed. *Theories of the Universe: From Babylonian Myth to Modern Science.* Free Press, Glencoe, Ill., 1957.

Nagel, F., and Newman, J. E. *Gödel's Proof.* New York University Press, New York, 1958.

Newton, I. *Opticks.* Reprint. Dover Publications, New York, 1952.

Pancheri, L. M. "Pierre Gassendi, a forgotten but important man in the history of physics." *American Journal of Physics 46,* 455 (May 1978).

Planck, M. *Survey of Physical Theory.* Dover Publications, New York, 1960.

Popper, K. *The Logic of Scientific Discovery.* Harper and Row, New York, 1965.

Sambursky, S. *The Physical World of the Greeks.* Routledge and Kegan Paul, London, 1956.

Sandage, A. R. "The time scale for creation," in *Galaxies and the Universe,* ed. L. Woltjer. Columbia University Press, New York, 1968.

Schrödinger, E. *Mind and Matter.* Cambridge University Press, Cambridge, 1958.

Smullyan, R. *What Is the Name of This Book?* Prentice-Hall, Englewood Cliffs, N.J., 1978.

Stapledon, O. *The Star Maker.* 1937. Reprinted in *Last and First Men and Star Maker.* Dover Publications, New York, 1968.

Stent, G. S. "Limits to the scientific understanding of man." *Science 187,* 1052 (March 21, 1975). "Human sciences face an impasse since their central concept of the self is transcendental."

Toulmin, S., and Goodfield, J. *The Fabric of the Heavens: The Development of Astronomy and Dynamics.* Harper and Row, New York, 1961.

Van Melsen, A. G. *From Atomos to Atom: The History of the Concept Atom.* Duquesne University Press, Pittsburgh, Pa., 1951. Reprint. Harper and Row, New York, 1960.

Whitehead, A. N. *Adventure of Ideas.* Macmillan, New York, 1933.

Wigner, E. F. *Symmetries and Reflections.* Indiana University Press, Bloomington, 1967.

6 SPACE AND TIME

I do not define time, space, place and motion, as being well known to all.
— Newton (1642–1726), *Principia*

SPACE, TIME, AND MOTION

Nothing puzzles me more than time and space; and yet nothing troubles me less.
— Charles Lamb, in a letter (1810)

CLOTHED AND UNCLOTHED SPACE. From the Heroic Age of Greece until modern times we see the development, side by side, of two views on the nature of space: "clothed" space and "unclothed" space.

Space as a void – existing in its own right and independently of the things contained – was at first a lofty concept that many could not take seriously. It seemed more natural to think of space as clothed and made real with a continuous distribution of material and ethereal elements. Aristotle, who believed in clothed space, regarded the notion of a vacuum as nonsense and said that what is empty is nothing and what is nothing has no existence. This was also the view of the Medievalists, of Descartes and many others in the Renaissance, and of nineteenth-century physicists who filled space with an electromagnetic ether because they thought that space by itself was incapable of transmitting light waves. An unclothed space – a void existing in its own right, absolute, and independent of contained things – is an idea that originated with the Atomists. It became the master concept of the Newtonian universe, and in this century has blossomed into the spacetime of relativity theory.

ARISTOTLE'S EQUATION OF MOTION. In the days of Aristotle, space and time were only vaguely defined and had not been sharpened into their modern forms. Distances were easily measurable because space was identified with the things that can be directly seen, but time was not easily measurable because its intervals are not directly seen. How intervals of space and time should be combined to determine the properties of motion was not at all clear, and motion and change of any kind were often not distinguished from each other.

Aristotle's equation of motion was

$$\text{force} = \text{resistance} \times \text{motion} \qquad (6.1)$$

But actually he had no general formula, nor any way of measuring force, resistance, or motion. He argued qualitatively, reasoning from the everyday experience that effort is needed to maintain a state of motion. "A body will move through a given medium in a given time, and through the same distance in a thinner medium in a shorter time," he said; "it will move through air faster than through water by so much as air is thinner and less corporeal than water." It was therefore natural to think that objects of different weights should fall through air at different speeds.

Aristotle argued that a vacuum could not exist in nature, because in a vacuum there would be no resistance and all forces would produce infinite speeds. Space was clothed with substance that not only gave it reality but also moderated the motion of moving objects. The Atomists had claimed that atoms moved freely and eternally through the void, but to Aristotle this idea was patently absurd; even if it were granted that the atoms had moderated motion, what then would be the cause of the motion? The Atomists "say there is always movement. But why and what this movement is they do not say, nor, if the world moves in this way or that, do they tell us the cause of the motion." Aristotle had common sense on his side, and his views remained common sense for two thousand years until the discovery of the Newtonian equation of motion:

$$\text{force} = \text{mass} \times \text{acceleration} \qquad (6.2)$$

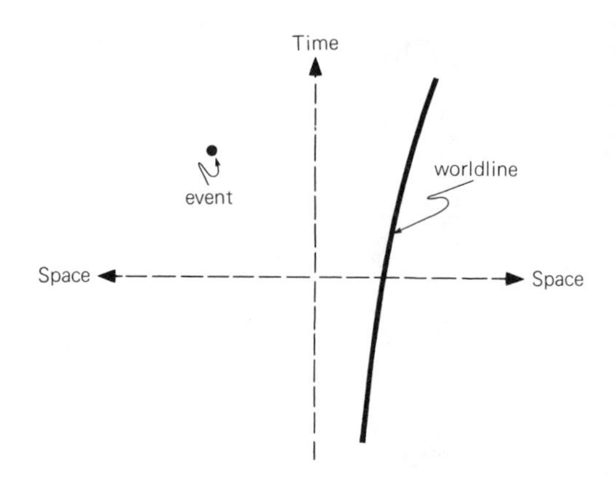

Figure 6.1. Space and time diagrams were used by the Medievalists and are not difficult to understand. A point is an event, something that happens at an instant in time at a position in space. A worldline *represents an object that endures; it has different positions at different instants of time and can be regarded as a string of events.*

FROM ARISTOTLE TO HOOKE. We must not think that the giant step from Aristotle to Newton was the result of the genius of Galileo and Newton alone. The notion of impetus – which we now call momentum – can be traced back to Hipparchus in the second century B.C. and was discussed by Philoponus in the sixth century. Medieval and Renaissance thinkers refined the notions of space and time; they developed the ideas that velocity is the change of distance with time, that acceleration is the change of velocity with time, and that falling objects have constant acceleration; and they used graphic methods of representing motion (see Figures 6.1 and 6.2).

In the fourteenth century, scholars of Merton College, Oxford, such as William Heytesbury, showed that if v is the velocity and a is a constant acceleration, then in time t the velocity attained is $v = at$ and the distance traveled is $x = \frac{1}{2}at^2$. Their results were accomplished by graphic means that anticipated the discovery of calculus. William Ockham, also of Merton College, argued that forces can act at a distance without any need for direct contact between objects. Jean Buridan, a French philosopher who was a student of Ockham and the teacher of Nicholas Oresme, argued that impetus is proportional to the velocity of a body and also to its quantity of matter. The planets are not continually pushed along in their orbits, he said, but move because of their initial impetus. The initial impetus of a thrown stone is sufficient to keep it moving, and its impetus is slowly lost as a result of the resistance of air. Aristotle, lacking the idea of impetus, had supposed that moving

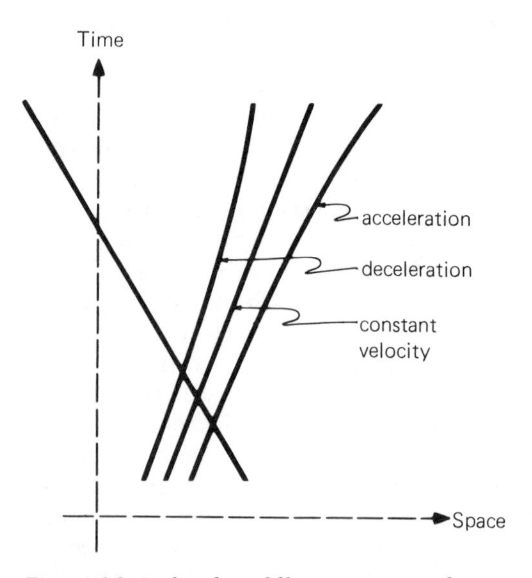

Figure 6.2. Inclined worldlines represent objects in relative motion. If the velocity changes, the worldline is curved.

objects were continually pushed: The air displaced at the front of a moving stone was replaced by air coming in from behind that pushed on the stone and maintained its motion. Buridan and his successors conceived the idea that falling bodies gain equal amounts of impetus in equal intervals of time.

In the fifteenth century, Leonardo da Vinci wrote on such subjects as the rotation of the Earth, the origin and antiquity of fossils, and the impossibility of perpetual motion, but his work was little known and was not influential. He, and others, also advocated the idea that falling bodies accelerate uniformly.

In the sixteenth century, Simon Stevinus, a Dutch-Belgian mathematician, performed the experiment that has wrongly been attributed to Galileo. He dropped objects of unequal weight from a high building and observed that they reached the ground at almost the same instant (see Figure 6.3).

Figure 6.3. Simon Stevinus (1548–1620) dropped objects of unequal weight and showed that they reach the ground at about the same time. That this result is to be expected is shown by the following thought experiment. We take a heavy object of weight W and divide it into two equal parts of weight ½W. Let us now tie them together with string and drop them. They will obviously fall in the same time as the original weight W. Now separate them and drop them individually and ask: Do the weights ½W fall slower than the weight W? If you think they do, then tie them together with a single hair and drop them again. Can the hair cause them to fall faster? (If we ignore the resistance of air, it cannot.)

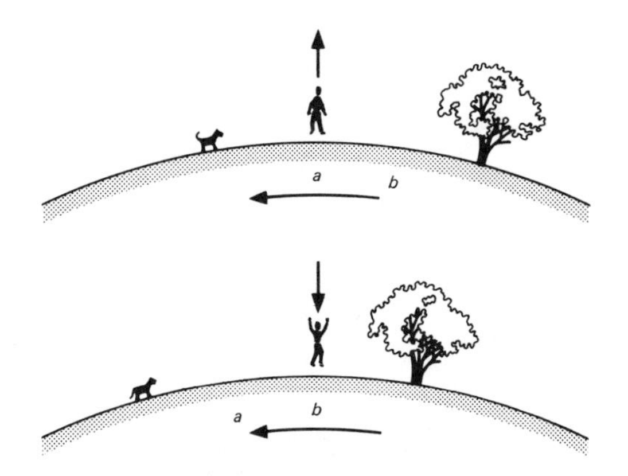

Figure 6.4. Ptolemy's proof that the Earth does not rotate. If a person on the Earth's surface at point a jumps up, and the Earth rotates, that person will fall back at point b. This proves, it was thought, that the heavens rotate and not the Earth.

In the early seventeenth century, Pierre Gassendi, a French philosopher, dropped stones from the top of the mast of a moving ship and showed that the stones landed at the foot of the mast, just as if the ship had been stationary. Ptolemy had said that if the Earth rotated then a person jumping up would not return to the ground at the same place (see Figure 6.4). This meant that a person who jumped up in a moving ship would not return to the deck at the same place. Gassendi's experiment proved that Ptolemy was wrong (see Figure 6.5).

Jeremiah Horrocks, a clergyman who died at the age of twenty-two, showed that the Moon moves in a Keplerian ellipse. He suggested that the disturbances in the Moon's motion are due to the influence of the distant Sun, and he anticipated universal gravity by suggesting that the planets perturb one another's motion.

Galileo brought together the medieval discoveries concerning space, time, and motion. He showed that impetus (still vaguely defined in various ways) is conserved in freely moving bodies; and by using balls that rolled down inclined planes he demonstrated that falling objects are uniformly accelerated. He showed that a pendulum has a period dependent on its length and not its weight (but he used his

Figure 6.5. Pierre Gassendi (1592–1655) of France dropped stones from the mast of a moving ship. He found that they fell at the foot of the mast when the ship was in motion, just as when it was stationary, and thus he demonstrated that Ptolemy was wrong. Newton later said, "The motions of bodies enclosed in a given space are the same relatively to each other whether that space is at rest or moving uniformly in a straight line without circular motion."

pulse and the flow of a jet of water to measure time and did not use the pendulum). Galileo failed to realize that circular motion is accelerated motion and thought that a planet moving in a circular orbit would continue to move in this fashion without the need of a constraining force. He was not a cosmologist and was silent on the problem of whether the universe is finite; furthermore, he disregarded Kepler's work and did not theorize about planetary motions other than to argue that Copernicus, and not Ptolemy, was correct.

Giovanni Borelli, a contemporary of Horrocks and Galileo, discussed planetary motions in the light of Galileo's work and speculated on the nature of universal gravity.

René Descartes believed that space existed only by virtue of a continuous distribution of matter, and that all forces acted only between adjacent things and could not affect distant bodies. He believed the planets were controlled by the vortical motion of material and ethereal elements in space, and that gravity was the result of these swirling motions creating pressure. (The "teacup effect" illustrates Descartes' idea: When tea is stirred in a cup, the tea leaves floating on the surface tend to concentrate in the center.) Descartes enunciated laws of motion that resembled the Newtonian laws. He clarified ideas on the nature of space, time, and motion and said, "Give me matter and motion and I will construct the universe."

Descartes combined geometry and algebra into *analytical geometry,* using coordinate systems in which the position of a point is determined by its three coordinate values x, y, and z. Measuring grids or graphs were quite familiar; they had been used by the ancient Egyptians for land surveying, they were employed in the measurement of latitude and longitude, and the Medievalists had used them. But Descartes was the first to use graphs to study geometry with the aid of algebra. Thus the distance between two points in the x and y plane is given by the Pythagorean rule

$$(\text{distance})^2 = (x \text{ interval})^2 + (y \text{ interval})^2 \quad (6.3)$$

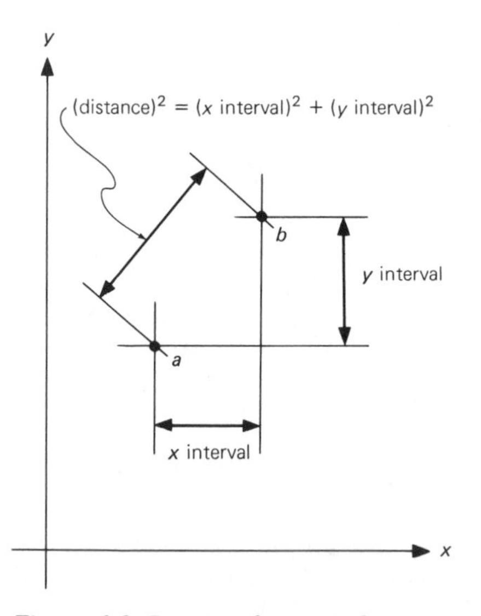

Figure 6.6. Distance between the two points a and b is given in terms of the x interval and y interval by the Pythagoras rule.

as in Figure 6.6; and in the three-dimensional x, y, and z space the distance between two points is

$$(\text{distance})^2 = (x\text{ interval})^2 + (y\text{ interval})^2 + (z\text{ interval})^2 \qquad (6.4)$$

The equation of circle A of radius r in Figure 6.7 is

$$x^2 + y^2 = r^2$$

and that of circle B, displaced a distance a in the x direction and a distance b in the y direction, is

$$(x - a)^2 + (y - b)^2 = r^2$$

Christiaan Huygens and Descartes realized what Galileo had failed to understand: that motion not in a straight line is accelerated motion (see Figure 6.8). Huygens showed that in circular motion there is an acceleration v^2/r toward the center of the circle, where v is the speed and r is the radius of the circle. A stone at the end of a piece of string, when whirled around, is continually accelerated toward the hand by the string pulling on the stone. Huygens, like Descartes, believed in clothed space and also thought that gravity was a force created by pressure. The Cartesians (those who followed Descartes) rejected the idea of a force acting at a distance as mystical and said, with some justice on their side, that it was of astrological origin.

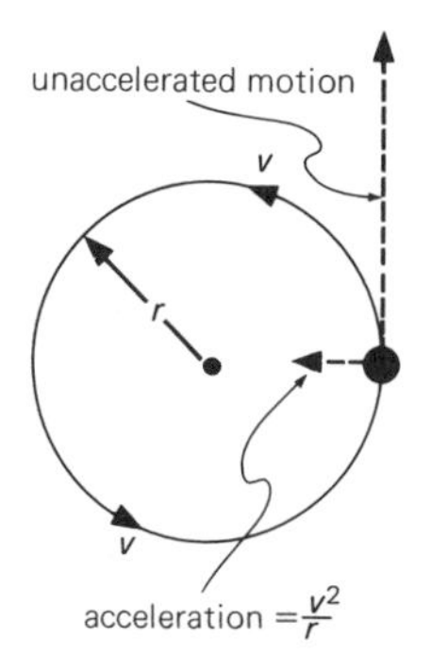

Figure 6.8. A body moves at constant speed v in a circular orbit of radius r. Although its speed is constant, its velocity (which has direction) continually changes, and the acceleration is v^2/r and is directed toward the center. The body is obviously accelerated toward the center, for otherwise it would move away in a straight line.

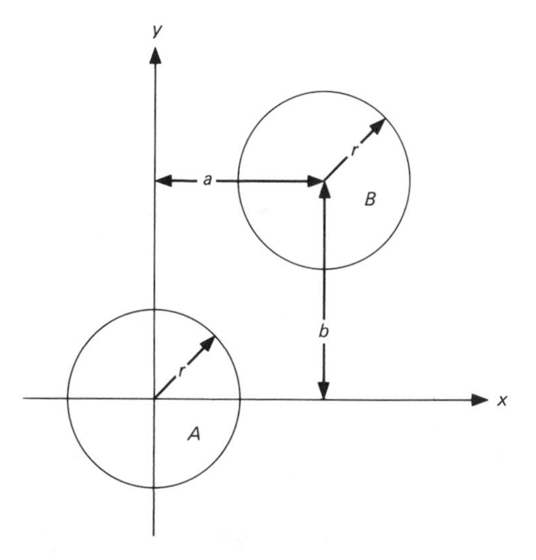

Figure 6.7 Diagram illustration the equations of circles in terms of the x and y coordinates.

Robert Hooke, Christopher Wren, and Edmund Halley outlined qualitatively what Newton later explained quantitatively. A freely moving object travels at constant speed in a straight line when there is nothing to force it from that natural state. Because the planets do not move in straight lines, but follow curved orbits about the Sun, they are therefore continually pulled toward the Sun by a gravitational force. Hooke demonstrated the idea with a conical pendulum and showed that planetary motions can be understood by mechanical principles (see Figure 6.9). At about the time when Newton was silently pondering these matters, Hooke realized that the forces that control the Solar System, that draw the planets to the Sun and the Moon to the Earth, are the same as that force which causes apples to fall off trees. "I shall explain," he announced, "a System of the World differing in many particulars from any yet known, answering in all things to the common rules of mechanical motions . . . that all celestial bodies whatsoever have an attraction or gravitating power to their own centers, whereby they attract not only their own parts, and keep them from flying from them, as we may observe the Earth to do,

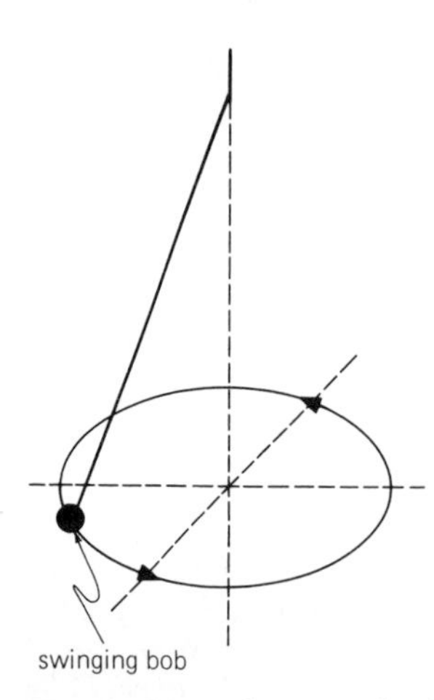

Figure 6.9. Robert Hooke (1635–1703) used a conical pendulum (the bob swings in a circle or an ellipse) to demonstrate the motion of the planets.

but that they do also attract all other celestial bodies that are within the sphere of their activity."

NEWTONIAN UNIVERSE. We have seen eminent scientists of the seventeenth century bringing together great thoughts that culminated in Hooke's mechanistic picture of the universe. It was the astounding genius of Newton, reflecting on these matters for many years, that converted pictorial descriptions into mathematical laws. Of time and space, Newton said: "Absolute, true, and mathematical time, of itself, and from its own nature, flows equably without relation to anything external . . . Absolute space, in its own nature, without relation to anything external, remains always similar and immovable." Unclothed space exists in its own right: That we can understand; but nobody has ever understood what Newton meant by his definition of time, except that time is objective and physical and is not purely subjective and psychical. Newton's three laws of motion state:

A body continues in its state of rest, or of uniform motion in a straight line, unless compelled to change that state by an impressed force. Momentum is conserved, where momentum is mass × velocity, and weight is the force that a mass experiences on a planet's surface.

The rate of change of momentum in time is equal to the impressed force and is in the direction of the force. If the mass of a body is constant, the rate of change of its momentum is mass × acceleration, and the law of motion is expressed as force = mass × acceleration.

To every force there is an equal and opposite force. This important law says that the sum of all forces at any place is zero (see Figure 6.10). For example, the weight of a person is balanced by a force in the ground that pushes upward. Notice that the equation of motion can be written

$$\text{force} - \text{mass} \times \text{acceleration} = 0 \qquad (6.5)$$

The first term on the left is the *applied force* and the second term is *inertial force* such that

$$\text{applied force} + \text{inertial force} = 0 \qquad (6.6)$$

The inertial force is the force experienced during acceleration, and is in the opposite direction to the acceleration. Thus in circular motion the acceleration is toward the center, but the inertial force (in this case the centrifugal force) is directed away from the center. A stone whirled around on the

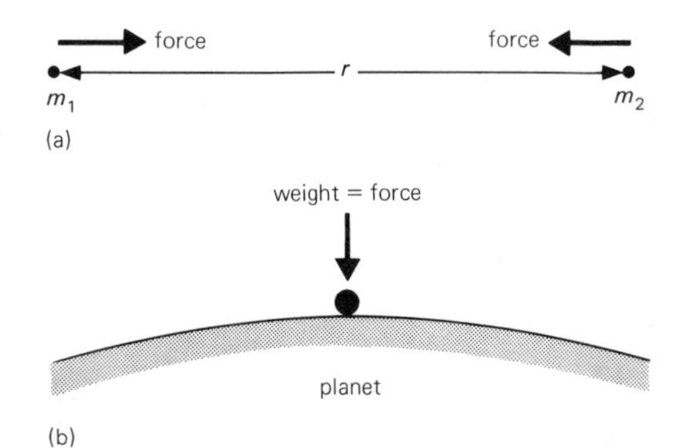

Figure 6.10. (a) Two bodies of mass m_1 and mass m_2, separated by a distance r, attract each other with a gravitational force Gm_1m_2/r^2, where G is the universal constant of gravity. (b) At the surface of a planet of mass M and radius R, a body of mass m has a weight GMm/R^2. Weight on a planetary surface is equal to the gravitational force.

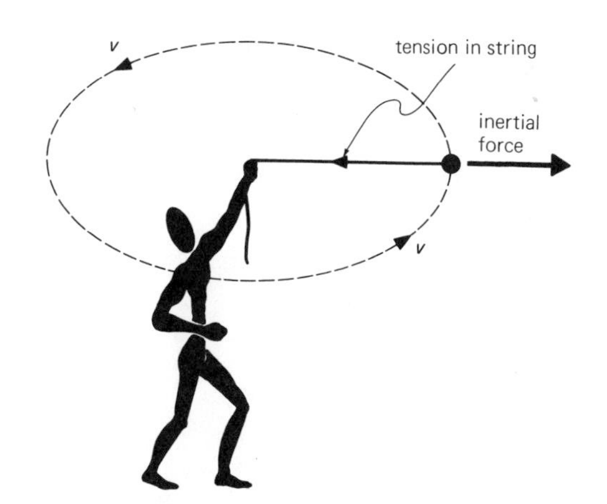

Figure 6.11. A body in circular motion at constant speed v is accelerated toward the center. The inertial force, which in this case is the centrifugal force, is in the opposite direction and equals mass × acceleration.

end of a string has a centrifugal force that pulls outward and is equal and opposite to the force in the string that pulls inward (see Figure 6.11). This explains why in a spaceship that orbits the Earth an astronaut does not feel the pull of the Earth's gravity: The applied force is gravity, and the inertial force resulting from the motion of the spaceship exactly cancels gravity. An object moves in such a fashion that its inertial force always cancels the applied force. A person moving with that object therefore experiences no force at all.

The strength of gravity varies as the inverse square of distance from a body such as the Sun. How was this discovered? Kepler's third law gave the answer. Suppose that a planet moves in a circular orbit of radius r at speed v. The acceleration to the center is v^2/r, and therefore

$$\text{gravitational force} = \text{mass of planet} \times \frac{v^2}{r} \qquad (6.7)$$

The circumference of the orbit is $2\pi r$, and because the period P is the time to revolve once, we have $P = 2\pi r/v$. Kepler's third law for all planets states that P^2 is proportional to r^3. We have just seen that P^2 is also proportional to r^2/v^2, and therefore it follows that v^2 is proportional to $1/r$. Hence the gravitational force is proportional to $1/r^2$ and must diminish as the inverse square of distance from the Sun. Newton showed that a spherical body exerts a gravitational attraction as if all its mass were concentrated at the center of the body; that the natural orbits of planets are ellipses; that, in general, the orbits of bodies in free fall (i.e., freely moving in gravitational fields) are either ellipses, parabolas, or hyperbolas; and that all bodies in the universe attract one another with gravitational forces that vary inversely as the square of their separating distances.

Isaac Newton, as professor of mathematics at Cambridge University, gave eight lectures a year, which few students attended. His great work in three volumes, the *Mathematical Principles of Natural Philosophy* (written in Latin and referred to as the *Principia*), was written in less than two years; finished in 1686, it was published in 1687 at Edmund Halley's expense and sold for seven shillings a set. From a few axioms and definitions, Newton developed an array of propositions and then proceeded to explain mathematically the twice-daily tides on Earth caused by the Sun and Moon, the flattening of the Earth at the poles owing to its rotation, the precession of the axis of the Earth's rotation once every 26,000 years, the motions and perturbations of the Moon's orbit, and the paths of the planets and comets. He, and Leibniz independently, also developed the mathematical art of calculus.

Newton wrote: "I do not know what I may appear to the world; but to myself I seem to have been only like a boy playing on the seashore, and diverting myself in now and then finding a smoother pebble or a prettier shell than ordinary, whilst the great ocean of truth lay all undiscovered before me." On Newton's tomb in Westminster Abbey are these words: "Mortals! Rejoice at so great an ornament to the human race!"

RELATIVE AND ABSOLUTE MOTION. In the Newtonian universe motion is both relative and absolute. We can understand the distinction by thinking of riding in an automobile. If the velocity is constant, there are no inertial forces, and the velocity must be measured relative to the ground or other bodies, such as passing cars. A passenger with closed eyes cannot determine the velocity because it is purely relative and has no absolute value. When the velocity changes, however, there is a force – the inertial force – experienced during acceleration, and this force is not produced by motion relative to anything. The passenger with closed eyes is able to estimate the acceleration and also the magnitude and direction of the inertial force experienced. In Newtonian space and time, and in the modern relativity theory of spacetime, velocities are relative but accelerations are absolute. In the Newtonian universe unclothed space is at last a reality in its own right: still relative in the Atomist sense that one position is measured relative to other positions and in the medieval sense that velocities are measured relative to other velocities, but absolute in its own right and in the new recognition that accelerations are absolute and are not measured relative to each other.

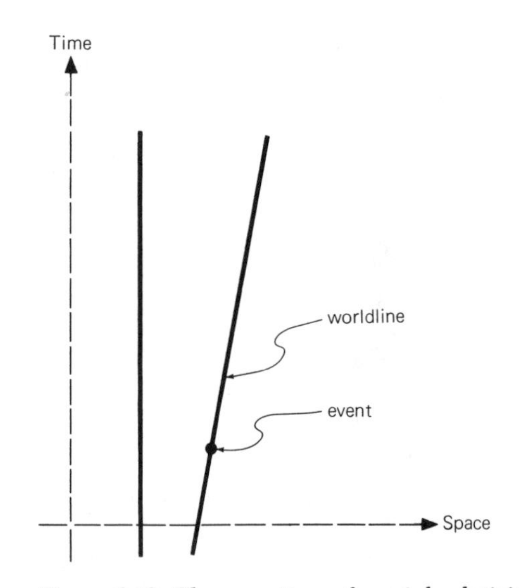

Figure 6.12. The spacetime of special relativity contains events and worldlines.

SPACETIME OF SPECIAL RELATIVITY

SPACE AND TIME DIAGRAMS. The restless world is in a state of continual change; at one instant of time there is a distribution of things in space, and at a later instant there is a different distribution. This variableness can be displayed by drawing a diagram in which things are distributed in both space and time. Time has only one dimension, in accordance with our experience that things in the past, present, and future form a continuous sequence in much the same way as the segments and points of a line. It is not easy to represent on paper three-dimensional space with time included as an additional fourth dimension – it is difficult enough to represent three-dimensional space on a two-dimensional piece of paper. Merely for the sake of simplicity we shall therefore show only one of the three dimensions of space. In this simple space and time diagram we can now draw points and lines (see Figure 6.12). A point is called an *event;* it is something at an instant in time and at a place in space. Any object, such as a stone or an atom, lasts for a long period of time and is represented by a line; this line shows its position in space at each instant of time and is called a *worldline.*

Newtonian space and time are both public property that all observers share in common, and intervals of space and time between any two events are the same for everybody. Suppose that an apple falls from a tree, and a person standing nearby observes that the apple drops a distance of 5 meters in a time of 1 second. A second person, in motion relative to the tree, also observes that the apple drops a distance of 5 meters in a time of 1 second. In the Newtonian universe the intervals of space and time between two events are said to be invariant (the same for everybody).

Space and time diagrams, with their events and worldlines, were used by the Medievalists, and there is nothing particularly difficult or novel about them. Until the beginning of this century they were regarded as a convenient graphic way of illustrating the way things change. Then came special relativity and pictures of this kind acquired a new physical meaning.

There is one thing we should notice about space and time diagrams: Nothing ever changes! Everything is shown as it was, as it is, and as it will be. We cannot speak of anything changing or moving in a space and time diagram because time has already been used and cannot be used twice.

SPECIAL RELATIVITY. The theory of special relativity emerged toward the end of the nineteenth century and was brought together in its final form in 1905 by the genius of Albert Einstein. It has withstood countless tests and is now in everyday use by physicists. Yet even nowadays, when we pause to reflect, the theory is just as startling as it was at the beginning of this century. Relativity requires that we abandon the belief that intervals of time and space are the same for everybody. We lose these two old invariants and are given instead two new invariants.

The first of the new invariants is the *speed of light* in empty space; it is the same for everybody and is independent of the velocity of the person measuring it. In 1887, Albert Michelson and Edward Morley showed that the speed of light c is the same in all directions on the Earth's surface. The Earth moves at a velocity $v = 30$ kilometers a second around the Sun, and Michelson and Morley expected to find that the measured speed of light was $c + v$ and $c - v$ in opposite directions parallel to the Earth's motion. Instead, they found that the speed of light is the same in both directions. The Sun, as we now know, moves at 300 kilometers per second around the Galaxy, and the Galaxy itself is also in motion in the Local Group of galaxies. But these velocities, and the much higher relative velocities of particles in modern experiments, do not affect the relative value of the speed of light. We are not greatly surprised to learn that light moves at speed c, and c is a speed limit for all particles moving at high energy. It is the constancy of c for everybody that is surprising. For instance, an object moves past us at half the speed of light, and relative to that object light has the same speed as that which we measure. Even if the object moves at $0.9c$, or $0.999c$, the speed of light relative to it is always c.

Physicists of the nineteenth century thought that space by itself was just an empty nothing and that electromagnetic waves such as light could not propagate unless this emptiness was filled with a medium called the electromagnetic ether. The invariance of the speed of light eventually killed the ether theory and led to the realization that space is more than mere emptiness: In conjunction with time, it possesses physical structure. It is not necessary to clothe space with a stressed ether to give it physical reality, for it has its own reality of a kind that makes the speed of light the same for all persons, independent of their relative motion.

The second invariant is the *spacetime interval.* Intervals of space and intervals of time by themselves are no longer invariant for all observers; instead, when combined in the following way, they form an invariant spacetime interval:

$$(\text{spacetime interval})^2 = (\text{time interval})^2 - (\text{space interval})^2 \qquad (6.8)$$

Note that if time is measured in seconds (or years), then distance is measured in light seconds (or light years); this is possible because the speed of light is the same for everybody. All observers, whatever their relative velocities, are in complete agreement on the value of the spacetime interval between any two events when it is determined in this way. Space and time are now fused together to form a united four-dimensional *spacetime* world. Space by itself and time by itself are peculiar to each observer, and only spacetime is the public thing that we all share in common. This was stressed by Hermann Minkowski in 1908 when he said, "Henceforth space by itself and time by itself are doomed to fade away into mere shadows, and only a kind of union of the two will preserve an independent reality."

Einstein had been a student of Minkowski's, and it was Minkowski who in 1908 showed that spacetime pictures do not just offer fictional ways of taming the restless world, but actually portray a four-dimensional physical reality.

A person, or a thing – let us call it an observer – is represented by a worldline that extends from the past to the future. The observer receives light signals that come in from the past and transmits light signals that go out into the future. These light signals come in and go out on the *lightcones,* as shown in Figure 6.13, and travel each second a distance of 1 light second. The backward lightcone, on which incoming light travels, extends outward into the past; the forward lightcone, on which outgoing light travels, extends outward into the future. All observers have similar lightcones, no matter how inclined their worldlines are.

Information of every kind comes to the observer at speeds less than or equal to the speed of light (sound waves, for example, travel at much less than the speed of light); therefore the events observed are either within or on the backward lightcone. Also, those events that are influenced by the observer, by means of outgoing signals, lie either within or on the forward lightcone. All observers agree that events within or on their backward lightcones belong to the past, and events within or on their forward lightcones belong to the future. But they do not agree on the past–future ordering of all events outside their lightcones.

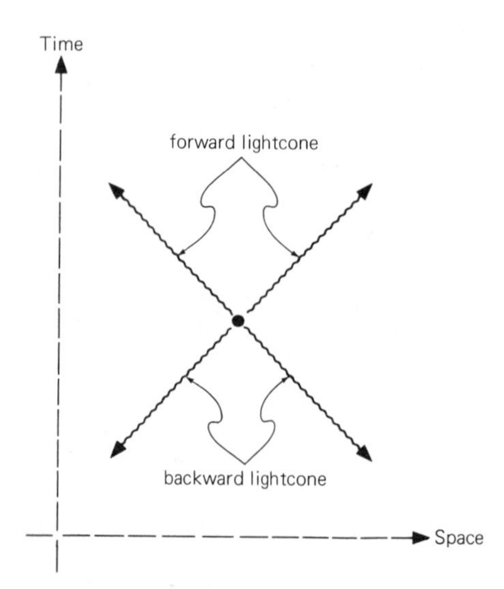

Figure 6.13. Lightcones at a point in spacetime. Light comes inward from the past on the backward lightcone and goes outward into the future on the forward lightcone.

Observers in relative motion have worldlines inclined to each other. Consider two observers A and B (Albert and Bertha) who are in relative motion. They have similar lightcones, yet they do not share the same space and the same time but have different spaces and different times within a common spacetime. Observer A measures time along his worldline (the intervals of length of his worldline are his actual intervals of time), and his space is perpendicular to his worldline; observer B measures time along her worldline, and her space is perpendicular to her worldline. Unfortunately, we can show that space is perpendicular to only one worldline, and this is because the pieces of paper on which we draw our diagrams do not have the same geometry as spacetime (see Figure 6.14). (This source of confusion is discussed in the next section.) From what we have said it is now clear that A's time is compounded of elements of B's time and space, and B's time is compounded of elements of A's time and space; furthermore, their spaces are compounded of elements of each other's space and time. Two events that are simultaneous (occurring at different places at the same instant of time) in A's space are not simultaneous in B's space, and two events occurring at the same place in A's space, but at different instants of time, do not occur at the same place in B's space.

What then, one might ask, is the true decomposition of spacetime into space and time? There is none: All decompositions into space and time are equally real, and hence the name *relativity*. The four-dimensional world of spacetime contains only events and worldlines; each event has its lightcones, and each worldline determines its own space and time.

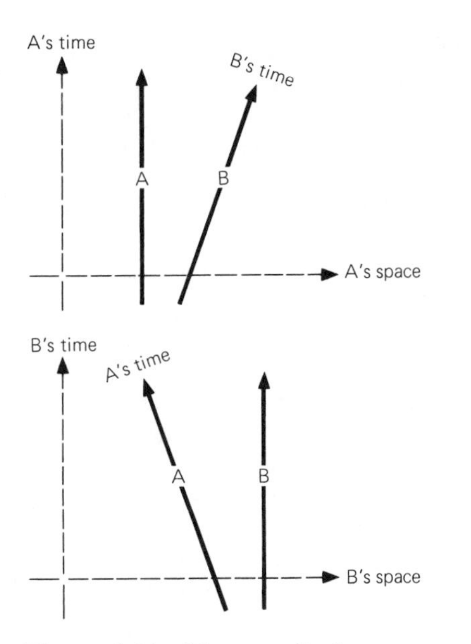

Figure 6.14. We can display a spacetime diagram so that A's space is perpendicular to A's worldline, or so that B's space is perpendicular to B's worldline, but we cannot show on paper the fact that both spaces are perpendicular to their worldlines. This is because the geometry of spacetime is unlike that of a blackboard or a sheet of paper.

What we have discussed is known as special relativity, and the algebraic details of the subject can be found in many books on elementary physics. A term that relates the intervals of time and space of observers in relative motion at speed v is quite easily derived: $(1 - v^2/c^2)^{1/2}$. This algebra of *Lorentz transformations* has been omitted because I think it is not essential for the purpose of understanding the concepts of relativity theory. Those familiar with algebra can easily acquire this extra command of the subject, but those who do not know algebra are equally capable of understanding the basic concepts.

INVARIANCE AND COVARIANCE. As we have already observed, quantities like the speed of light and the spacetime intervals between events, which are the same for all observers, independent of their relative velocities, are said to be *invariant*. Laws, and also equations, that are the same for all observers and are independent of their velocities are said to be *covariant*. The Newtonian universe provides us with simple examples of covariant laws. If we are on a ship moving at sea, or in a plane flying in the air, and toss a coin in the air, it falls back into the hand in exactly the same way as when we are standing on the ground. The laws governing the motion of the coin are the same, independent of our velocity, and are therefore covariant. In special relativity all laws of nature are covariant in a special sense. Consider a windowless laboratory that moves freely and at constant velocity. If, in this laboratory, we measure the speed of light, or perform any experiment whatever, we get always the same results, independent of the velocity of the laboratory (see Figure 6.15). In other words, no experiments that can be performed will tell us the velocity of the laboratory. The laws of nature are covariant and the same for observers in all laboratories moving at constant velocity.

In special relativity the laws of nature are covariant if the observer's velocity is constant. Despite the physical structure of spacetime, with its speed limit that is the same for everybody, we are still not very far away from Newtonian ideas. Velocity is

Figure 6.15. In a laboratory moving at constant velocity the laws of nature are covariant; they are independent of its velocity, and no experiment that can be performed in a windowless laboratory will reveal the velocity.

relative but acceleration is absolute. An accelerated observer experiences a force known as an inertial force, which is peculiar to the observer and is not shared by other observers of different motion. The laws of nature in special relativity are therefore not covariant for accelerated observers. The laws in *special* relativity are covariant only for a *special* class of observers whose motion is inertial (at constant velocity). We shall later see that the laws in *general* relativity are covariant for a more *general* class of observers who are in free fall.

Despite its limitations, physicists use special relativity all the time for accelerated (noninertial) motions. The trick is quite simple. Suppose we are in a laboratory that is accelerated. In our experiments we obtain results that are not strictly in accord with the predictions of our covariant equations; but we do not throw away the results and give up in despair. (Remember, laboratories on the Earth's surface are noninertial because of the Earth's rotation and its revolution about the Sun and the center of the Galaxy.) Instead, we imagine ourselves moving inertially (at constant velocity), and we study the events that are occurring in the accelerated laboratory. The covariant laws, which now hold, tell us what are the true results and what are the corrections that must be made to the observations obtained in the accelerated laboratory. The physicist therefore uses imaginary inertial motion to explain the results obtained in a state of noninertial motion.

SPACE TRAVEL

The geometry of spacetime is not the same as that of a piece of paper, and the shortest distance between two events is generally not a straight line.

Spacetime reveals its most interesting properties when bodies have relative speeds close to the speed of light. Most people find the results surprising and difficult to understand. How is it possible for a space traveler to journey in a lifetime to galaxies millions of light years away? How is it possible that a space traveler can return to Earth and find that his twin sister has grown old while he is still young? The answer is actually quite simple and springs from the fact that the geometry of spacetime is not the same as the geometry of the blackboard or the piece of paper on which we draw space time diagrams. You don't like geometry? Never mind – the geometry of spacetime is remarkably simple. All the confusion surrounding special relativity comes from the fact that the distance between two points in spacetime is not measured in the same way as the distance between two points on a piece of paper.

The distance between two points on a piece of paper is given by the Pythagorean rule:

$$(\text{space interval})^2 = (x \text{ interval})^2 + (y \text{ interval})^2$$

and this can be extended to space intervals in three dimensions, as shown by Descartes. If we go to four-dimensional space, in which the extra dimension is time, we might expect that distance in space and time would be measured in the same way, according to the Pythagorean rule:

$$(\text{space-and-time interval})^2 = (\text{time interval})^2 + (\text{space interval})^2 \qquad (6.9)$$

This would give us the space-and-time interval in Newtonian space and time, but it does not give the spacetime interval in special relativity. The spacetime interval that is the same for all observers, independent of their velocity, is actually

$$(\text{spacetime interval})^2 = (\text{time interval})^2 - (\text{space interval})^2 \qquad (6.10)$$

The minus sign (instead of a plus sign) is contrary to our common sense and accounts for the bewildering properties of spacetime.

We are accustomed to the idea that the shortest distance between two points is a straight line. This is true in ordinary space but is not true in spacetime. To understand why this is so let us consider the lightcones that are the paths on which lightrays travel.

In each second of time light travels 1 light second in space; for a lightray this means

interval of time = interval of space (6.11)

and therefore the spacetime interval between any two events on a lightcone, given in equation (6.10), is always zero. Consider a star at a distance of 1000 light years. Light from this star hurries to us at great speed across a wide gulf of space, and after 1000 years it enters the eye. No one can disagree that light has traversed immense intervals of space and time. And yet the spacetime interval between the eye and the star is zero! Spacetime distances along the backward lightcone to all events that send us signals, and along the forward lightcones to all events that receive our signals, are always zero.

We refer normally to the different regions of space and time as *here–now, there–now, here–then,* and *there–then* (see Figure 6.16). An observer is represented by a worldline, and a point on this worldline, at a given instant, is referred to by the observer as the here–now. On the backward lightcone are the observed events of the there–then. But these observed events of the there–then are separated from the here–now by spacetime distances of zero length. Hence, in spacetime, the observer is not confined to a point on a worldline, but instead reaches out along the lightcones that stretch through all of space and time. In spacetime we are at zero distance from the stars.

Consider a straight worldline connecting two events labeled *a* and *b*. Now consider the line *acb* (i.e., *a* to *c* and then *c* to *b*) along the lightcones intersecting *a* and *b*, as shown in Figure 6.17. The length of *acb,* as drawn on paper, is longer than the straight line *ab*. But in spacetime the distance *acb* is zero. The longest spacetime distance between *a* and *b* is measured along the straight line that passes through *a* and *b*; all paths

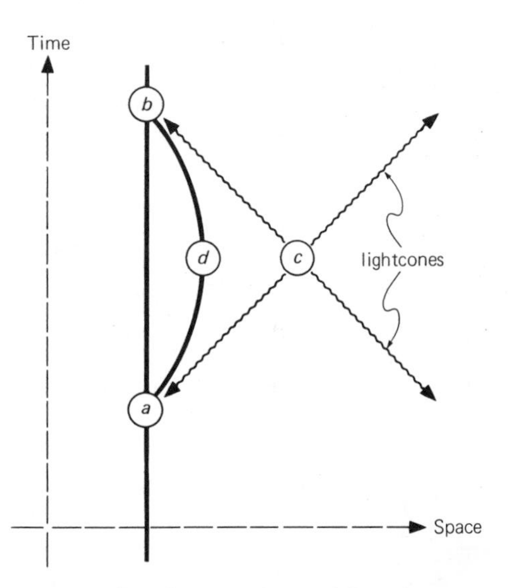

Figure 6.17. The straight worldline between a and b is the longest distance. The lightcone distance acb (i.e., a to c and then c to b) is of zero length. The nearer the bent worldline adb approaches the lightcones acb, the shorter its length. A person's time is measured along his or her worldline (intervals of experienced time are actually equal to intervals of length of the worldline), and the time taken to go from a to b is the length of the worldline. The longest time is along the straight worldline, and the time taken gets shorter as the worldline adb approaches acb. This explains the twin paradox.

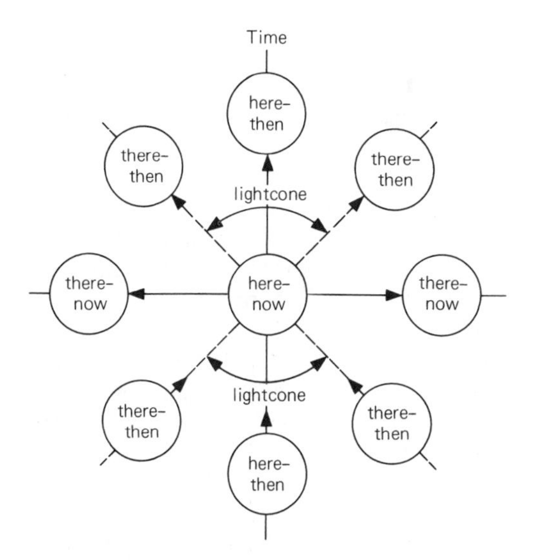

Figure 6.16. This illustrates the way we normally think of space and time. But in spacetime the distance on the lightcones from the here–now to the there–then is zero.

between a and b that bend out toward the lightcones are of shorter length. Spacetime contradicts ordinary common sense, and this has led to the so-called twin paradox.

THE TWIN PARADOX. Time is measured along a worldline and the age of a person (or an object) is the length of that person's (or that object's) worldline. But bent and curved worldlines, connecting any two points in spacetime, are shorter than straight worldlines. A person who goes from a to b via a bent worldline such as adb will arrive more quickly than another person who goes via a straight worldline. Both measure time with a wristwatch, or by the number of heartbeats, or with any other transported horological device, and the lapsed time in each case is equal to the length of the worldline measured in spacetime.

Twins A and B are born together and their worldlines begin at the same event. But for the rest of their lives their worldlines will follow different paths and whenever they meet they will have slightly different ages as

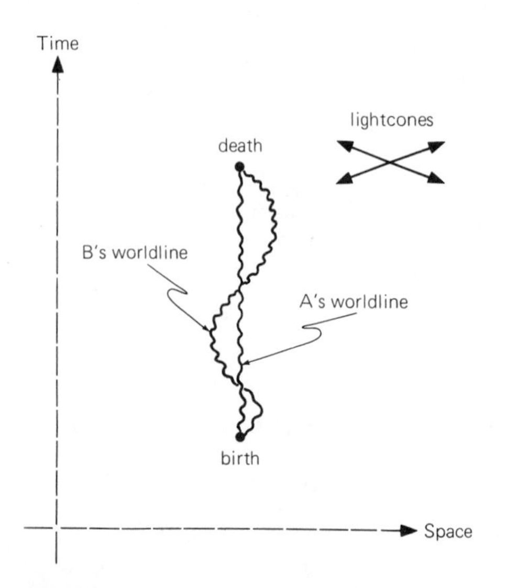

Figure 6.18. Crinkled worldlines of the twins A and B. Their worldlines follow different paths in spacetime, as they travel around, and are of slightly unequal length. It follows that the twins have slightly different ages whenever they meet. This difference in age becomes large if one twin travels on a spaceship at high speed.

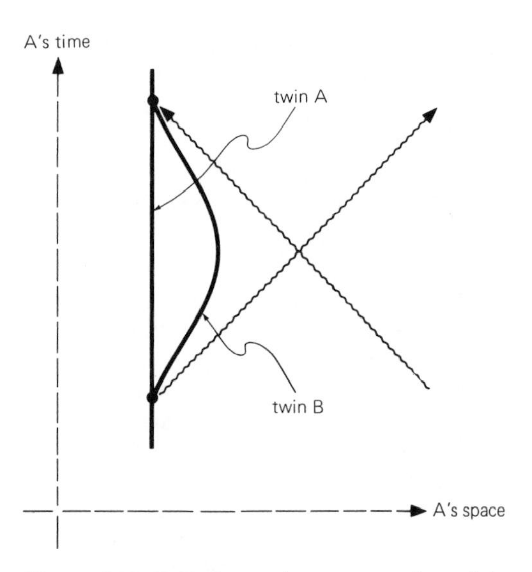

Figure 6.19. Twin B travels in a spaceship while twin A stays at home.

measured by the clocks they carry (see Figure 6.18). Suppose that at a reunion both die simultaneously in the same house. Let us suppose that A has stayed at home all his life and dies at the age of eighty. The other, B, has traveled a great deal by plane during her life and has a more crooked worldline. When she dies at the same instant as A, her worldline is slightly shorter than A's, and she has therefore lived a slightly shorter life. The difference in ages in this example is very small, perhaps as small as 1/10,000 ot a second.

Now suppose that B travels in a spaceship at extremely high speed and eventually returns to Earth (see Figure 6.19). Initially, the twins A and B were the same age, but when B returns, it is immediately apparent that A is older than B. "You have cheated!" says A; "you have gained and I have lost time!" "No I haven't," says B; "I am younger than you, my heart has beat fewer times than yours, and you have done more than I have. You have brushed your teeth more times and you have read more books than I have."

The space-traveling twin B experiences periods of acceleration in her travels (acceleration to get started, acceleration to turn

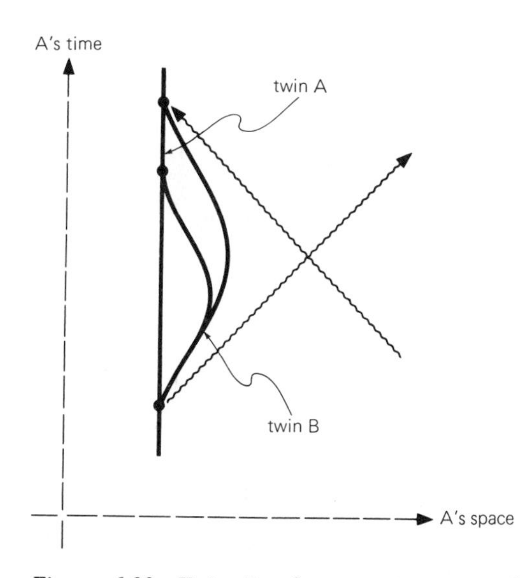

Figure 6.20. Twin B takes two journeys of unequal distance in which the accelerations are the same. The age difference between the twins increases with the length of the journey and has therefore nothing to do with the actual experience of acceleration.

around to come back, and acceleration – or rather deceleration – to return to Earth), but acceleration by itself is not the cause of the asymmetric aging of the twins. If B takes a longer journey (see Figure 6.20), but with exactly the same acceleration as in the journey illustrated in Figure 6.19, the difference in the ages of A and B will be even greater. Twin B ages less because of travel near the lightcones; and the greater the distance traveled, the greater the difference in aging.

Consider the outward-bound trip and let us view A in B's space and time, as shown in Figure 6.21. Twin A is now seen to be traveling away close to the lightcone. Surely this proves that A should be younger than B? But B must eventually change direction, and in order to catch up with A, must travel even closer to the lightcone. This makes B younger than A when they meet.

The technical problems of space travel at speeds close to that of light are very considerable, and will not be solved in the immediate future. We shall probably not encounter in any noticeable way the bizarre phenomenon of asymmetric aging for many centuries to come. But asymmetric aging occurs repeatedly with high-energy particles. A subatomic particle known as the muon decays in about a millionth of a second, and this lifetime is measured along its worldline. When a muon moves close to the speed of light it travels a large distance before it is seen to decay into other particles. In its own time, measured along its worldline, it decays in only a millionth of a second; but to us, who are not traveling with it, it is seen to decay much more slowly. The decay time is multiplied by $1/(1 - v^2/c^2)^{1/2}$, where v is the relative speed. Thus, if $v = 0.9998c$, then the observed decay time is increased by 50. Similarly, if space travelers journeyed away and then returned at this high speed for 1 year, on their arrival they would find that everybody on Earth had aged by 50 years (see Figure 6.22).

WHAT IS TIME?

THE AUGUSTINIAN DILEMMA. We think we know what space is like: It is that thing all around us that stretches away, in which objects are visibly distributed. It is spanned

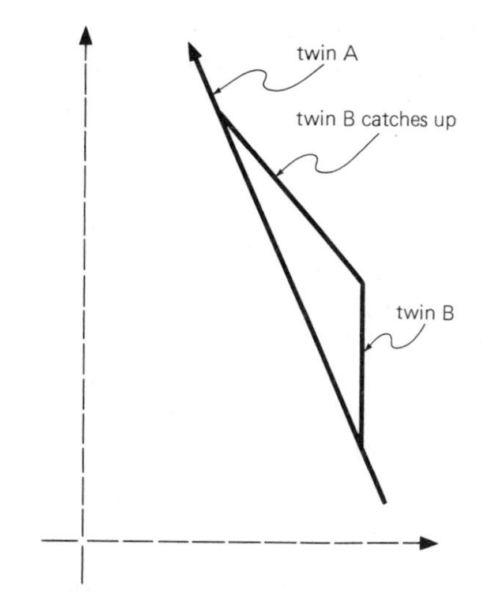

Figure 6.21. Outward-bound trip of B as viewed in B's space. B sees A moving away. To catch up with A, B must later travel even faster than A, and hence at their reunion B is again younger than A.

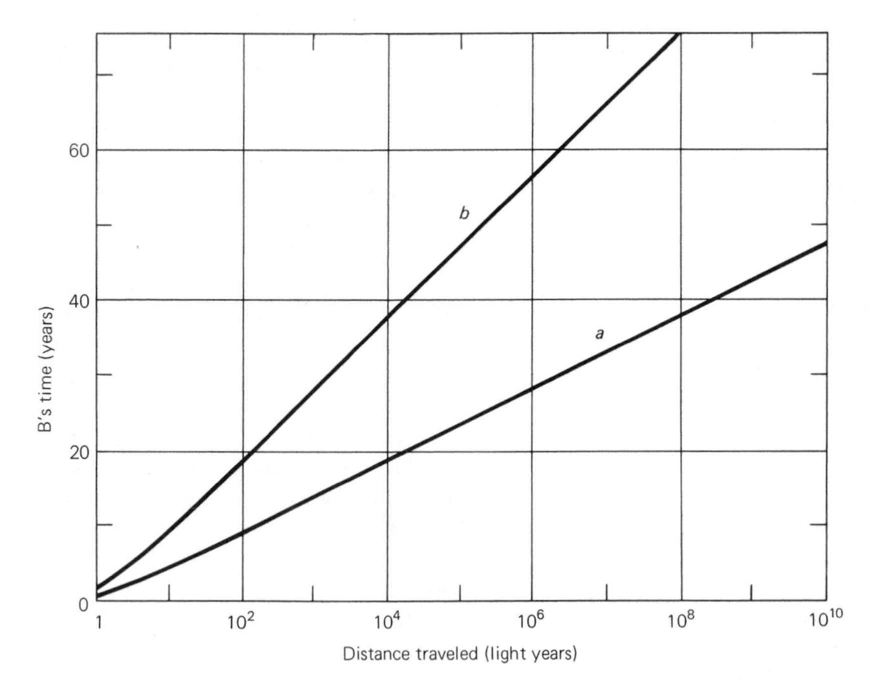

Figure 6.22. Twin B travels away at a constant acceleration of 1 g (the inertial force is equal to weight on the Earth's surface), and at half way B then decelerates at 1 g. The total distance traveled, in B's time, when B is once again stationary (relative to A), is shown in curve a. If B returns to Earth, by accelerating and decelerating as before, the trip takes twice as long, as shown by curve b. In 60 years of your own time you can travel to the Andromeda galaxy and back, and the Earth will be 4 million years older on your return.

by distances and measured in units such as centimeters, which can be directly perceived by looking at the size and separation of objects. But time is not quite so simple because we cannot observe objects distributed within it, and we cannot directly perceive with the five senses intervals of time such as seconds. It seems that we experience intervals of time subjectively but cannot directly observe them objectively outside ourselves. The impression gained from everyday life is that all experience is of two kinds: objective things diversified in space, and subjective things (ideas, emotions, self-awareness, and so forth) diversified in time. Somehow these two sides of experience, which supplement each other, are brought together to make up our world pictures.

The time that we experience as human beings, the "time that devours all things," according to Ovid, is not quite the same as that used in science. Our experienced time has been simplified by science and made into a continuous one-dimensional space that obeys the Hausdorff axioms (see the Reflections section of this chapter). We must therefore not be surprised if physical time lacks many of the temporal characteristics we experience. The reason for this difference between theoretical time and experienced time may be that science is still trying to catch up with the real world.

The nature of time is a difficult and perplexing subject that provokes endless philosophical discussion, and the words of St. Augustine in the fifth century still strike a responsive chord: "What, then, is time? If no one asks me, I know what it is. If I wish to explain what it is to him who asks me, I do not know."

FROZEN SPACETIME. Science has seized time, stripped away many of the characteristics ascribed to it in everyday life, and made it akin to space. Our physical world consists of four-dimensional spacetime that decomposes into a three-dimensional space and a one-dimensional time relative to each worldline. Time is measured by some device that follows the worldline, such as an atomic clock, a lighted candle, or a beating heart, and the worldline extends from the past to the future. Spacetime is in a frozen, unchanging state, and we refer to all its regions in a common tense. It is misleading to say of spacetime that an event has happened, another is happening, and yet another will happen, for all are present and displayed together. It is also misleading to say that an object moves along a worldline from the past to the future, for the object exists simultaneously at all points along the worldline. We must guard our tongues and continually remember that time is already contained within spacetime and cannot be used twice. If we make a mistake and say that a particle moves along a worldline, we are confronted by the question of the speed at which it moves through time. This compels us to invent a second time, and having admitted the possibility of movement through time, we must decide at what speed the particle moves through this second time. Motion through time opens up the possibility of serial time consisting of an infinite regression of timelike dimensions. We avoid this absurdity by realizing that spacetime does not endure in time, any more than it is dispersed in space, and that by bringing in extraneous time we violate the idea of containment.

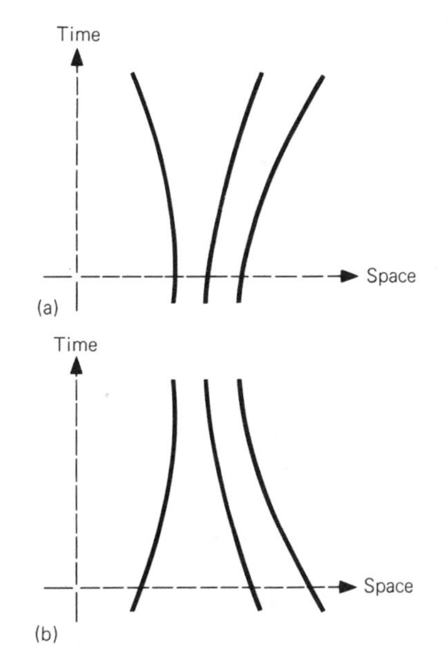

Figure 6.23. A spacetime picture (a) turned upside down (b) has the past and future reversed. In the picture there is often little to indicate which is the real past and future.

THE ARROW OF TIME. The world of spacetime contains worldlines, and these worldlines do not have any arrows affixed to them pointing in the direction of the future. In the spacetime picture it is therefore difficult to identify what determines the past and the future. When a spacetime picture is turned upside down, there is little or nothing in the picture that prevents us from relabeling the top as the future and the bottom as the past (see Figure 6.23). Past and future are labels that often have no spacetime meaning. Yet obviously a universe that is time-reversed becomes utterly different: Life begins in the grave and ends in the cradle, cool objects grow hot, candles and stars absorb radiation, and everything is in a hilarious topsy-turvey state. One is inclined to think that such time-reversed universes, if they exist, do not contain life and are rejected by the anthropic principle.

Most of the laws of physics do not distinguish between the past and the future and are therefore said to be *invariant to time reversal*. The laws of microscopic physics, such as the equations that govern the motions of particles, are reversible in time; they remain unchanged when we turn the spacetime picture upside down and relabel the past and the future. Even the equations that govern the propagation of light are time-reversible. We have supposed, in accordance with experience, that light comes toward us on the backward lightcone and leaves us on the forward lightcone. When the spacetime picture is turned upside down we find that light comes toward us from the future and goes outward into the past. This strange situation is not forbidden by the theory of electromagnetism, and nothing in the spacetime picture says that lightrays must propagate only into the future.

Thermodynamics is the branch of macroscopic physics that deals with the properties of heat. The second law of thermodynamics says that the behavior of heat is not time-reversible. No scientist would dream of contradicting this sacrosanct law. A cup of coffee on my desk gets cooler in the direction of the future and gets hotter in the direction of the past. If that did not happen then you and I would not exist. The laws of thermodynamics that govern the cup of coffee and its environment are not time-reversible; yet the laws that govern the behavior of the individual particles that make up the cup of coffee are time-reversible. Heat must always flow from hotter to cooler nearby regions, even when we reverse time in the equations that control the motions and behavior of individual particles. The behavior of heat (thermodynamics) provides us with an arrow of time, whereas the behavior of individual particles (particle physics) has no arrow of time (see Figure 6.24). This is rather puzzling, and there are still several aspects of the subject that are not fully understood. Let us try to understand in an elementary way what happens in the spacetime picture when heat flows from hotter to cooler regions. We can regard heat in this case as an agitation of particles: The hotter the particles, the more agitated is their motion. Now an agitated particle moves about and has a wavy or crinkled worldline, and crinkled worldlines communicate their agitation, by collisions and the emission of radiation, to their neighboring less agitated worldlines. Hence hot (or crinkled) worldlines, surrounded by less hot (less crinkled) worldlines, as shown in Figure 6.25, are cooler in the direction of the future. This, in a nutshell, is the second law of thermodynamics. If you are shown a spacetime diagram with many crinkled worldlines, you immediately know which is the past and which is the future. The agitation itself is governed by the laws of particle physics and is unaffected

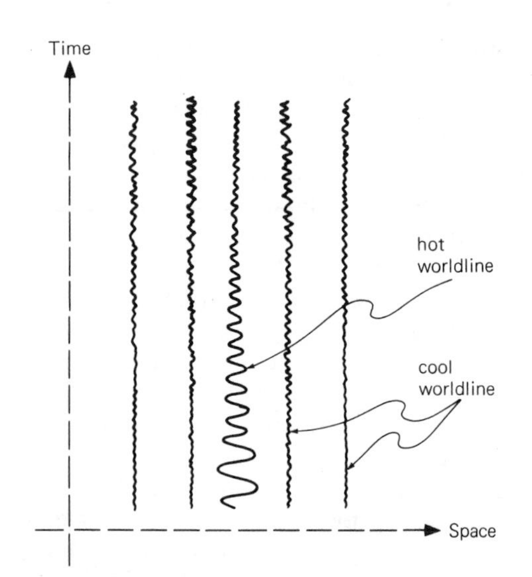

Figure 6.25. A hot (agitated) worldline surrounded by cool (less agitated) worldlines. Agitation is communicated to neighboring worldlines by particle collisions. Therefore the hot worldline is cooler in the direction of the future. In other words, heat diverges from hot to cool regions in the course of time.

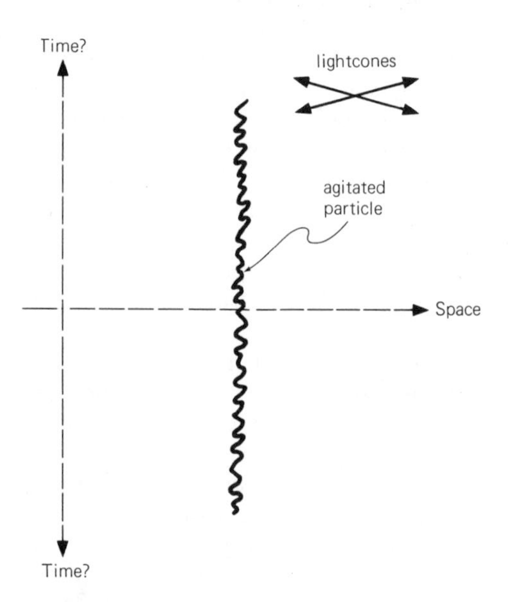

Figure 6.24. Heat is particle agitation, and an agitated particle is represented by a crinkled worldline. Note that the spacetime picture looks similar when turned upside down; this indicates that time reversal does not affect the behavior of particles.

by time reversal, but the diffusion and spread of agitation, from hot to cool regions, is governed by the second law of thermodynamics, which is not time-reversible. Single particles are time-reversible, but systems of particles are not. It is like hanging a picture on a wall: If you look only at the dabs of paint, you will not know which way to hang it; you have to look at large areas of the picture to determine whether it is the right way up.

The universe consists of numerous many-particle systems (living creatures, planets, stars, galaxies), and most scientists believe that the arrow of time is determined, either partly or totally, by the statistical behavior of these many-particle systems. The organized energy of these systems becomes disorganized in the course of time, and order always tends toward a state of disorder. In trillions of years' time, when all stars have died and all systems have attained their lowest accessible energy states, it will be difficult to determine the direction of time. In our universe, where matter is distributed irregularly in the form of stars and galaxies, and where energy continually cascades into dispersed states of lower energy, we are provided with a cosmic arrow of time. Is there then no arrow of time in a universe that is perfectly uniform and without irregularities of any kind? This is rather difficult to answer, because such a universe might be expanding and the change in its density would identify which was the past and which the future. But how would we know that it was expanding? – it might be contracting, so that higher density would lie in the future and not the past (see Figure 6.26). Thomas Gold has championed the idea that the universe gets a sense of time direction entirely from its state of expansion, and therefore the future always lies in the direction of diverging worldlines of the galaxies. It is not clear in this case what would happen if the universe ceased to expand and began to collapse.

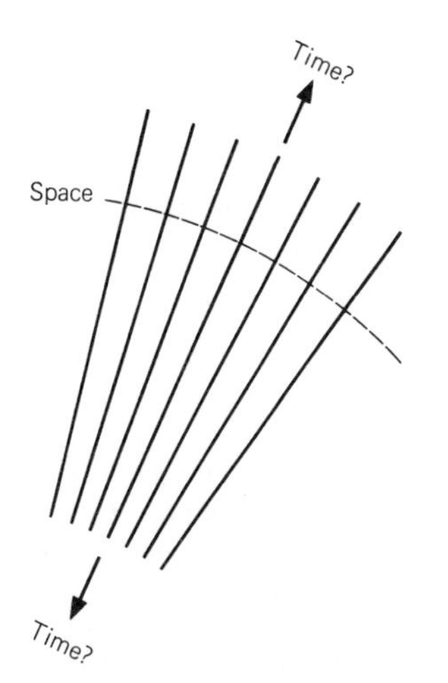

Figure 6.26. Spacetime picture of diverging worldlines in an expanding or contracting universe. Does this picture fix the direction of time?

THE CONFLICT OF PHYSICAL AND EXPERIENCED TIME. In our everyday experiences we are acutely aware of the "now" as a moment of enhanced vividity that divides the known past from the unknown future. We have an awareness of transience: of things becoming, bursting forth into actuality, and then fading into a limbo of memory. We think of our awareness as if it were an asymmetric wave of vividity advancing into the future, whose intensity is almost zero ahead, rises sharply to a maximum at the present moment, and then falls off slowly into the past and is extinguished at birth, as illustrated in Figure 6.27. Our thought and language patterns contain a mixture of primitive and sophisticated expressions concerning the nature of time, and their conflict leads to endless confusion. In the midst of this confusion we debate such subjects as the nature of time, the arrow of time, and free will versus determinism. We imagine time flowing past us, ourselves moving through time, the future approaching and the past receding; and our thoughts embrace the discordant concepts of transience and continuity that we have so far failed to harmonize. We have only to ask at what speed the "now" moves through time, and how a sense of transience can be reconciled with continuity, to realize that we still do not understand time.

The physical time of spacetime is an

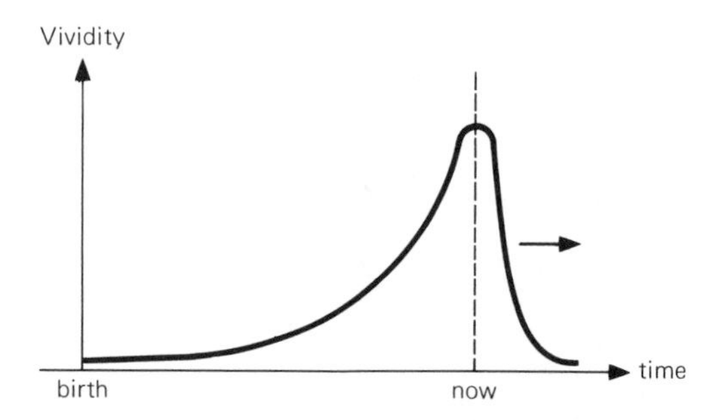

Figure 6.27. A wave of vividity advancing into the future. This picture illustrates our usual way of thinking and speaking about time. And yet it is nonsensical: It implies the existence of a second time that determines the speed of movement through time. Also, why is it that the vividity wave of awareness moves in only one direction?

abstract refinement of Newtonian time; and Newtonian time, which is still used in sciences other than physics, is itself an abstract refinement of the colloquial time used in everyday affairs. Some people have said that spacetime contains real time, and the transience of our experiences is only an illusion peculiar to living creatures. This philosophy revives the Parmenidean doctrine according to which reality is changeless and our experiences are illusions. Which is the real thing – the idea or the experience? – is a matter open to endless debate. If ideas derive from experiences, and experiences are illusory, how can ideas be more real than the experiences? If living creatures are nothing but bundles of worldlines in spacetime, how can worldlines have an illusion of transience when nothing in spacetime changes?

At this point we should remember that physics advances by modifying and discarding old ideas that were once mistaken for reality. Physicists are aware of the impermanence of our basic theories, and many feel uneasy when they hear philosophers and biologists discuss the nature of time in the physical world. If, in the future, we once again succeed in changing the nature of physical time in such a way as to incorporate some of the more subtle properties of experienced time, then the physical universe and cosmology will be fundamentally transformed.

REFLECTIONS

1 *"Where I am not understood, it shall be concluded that something useful and profound is couched underneath" (Jonathan Swift [1667–1745],* Tale of a Tub).

2 *Discuss: "clothed" and "unclothed" space; nature abhors a vacuum (Latin proverb); "a vacuum is repugnant to reason" (René Descartes).*

* *"I don't say that matter and space are the same thing. I only say, there is not space, where there is no matter; and that space in itself is not an absolute reality." This was written by Gottfried Leibniz (1646–1716) in a letter to Samuel Clarke, who argued in defense of Newton's ideas and of the reality of an absolute space that is independent of the existence of matter. Leibniz shared the views of many Continental philosophers, who believed that unclothed space was meaningless. He also shared the view that forces of any kind could not act at a distance unless they were conveyed by a material medium.*

* *"A Frenchman coming to London finds matters considerably changed, in philosophy as in everything else. He left the world filled, he finds it here empty. In Paris you see the universe consisting of vortices of a subtle matter; in London nothing is seen of this. With us it is the pressure of the moon that causes the tides of the sea; with the English it is the sea that gravitates toward the moon . . . Moreover, you may perceive that the sun, which in France is not at all involved in the affair, here has to contribute by nearly one quarter. With your Cartesians everything takes place through pressure, which is not easily comprehensible; with Monsieur Newton it takes place through attraction, the cause of which is not better known either" (Voltaire [1694–1778],* Letters from London on the English).

3 *The Hausdorff axioms (Felix Hausdorff,* Basic Features of Set Theory, *1914) express many of our intuitions about space (see Figure 6.28). We consider dots, and imagine them surrounded by circles, called neighborhoods, which can expand and contract. The axioms are:*

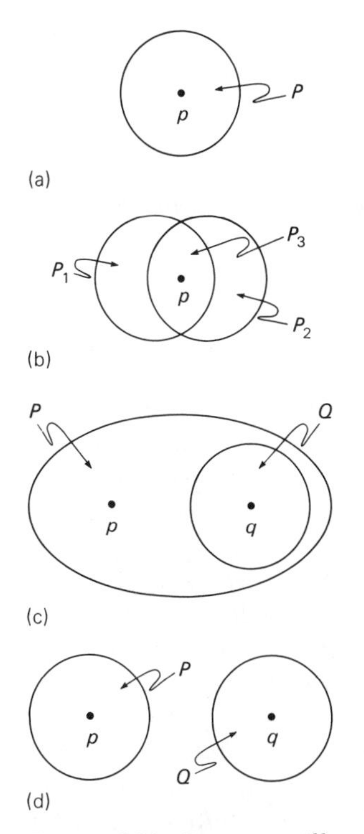

Figure 6.28. Diagrams illustrating the Hausdorff axioms of space.

(a) To a point p there is at least one neighborhood P that contains p.
(b) If P_1 and P_2 are two neighborhoods of the same point p, there exists a neighborhood P_3 that is contained within P_1 and P_2.
(c) If point p has a neighborhood P, and point q *is contained within P, there exists a neighborhood Q that is contained within P.*
(d) The points p and q have neighborhoods P and Q with no points in common.

From these axioms, Hausdorff developed a formal treatment of the concept of spatial continuity.

Use the Hausdorff axioms to discuss one-dimensional continuous time. Do the axioms assign an arrow to time, and do they satisfactorily account for our experience of time?

4 *In the Cartesian coordinate system shown in Figure 6.29, of origin O, the point A is the position denoted by the coordinate distances x, y, and z. Note that with the three perpendicular axes* X, Y, *and* Z *we can construct the three perpendicular planes XY, YZ, and ZX. Consider the following reflection symmetries:*

Reflection through the axes: Point A at position x,y,z becomes the point C at x, −y, −z, the point D at −x, y, −z, and the point E at −x, −y, z.

Reflection through the planes: Point A at position x,y,z becomes the point F at x,y,−z, the point G at −x,y,z, and the point H at x, −y, z.

Reflection through the origin: Point A at position x, y, z becomes the point B at −x, −y, −z.

Starting from A, show that the points C, D, and E can be obtained by reflecting twice through the planes. Show that the point B can be obtained by reflecting thrice through the planes but cannot be obtained by reflections through the axes. Note that all points can be obtained from A by multiple reflections through planes, or by a combination of reflections through axes and the origin.

The above geometric reflections are known as discrete isotropies. If the point A, for all values of x, y, and z, is reflection symmetric through an axis, then we have cylindrical symmetry about that axis. If the point A, for all values of x, y, and z, is reflection symmetric through the origin, we have spherical symmetry about the origin. These last two cases are examples of continuous isotropies, and spherical symmetry is

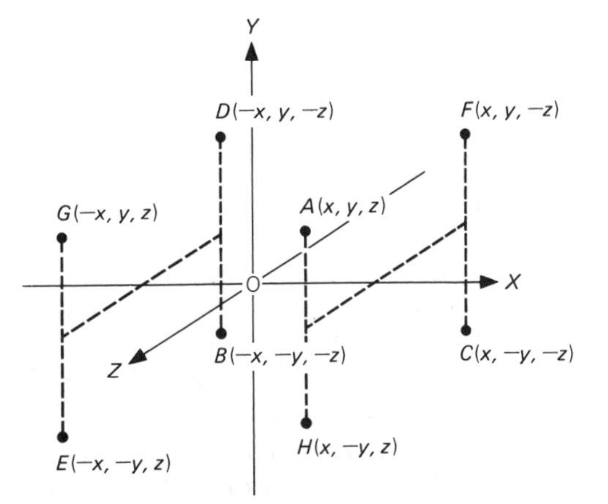

Figure 6.29. This shows how the point A at x, y, and z is reflected through the origin O; the axes X, Y, and Z; and the planes XY, YZ, and ZX.

the isotropy in which all directions are alike. A uniform (homogeneous and isotropic) space is reflection symmetric through all points everywhere (the origin O can be anywhere).

5 *Why is the twin paradox not a paradox?*

* *Traveling at 0.99 times the speed of light during your entire lifetime, how far can you get in the universe? If at half way you turn around and return to Earth, how long will the journey have taken in Earth time?*

Traveling with an acceleration and deceleration of 1 g, *how long in your time would it take to reach the center of the Galaxy? If you make a round trip, accelerating and decelerating at 1* g, *how far will you have traveled in half your lifetime, and how much older will the Earth be when you step out of your spaceship? (Use Figure 6.22.)*

6 *Time goes, you say? Ah no!*
Alas, Time stays, we *go.*
— Austin Dobson (1840–1921), The Paradox of Time

The idea of time as a fourth dimension was first proposed by Charles Hinton in What Is the Fourth Dimension? *(1887), and was the theme of H. G. Wells's* The Time Machine *(1895). To account for alleged precognition in dreams, John Dunne in* An Experiment with Time *(1927) formulated a theory of serial time in which it is possible for consciousness to travel backward and forward in other dimensions of time.*

* *Could life exist in a time-reversed universe? What is it in physics that distinguishes the past from the future?*

* *Arthur Eddington in* The Nature of the Physical World *(1928) introduced the term* arrow of time. *He wrote: "The great thing about time is that it goes on. But this is an aspect of it which the physicist seems inclined to neglect . . . I shall use the phrase 'time's arrow' to express this one-way property of time which has no analogue in space. It is a singularly interesting property from a philosophical standpoint. We must note that:*

(1) It is vividly recognized by consciousness.

(2) It is equally insisted on by our reasoning faculty, which tells us that a reversal of the arrow would render the external world nonsensical.

(3) It makes no appearance in physical science except in the study of the organization of a number of individuals. Here the arrow indicates the direction of progressive increase of the random element."

7 *The age-old problem of free will and determinism deserves thought. Human beings believe that they control their lives and have the freedom to influence at will the events of the surrounding world. But if human beings are part of the physical world and are completely contained within it, then they are also controlled by physical laws, and their belief in free will is an illusion. Either we have no free will – we are puppets – and are not responsible for the things we do, or we have free will according to mystical principles not yet accessible to rational thought. The argument applies not only to the physical universe but to all universes. Even if we believe in a rational spiritual realm (a view promoted by Saint Augustine, who believed in predestination), we cannot escape the free will–versus–determinism dilemma, for our actions are then controlled by nonphysical laws equally implacable, as was argued by Pelagius.*

* *"Only two possibilities exist: either one must believe in determinism and regard free will as a subjective illusion, or one must become a mystic and regard the discovery of natural laws as a meaningless intellectual game" (Max Born, "Man and the atom," 1957).*

* *"If it is necessary, then it is not a sin; if it is optional, then it can be avoided" (Pelagius [360–420]). Pelagius was a British theologian who lived at the time of Saint Augustine, and his argument against original sin is known as the Pelagian heresy.*

FURTHER READING

Abers, E. S., and Kennel, C. F. *Matter in Motion: The Spirit and Evolution of Physics.* Allyn and Bacon, Boston, 1977.

Bondi, H. *Relativity and Common Sense.* Doubleday, Anchor Books, New York, 1964.

Born, M, *Einstein's Theory of Relativity*. 1924. Reprint. Dover Publications, New York, 1962.

Boyer, C. B. *A History of Mathematics*. John Wiley, New York, 1968. Stresses the early developments and is slightly technical. An elementary discussion of the Hausdorff axioms is given on p. 668.

French, A. P. *Newtonian Mechanics*. W. W. Norton, New York, 1971. A clear exposition with historical comments.

Gardner, M. *The Ambidextrous Universe*. Basic Books, New York, 1964.

Gardner, M. *The Relativity Explosion*. Random House, New York, 1976.

Kaufmann, W. J. "Traveling near the speed of light." *Mercury,* January–February 1976.

Koyré, A. *Newtonian Studies*. Chapman and Hall, London, 1965.

Taylor, E. F., and Wheeler, J. A. *Spacetime Physics*. W. H. Freeman, San Francisco, 1966.

Toulmin, S., and Goodfield, J. *The Fabric of the Heavens: The Development of Astronomy and Dynamics*. Harper and Row, New York, 1961.

ON THE NATURE OF TIME

Butler, S. T., and Messel, H., ed. *Time: Selected Lectures on Time and Relativity, the Arrow of Time and the Relation of Geological and Biological Time*. Pergamon Press, Oxford, 1965.

Dunne, J. W. *An Experiment with Time*. Macmillan, New York, 1927.

Gold, T. "The arrow of time." *American Journal of Physics 30,* 403 (June 1962).

Haber, F. C. *The Age of the World, Moses to Darwin*. Johns Hopkins Press, Baltimore, 1959.

Layzer, D. "The arrow of time." *Scientific American,* April 1975.

Priestley, J. B. *Man and Time*. Aldus Books, London, 1964.

Schlegel, R. *Time and the Physical World*. Dover Publications, New York, 1968.

Wells, H. G. *The Time Machine*. Random House, New York, 1931. This edition has an interesting preface by Wells, written 36 years after the book was first published.

Whitrow, G. J. *What Is Time?* Thames and Hudson, London, 1972.

SOURCES

Alexandroff, P. *Elementary Concepts of Topology*. Dover Publications, New York, 1961.

Borel, E. *Space and Time*. 1922. Reprint. Dover Publications, New York, 1960. Still a useful book.

Born, M. "Man and the atom." *Bulletin of the Atomic Scientists,* June 1957.

Butterfield, H. *The Origins of Modern Science, 1300–1800*. Bell & Sons, London, 1957. Rev. ed. Free Press, New York, 1965.

Chew, G. F. "The dubious role of the space–time continuum in microscopic physics." *Science Progress,* October 1963.

Ellis, G. F. R., and Harrison, E. R. "Cosmological principles I: symmetry principles." *Comments on Astrophysics and Space Physics 6,* 23 (1974).

French, A. P. *Special Relativity*. W. W. Norton, New York, 1968. A clear mathematical treatment of special relativity.

Hinckfuss I. *The Existence of Space and Time*. Oxford University Press, Clarendon Press, Oxford, 1975.

Holton, G. "On the origin of the special theory of relativity." *American Journal of Physics 28,* 627 (October 1960).

Jammer, M. *Concepts of Space: The History of the Theories of Space in Physics*. Harvard University Press, Cambridge, Mass., 1954. Reprint. Harper and Row, New York, 1960.

Jammer, M. *Concepts of Force: A Study in the Foundations of Dynamics*. Harvard University Press, Cambridge, Mass., 1957. Reprint. Harper and Row, New York, 1962.

Kline, M. *Mathematics and the Physical World*. Thomas Y. Crowell, New York, 1969.

Lanczos, C. *Space through the Ages*. Academic Press, New York, 1970.

Lucretius. *The Nature of the Universe*. Trans. in prose by R. E. Latham. Penguin Books, Harmondsworth, Middlesex, 1951.

Munitz, M. K. *Space, Time and Creation: Philosophical Aspects of Scientific Cosmology*. Free Press, Glencoe, Ill., 1957.

Reichenbach, H. *The Philosophy of Space and Time*. Dover Publications, New York, 1957.

Sambursky, S. *The Physical World of the Greeks*. Routledge and Kegan Paul, London, 1956.

Smart, J. J. C., ed. *Problems of Space and Time: From Augustine to Albert Einstein*. Macmillan, New York, 1964.

Waerden, B. L. van der. *Science Awakening*. Noordhoff, Groningen, Holland, 1954.

Whittaker, E. *Space and Time: Theories of the Universe and the Arguments for the Existence of God*. Thomas Nelson, London, 1946.

Wigner, E. P. "Violations of symmetry in physics." *Scientific American,* December 1965.

ON THE NATURE OF TIME

Augustine. *Confessions.* Book Eleven. Trans. and ed. A. C. Outler. Library of Christian Classics, Westminster Press, London, 1955. Reprinted in *Problems of Space and Time. From Augustine to Albert Einstein,* ed. J. J. C. Smart. Macmillan, New York, 1964.

Davies, P. C. W. *The Physics of Time Asymmetry.* University of California Press, Berkeley, 1974. A technical treatment of time in cosmology and thermodynamics.

Eddington, A. S. *The Nature of the Physical World.* Cambridge University Press, Cambridge, 1928.

Gold, T., ed. *The Nature of Time.* Cornell University Press, Ithaca, N. Y., 1967.

Hook, S., ed. *Determinism and Freedom in the Age of Modern Science.* New York University Press, New York, 1958. Reprint. Macmillan, New York, 1961.

Huxley, A. *Tomorrow and Tomorrow and Tomorrow.* Harper and Row, New York, 1956.

Sachs, R. G. "Time reversal." *Science 176,* 587 (1972).

Whitrow, G. J. *The Natural Philosophy of Time.* Thomas Nelson, London, 1961.

7 CURVED SPACE

Man has weav'd out a net, and this net throwne
Upon the Heavens, and now they are his owne.
— Donne (1571–1631), *Ignatius His Conclave*

NON-EUCLIDEAN GEOMETRY

THE PARALLEL POSTULATE. Geometry was the art of land measurement in the ancient delta civilizations and was indispensable in the construction of such mammoth works as Stonehenge and the Great Pyramid of Giza. It consisted mainly of rule-of-thumb and trial-by-error methods. According to the sacred Rhind Papyrus, the Egyptians of 1800 B.C. used for π the value $(16/9)^2 = 3.1605$, as compared with its more exact value 3.1416. The Babylonians of 2000 B.C. and the Chinese of 300 B.C. used the rule that the circumference of a circle is three times its diameter, and this value for π is sanctioned by Hebraic scripture. The Greeks, in their thorough fashion, developed geometry into a science that climaxed with the axiomatic and definitive treatment provided by Euclid in the third century B.C.

What is the axiomatic method? Suppose that you wish to persuade someone that a statement S is true. You can show that this statement follows logically from another statement T that the person already accepts. But if the person is unconvinced of the truth of T it will be necessary to show that T follows logically from yet another statement U. This process might have to be repeated many times until eventually a statement Z is reached that is accepted as obviously true and without further need of logical justification. This basic statement is called an axiom or a postulate. Euclid used 5 intuitionally obvious axioms and from these, with numerous definitions, derived all that was then known of geometry in 465 theorems.

The axiom of most interest to us, which remained controversial until the nineteenth century, is the Euclidean parallel postulate. This postulate asserts that through any point there is one and only one parallel to a given straight line. The definition of *parallel* states that two straight lines in the same plane are parallel if they do not intersect (see Figure 7.1).

For more than 2000 years most people acquainted with geometry were willing to accept the parallel postulate as intuitionally obvious. A few exceptional people, including Euclid himself, confessed uneasiness because the parallel postulate cannot be verified by direct appeal to experience. We always encounter only segments of straight lines, never straight lines of infinite length, and hence we cannot assert with utmost confidence that two straight lines will always remain equidistant when extended to unlimited distances. Through the ages the few who felt uneasy sought for a more basic axiom from which the parallel postulate could be derived. But all attempts failed. John Wallis of the seventeenth century thought that our ordinary senses assure us that the geometric relations of a figure are

1 Kings 7:23
π = 3.

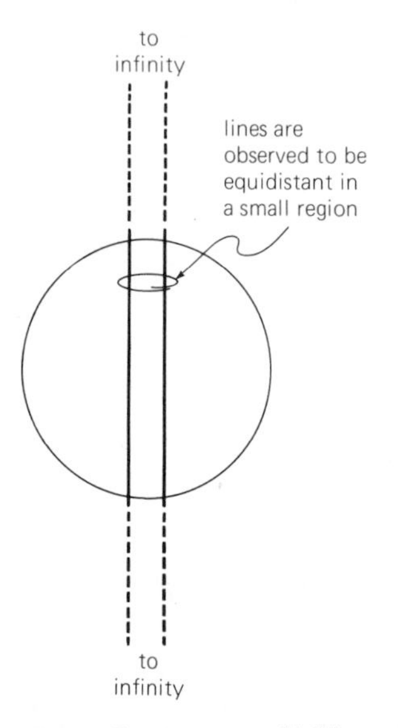

Figure 7.1. Two parallel lines, when extended to unlimited distances, remain equidistant. Do they? How can we know if we have no experience with lines of unlimited length?

unchanged when the figure is scaled in size. He therefore postulated: "If the sides of a triangle are changed in the same ratio, the angles of the triangle remain unchanged." From this axiom it can be shown that the parallel postulate follows immediately. But like all other attempts to find a more acceptable statement, it fails because it is merely the Euclidean postulate restated in an alternative form. Clearly, on the basis of experience, it is impossible to declare that the geometric relations of a triangle are unchanged when the triangle becomes of unlimited size. Geralamo Saccheri was also convinced that the parallel postulate could be proved by appeal to more obvious truths. In his work *Euclid Vindicated,* published in 1733, he wrongly thought that he had at last established the parallel postulate as a transparent truth. He showed that two straight lines, intersecting at an infinite distance, have an angle of intersection that is zero. But if the angle of intersection is zero, he argued, the lines are indistinguishable; and straight lines that are distinguishable, and do not intersect, are necessarily parallel. In this work, Saccheri derived and discussed many non-Euclidean theorems but failed to realize that non-Euclidean geometry can have a theoretical validity equal to that of Euclidean geometry.

Immanual Kant shared the prevailing belief that Euclidean geometry was transparently true and that no alternative system of geometry was conceivable by the human mind. In the *Critique of Pure Reason* he tried to place Euclidean geometry on a secure foundation by arguing that its axioms were a priori (prior to experience) and were therefore "an inevitable necessity of thought." Kant made the mistake of supposing that what is unimaginable is automatically impossible: In mathematics and physics what is possible today was often unimaginable yesterday.

We now know that the parallel postulate is fundamental and cannot be reduced to a more basic axiom. It is fundamental to Euclidean space and singles out Euclidean space from all other possible spaces. In Euclidean space the circumference of a circle is π times its diameter and the sum of the interior angles of a triangle is equal to two right angles. In other spaces these relations are not true.

THE THREE UNIFORM SPACES. There are countless possible spaces with their own geometries, and all are equally valid and self-consistent. Euclidean space, however, is uniform; it is homogeneous and isotropic (all places and all directions are alike), and it has a *congruence geometry.* In a congruence geometry all spatial forms are invariant under translations and rotations; thus, if the ratio of the circumference and diameter of a circle is π, this ratio is the same everywhere for all circles.

Of all possible non-Euclidean spaces there are only two that are also uniform in the same way as Euclidean space. Both were discovered in the nineteenth century. The first has *hyperbolic geometry* and was discovered by Johann Gauss, Nikolai Lobachevski, and Janos Bolyai; the second has *spherical geometry* and was discovered by Georg Riemann.

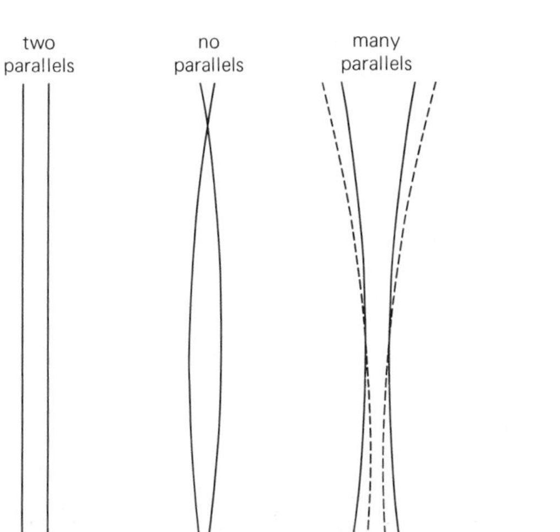

Figure 7.2. Parallel lines in uniform spaces. In flat space there is only one parallel to a straight line; in spherical space there are no parallels to a straight line; and in hyperbolic space there are many parallels to a straight line.

Hyperbolic and spherical spaces, like Euclidean space, are uniform and have congruence geometries. But unlike Euclidean space, they have an intrinsic scale length that is denoted by R. In all regions of space small in size compared with the length R, their geometries closely resemble Euclidean geometry. In other words, when R is large, it is not easy to distinguish between the three uniform spaces. We now see why the axioms of geometry must contain a postulate, such as the parallel postulate, that refers to what happens at large distances. It is the only way in which the congruence geometries of uniform spaces can be distinguished.

Our knowledge of the world derives directly from our experience with small-scale phenomena (that is, small on the cosmic scale), and this local experience contains no apparent information concerning a large intrinsic scale length R. This explains our predilection for the Euclidean geometry that has no intrinsic scale length. All people who live in hyperbolic and spherical spaces, and have experience of only small-scale local phenomena, think automatically in terms of Euclidean geometry.

The three uniform spaces are distinguished by the following postulates: In hyperbolic space there is more than one parallel to a straight line through a given point; in Euclidean space there is one parallel to a straight line through a given point; and in spherical space there is no parallel to a straight line through a given point (see Figure 7.2). In hyperbolic space the circumference of a circle is greater than π times its diameter and the sum of the interior angles of a triangle is less than two right angles; in spherical space the circumference of a circle is less than π times its diameter and the sum of the interior angles of a triangle is greater than two right angles (see Figure 7.3).

CURVATURE OF SPACE. We are familiar with two-dimensional surfaces in the three-dimensional space of our world, and we have no difficulty in visualizing uniform surfaces that are homogeneous and isotropic. A uniformly flat and infinitely large surface illustrates the nature of infinite Euclidean space, and the uniformly curved surface of a sphere illustrates the nature of non-Euclid-

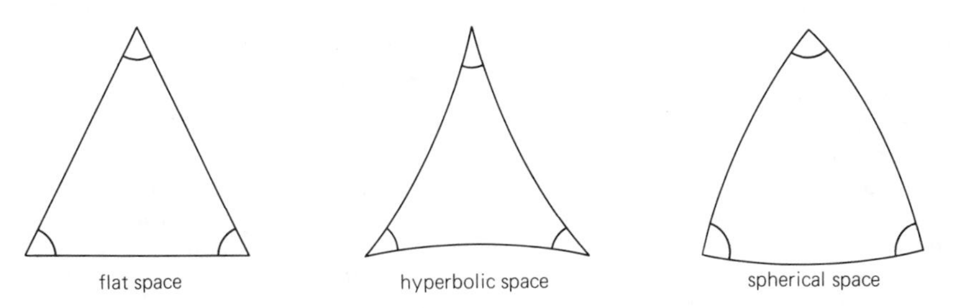

Figure 7.3. Triangles in flat, hyperbolic, and spherical spaces.

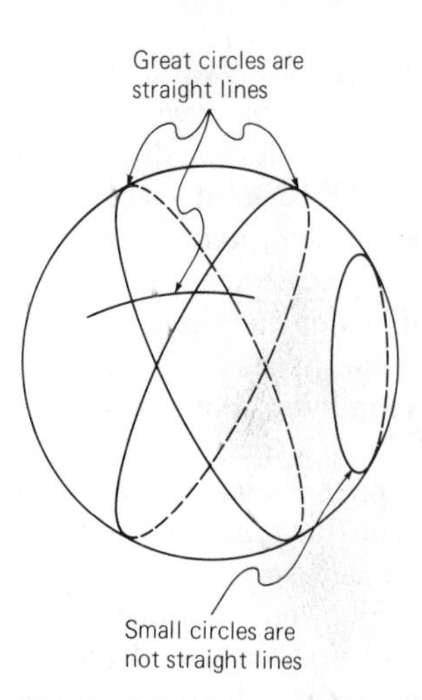

Figure 7.4. Great circles on the surface of a sphere. Great circles are straight lines; they always intersect each other, and it is therefore impossible to draw parallel straight lines on the surface of a sphere.

ean spherical space. Straight lines drawn on the surface of a sphere are called great circles. These great circles always intersect each other at finite distances, and it is therefore apparent that parallel straight lines do not exist on the surface of a sphere (see Figure 7.4). Furthermore, the sum of the interior angles of a triangle on the surface of a sphere always exceeds two right angles

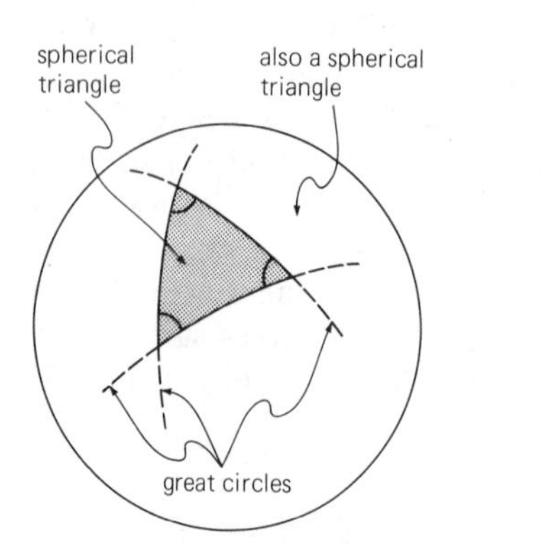

Figure 7.5. Triangles on the surface of a sphere. The sum of the interior angles is greater than two right angles. Note that the surface outside a small triangle is itself a large triangle. This can be demonstrated by allowing the small triangle to expand; the large external triangle then contracts and becomes a small triangle on the other side of the sphere.

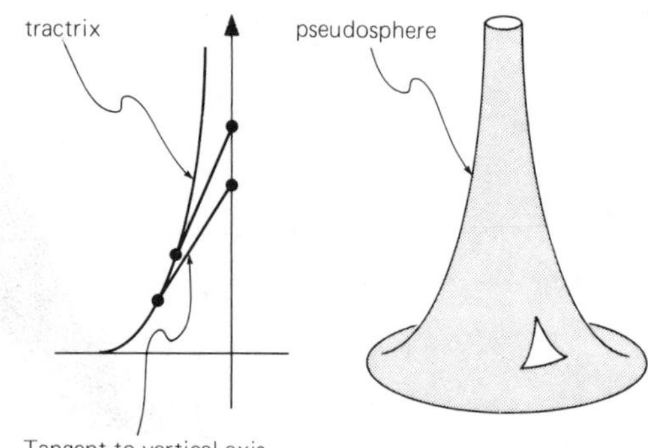

Figure 7.6. A pseudosphere has negative curvature and is an approximate representation of hyperbolic space. Note that the sum of the interior angles of a triangle on its surface is less than two right angles.

(see Figure 7.5). A spherical surface illustrates rather nicely the geometry of finite spherical space.

But what kind of surface illustrates the geometry of hyperbolic space? It was shown by the mathematician David Hilbert that we cannot construct in Euclidean space a two-dimensional surface that everywhere represents the geometry of uniform hyperbolic space. The surface of a pseudosphere (Figure 7.6) has hyperbolic geometry, but it is not uniform in the sense of being homogeneous (all places are alike) and isotropic (all directions are alike) at any place. A saddle-shaped surface (Figure 7.7) is homogeneous

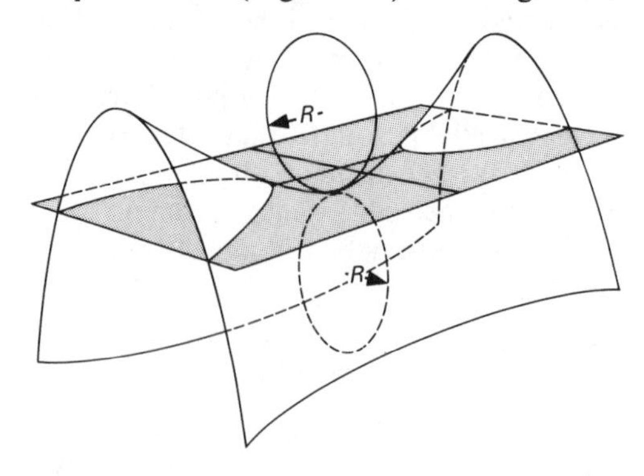

Figure 7.7. A saddle-shaped surface also has negative curvature and, in the central region of the saddle, is a representation of hyperbolic space. Note that the radii of curvature R are on opposite sides of the surface, and hence the curvature K is negative and equal to $-1/R^2$.

and isotropic in a small central region, but obviously all places are not alike and the surface is therefore not uniform. On such surfaces, which are curved oppositely in different directions (and which illustrate the properties of hyperbolic geometry), the sum of the interior angles of a triangle is less than two right angles. A saddle-shaped surface has the advantage of demonstrating that hyperbolic space is "open" and of infinite extent (like Euclidean space), whereas a spherical surface demonstrates that spherical space is "closed" and of finite extent.

The three uniform spaces are defined in the following way:

spherical space: K is positive
Euclidean space: K is zero
hyperbolic space: K is negative

where the quantity K is called *curvature*. Curvature of a uniform surface is numerically equal to $1/R^2$, where R is a *radius of curvature* and is the intrinsic scale length previously mentioned. A surface has two radii of curvature measured in directions perpendicular to each other, and in a uniform surface they are equal and everywhere the same. When both radii of curvature are on the same side of the surface the curvature K is positive, and when they are on opposite sides of the surface the curvature K is negative. A flat surface has an infinite radius (R = infinity) and the curvature is therefore zero. This is why Euclidean space, which has zero curvature, is often said to be flat. A spherical surface has its two radii of curvature on the same side, and therefore its curvature $K = 1/R^2$ is positive. A saddle-shaped surface has its two radii of curvature on opposite sides, and the curvature $K = -1/R^2$ of hyperbolic space is negative. A three-dimensional space has six radii of curvature, because it contains three perpendicular surfaces, each of which has two radii; when the space is uniform there are only three radii, and the curvature K, equal to $1/R^2$, is the same everywhere. In flat three-dimensional space, K is zero at all points.

We have considered two-dimensional surfaces embedded in three-dimensional space and have found that they illustrate the properties of non-Euclidean geometry. But it is not necessary to adopt a three-dimensional view in order to investigate the geometry of two-dimensional surfaces. Two-dimensional creatures, such as "Flatlanders" who live in flat surfaces, "Spherelanders" who live in spherical surfaces, and "Hyperlanders" who live in hyperbolic surfaces, are unaware of a third dimension and yet are able to survey their two-dimensional worlds and determine their appropriate geometry. Similarly, we do not have to think of our three-dimensional space embedded in a higher-dimensional space in order to determine its appropriate geometry. In a uniform space the sum of the three interior angles of a triangle, minus two right angles, is equal to the curvature K multiplied by the area of the triangle. Because the sum of two right angles equals π radians, we have the rule

$$\text{sum of interior angles} - \pi = K \times \text{area}$$

and this enables us to determine the curvature.

Curvature of space is a loose expression that means space has non-Euclidean geometry. It originated in the last century when the non-Euclidean geometries of curved surfaces were studied, and it is picturesque and harmless provided we do not take it too literally. Our universe contains three-dimensional space, or rather four-dimensional spacetime, which is not embedded and curved within a higher-dimensional space. Curvature must be understood as an intrinsic geometric property: We should not try to visualize it as having an extrinsic significance.

CURVATURE OF SPACE

DENSE TRIANGULATION. Let us take a flat surface and draw on it numerous small triangles. The interior angles of each triangle have everywhere a sum equal to two right angles. We now bend and stretch the surface into any desired smooth shape (see Figure 7.8). The triangles remain triangular, because they are very small, and their

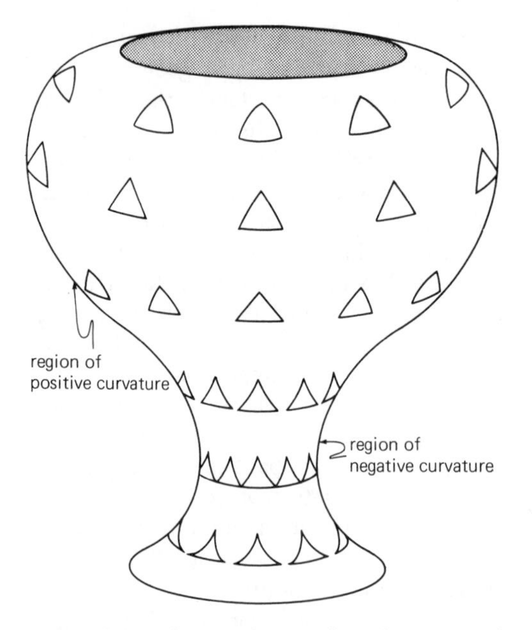

Figure 7.8. A bent and warped surface covered with numerous triangles. Each triangle tells us the value of the local curvature.

sides remain approximately straight. The change in the shape of the triangles is an indication of the change in the shape of the surface. Actually we can do better than just look at the shapes of triangles. We know that the sum of the interior angles of each triangle is equal to two right angles plus the curvature multiplied by the area of the triangle:

$$\text{sum of interior angles} = \pi + K \times \text{area}$$

where π is the sum of two right angles and the area is that of a triangle at the place of curvature K. In regions of positive curvature the interior angles exceed two right angles, and in regions of negative curvature (as when the surface is saddle-shaped) the interior angles are less than two right angles. This gives us a way of determining the curvature that does not require measurements made outside the surface and does not require that we know the original shapes of the triangles when the surface was flat.

Unfortunately, measuring the angles and areas of triangles has a disadvantage. To determine the shape of the surface we must know the curvature at each point. We must therefore shrink the triangles down to a very small size and cover the whole surface with dense triangulation. Ideally, the triangles should be infinitesimally small. Triangles of infinitesimal size, however, have infinitesimal areas, and have interior angles whose sum is two right angles. In other words, vanishingly small triangles have Euclidean geometry, and cannot easily be used for measuring curvature.

THE "OUTSTANDING THEOREM." The great mathematician Johann Gauss was the first to use *differential geometry* to study the geometry of space. When Gauss participated in large land survey at the invitation of the Hanoverian government, he was confronted with a surface deformed into hills and valleys. Ordinary geometry is of little help in the theoretical study of an inhomogeneous surface and Gauss had a brilliant idea.

To explain Gauss's discovery we shall start with a flat surface and lay out a network of imaginary lines that form what is called a *coordinate system* (see Figure 7.9). If the coordinates are perpendicular to each other, and labeled x and y, the distance between any two points is given by the Pythagorean rule:

$$(\text{space interval})^2 = (x \text{ interval})^2 + (y \text{ interval})^2$$

Another person might choose a different set of perpendicular coordinates x' and y' and the same distance between the same two points would be given by

$$(\text{space interval})^2 = (x' \text{ interval})^2 + (y' \text{ interval})^2$$

Coordinates are obviously just a convenient way of marking out a surface and do not affect the actual distances between the points in the surface. If we wish, we can use any coordinate system consisting of a network of arbitrarily curved lines (not necessarily intersecting perpendicularly), and the general Pythagorean rule becomes

$$\begin{aligned}(\text{space interval})^2 = {} & f(x \text{ interval})^2 \\ & + 2g(x \text{ interval} \times y \text{ interval}) \\ & + h(y \text{ interval})^2\end{aligned}$$

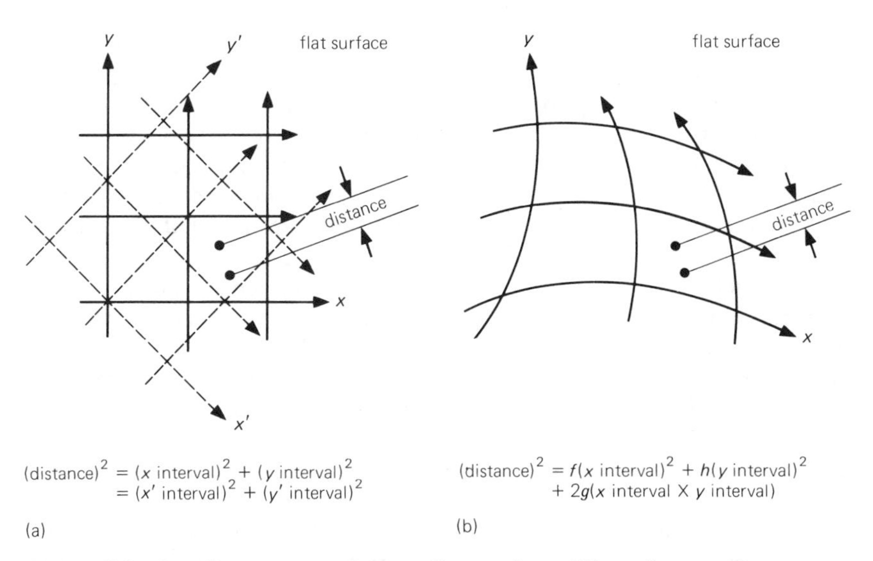

Figure 7.9. Coordinate systems on a flat surface. When the coordinates are perpendicular, as x and y or x' and y' are in (a), the distance between two points is given by the Pythagorean rule. When the coordinates are curvilinear (b), and not necessarily perpendicular, the distance between two points is given by a more general rule.

An equation of this kind, which gives the distance between two points in terms of arbitrary coordinates, is known as a *metric equation,* and *f, g,* and *h* are the *metric coefficients*. In a flat surface the metric coefficients depend only on the coordinates chosen, and we are free to choose coordinates such that $f = 1$, $g = 0$, and $h = 1$, as in the ordinary Pythagorean rule.

We now deform the surface into any desired smooth shape (see Figure 7.10). The *x* and *y* coordinate lines marked on the surface also change with the surface. The distance between two points that are close together is still given by the metric equation just stated, but the metric coefficients *f, g,* and *h* have changed. Altering the coordinates changes the metric coefficients, as does altering the shape of the surface.

A surveyor confronted with an undulating landscape does not start with a flat surface. As it happens, however, this is not a great disadvantage. The surveyor lays out on the surface an imaginary network of intersecting lines and from the beginning has a metric equation in which the metric coefficients *f, g,* and *h* depend on the coordinates chosen and the shape of the surface. Different surveyors will use different coordinates, but their metric equations will always agree on the space intervals between adjacent points. Gauss discovered that despite the arbitrary nature of the coordinates cho-

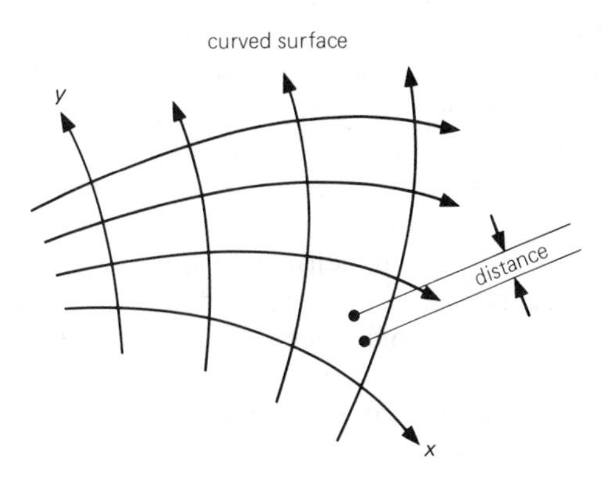

(distance)^2 = f(x interval)^2 + 2g(x interval X y interval) + h(y interval)^2

Figure 7.10. Distance between two adjacent points on a curved surface. The metric coefficients f, g, and h have values that depend on the arbitrarily chosen coordinates, and also on the geometry of the surface. The curvature can be determined from the way the metric coefficients vary from point to point in the surface, as shown by Gauss.

sen, and despite the dependence of the metric coefficients on the chosen coordinates, it is possible to determine the shape of the surface from the way in which the metric coefficients vary from place to place in the surface. The actual values of the metric coefficients are not important – they can always be changed by altering the coordinates – but the way in which they vary in the surface contains all the information that is needed to determine its geometry. Gauss was excited by his discovery and called it the *theorema egregium* – or the "outstanding theorem." He wrote in a letter, "These investigations deeply affect many other things; I would go so far as to say they are involved in the metaphysics of the geometry of space."

RIEMANNIAN SPACES

THE GENIUS OF RIEMANN AND CLIFFORD'S DREAM. Georg Riemann saw that the work done by Gauss had opened up an entirely new approach to the study of geometry. Riemann's great achievement was to show how this new approach could be generalized and applied to three-dimensional and higher-dimensional spaces. The distance between adjacent points is given by a metric equation that is nothing but a glorified Pythagorean rule; it contains the metric coefficients and is expressed in terms of arbitrary coordinates extending throughout space. In a two-dimensional space there are in general three metric coefficients; in a three-dimensional space there are in general six metric coefficients; and in a four-dimensional space there are in general ten metric coefficients.

Riemann derived differential equations for the variations of the metric coefficients. One of these equations gives us what is known as the *Riemann curvature,* which is the curvature discussed in an earlier section of this chapter. In a two-dimensional space the curvature has a single value K; and if the space is inhomogeneous, K varies from place to place. In a three-dimensional space the curvature is a more complicated expression containing six components that in general have different values; if the space is uniform (homogeneous and isotropic), three of the components are zero and the other three are everywhere constant and equal to K; when the space is flat, and hence Euclidean, all components are zero. In a four-dimensional space the curvature has twenty components, and in a five-dimensional space it has fifty components, and all components are zero if these spaces are flat.

If we wish to think of curvature as the deformation of a space that is embedded within a higher dimensional flat space, then for a curved two-dimensional space we need a flat space of three dimensions (this is the case with which we are familiar); for a curved three-dimensional space we need a flat space of six dimensions; and for a curved four-dimensional space we need a flat space of ten dimensions. Spacetime is four-dimensional, but when speaking of its curvature we do not attempt to visualize it embedded in a ten-dimensional flat space. Such a feat of the imagination is beyond human power and is quite unnecessary because our physical world consists of four-dimensional curved spacetime and not of flat ten-dimensional space. Moreover, flat space is not in any way more fundamental than curved space.

Riemann foresaw the possibility that science would evolve beyond Newtonian theory, with its three-dimensional Euclidean space, and might one day draw on the more general theory of space that he had developed. At the conclusion of his inaugural doctoral lecture "On the hypotheses that form the foundation for geometry," delivered in 1854, he said, "This leads us into the domain of another science, that of physics, into which the object of this work does not allow us to go today." Einstein paid to Riemann this tribute: "Only the genius of Riemann, solitary and uncomprehended, had already won its way by the middle of the last century to a new conception of space, in which space was deprived of its rigidity, and in which its power to take part in physical events was recognized as possible."

William Clifford, a young mathematician who translated Riemann's work into English, was greatly influenced, and he enthusiastically championed the idea of a fusion of geometry and physics. In the *Common Sense of the Exact Sciences,* published posthumously in 1885, he made the prophetic remarks: "Our space may be really the same (of equal curvature), but its degree of curvature may change as a whole with time. In this way our geometry based on the sameness of space would still hold good for all parts of space, but the change of curvature might produce in space a succession of apparent physical changes." This was a remarkable anticipation of the expanding space of modern cosmology. He continued: "We may conceive our space to have everywhere a nearly uniform curvature, but that slight variations of the curvature may occur from point to point, and themselves vary with time. These variations of the curvature with time may produce effects which we not unnaturally attribute to physical causes independent of the geometry of our space. We might even go so far as to assign to this variation of the curvature of space 'what really happens in that phenomena which we term the motion of matter.' " In 1876, Clifford wrote: "I wish here to indicate a manner in which these speculations may be applied to the investigation of physical phenomena. I hold in fact

(i) That small portions of space *are* in fact of a nature analogous to little hills on a surface which is on the average flat; namely, that the ordinary laws of geometry are not valid in them.

(ii) That this property of being curved or distorted is continually being passed on from one portion of space to another after the manner of a wave.

(iii) That this variation of the curvature of space is what really happens in that phenomena which we call the *motion of matter,* whether ponderable or ethereal.

(iv) That in the physical world nothing else takes place but this variation, subject (possibly) to the laws of continuity."

Clifford was an outstanding mathematician whose speculative ideas, outrageous in their day, anticipated general relativity by 40 years. He died at the age of 34 in the year in which Einstein was born.

REFLECTIONS

1 *Find out what the Euclidean axioms are. The Euclidean parallel postulate, as stated in the text, is referred to as Playfair's axiom. John Playfair (1748–1819) expressed the parallel postulate in this more acceptable form which we now know was actually used shortly after Euclid's death.*

* *David Hilbert (1862–1943) of Germany, the world's leading mathematician in the early decades of this century, championed the axiomatic method. Among his many accomplishments, in this field alone, he showed that the Euclidean axioms are self-consistent. The axiomatic method no longer requires that axioms be transparent truths. It is necessary only that the axioms be free of contradiction (or be self-consistent) and be sufficient to construct a theoretical system. Some axioms may at first seem strange, but common sense is not an infallible guide, and it is the theoretical consequence of the axioms that determines their validity.*

2 *Discuss the statement that Euclidean geometry is the only geometry without an intrinsic standard length.*

* *Why do people who live in spherical and hyperbolic spaces of small curvature think always in terms of Euclidean geometry?*

* *What is the sum of the interior angles of the smallest and largest triangles on the surface of a sphere? The curvature is $K = 1/R^2$ and the total surface is $4\pi R^2$. Note that the outside of a triangle is also a triangle. Show that a triangle becomes a hemisphere when the sum of the interior angles is 3π.*

* *Draw a coordinate system of lines on a thin sheet of rubber. Now stretch the rubber sheet over various objects that are smooth and curved, and notice how the coordinate lines change. Next draw a triangle on the rubber sheet when it is flat, and use it to estimate curvature when the rubber sheet is stretched over a curved surface.*

3 *"Space-curvature is something found in nature with which we are beginning to be familiar, recognizable by certain tests, for which ordinarily we need not a picture but a name" (Arthur Eddington,* The Expanding Universe, *1933).*

* *On the word* curvature, *introduced by Riemann, Howard Robertson wrote: "This name and this representation are for our purpose at least psychologically unfortunate, for we propose ultimately to deal exclusively with properties intrinsic to the space under consideration – properties which in the later physical applications can be measured within the space itself – and are not dependent upon some extrinsic construction, such as its relation to an hypothesized higher dimensional embedding space. We must accordingly seek some determination of K – which we nevertheless continue to call curvature – in terms of such inner properties" ("Geometry as a branch of physics," 1949).*

4 *"Now, what I want is Facts . . . Facts alone are wanted in life" (Mr. Gradgrind in* Hard Times *by Charles Dickens [1812–1870]).*

* *Consider circles and spheres of radius r (as measured by a stretched tape) in a uniform space of curvature K. If r is small compared with R, the circumference of a circle is given by the expression*

$$\text{circumference} = 2\pi r(1 - Kr^2/6)$$

and the area of the circle is expressed in this way:

$$\text{area} = \pi r^2(1 - Kr^2/12)$$

The equation for the surface area of a sphere is

$$\text{surface area} = 4\pi r^2(1 - Kr^2/3)$$

and that for the volume of a sphere is

$$\text{volume} = \frac{4}{3}\pi r^3(1 - Kr^2/5)$$

The departure from Euclidean geometry in all cases is approximately $(r/R)^2$. In our universe, R has a value of perhaps about 10^{10} light years. Even with figures that have a value of r as large as a million light years, the departure is only about 1 part in 10^8.

* *The surface of a cylinder is homogeneous (all places are alike). Show that the surface is not isotropic and that it has zero curvature (see Figure 7.11). Do triangles and circles on a cylindrical surface have Euclidean geometry? (Hint: Draw a triangle and a circle on a sheet of paper, and then wrap the sheet around the cylinder.)*

5 *Nikolai Lobachevski (1793–1856), an outstanding Russian mathematician of the University of Kazan, developed hyperbolic geometry, which he referred to as "imaginary geometry"; he published his results in 1829. Lobachevski said, "There is no branch of mathematics, however abstract, that may not someday be applied to phenomena of the real world." Although an outstanding teacher and university administrator, he was fired from his academic position in 1846, presumably for his unorthodox work, but no reason was given.*

* *Janos Bolyai (1802–1860) of Hungary, unaware of Lobachevski's recently published work, also developed the theory of hyperbolic geometry. "Out of nothing I have created a strange new universe," he said. His enthusiasm was quenched, however, when Gauss replied in a letter that he himself had discovered similar results many years previously.*

* *Johann Karl Friedrich Gauss (1777–1855), of humble origin, was an infant prodigy in mathematics who rose to become the*

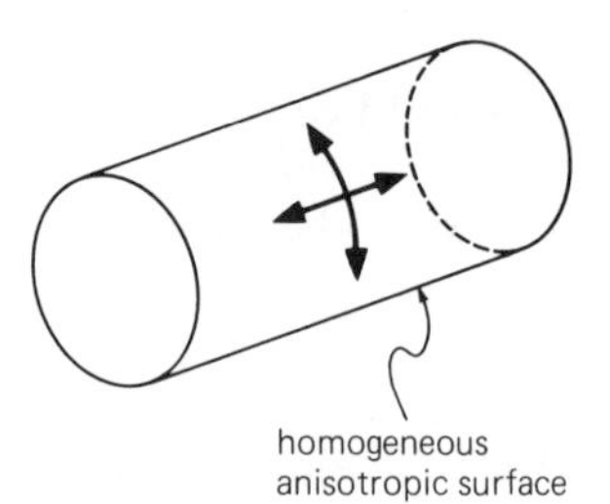

Figure 7.11. The surface of a cylinder has zero curvature, but is anisotropic and is topologically different from Euclidean space. Similarly, a pseudosphere has negative constant curvature, but is anisotropic and is topologically different from hyperbolic space.

prince of nineteenth-century mathematicians. In addition, he made fundamental advances in physics and astronomy, and while he was the leading professor at the University of Göttingen, had Riemann for a brief period as one of his students. Despite his reputation, Gauss was apparently afraid to publish his researches in non-Euclidean geometry; in a letter he noted that he had "a great antipathy against being drawn into any sort of polemic." In 1817 he said: "Perhaps in another world we may gain other insights into the nature of space which at present are unattainable to us. Until then we must consider geometry of equal rank not with arithmetic, which is purely logical, but with mechanics, which is empirical."

* *Georg Friedrich Bernhard Riemann (1826–1866) was the son of a Lutheran minister. He made numerous advances in several fields, mainly in mathematics, and initiated the subject of tensor calculus. (A tensor is a term having many components that vary in value from place to place.) Riemann's investigations into the structure of space were ignored by his contemporaries, who regarded this work as excessively theoretical and speculative. Riemann suffered from poor health and died at the age of 39.*

A zero-order tensor, known as a scalar, has a single value at each point in space. Temperature is an example of a zero-order tensor. A first-order tensor is a vector with n components at each point, where n is the number of dimensions of space. The motion of a simple fluid has at each point 3 components of velocity and is an example of a vector in three-dimensional space. A second-order tensor has n^2 components; many fluids have complex motions that require the descriptive power of second-order tensors. The metric tensor is second order, but owing to certain symmetries in our conception of space, it contains not n^2 components (or metric coefficients), but $n(n + 1)/2$ independent components. There are 3 components (as used by Gauss) when $n = 2$, 6 components when $n = 3$, and 10 components when $n = 4$, as in spacetime. The Riemann curvature is a fourth-order tensor with n^4 components, and in four-dimensional spacetime it has 256 components at each point. Symmetries in this tensor reduce the number of independent components to $n^2(n^2 - 1)/12$, so that when $n = 2$ there is 1 component, when $n = 3$ there are 6 components, and when $n = 4$ there are 20 components.

The theory of space curvature and tensor calculus was further developed by the mathematicians Elwin Christoffel (1829–1900) of Germany and Giuseppi Ricci (1811–1881) and Tullio Levi-Civita (1873–1941) of Italy. Einstein combined space and time into spacetime, drawing on the work of Riemann, Christoffel, Ricci, and Levi-Civita, and endowed spacetime with a varying curvature that explains gravity.

6 *In non-Euclidean space the sum of the interior angles of a large triangle will not equal 180 degrees (or π radians). Gauss performed an experiment, using three mountain peaks as the vertices of a triangle, and found that the interior angles added up to 180 degrees within the uncertainties of the measurements. His triangle was far too small, and furthermore, such an experiment is impossible without knowing how lightrays travel in curved space.*

In 1900, Karl Schwarzschild, a German astronomer, tried to measure the curvature of space. Lightrays from a star, intercepting the Earth's orbit at two widely separated points, form a triangle. By measuring the angles of such triangles, Schwarzschild attempted to determine the curvature of space. He concluded that if space is indeed curved, it has an extremely large radius of curvature, and he wrote: "One finds oneself here, if one will, in a geometrical fairyland, but the beauty of this fairy tale is that one does not know but that it may be true. We accordingly address the question here of how far we must push back the frontiers of this fairyland; of how small we must choose the curvature of space, how great its radius of curvature." In 1916, the year he died as a soldier of World War I, he found the first exact solution of Einstein's equation for the exterior geometry of spherical body such as a star. The Schwarzschild solution is of fundamental importance and is the equivalent of the Newtonian inverse-square law of gravity.

* *William Clifford (1845–1879) was a brilliant mathematician who died of tuberculosis while still relatively young. In the introduction to Clifford's book* The Common Sense of the Exact Sciences, *James Newman wrote: "All his life he had burdened his physical powers. The abundant but self-consuming nervous energy, the warfare against false beliefs, the self-goading search for new riddles and new challenges, the full submission to the demands of his intellect, were altogether out of proportion to what the physical machine could endure."*

7 *The following experiment demonstrates the nature of a* congruence geometry *and requires only a wall and a flashlight (see Figure 7.12). On the glass cover at the front of the flashlight is attached a transparent piece of paper on which a figure, such as a triangle or a circle, is heavily inked. The flashlight is held at a constant distance from the wall, and is moved around, always with the beam perpendicular to the wall. The figure seen on the wall preserves its*

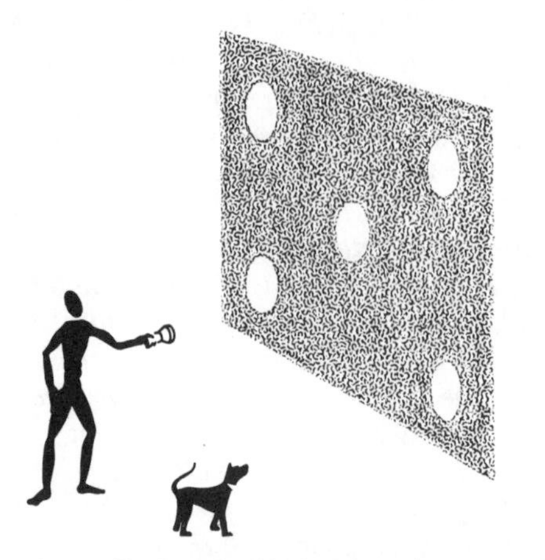

Figure 7.12. A flashlight beam is projected onto a wall. If the flashlight is moved about at constant distance from the wall, with the beam perpendicular to the wall, the projected figure demonstrates the nature of a congruence geometry.

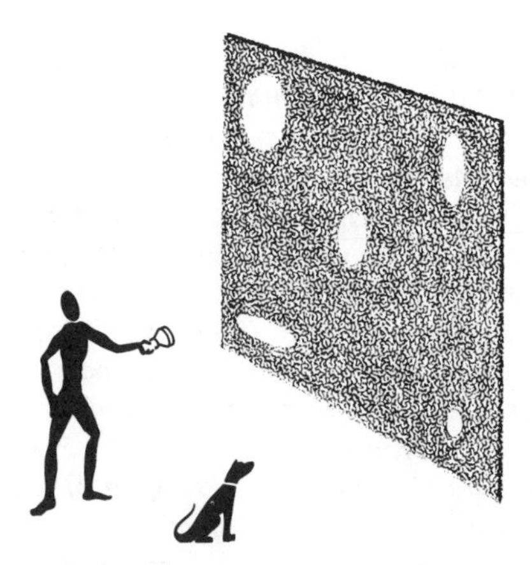

Figure 7.13. Altering the distance of the flashlight from the wall and varying the angle of incidence as the beam scans the wall demonstrates the nature of an affine geometry, which is more general than a congruence geometry.

shape and size and is "invariant to translations and rotations." This illustrates the properties of a uniform space.

With this simple apparatus we can also demonstrate some of the properties of an affine geometry *(see Figure 7.13). In this kind of geometry a point remains always a point and a straight line is always the shortest distance between two points Riemannian spaces have affine geometry. As the incident flashlight beam moves about on the wall, we are free to do two things: vary the distance of the flashlight from the wall and vary the angle of incidence; the first causes the projected figure to alter in size, and the second changes its shape. In this second experiment there is one thing we should not do: increase the angle of incidence to the extent that a point on the cover glass of the flashlight is projected as a line. This happens when a circle on the cover glass becomes a parabola on the wall. At a greater angle of incidence the circle becomes a hyperbola. When we are free to use all angles of incidence, so that circles are projected not only as ellipses but also as parabolas and hyperbolas, we have an example of a* projective geometry *(see Figure 7.14). An affine geometry is more general than a congruence geometry, and a projective geometry is more general still.*

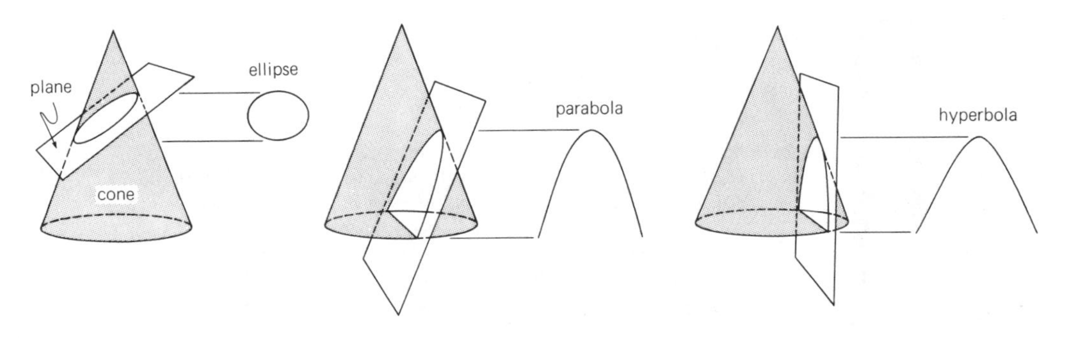

Figure 7.14. Ellipses, parabolas, and hyperbolas are conic sections *produced by the intersection of a plane and a circular cone.*

FURTHER READING

Abbott, E. A. (1838–1926), *Flatland.* Dover Publications, New York, 1952.

Bell, E. T. *Men of Mathematics: The Lives and Achievements of the Great Mathematicians from Zeno to Poincaré.* Simon and Schuster, New York, 1937.

Burger, D. *Sphereland: A Fantasy about Curved Spaces and an Expanding Universe.* Thomas Y. Crowell, Apollo ed., New York, 1969.

Clifford, W. *The Common Sense of the Exact Sciences.* 1885. Reprint. Dover Publications, New York, 1955.

Greenberg, M. J. *Euclidean and Non-Euclidean Geometries: Development and History.* W. H. Freeman, San Francisco, 1973.

Hall, R. S. *About Mathematics.* Prentice-Hall, Englewood Cliffs, N. J. 1973.

Jaggi, M. P. "The visionary ideas of Bernhard Riemann." *Physics Today,* December 1967.

Kline, M. *Mathematics and the Physical World.* Thomas Y. Crowell, New York, 1969.

Le Lionnais, F. *Great Currents of Mathematical Thought; Vol. I, Mathematics: Concepts and Developments. Vol. II, Mathematics in the Arts and Sciences.* Dover Publications, New York, 1871.

Manning, H. P. *Geometry of Four Dimensions.* Macmillan, London, 1914. Reprint. Dover Publications, New York, 1956.

Resnikoff, H. L., and Wells, R. O. *Mathematics in Civilization.* Holt, Rinehart and Winston, New York, 1973.

Rucker, R. B. *Geometry, Relativity and the Fourth Dimension.* Dover Publications, New York, 1977.

Stewart, I. "Gauss." *Scientific American,* July 1977.

Waerden, B. L. van der, *Science Awakening.* Noordhoff, Groningen, Holland, 1954.

SOURCES

Eddington, A. S. *The Expanding Universe.* Cambridge University Press, Cambridge, 1933. Reprint. University of Michigan Press, Ann Arbor Paperback, Ann Arbor, 1958.

Robertson, H. P. "Geometry as a branch of physics," in *Albert Einstein: Philosopher–Scientist,* ed. P. A. Schilpp. Library of Living Philosophers, Evanston, Ill., 1949.

GENERAL RELATIVITY

It is as if a wall which separated us from the truth has collapsed. Wider expanses and greater depths are now exposed to the searching eye of knowledge, regions of which we had not even a presentiment. It has brought us much nearer to grasping the plan that underlies all physical happening.
— Herman Weyl (1885–1955), *Space, Time, and Matter*

PRINCIPLE OF EQUIVALENCE

Gravitational and inertial forces produce effects that are indistinguishable: This principle of equivalence is a stepping-stone to the theory of general relativity. It brings into focus an important relation between motion and gravity and leads to a second stepping-stone, which is the realization that gravity and geometry have much in common. Then, in a flight of inspiration, we leap across a gulf that is spanned by non-Euclidean geometry; on the other side lies an entrancing land into which comparatively few explorers have penetrated. No one can claim to have a liberal education who has not glimpsed, even vaguely, the universe of general relativity.

Figure 8.1. "Let us then take a leap over a precipice so that we may contemplate Nature undisturbed" (Arthur Eddington, The Nature of the Physical World, *1928*).

We know that an inertial force, such as centrifugal force, always exists when a body is accelerated. We recall also that when a body is in free fall, and hence moves freely in space under the influence of gravity, it follows a path of such a kind that the sum of the inertial and gravitational forces is zero. With such items of knowledge, sufficient to land men on the Moon, we have made our first step toward the theory of general relativity.

We begin by considering an imaginary laboratory that is free to roam in space. It is equipped with apparatus and contains experimenters who conduct various investigations. The experimenters are unable to look outside to see what is happening because the laboratory has no windows.

The laboratory is out in space, far from the nearest stars, where gravity is virtually zero (see Figure 8.2). It moves freely, and because there is nothing to accelerate it, the inertial force is zero. This kind of motion, free and unaccelerated, is known as inertial motion, and the laboratory is said to be in an inertial state. The experimenters perform tests designed to detect acceleration and

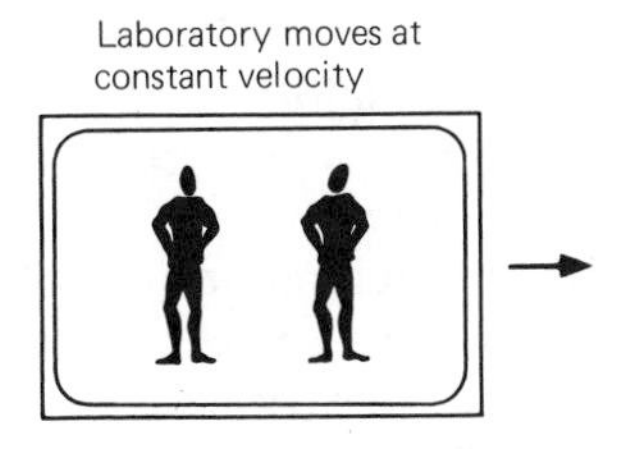

Figure 8.2. A windowless laboratory moves freely and at constant velocity in space. This is an inertial system, and the experimenters inside cannot determine its velocity.

announce that their laboratory is unaccelerated and is therefore in an inertial state. After a period of time the laboratory approaches a star, swings around the star in a curved orbit, and then moves away (see Figure 8.3). While this is happening the experimenters, unaware of the presence of a nearby star, continue to make measurements and continue to announce that the laboratory is unaccelerated and in an inertial state. This is because the laboratory follows a free-fall orbit of a kind such that the inertial force resulting from its acceleration exactly cancels the gravitational force of the star. In a windowless laboratory the principle of equivalence is a postulate of impotence which asserts that it is impossible to distinguish between gravitational and inertial forces.

Our inability to distinguish between inertial and gravitational forces is also shown in the following way (see Figure 8.4). We suppose the laboratory is accelerated by an applied force such as the thrust of a rocket engine. All objects inside the laboratory now experience an inertial force owing to the acceleration. With a piece of apparatus, such as a pendulum, the experimenters are able to measure the acceleration. Let the thrust of the rocket engine be adjusted so that the acceleration equals g at the Earth's surface; this means that the velocity increases 9.8 meters a second every second. The forces now acting on the experimenters and their apparatus are the same in magnitude as when the laboratory is at rest on the surface of the Earth. The principle of equivalence states that the experimenters cannot,

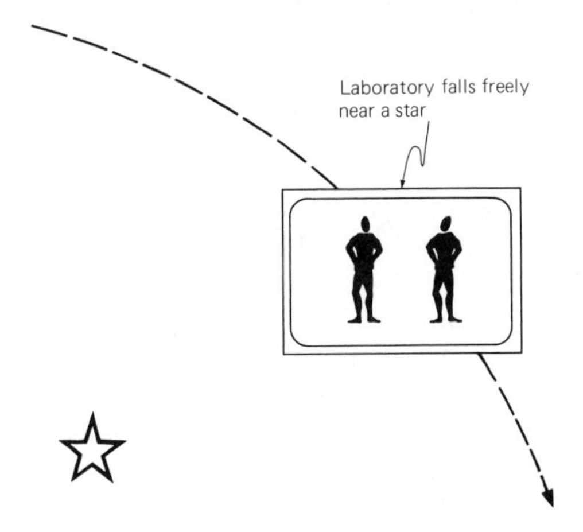

Figure 8.3. The laboratory now falls freely in the vicinity of a star. The inertial force resulting from acceleration cancels the effect of gravity, and the experimenters think that they still have inertial motion and that they are still moving at their previous constant velocity.

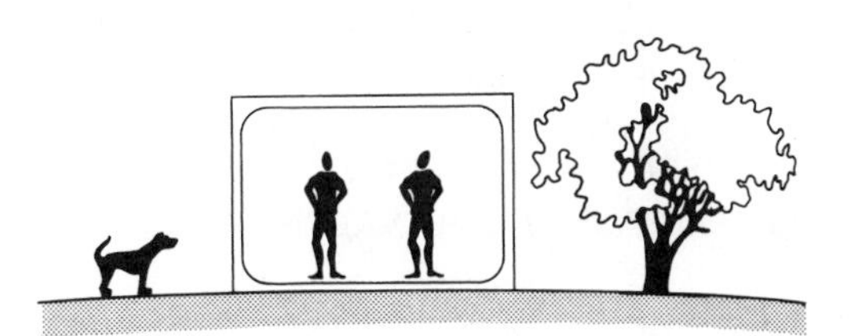

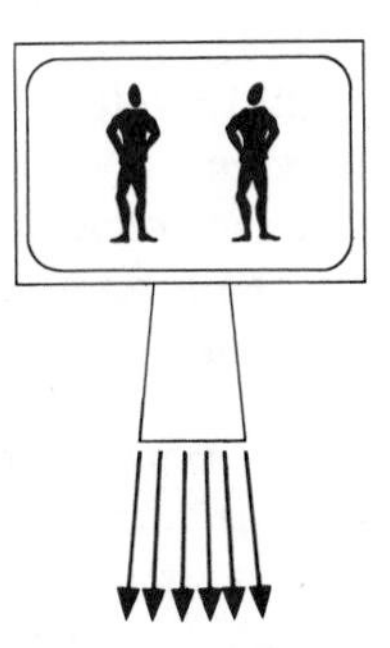

Figure 8.4. Objects in a laboratory resting on the Earth's surface are subject to a gravitational force. Objects in an accelerated laboratory out in space are subject to an inertial force. The experimenters inside the laboratories cannot distinguish between the two forces.

with experiments of any kind, determine whether the laboratory is accelerating in space or is at rest on the Earth's surface. In one case the forces are inertial and in the other case gravitational, and the experimenters are unable to distinguish between the two.

According to special relativity the laws of nature and the equations of physics are the same in all inertial systems (that is, in systems that are not accelerated). Consequently, in laboratories moving at arbitrary but constant velocities, all experimenters obtain identical results when performing the same tests. Yet we have just seen that these experimenters are not able to tell whether their laboratories are inertial or free falling. Therefore, the laws of nature are the same in inertial and free-falling laboratories, and all experimenters are able to use special relativity to explain the results of their experiments. That special relativity can be used in windowless laboratories in free fall indicates an amazing state of affairs in nature.

CLOSER LOOK AT THE PRINCIPLE OF EQUIVALENCE

The principle of equivalence can be broken down into what Robert Dicke at Princeton University has described as the weak and strong principles of equivalence. The weak principle can be traced back to the Middle Ages; it lies at the heart of Newtonian theory, and we shall therefore refer to it as the *Newtonian principle of equivalence*. The strong principle was introduced by Einstein in 1911, and we shall refer to it as the *Einstein principle of equivalence*.

THE NEWTONIAN PRINCIPLE OF EQUIVALENCE. The Newtonian form of the principle of equivalence states that all bodies fall freely in the same way, no matter what their mass. We see this idea emerging in the late Middle Ages in the work of the scholars of Merton College and the University of Paris, and in the demonstration by Simon Stevinus that bodies of unequal weight, when dropped, tend to reach the ground in equal time. Galileo rolled balls of different weights down an inclined plane and observed that they accelerate in the same way. Newton experimented with pendulums of equal lengths and different weights and confirmed that they swing with the same period.

For a body in free fall, Newton's equation of motion is

$$\text{mass} \times \text{acceleration} = \text{gravitational force} \quad (8.1)$$

where the mass is that of the accelerated body. The gravitational force acting on the body is proportional to its mass, and we have

$$\text{mass} \times \text{gravity} = \text{gravitational force}$$

where gravity (or gravitational field) is the force acting on a unit of mass. The mass of the moving body appears on both sides of the equation of motion (8.1) and can therefore can be canceled, giving

$$\text{acceleration} = \text{gravity} \quad (8.2)$$

This is the Newtonian equation of motion for a body in free fall and is independent of the mass of the body. It explains why bodies of different masses accelerate in the same way and why, when they commence at the same place with the same velocity, they follow identical orbits. The fact that an acceleration of 9.8 meters a second every second equals g, where g is the gravity at the Earth's surface, is an illustration of equation (8.2).

These thoughts can be expressed in a different fashion. The mass of a body in motion acts in two ways: First, it has an inertial property; and second, it responds to gravity. We may say that the body has an *inertial mass* and a *gravitational mass*. When the body is accelerated we are given

$$\text{inertial mass} \times \text{acceleration} = \text{gravitational force}$$

and the gravitational force acting on the body is given by

$$\text{gravitational mass} \times \text{gravity} = \text{gravitational force}$$

These last two equations show that mass is both inertial and gravitational. By equating the two masses,

$$\text{inertial mass} = \text{gravitational mass} \qquad (8.3)$$

we obtain the Newtonian equation of motion for all bodies in free fall. The Newtonian principle of equivalence, enshrined in equation (8.2), which states that all free-fall motion is independent of mass, or alternatively, that the sum of the inertial and gravitational forces in free fall is zero, can be interpreted to mean that the inertial and gravitational masses are equal. Equation (8.1) thereby becomes equation (8.2) because of equation (8.3). The equality of inertial and gravitational masses is an expression of the fact that inertial and gravitational forces cancel exactly in a free-falling laboratory, and that the inertial force in an accelerated laboratory is indistinguishable from the gravitational force in a laboratory resting on the surface of a planet.

Astronauts in a spaceship orbiting the Earth are in free fall and follow the same orbit as their spaceship. All objects inside their spaceship are also in free fall and follow the same orbit. As an example, we consider two spheres, one of gold and the other of aluminum (see Figure 8.5). When placed side by side in the spaceship they float together because they are following similar free-fall orbits. If their diameters are equal, the gold sphere has a mass slightly more than seven times that of the aluminum sphere, and yet both fall freely in exactly the same way.

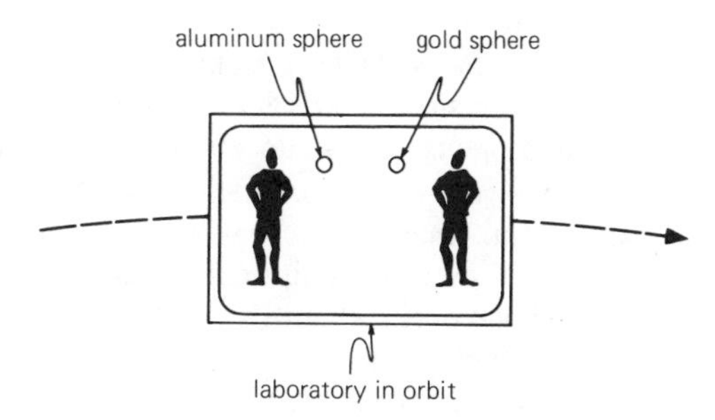

Figure 8.5. Two spheres float side by side in a space vehicle orbiting the Earth. One sphere is made of gold and the other is made of aluminum, yet they fall freely about the Earth in similar orbits. This is because gravitational mass is always equal to inertial mass.

The Newtonian principle of equivalence comes as no surprise to those of us accustomed to using the equation of motion in a gravitational field. Yet despite our contempt bred of long familiarity, the principle is nonetheless still surprising. The nucleus of an atom has a positive charge that produces an intense electric field inside the atom. An electric field has energy, and energy has mass, and therefore electric fields have mass. In the case of gold the subatomic electric fields have a mass that is approximately 0.5 percent of the total mass. Put differently, 1 part in 200 of the weight of gold is due solely to the weight of subatomic electric fields. The nucleus of an atom of aluminum has a smaller positive charge, and the weight of the subatomic electric fields of aluminum is accordingly a much smaller fraction of the total weight. Yet we find that gold and aluminum bodies behave in exactly the same way in free fall. Electric fields are therefore subject to the same inertial and gravitational forces as all other forms of mass; in other words, the inertial and gravitational masses of electric fields are equal. Consider also two bodies of the same material floating side by side in free fall. One of the bodies is heated while the other remains cool, and we observe that they both continue to float side by side. Heat energy has mass, yet the hot and cool bodies have similar free-fall orbits, and the mass associated with heat energy is therefore subject to the same gravitational and inertial forces as all other forms of mass. Considerations of this kind lead to the general conclusion that all forms of energy have an associated mass that obeys the Newtonian principle of equivalence.

Newton experimented with pendulums and found that their gravitational and inertial masses are equal to within 1 part in 10^3 (see Figure 8.6). Friedrich Bessel, who was the first to succeed in measuring the parallax of a star, showed in 1827 that the gravitational and inertial masses of pendulums

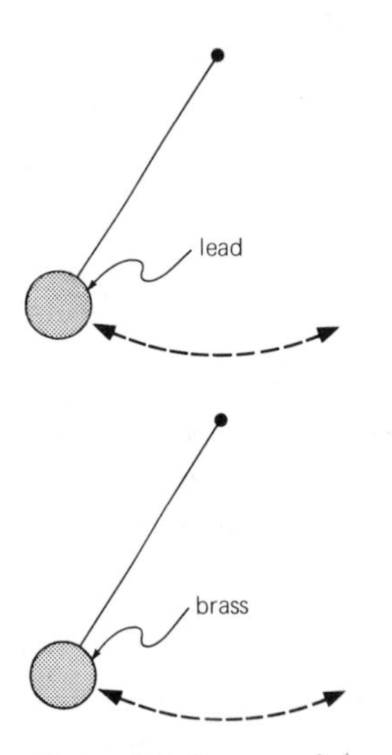

Figure 8.6. Two pendulums of the same length, made of different materials, have equal periods. Why? Because gravitational and inertial masses are equal.

are equal to within 2 parts in 10^5. The Hungarian nobleman Roland von Eötvös greatly improved on this result in 1890 by using a torsion balance, and found for different substances – such as snakewood and platinum – that their gravitational and inertial masses are equal to a few parts in a billion. In more recent years the Eötvös experiment has been greatly refined, and it has been shown that the gravitational and inertial masses of gold and aluminum are equal to within a few parts in a trillion.

THE EINSTEIN PRINCIPLE OF EQUIVALENCE. The Einstein form of the principle of equivalence states that inertial and free-falling systems are entirely equivalent. In inertial and free-falling laboratories there are no experiments of any kind that are capable of distinguishing between inertial and free-falling motion. Such experiments might involve light beams, nongravitational forces, electric and magnetic fields, or anything else. The Einstein principle of equivalence declares that the acceleration of a free-falling laboratory cancels completely the effect of gravity, not only dynamically, as in the weaker form of the principle, but also in all conceivable experiments in every branch of science. Special relativity, and not just Newtonian mechanics, can therefore be used in free-falling systems as well as in inertial systems, and this is the essence of the principle of equivalence in its strong form.

GEOMETRY AND GRAVITY

CURVED SURFACES HAVE PROPERTIES ANALOGOUS TO THOSE OF GRAVITY. If the principle of equivalence is the first stepping-stone to general relativity, the second is the realization that geometry and gravity have much in common. To illustrate the similarity of geometric curvature and gravity, let us consider a large rubber sheet that initially is stretched flat. The curvature is everywhere zero and the surface is like the flat spacetime that exists far from a star where gravity is essentially zero, and where a laboratory moves inertially at constant velocity. If we take a small ball, such as a ball bearing, and roll it on the surface of the sheet, it will also move inertially at constant velocity. The ball bearing moves in a straight line at constant speed just like the laboratory that is far from a star. In the center of the sheet we now place a heavy ball that produces a large depression (see Figure 8.7). Far from the central body the surface is almost flat and its curvature is very small. This is like the almost-flat spacetime that exists far from a star. Close to the central body the curvature of the surface is large. A ball bearing moving on the surface accelerates in a way that is analogous to the acceleration of a laboratory in the vicinity of a star. By altering the initial speed of the ball bearing we can make it describe elliptical, parabolic, and hyperbolic orbits about the central body that is producing the curvature. The ball bearing moves in the same way as the free-falling laboratory, and the curvature of the surface mimics the properties of gravity. Where the curvature is zero, gravity vanishes, and the ball bearing and the laboratory move inertially; where the curvature

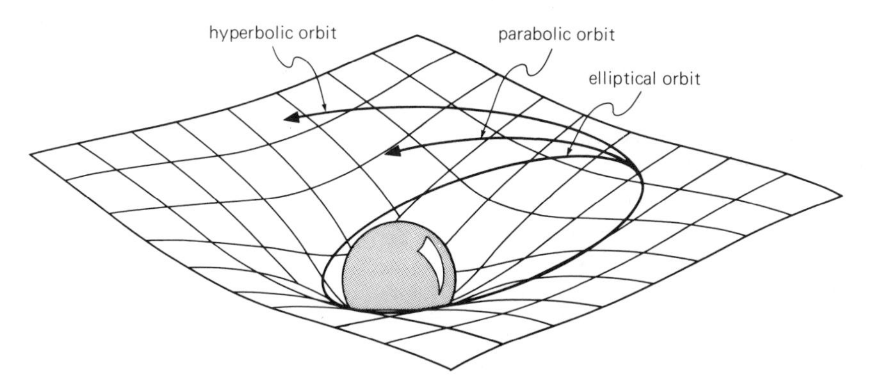

Figure 8.7. A stretched rubber sheet is depressed by a heavy spherical body. The curvature of the sheet mimics the effect of gravity, and a ball bearing follows an orbit that is either elliptical, parabolic, or hyperbolic.

is not zero, gravity exists, and the ball bearing and the laboratory follow similar orbits.

Bodies moving inertially have constant velocities and straight worldlines in the flat spacetime of special relativity. Bodies in free fall have acceleration and have therefore curved worldlines in the flat spacetime of special relativity. An experimenter in a free-falling laboratory does not know that the laboratory is accelerating but thinks that it has a straight worldline. By peeping through a hole cut in the wall of the laboratory, however, the experimenter sees a nearby star. This discovery reveals that the laboratory is actually accelerating and has a curved worldline in the spacetime of special relativity that is used inside the laboratory. The experimenter might wonder whether there is perhaps some grand theory of gravity that reduces locally to special relativity in a state of free fall. What would this grand theory be? If we think about it for some time, we might stumble on the ideas that free-falling bodies have straight worldlines, just as inertial bodies do when there is no gravity. This would mean abandoning flat spacetime and finding a way in which gravity alters the geometry of spacetime so that free-falling bodies have straight worldlines – a theory that would take us from the old picture of curved worldlines in flat spacetime to a new picture of straight worldlines in curved spacetime. And if we were like Einstein, we might eventually succeed in discovering the grand theory that governs the geometry of spacetime, and we would then have accomplished the most imaginative feat in the history of science.

TIDAL FORCES

TIDAL FORCES ARE COMPLICATIONS THAT NONETHELESS HELP US TO UNDERSTAND GRAVITY. Before proceeding with our main theme – the theory of general relativity – we must turn aside and consider tidal forces. To illustrate the principle of equivalence we have used imaginary laboratories. Thought experiments of this kind are often used, usually consisting of idealizations that enable us to understand basic principles. In special relativity, for example, we perform thought experiments in systems having inertial motion; yet this kind of motion is an idealization because it rarely exists in nature. However far we escape into the depths of intergalactic space, the gravitational pull of the nearest galaxies will produce acceleration and destroy an assumed state of inertial motion. There are certain places, of course, hardly larger than points, where the pull and counterpull of nearby galaxies just balance and the gravitational force is zero, and at these places it can be said that motion is truly inertial. But bodies in motion occupy these places only momentarily, and their motion, like that of all other bodies, is generally noninertial. Inertial motion, although it rarely exists, is nevertheless a very useful idealization.

The principle of equivalence, as we have presented it, is also an idealization. When we perform experiments in imaginary laboratories and other spatial regions of finite volume, the principle is strictly true only when the gravitational field is the same everywhere in the region of finite volume. This actually was the way in which Einstein defined the equivalence principle: in terms of a uniform gravitational field that does not vary from place to place. But a uniform gravitational field is an idealization, because gravity always varies in space and is never everywhere the same. A spherical body produces a gravitational field that decreases as the inverse square of distance; consequently, finite regions of space in which the field is constant cannot exist.

Inside a laboratory resting on the Earth's surface gravity is not everywhere the same: When an object is raised 3 meters above the floor its weight is reduced by 1 part in a million. A variation of weight of this kind does not exist in a laboratory in space that is accelerated at a constant rate by a rocket engine, for all parts of the interior are equally accelerated, and the inertial force acting on any object is the same wherever that object is placed in the laboratory. It follows, despite what has been previously said, that there are experiments capable of distinguishing between stationary laboratories in gravitational fields and accelerated laboratories out in space. All we have to do is look for a variation in the force acting on an object as its position is changed; and the larger the laboratory, the easier it is to detect such a variation.

Consider a body moving freely under the influence of gravity (see Figure 8.8). Its many atoms do not occupy the same position, and therefore, because gravity is not everywhere the same, they are pulled by slightly different gravitational forces. The atoms, however, are all stuck together and follow the same orbit. If they were not stuck together, but could move freely and independently, they would all follow slightly different orbits. The force that tends to distort and even to tear a body apart, because each atom tries to follow its own free-fall orbit, is known as the *tidal force*. The center of mass of a spherical body is the only point actually in a state of free fall, and at this point the tidal force vanishes. All other points of the body are not in perfect free fall; at these points there exists a detectable tidal force. What happens is obvious: All atoms are constrained to follow the center of mass, and the inertial force owing to this motion does not exactly cancel the gravitational force at positions other than the center of mass. The tidal force that tends to tear a body apart is the result of the variation of gravity, and at any point in a body it is approximately equal to the gravitational force at that point minus the gravitational force at the center of mass.

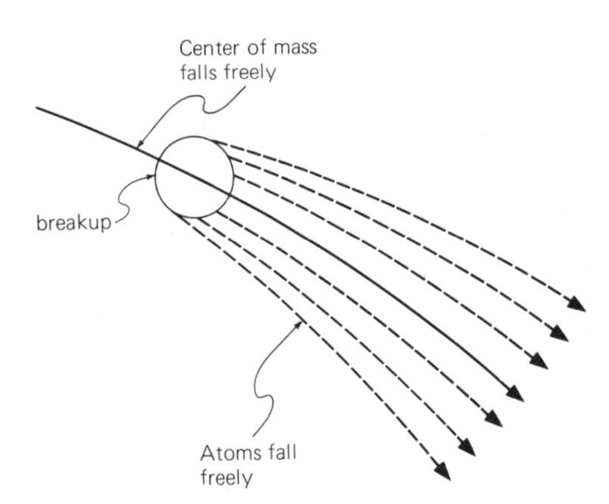

Figure 8.8. A spherical body falls freely under the influence of gravity, but only its center of mass follows a perfect free-fall trajectory. All its atoms are stuck together and are compelled to follow the center of mass. If the atoms were free, and not stuck together, they would follow independent and slightly different trajectories because gravity is not exactly the same everywhere within the body.

Let us return to our imaginary laboratory in free fall out in space. By searching for and detecting the presence of a tidal force, the experimenters can tell whether their labora-

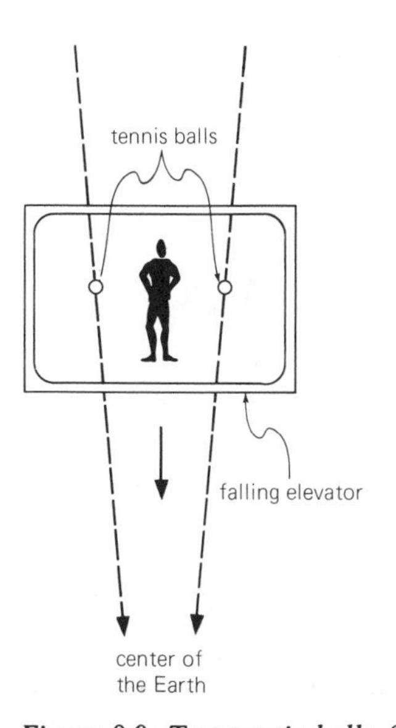

Figure 8.9. Two tennis balls floating in a free-falling elevator. As the elevator plunges toward the center of the Earth the tennis balls move toward each other. They actually accelerate toward each other, and to an observer in the elevator it seems that a force is acting on the balls; this force, a result of the nonuniformity of gravity, is known as the tidal force.

tory is in an inertial state or is free falling under the influence of gravity. Our imaginary laboratory could be an elevator that falls freely in a vertical shaft penetrating deep into the Earth. We can place in this elevator two tennis balls that float side by side at the same height above the floor (see Figure 8.9). As the elevator plunges toward the center of the Earth, we notice that the tennis balls accelerate slowly toward each other. Both are in free fall, each traveling toward the center of the Earth, and naturally their separating distance decreases. By observing the way in which floating objects move relative to each other, as in this case, we are able to determine whether a laboratory is inertial or free falling.

Tidal forces are normally small and can often be neglected. Scientists rarely pay attention to the tidal forces within their laboratories on Earth. A principle, however, is a principle, and if equivalence is fundamental there should be some way of stating it without the bother of tidal forces. The time has come to let the reader into a secret: All our talk about experiments in imaginary laboratories was for the sake of illustrating equivalence in a vivid and easily understandable way. Equivalence of results obtained by experimenters in inertial and free-falling laboratories is actually not of fundamental theoretical importance. What is important is to have equivalence as a property of the equations of physics.

The principle of equivalence cannot apply everywhere in a free-falling laboratory simply because all parts are not in perfect free fall. Gravitational and inertial forces are actually indistinguishable only in free-falling regions of extremely small volume. We are unable therefore to perform experiments with apparatus of finite size to establish with utmost precision the truth of the equivalence principle. But this is not terribly important. The equation of motion and other equations of physics are statements about what happens in regions of extremely small size. Let dx, dy, dz indicate the size of an extremely small region, measured as intervals in the x, y, z directions, as shown in Figure 8.10. (Here dx means difference or differential in x, and dy and dz are defined similarly.) The equations of physics are cast into a form that tells us what happens when dx, dy, dz, and other differences, such as dt in time, all shrink to zero. This is why they are called differential equations. The basic laws are expressed in terms of differential equations that refer to what happens in infinitesimal regions. The principle of equivalence now means, in its weak form, that the inertial and gravitational forces cancel each other in a free-falling state, and in its strong form, that the equations of physics do not distinguish between inertial and free-falling states.

Let us consider a surface, such as that of a vase, having a curvature K that varies from place to place. On the surface we draw a figure, such as a circle, and notice that inside this figure the curvature K varies by small amounts. This closed region is analogous to a laboratory in free fall, and the

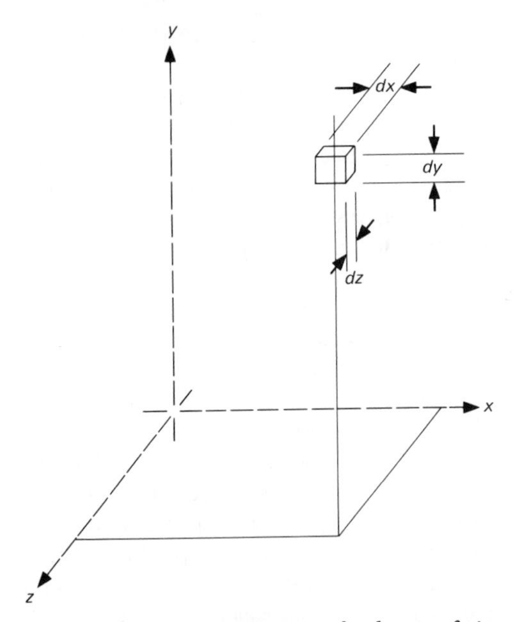

Figure 8.10. *A small element of volume, of size dx, dy, and dz. The equations of physics refer to what happens in such small elements when dx, dy, and dz shrink to zero. The principle of equivalence, free of the bother of tidal forces, states that the equations of physics do not distinguish between inertial and free-falling elements of infinitesimal volume, and the equations are the same in either case.*

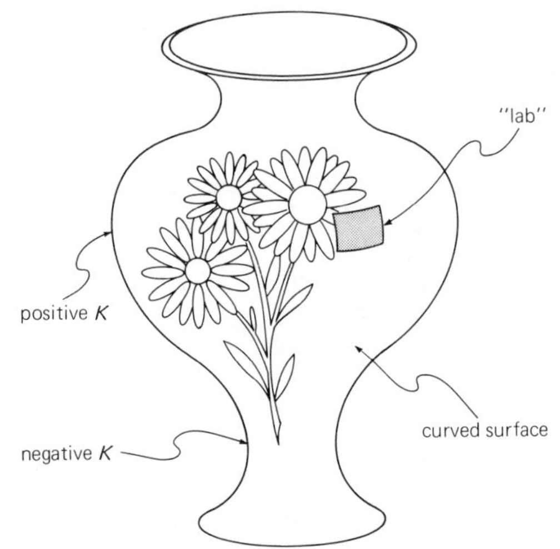

Figure 8.11. *A thin and flexible piece of material, labeled "lab," fits snugly against a curved surface. The variation of curvature within the lab is analogous to the variation of gravity (the tidal force) within a free-falling laboratory.*

variation of K is similar to the tidal force that varies from place to place inside the laboratory. We next take a small and flexible piece of thin material that has been cut into the shape of a square. This piece of material, labeled "lab" in Figure 8.11, always fits snugly on the surface. We notice that the curvature K not only varies from place to place in the lab, but also varies in time as we move the lab around on the surface. This again is analogous to what happens in a free-falling laboratory: The tidal forces vary from place to place within the laboratory, and they also vary with time. The principle of equivalence applies only to infinitesimal regions (in laboratories of extremely small size); similarly, Euclidean geometry can be used only in extremely small regions on a curved surface (in labs of extremely small size). Thus tidal forces, which at first seemed to be a bothersome complication, do not detract from the beauty of the principle of equivalence; they in fact help us to understand the relation between gravity and geometry. On the one hand we have tidal forces, and on the other hand we have a variation of curvature, both in regions of finite size; whereas in extremely small regions the tidal forces vanish and we are able to use Euclidean geometry.

Although experiments performed in laboratories do not strictly obey the equivalence principle under all circumstances, we shall follow the usual practice and continue to use this picturesque and convenient way of performing thought experiments.

THEORY OF GENERAL RELATIVITY

EINSTEIN'S EQUATION. Einstein's theory of general relativity was developed in the early years of this century and reached its final form in 1916. The Newtonian universe with its Euclidean geometry and gravitational forces was at last overthrown and replaced with a universe consisting of spacetime of varying curvature. The curved orbits of freely moving bodies in the Newtonian universe became straight orbits in the curved spacetime of the Einstein universe.

We should mention that straight orbits are referred to as *geodesics*. They are straight within the local geometry, but to an observer elsewhere, whose local geometry is different, they appear curved. A geodesic in flat spacetime (as in special relativity) is the familiar straight line, and a particle following such a geodesic is unaccelerated and has a straight worldline. A geodesic on the surface of a sphere is a great circle. In general relativity all free-fall motion including inertial motion – follows geodesic paths.

The Einstein equation of general relativity states that the curvature of spacetime is

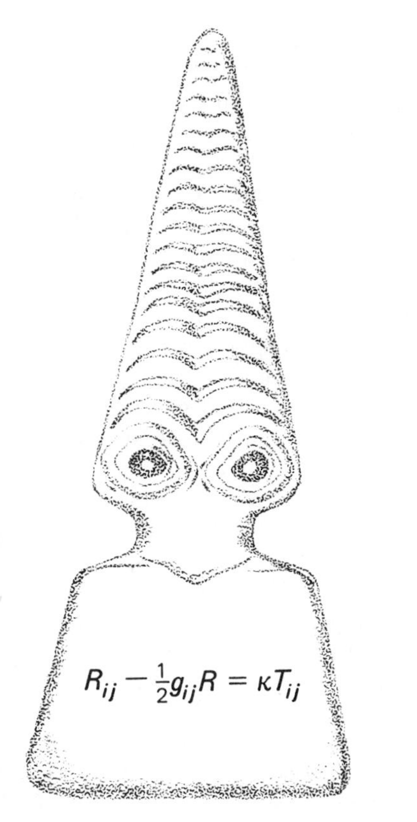

Figure 8.12. The Eye Goddess (Syria, 2800 B.C.) displaying the Einstein equation as a modern addition. The Einstein equation is explained in the text in a qualitative manner.

influenced by matter; or the strain of spacetime is related to the stress produced by matter. Expressed in the simplest manner possible, the equation is

curvature of spacetime = constant × matter

The "matter" on the right side includes all forms of energy that have mass. This is a mathematical equation that breaks down into 10 separate equations, and so far very few exact solutions have been discovered. We can interpret the Einstein equation to mean that curvature is equivalent to gravity. When the curvature is only slight and gravity is therefore weak, as in the Solar System, the Einstein equation simplifies and becomes the Newtonian law of gravity and the Newtonian law of motion. General relativity by itself does not tell us the value of the "constant" that couples curvature and matter; but by making a comparison with Newtonian theory it is found to contain the universal gravitational constant G, which must be determined by measurements.

The Einstein equation does not say that curvature and matter are the same. They are, of course, distinctly different. Instead, it shows how curvature and matter influence each other. The Riemannian curvature of four-dimensional spacetime, discussed in Chapter 7, has 20 components, and therefore the Einstein equation, which has only 10 separate equations built into it, cannot determine all components of the curvature at each point of spacetime. This is just as well, because in the empty space around the Sun there is no matter, and yet a gravitational field exists that controls the motions of the planets. In the absence of matter, spacetime is therefore not necessarily flat, as in special relativity; it may have curvature determined by the presence of distant bodies such as the Sun. All components of the Riemannian curvature are determined when we take into account distant as well as local matter. A rubber sheet, stretched and initially flat, illustrates what happens in spacetime (see Figure 8.13). A heavy ball placed on the sheet produces a depression and the curvature of the sheet diminishes

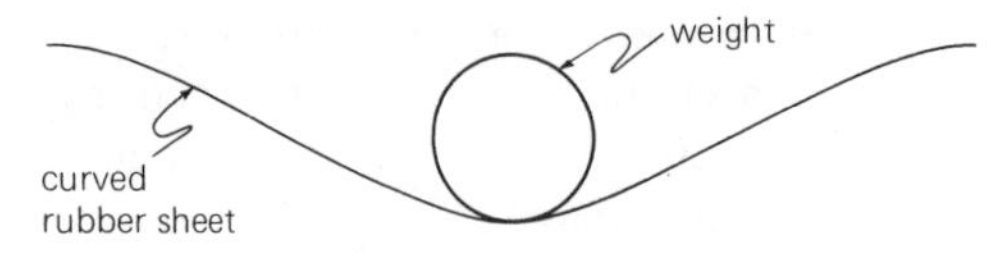

Figure 8.13. Spacetime is curved not only by local matter, but also by matter at a distance, as shown by the curved rubber sheet.

with distance from the ball. This demonstrates how curvature is produced by distant as well as local matter.

Gravity, acting in a mysterious way across a vacuum, has vanished and has been replaced with the geometrical curvature of physical spacetime. Yet we still persist in using the old Newtonian language of gravity. One reason is that the language of gravity is more vivid and familiar than the language of differential geometry; and another is that it helps us to distinguish gravity from other forces, such as electromagnetic, strong, and weak forces, that are not fully geometrized in general relativity.

Gravity in the Newtonian universe obeys what is called the *principle of superposition.* This means that the gravitational force at any point is the sum of the forces produced by all bodies, and the force produced by each body is unaffected by the presence of the other bodies. If one body by itself produces a force F_1, and another body by itself produces a force F_2, the two bodies existing at the same time in their same places produce a combined force of $F_1 + F_2$. The force each produces is independent of the presence of the other. But in the Einstein equation the geometric curvature produced jointly by two or more bodies is not exactly the sum of the curvatures produced by the bodies when taken separately. Hence the Einstein equation does not obey the principle of superposition, and we cannot add together the curvatures of simple configurations to find the curvature of a complex configuration. This, incidentally, is also true of a stretched rubber sheet that is depressed at several places by different weights: The total depression at any point is not exactly the sum of the depressions taken separately. The amazing feature of general relativity is that not only do bodies curve spacetime, but the separate curvatures they produce also act on each other, and this self-interaction of spacetime is what is so important about general relativity. This self-interaction exists because the curvature of spacetime is itself a form of energy, which produces its own gravitational field, and is hence the source of further curvature. This explains why gravity in the vicinity of a body such as the Sun does not obey precisely the inverse-square law: The energy that resides in the spacetime curvature outside the body makes its own contribution to the distant gravitational field. Thus curvature generates curvature, whereas in the Newtonian universe gravity does not generate gravity. In the case of comparatively weak gravitational fields, such as that of the Sun, the Einstein equation simplifies to the Newtonian equations with the addition of very small corrections. Tests of general relativity are usually difficult because regions of strong gravity are not readily available to us.

Gravity in the Newtonian universe propagates at infinite speed. Thus a body, such as a star, creates a gravitational field that is instantaneously everywhere. But in the Einstein universe, gravity – or rather spacetime curvature – propagates at the speed of light. The Einstein equation is in fact a dynamic wave equation that generates and propagates the curved deformations of spacetime. Two stars in orbit about each other have a combined gravitational field that varies periodically with time (see Figure 8.14). But gravity is spacetime curvature that contains energy; hence, if gravity varies periodically, the spacetime curvature varies periodically, and energy is continually redistributed in the surrounding region of spacetime. This cyclic bending and warping of spacetime streams away in all directions at the speed of light, and energy is lost from the orbiting stars in the form of gravitational waves. This loss of energy means that the two stars are slowly spiraling toward each other on a time scale of billions and even trillions of years. Energy is lost

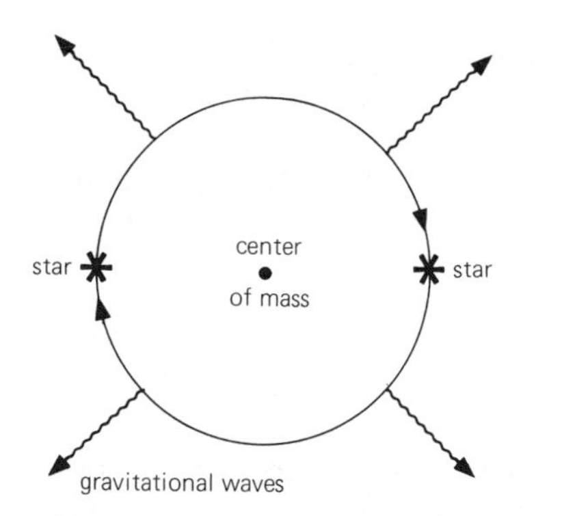

Figure 8.14. Two stars in orbit about each other radiate gravitational waves. The waves are ripples in spacetime and are the result of the periodic variation of the curvature of spacetime. Spacetime curvature contains energy, and hence the outward-traveling ripples, or gravitational waves, carry energy away at the speed of light. As a result, the stars slowly spiral toward each other.

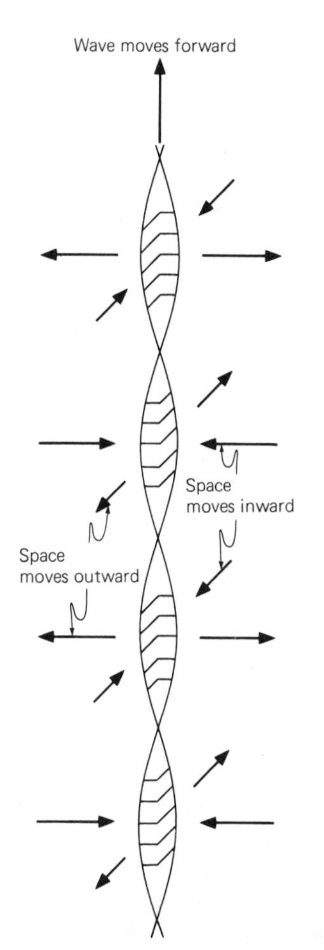

Figure 8.15. A gravitational wave is a traveling ripple of spacetime curvature. This picture illustrates the way space is periodically changed by the passage of a gravitational wave.

very slowly and therefore gravitational waves are extremely difficult to detect. Joseph Weber of Maryland and several other physicists have attempted for many years the arduous task of detecting gravitational waves generated by events of a catastrophic nature in the Galaxy. Such waves, because of their extreme weakness, have so far not been observed, at least not in a manner that is beyond doubt. The problem is not just to observe the time-variation of gravity – the spring and neap tides reveal that – but to detect the energy in spacetime that travels as waves at the speed of light (see Figure 8.15).

LABORATORIES WITH WINDOWS. We have seen that in free fall the effect of gravity is abolished locally and that all experiments performed in free-falling laboratories give results that are the same as those obtained in the flat spacetime of special relativity. (Notice that we are not bothering with tidal forces or, equivalently, the curvature variations within the laboratory.) If we use the Einstein equation in a state of free fall, we find that it reduces to the same equations that are used in special relativity. It obeys the principle of equivalence by abolishing local spacetime curvature in a free-falling system.

Let our experimenters now cut holes in the walls of their laboratory and observe all that is happening outside. What they see is in accordance, not with special relativity, which applies only locally inside the laboratory, but with general relativity, which applies globally. Everything of a dynamic nature observed from their laboratory with windows is explained with the Einstein equation. The Einstein equation is covariant (the same) for all free-falling observers and is hence covariant for a class of observers more general than the inertial class of observers in special relativity.

The Sun moves freely in the Galaxy and the Earth moves freely about the Sun, yet we ourselves on the surface of the Earth are not in free fall within the gravitational field of the Earth. This does not mean that we on Earth cannot use the Einstein equation. Some mathematicians and physicists use it all the time. What they do is quite simple: They imagine themselves in a free-falling system and are then able to calculate with the Einstein equation the corrections that must be used in a state that is not free falling, like that on Earth.

EINSTEIN'S QUEST FOR A UNIFIED THEORY. Einstein once said that the Riemannian geometry on the left side of the Einstein equation is like an elegant marble hall. On the right side is what might be called an outside yard into which is put almost anything we choose, subject to certain elementary constraints. It has been the custom, since the time when Einstein first formulated the theory, to place on the right side "matter" and other things that have an energy content. But Einstein was cautious on this aspect of the theory and at one time allowed for a more general interpretation of the meaning of "matter." General relativity by itself does not tell us what the fundamental nature of matter is. Einstein was therefore not entirely satisfied. He regarded the right side as incomplete, and in the later years of his life he said: "The right side is a formal condensation of all things whose comprehension in the sense of a field theory is still problematic. Not for a moment, of course, did I doubt that this formulation was merely a makeshift in order to give the general principle of relativity a preliminary expression. For it was essentially not anything more than a theory of the gravitational field, which was somewhat artificially isolated from a total field of as yet unknown structure."

Because Einstein viewed his great theory as incomplete, he sought for many years a way in which geometry could be equated more directly with the fundamental properties of matter. Maxwell had previously unified electricity and magnetism into electromagnetism, and Einstein sought to unify gravity with electromagnetism, but his quest for a unified theory never succeeded. Recently, eminent physicists have combined the electromagnetic and weak forces into an *electroweak* force. Grand unified theories have also been proposed that unify the electroweak and strong forces into a single *hyperweak* force. The hyperweak force ruled in the very early universe, when the energy concentration was extremely high, but in the present universe it operates in different ways that appear to us as three separate forces. The hyperweak force is still feebly active in its unified form, and as a result it is possible that all matter is decaying into radiation on a time scale of 10^{30} or so years.

TESTS OF GENERAL RELATIVITY

What is needed is a homely experiment which could be carried out in the basement with parts from an old sewing machine and an Ingersoll watch, with an old file of Popular Mechanics *standing by for reference.*
— Howard Robertson, in *Albert Einstein: Philosopher–Scientist* (1949)

STRENGTH OF GRAVITY. Various tests of general relativity have been successfully conducted and have yielded results in close agreement with the predictions of the theory. Such tests are usually not easy to perform, because, with the weak gravitational fields available to us, the difference between the Einstein and Newtonian theories is quite small.

Gravity is weak, generally speaking, when the escape speed from a body is small compared with the speed of light. We can say that

$$\text{strength of gravity} = \left(\frac{v}{c}\right)^2$$

where v is the escape speed and c is the speed of light. At the surface of the Sun the strength of gravity is 4.24×10^{-6}, or roughly 4 parts in a million. Hence gravity is weak not only at the surface of the Sun, but also throughout the Solar System, and consequently the effects peculiar to general relativity are exceedingly small. When the strength of gravity approaches unity, as in the vicinity of black holes, gravity is strong and the full effects of general relativity become apparent.

Einstein proposed three famous tests of general relativity, which will now be discussed briefly.

DEFLECTION OF STARLIGHT. The first of Einstein's tests is the deflection of starlight in the Sun's gravitational field (see Figure 8.16). Light from a distant star, when it passes close to the Sun's disk, should be deflected through a small angle that is twice the amount predicted by the Newtonian theory of gravity. General relativity theory predicts a deflection of 1.75 seconds of arc for starlight grazing the edge of the Sun's disk – a angle about equal to that subtended by a small finger at a distance of 1 kilometer. The deflection in radians is equal to twice the strength of gravity, where 1 radian is 57.3 degrees of arc. There are 206,265 seconds of arc in 1 radian, and if this is multiplied by twice the Sun's strength of gravity, we obtain 1.75 seconds of arc.

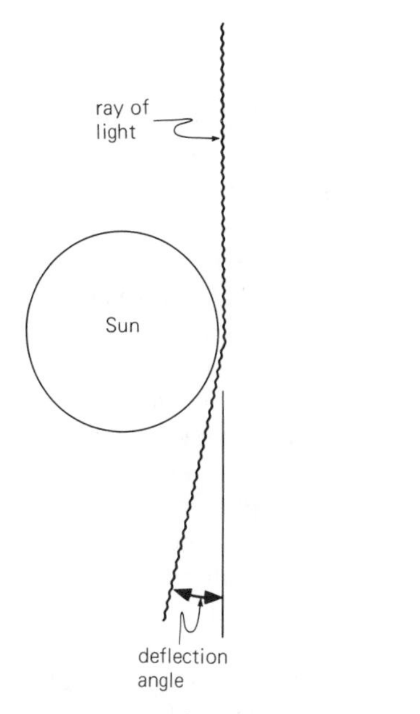

Figure 8.16. The deflection of a lightray, grazing the Sun's disk, is 1.75 seconds of arc.

The light-deflection test was proposed by Einstein in 1916 and was not performed until after World War I, in 1919. Stars are seen close to the Sun only for a short time during a solar eclipse, and two eclipse expeditions were organized at the suggestion of Frank Dyson, the astronomer royal of Britain, and of Arthur Eddington, the first person outside Germany to champion the theory of general relativity. The expeditions observed the eclipse from Sobral in Brazil and from the island of Principe off the coast of West Africa. The Sobral expedition observed a displacement of 1.98 seconds of arc and the expedition to Principe, one of 1.61 seconds of arc. Observations of this kind are difficult to make in the short time available during an eclipse, and the results were considered to be in reasonable agreement with the Einstein prediction. Both measurements were significantly greater than the Newtonian prediction of 0.87 seconds of arc.

More precise measurements have since been made in other eclipse expeditions. The most precise results, which fully confirm the Einstein prediction, have been made in recent years by radioastronomers. Radio waves from distant radio sources are deflected in the same way as starlight, and radioastronomy has the advantage that observations close to the Sun can be made continually, without waiting for an occasional eclipse of short duration.

PRECESSION OF PLANETARY ORBITS. The second test is the precession – or slow drift – of a planetary orbit such as that of Mercury (see Figure 8.17). Whenever gravity fails to obey exactly the inverse-square law, a precession of an elliptical orbit always occurs. The gravitational field of the Sun, according to general relativity theory, is not exactly of the inverse-square law form

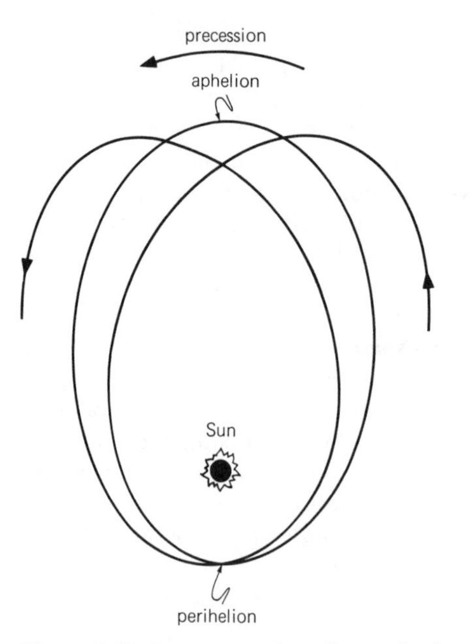

Figure 8.17. Precession of an elliptical orbit of a planet about the Sun.

except at very large distances from the Sun. The precession owing to the general relativity effect is therefore most obvious in the case of Mercury, the planet closest to the Sun. For planetary orbits that are almost circular the precession is independent of the eccentricity of the orbit. The amount of the precession, measured in radians per revolution, is equal to the strength of gravity. In this case, however, the strength of gravity is determined at the planetary orbit and not at the surface of the Sun. If u is the orbital speed of the planet, then at the radius of the planetary orbit, we have

$$\text{strength of gravity} = 2\left(\frac{u}{c}\right)^2$$

The precession of planetary orbits in the Solar System resulting from this effect is quite small. For Mercury, which has an orbital speed of 48 kilometers per second and an orbital period of 80 days, the precession is equal to 43 seconds of arc per century. This amounts to a complete revolution once every 3 million years. Perturbations by other planets, which also cause deviations from the inverse-square law of the Sun's gravity, produce a much larger precession, which must first be subtracted from the observed precession of Mercury's orbit. What remains after this subtraction is in close agreement with the predicted precession of 43 seconds of arc per century.

GRAVITATIONAL REDSHIFT. The third test is the gravitational redshift effect. Radiation escaping from a body like a star or a planet loses energy because of the pull of gravity. The energy of a lightray is proportional to its frequency, and as the energy decreases, the frequency also decreases. It follows that the wavelength of a lightray increases as it travels away from a gravitating body. A lightray falling on a gravitating body like a star or a planet gains energy because of the pull of gravity, and its wavelength decreases.

Let λ be the wavelength of the radiation emitted from the surface of a body, and λ_0 be the wavelength when the radiation has escaped to a great distance; the redshift is the fractional amount by which the wavelength has increased:

$$\text{redshift} = \frac{\lambda_0 - \lambda}{\lambda}$$

When gravity is weak, the redshift is equal to half the strength of gravity; it is therefore 2.12×10^{-6} for radiation escaping from the Sun and 7×10^{-10} for radiation escaping from the Earth. The redshift that occurs when radiation travels vertical distances of only tens of meters at the Earth's surface has been observed using methods developed in nuclear physics. Slow-decaying nuclei of certain atoms embedded in crystals emit short-wavelength radiation of sharply defined frequency (this is known as the Mössbauer effect), and the shift in frequency owing to a change in height at the Earth's surface has been successfully measured and found to be in agreement with prediction.

The gravitational redshift is a direct consequence of the principle of equivalence. This is shown by the following thought experiment. We consider a lightray of increasing wavelength that is escaping verti-

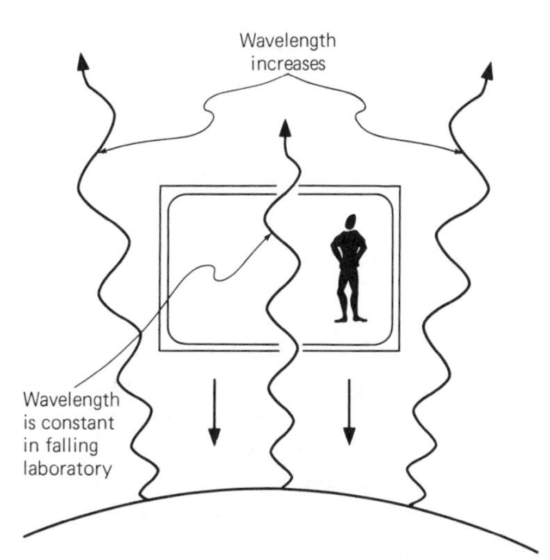

Figure 8.18. A lightray loses energy as it moves away from a star or a planet, and its wavelength steadily increases. Inside the free-falling laboratory the wavelength is constant. This enables us to calculate the gravitational redshift from the known acceleration of the laboratory.

cally from a gravitating body, as shown in Figure 8.18. We suppose that an experimenter is in a free-falling laboratory and the lightray enters through a hole in the floor and passes upward and out through a hole in the ceiling. Within this laboratory the effect of gravity is abolished, and the experimenter therefore sees the lightray enter and leave with its wavelength unchanged. We must now take into account the acceleration of the laboratory. While the lightray passes upward from the floor to the ceiling, the velocity of the falling laboratory increases. The wavelength is therefore, in effect progressively squeezed by the increasing velocity of the laboratory, and it remains of constant value to the experimenter who is inside. Because the wavelength is constant in the free-falling laboratory, in accordance with the principle of equivalence, it follows that to an outside observer, standing on the surface of the body, the wavelength increases as the lightray travels upward.

Imagine that we are out in space observing what happens on the surface of a gravitating body. All wavelengths of radiation received from the surface are increased by the same fractional amount and have the same redshift. Similarly, all frequencies are decreased by the same fractional amount. As a consequence, we see everything happening more slowly on the surface of the body. Let us ignore the practical aspects of the situation and take an extreme case in which the redshift is unity. The atoms on the surface now appear to vibrate twice as slow, as the same atoms around us. Everything else on the surface also appears to happen twice as slowly as out in our region of space. A person on the surface communicating to us by radio has a deep voice and talks very sluggishly. That person, instead of living for 70 years, lives for 140 years in our time. The redshift not only increases wavelengths and decreases frequencies; it also slows up the apparent rate at which everything happens. All intervals of time are increased in the same way as wavelengths, and two events separated by 1 second on the surface are seen to be separated by 2 seconds out in space.

Let us reverse the situation and imagine that we are living on the surface. Everything around us happens at a normal rate, as measured by our pulse rate, and if we are fortunate we shall live for the usual 70 years. (We are of course ignoring the practical aspects of living on the surface of a body of very strong gravity.) Incoming radiation from space is now blueshifted; all wavelengths are decreased by the same fractional amount and shifted toward the blue end of the spectrum. A person out in space, communicating with us by radio, has a high-pitched voice and talks very rapidly. Measured in our time, for a blueshift of one half, that person lives for only 35 years.

Summing up: To an observer out in space, all things in the vicinity of gravitating bodies are redshifted and appear to happen more slowly; to an observer in the vicinity of a gravitating body, all things out in space are blueshifted and appear to happen more rapidly.

OTHER TESTS OF GENERAL RELATIVITY. Various tests of general relativity have been suggested in addition to those proposed by Einstein, and several of these tests, such as the measurement of time delays of radar signals in the Solar System, have been successfully accomplished. As experimental ingenuity and observational precision have advanced, the results have become progressively more reassuring concerning the validity of general relativity. The discovery by Joseph Taylor and Russell Hulse, at the University of Massachusetts, of a pulsar in a close binary system now provides radioastronomers with an opportunity to observe the effects of gravitational fields much stronger than those previously available. In this binary system the precession of the pulsar orbit about the companion star is approximately 4 degrees of arc per year, which is about 35 thousand times faster than the precession of Mercury's orbit. The effect of gravitational radiation from this binary system has already been detected, and in the future we can look forward to more exacting tests of the theory of general relativity.

MACH'S PRINCIPLE

Mach's principle states that all inertial forces are due to the distribution of matter in the universe, and this principle was of help to Einstein in the development of general relativity.

ABSOLUTE SPACE. Newton said, "Absolute space in its own nature, without relation to anything external, remains similar and immovable." This abstract idea of an absolute space, of a space existing in its own right and without the need of material raiment, was contrary to common sense and was challenged by many philosophers and scientists. Gottfried Leibniz regarded Newton's ideas about space as outrageous and responded by asserting, "There is no space, where there is no matter." But Newton believed that he had proof of the absolute nature of space, and in the controversy that raged he was the only person who submitted his ideas to experimental tests.

Everybody agreed that uniform motion (motion of constant velocity) is relative, and to many people it seemed natural therefore that nonuniform motion (accelerated motion) would also be relative. Newton declared, however, that accelerated motion is absolute and is relative only to absolute space. We can understand what Newton had in mind by the following thought experiment. Let us imagine that only a single body exists in the whole of space. When this body has uniform motion it is impossible to determine how fast it is moving or the direction in which it moves. But when the body has nonuniform motion, then it is possible to ascertain how fast it is accelerated and also the direction in which it accelerates, because of the existence of inertial forces. Consider rotation: It is a form of acceleration, and in a rotating body there exists a centrifugal force that produces observable effects. For instance, a rotating planet or star bulges at the equator and is flattened at the poles. The measurement of effects of this kind enables us to determine how fast the body rotates, and this rotation, said Newton, is relative not to other bodies but to absolute space.

Absolute space, existing in its own right, was the overarching concept of the Newtonian universe. It was implicit in the Newtonian equation of motion, and no rival philosopher or scientist was able to suggest an alternative idea that could match it. Of the experiments dealing with the absolute nature of rotation that were discussed by Newton, the most famous is the rotating-bucket-of-water experiment, discussed in the Reflections at the end of this chapter.

BISHOP BERKELEY. The idea of absolute space was vigorously attacked by the Irish philosopher George Berkeley in a work entitled *Motion,* published in 1721. Berkeley's dislike of absolute space stemmed from the Aristotelian belief that space existed by virtue of its association with matter, and that unclothed space had no physical properties of its own; it was a sideless box that vanished entirely when nothing was contained. Berkeley said that space by itself was emptiness and was therefore nothing; its

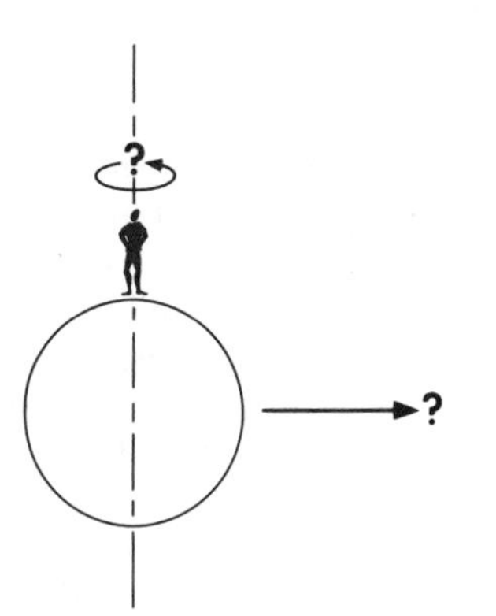

Figure 8.19. Berkeley's "solitary globe" in an empty universe. All forms of motion, uniform and accelerated, are unobservable and meaningless, according to Berkeley.

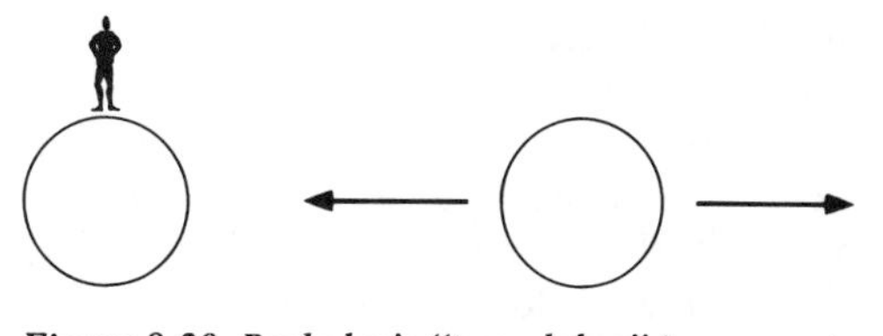

Figure 8.20. Berkeley's "two globes" in an empty universe. The only detectable motion, according to Berkeley, is their motion toward and away from each other, and their revolution about a common axis is unobservable and meaningless.

only property was extension, and without a supporting distribution of matter this property was meaningless.

Berkeley's principle, which lies at the heart of his argument, can be stated as follows: A single body in an otherwise empty universe has no measurable motion of any kind (see Figure 8.19). It is a postulate of impotence that unfortunately cannot be verified by observation. If we were to suppose, said Berkeley, that "the other bodies were annihilated and, for example, a globe were to exist alone, no motion could be conceived in it; so necessary is it that another body should be given by whose situation the motion should be understood to be determined. The truth of this opinion will be very clearly seen if we shall have carried out thoroughly the supposed annihilation of all bodies, our own and that of others, except that solitary globe." A single body, all alone, is thus denied conceivable motion of any kind, and hence, according to Berkeley, there is no way of determining whether it has rotation or not. If two bodies alone exist, then only their relative motion toward and away from each other can be measured, and we cannot therefore determine if they have revolution about each other. In the words of Berkeley: "Let two globes be conceived to exist and nothing corporeal besides them. Let forces then be conceived to be applied in some way; whatever we may understand by the application of forces, a circular motion of the two globes around a common center cannot be conceived by the imagination" (see Figure 8.20). If there are three bodies alone, then only their relative motion within a common plane is measurable, and we cannot determine if they have revolution about a common point. If there are four bodies alone, then their relative motion in three dimensions is measurable, but we cannot determine if they have revolution about a common axis. "Then let us suppose that the sky of the fixed stars is created; suddenly from the conception of the approach of the globes to the different parts of the sky the motion will be conceived."

Instead of an absolute space of independent existence, Berkeley invoked a "sky of fixed stars." The fixed stars became the reference points and all motion, uniform or nonuniform, was relative to the distant stars. To Berkeley, and to many other philosophers, it seemed natural to suppose that space in all respects was subordinate to matter and that the properties attributed by Newton to absolute space were in fact the result of the material content of the universe.

ERNST MACH. Ernst Mach, an Austrian physicist of the nineteenth century, expressed ideas essentially similar to those of Bishop Berkeley. Berkeley had stressed the relativity of all motion, uniform and accelerated, and Mach developed this theme and stressed the relativity of inertial forces.

Once again let us suppose there is only a single body in an empty universe. Motion of any kind, according to Mach, is inconceivable and therefore inertial forces cannot exist. When such a body rotates it does not have a centrifugal force. The addition of one, two, three, or more bodies now allows relative motion to be determined. But this leads to a dilemma: The additional bodies merely serve as reference points and can be made as small as specks of dust; how then can they account for the sudden creation of inertial forces? Having denied inertial force to the first body, it becomes absurd to suppose that such a force is suddenly acquired in full strength by the addition of a few small bodies. How then, at the same time that rotation becomes measurable, can centrifugal force be created? Mach's answer was that the inertial forces (or rather, the inertial masses) increase only slightly and that all inertial forces are determined by and are proportional to the total amount of matter in the universe. Hence, in some way, the universe of stars is responsible for the inertial forces of accelerated motion. A single body by itself has no measurable rotation and no detectable centrifugal force, but when the sky of fixed stars is created, the body has measurable rotation relative to the stars and acquires a centrifugal force because of the stars.

MACH'S INFLUENCE ON EINSTEIN. Einstein found Mach's argument suggestive and helpful during the years in which the theory of general relativity was being developed. Mach apparently was not aware of Berkeley's earlier work and did not refer to it in his publications; Einstein therefore attributed these views entirely to Mach and coined the term *Mach's principle.* This principle, in effect, states that local inertial forces are determined by the distribution and quantity of matter in the universe. It is a fascinating bootstrap conception that in some unknown way connects local inertial forces with the global properties of the universe.

The notion that inertial forces are determined by distant matter was for Einstein a signpost pointing to a possible connection between geometry and matter, in the sense that distant matter affects geometry and geometry affects local motion. If the geodesics (straight lines) of Riemannian spacetime are akin to the orbits of free-falling bodies in the Newtonian universe, and these orbits are affected by inertial forces, then according to Mach's principle the distant matter that affects geometry should also account for inertial forces. Einstein initially took the tentative view that spacetime was fully determined by matter and all inertial forces were the consequence of matter interacting with matter. "In a consistent theory of relativity," said Einstein as late as 1917, "there can be no inertia relative to 'space,' but only an inertia of masses relative to one another. If, therefore, I have a mass at a sufficient distance from all other masses in the universe, its inertia must fall to zero." At that time it was thought that the equation of motion of a body had to be postulated as a geodesic equation, as something in addition to the Einstein equation, and only later was it realized that the equation of motion, as a geodesic equation, was already implicit in the theory of general relativity.

Very soon it became clear to Einstein and other scientists that general relativity had outgrown the suggestive ideas of Mach's principle. Spacetime had achieved a physical reality of its own: Although its geometry is influenced by matter, and motion is controlled by geometry, the nature and existence of spacetime are not dependent on the existence of matter. Einstein then abandoned Mach's principle. The principle had served its purpose, and instead of continuing to seek for a way to materialize spacetime, Einstein took a new departure and sought for a way to geometrize matter. Some physicists have advanced the view that perhaps matter exists only by virtue of the geometrical properties of spacetime. Charles Misner and John Wheeler stressed in 1957 the contrasting views: Either the spacetime "continuum serves only as an *arena* for the struggles of fields and particles," or there "is nothing in the world except empty curved space. Matter, charge, electromagnetism, and other fields are only manifestations of the bending of space. *Physics is geometry.*" We can at least say that spacetime, as a result of general relativity, is physically real and does not exist by virtue of matter alone.

REFLECTIONS

1 *Albert Einstein (1879–1955) was awarded the Nobel Prize in 1922, not for his work on relativity theory, but for work published in 1905 on the photoelectric effect. (In the photoelectric effect light ejects electrons from metal surfaces, and the energy of the electrons depends on the frequency of the radiation, whereas the number of ejected electrons depends on the intensity of the radiation.) Einstein was a humble man who was indifferent to prizes and other honors. Yet many if not most scientists have a strong desire to win prizes and receive academic distinctions, even though we teach students that science is motivated by a spirit of inquiry and a desire to probe the mysteries of nature. Motivation in science is an interesting psychological study.*

* *Albert Einstein, son of an unsuccessful businessman, was a backward child, an inattentive pupil, and a dropout from high school. He was mainly self-taught; after getting his doctorate degree in 1905 he sought in vain for an academic appointment, and accepted a clerical position in a Swiss patent office. In that same year, 1905, he published three pioneering scientific papers, in one of which he advanced the theory of special relativity. Four years later he secured an academic position, and thereafter his progress was rapid. Einstein possessed a great power for sustained concentration, a penetrating grasp of the fundamentals of physics, and an immensely creative and disciplined imagination. One of his most famous sayings is that "God does not play with dice." Several aspects of the theory of quantum mechanics were developed by Einstein, but in his later years he grew aloof and skeptical, and viewed modern quantum theory as insufficiently profound. He believed that beyond its uncertainties there are undiscovered deterministic laws.*

* *In 1930, in an entertaining after-dinner toast to Einstein, who was present, George Bernard Shaw made the following remarks: "Religion is always right. Religion solves every problem and thereby abolishes problems from the universe. Religion gives us certainty, stability, peace and the absolute. It protects us against progress which we all dread. Science is the very opposite. Science is always wrong. It never solves a problem without raising ten more problems." Shaw then continued: "Copernicus proved that Ptolemy was wrong. Kepler proved that Copernicus was wrong. Galileo proved that Aristotle was wrong. But at that point the sequence broke down, because science then came up for the first time against that incalculable phenomenon, an Englishman. As an Englishman, Newton was able to combine a prodigious mental faculty with the credulities and delusions that would disgrace a rabbit. As an Englishman, he postulated a rectilinear universe because the English always use the word 'square' to denote honesty, truthfulness, in short: rectitude. Newton knew that the universe consisted of bodies in motion, and that none of them moved in straight lines, nor ever could. But an Englishman was not daunted by the facts. To explain why all the lines in his rectilinear universe were bent, he invented a force called gravitation and then erected a complex British universe and established it as a religion which was devoutly believed in for 300 years. The book of this Newtonian religion was not that oriental magic thing, the Bible. It was that British and matter-of-fact-thing, a Bradshaw [railway timetable]. It gives the stations of all the heavenly bodies, their distances, the rates at which they are travelling, and the hour at which they reach eclipsing points or crash into the earth. Every item is precise, ascertained, absolute and English.*

"Three hundred years after its establishment a young professor rises calmly in the middle of Europe and says to our astronomers: 'Gentlemen: if you will observe the next eclipse of the sun carefully, you will be able to explain what is wrong with the perihelion of Mercury.' The civilized Newtonian world replies that, if the dreadful thing is true, if the eclipse makes good the blasphemy, the next thing the young professor will do is to question the existence of gravity. The young professor smiles and says that gravitation is a very useful hypothesis and gives fairly close results in most cases, but that personally he can do without it. He is asked to explain how, if there is no gravitation, the heavenly bodies do not move in straight lines and run clear out of the universe. He replies that no explanation is needed because the universe is not rectilinear and exclusively British; it is curvilinear. The Newtonian universe thereupon drops dead and is supplanted by the Einstein universe. Einstein has not challenged the facts of science but the axioms of science, and science has surrendered to the challenge" (Blanche Patch [Shaw's secretary], Thirty Years with G.B.S., *Gollancz, London, 1951).*

2 *Discuss the principle of equivalence.*

* *Suggest an experiment that can be performed in a free-falling laboratory to check the equivalence of inertial and gravitational mass. How would you verify that the energy in a magnetic field has an inertial mass equal to its gravitational mass?*

* *What is the value (dollars per ounce) of the atomic electric fields of gold? The mass contribution of atomic electric fields varies roughly as the atomic number squared. What is the mass contribution of the atomic electric fields of aluminum?*

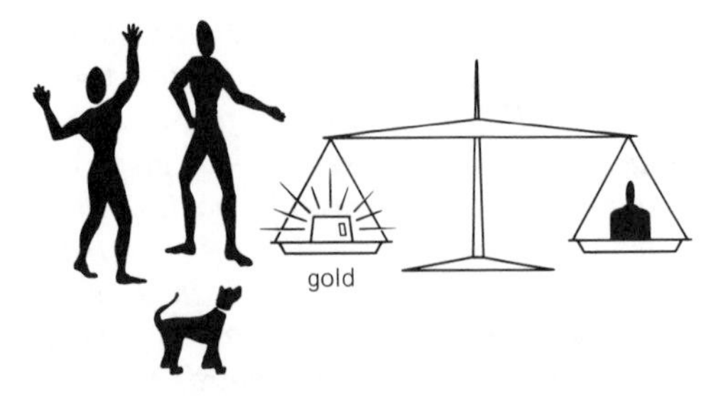

Figure 8.21. "You can take out the electric fields, they cost too much!"

* *Gravitational mass acts in two ways: actively as a source of gravity, and passively in response to gravity. The bodies of masses M_1 and M_2 attract each other gravitationally; discuss the active and passive roles that each plays. Notice that each pulls and is pulled by the other.*

* *What are tidal forces, and why are there tides twice daily on the Earth's surface?*

* *In what way are gravity and the geometry of curved surfaces similar to each other?*

* *What is the difference between* special *and* general *relativity?*

* *Discuss the Einstein equation.*

* *In the Newtonian picture gravity acts mysteriously and instantaneously across a vacuum. In the Einstein picture the curvature of physical spacetime is produced by matter and controls the motion of matter. Which picture do you find the more sensible?*

* *Stretch a rubber sheet in a frame in the horizontal plane. Now place a spherical and sufficiently heavy body in the center and roll small ball bearings in the saucer-shaped depression of the rubber sheet. Notice that the ball bearings follow trajectories similar to those of bodies moving in the gravitational field of the Sun.*

3 *The deflection of starlight by the Sun was predicted in 1801 by the German mathematician Johann von Soldner. He used the idea that light consists of particles that obey the Newtonian equations and estimated a deflection half that given by the theory of general relativity.*

The principle of equivalence, when considered by itself, is unable to explain why general relativity gives a deflection twice that expected from Newtonian theory. We can understand why this is so in a qualitative sort of way. Consider a laboratory on the Earth's surface in which a beam of light travels horizontally from one side to the other and is deflected downward slightly by gravity (see Figure 8.22). This suggests, according to the principle of equivalence, that in a free-falling laboratory the same beam of light will travel in a straight line and not be deflected. This would be true if the gravitational field were uniform and everywhere the same. But it is not, and therefore all parts of the laboratory, and all parts of the beam of light, are not simultanteously in a perfect state of free fall. Here is one instance where we cannot ignore tidal forces, or equivalently, we cannot ignore the variation of curvature of spacetime within the laboratory. A beam of light in a free falling laboratory is curved because the laboratory has finite size. When we take into account the variation of curvature within the laboratory we obtain a deflection that is twice what the principle of equivalence, without tidal forces, predicts.

4 *Newton wrote: "The effects which distinguish absolute from relative motion are centrifugal forces, or those forces in circular motion which produce a tendency of recession from the axis. For in a circular motion which is purely relative no such forces exist, but in a true and absolute circular motion they do exist, and are greater or less according to the quantity of the absolute motion."*

He then described the rotating-bucket-of-water experiment (see Figure 8.23): "For instance: If a bucket, suspended by a long

Figure 8.22. A beam of light in a laboratory on the Earth's surface is deflected downward by gravity. In a free-falling laboratory in a gravitational field the beam of light is also deflected. This is because there is spacetime curvature within the laboratory of finite size, and not all its parts are falling freely. (The deflections shown are exaggerated.)

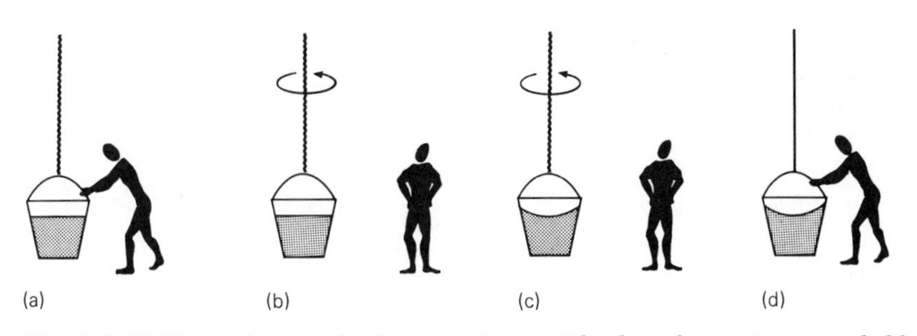

Figure 8.23. Newton's water-bucket experiment. A bucket of water is suspended by a rope and is rotated until the rope is tightly twisted. In (a) we see that the surface of the water is level and there is no centrifugal force. In (b) the bucket is released and it begins to rotate; the surface of the water is still level, indicating that centrifugal force is not created by the rotation of the bucket. In (c) the water is now rotating with the bucket and its surface is concave, thus indicating the existence of a centrifugal force acting on the water. In (d) the bucket is brought to rest; but the water still rotates and has a concave surface. This experiment, said Newton, indicates that centrifugal force is the result of absolute motion, and not relative motion.

cord, is so often turned about that finally the cord is strongly twisted, then is filled with water, and held at rest together with the water; and afterwards, by the action of a second force, it is suddenly set whirling about the contrary way, and continues, while the cord is untwisting itself, for some time in this motion; the surface of the water will at first be level, just as it was before the vessel began to move; but, subsequently, the vessel, by gradually communicating its motion to the water, will make it begin sensibly to rotate, and the water will recede little by little from the middle and rise up at the sides of the vessel, its surface assuming a concave form (as I have experienced) and the swifter the motion becomes, the higher will the water rise, till at last, performing its revolutions in the same times as the vessel, it becomes relatively at rest to it." In this experiment the surface of the rotating water is depressed in the center, and the amount of depression depends on how fast the water rotates in an absolute sense, not on the rotation of the water relative to the bucket. Newton therefore concluded that rotation can be determined absolutely, without reference to other bodies.

5 *Ernst Mach (1838–1916), a physicist and leading philosopher of science, believed that observations are of primary importance in science, and that all things that cannot be directly perceived exist only in our minds and have no objective reality. This positivist philosophy led him to a rejection of the atomic theory, because atoms are not directly observable. He also rejected the notion of space and time existing independently of observed material things, and he did not accept the special theory of relativity. Mach is best known for his study of supersonic flight. When a plane flies at the speed of sound in air we say that it moves at Mach 1. A Mach number x means that an object moves through a medium at x times the speed of sound in that medium. In 1872, Mach wrote: "For me only relative motion exists . . . When a body rotates relative to the fixed stars, centrifugal forces are produced; when it rotates relative to some different body and not relative to the fixed stars, no centrifugal forces are produced. I have no objection to just calling the first rotation so long as it be remembered that nothing is meant except relative rotation with respect to the fixed stars."*

6 *In the neighborhood of a rotating body a particle tends to be dragged around with the body. This sharing of rotation, or* inertial dragging, *was first discovered in 1918 by H. Thirring and J. Lense and is an effect owing solely to general relativity. Consider a shell – a hollow sphere – of matter of mass M, as shown in Figure 8.24, which contains a suspended rod. The suspended rod acts as an* inertial compass. *Normally, when the shell rotates, we are able to detect its rotation relative to the suspended rod. When, however, the mass M of the shell is large, the rod tends to be dragged around. In the limit, when M becomes exceedingly large, there is no apparent rotation because the compass of inertia rotates with the shell. This Thirring–Lense effect indicates that general relativity does possess a Machian*

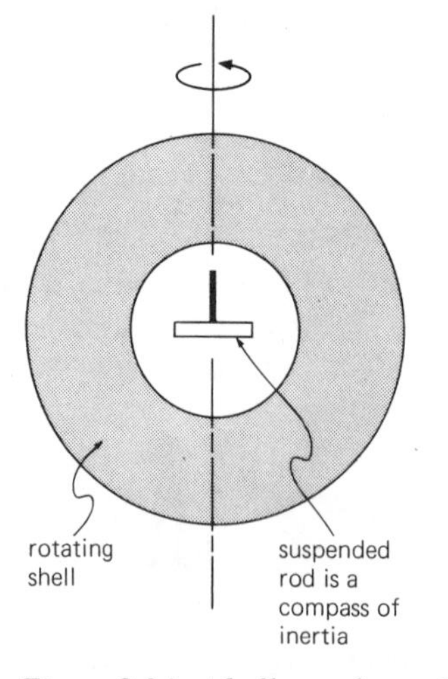

Figure 8.24. A hollow sphere of mass M rotates. Inside is suspended a rod, free to rotate independently, which acts as a compass of inertia. The rod tends to be dragged around by the rotating sphere. This is the Thirring–Lense effect. When M is exceedingly large, the compass of inertia rotates with the sphere, and rotation cannot be detected from inside.

property known as the dragging of the inertial frame. *But many exceptions can be found that do not conform to Mach's principle. In a rotating universe the compass of inertia should corotate, and cosmic rotation should therefore be undetectable and meaningless, according to Mach. But in all rotating models so far constructed with general relativity, the compass of inertia does not corotate. In other words, if we have a body in a state of inertial motion, any rotation of the universe can be detected relative to that body. This theoretical discovery, first made by Kurt Gödel in 1949, has delivered a mortal blow to Mach's principle and shown that it is not an indispensable part of general relativity.*

FURTHER READING

Barnett, L. *The Universe and Dr. Einstein.* William Sloane Associates, New York, 1948.

Bergmann, P. G. *The Riddle of Gravitation: From Newton to Einstein to Today's Exciting Theories.* Charles Scribner's Sons, New York, 1968.

Bernstein, J. *Einstein.* Viking Press, New York, 1973.

Chandrasekhar, S. "Einstein and general relativity: historical perspectives." *American Journal of Physics 47,* 212 (May 1979).

Davies, P. C. W. *Space and Time in the Modern Universe.* Cambridge University Press, Cambridge, 1977.

Eddington, A. S. *Space, Time, and Gravitation.* Cambridge University Press, Cambridge, 1920. Reprint. Harper Torchbooks, New York, 1959.

Einstein, A. *Out of My Later Years.* Philosophical Library, New York, 1950.

Fletcher, J. G. "Geometrodynamics: the geometry of space–time." *Discovery.* November 1964.

Frank, P. *Einstein: His Life and Times.* Alfred A. Knopf, New York, 1947.

Gamow, G. *Gravity.* Doubleday, Anchor Books, New York, 1962.

Hoffman, B. *Albert Einstein: Creator and Rebel.* Viking Press, New York, 1972.

Infeld, L., *Albert Einstein: His Work and Influence on Our World Order.* Charles Scribner's Sons, New York, 1950.

Kaufman, W. J. *Relativity and Cosmology.* Harper and Row, New York, 1973.

Lanczos, C. *Albert Einstein and the Cosmic World Order.* John Wiley, New York, 1965.

Sciama, D. W. *The Unity of the Universe.* Faber and Faber, London, 1959.

Sciama, D. W. *The Physical Foundations of General Relativity.* Doubleday, New York, 1969.

Wills, C. "Gravitational theory." *Scientific American,* November 1974.

SOURCES

Born, M. *Einstein's Theory of Relativity.* 1924. Reprint. Dover Publications, New York, 1962.

Brill, D. R., and Perisho, R. C. "General relativity: resource letter." *American Journal of Physics 36,* 1 (February 1968).

Eddington, A. S. *The Nature of the Physical World.* Cambridge University Press, Cambridge, 1928.

Einstein, A. *Relativity: The Special and the General: A Popular Exposition.* 1916. Reprint. Methuen, London, 1954.

Einstein, A. *The Meaning of Relativity.* Princeton University Press, Princeton, N.J., 1955.

Einstein, A., Lorentz, H. A., Weyl, H., and Minkowski, H. *The Principle of Relativity.* Dover Publications, New York, 1952. Some of Einstein's important papers on relativity are collected in this book.

Geroch, R. *General Relativity from A to B.* University of Chicago Press, Chicago, 1978.

Hilbert, D., and Cohn-Vossen, S. *Geometry and the Imagination.* Chelsea Publishing Co., New York, 1952.

Holton, G. "Einstein and the crucial experiment." *American Journal of Physics 37,* 968 (October 1969).

Jaki, S. L. "A forgotten bicentenary: Johann Georg Soldner." *Sky and Telescope,* June 1978.

Klein, H. A. *The New Gravitation.* J. B. Lippincott, Philadelphia, 1971.

Mach, E. *The Science of Mechanics.* Open Court, LaSalle, Ill., 1942.

Misner, C. W., and Wheeler, J. A. "Classical physics as geometry," *Annals of Physics 2,* 525 (1957)

Pearce, W. L., ed. *Relativity Theory: Its Origin and Impact on Modern Thought.* John Wiley, New York, 1968.

Richardson, R. S. *The Star Lovers.* Macmillan,

New York, 1967. Contains an informative chapter on Einstein.
Rindler, W. *Essential Relativity: Special, General, and Cosmological.* Van Nostrand Reinhold, New York, 1969.
Schilpp, P. A., ed. *Albert Einstein: Philosopher–Scientist.* Library of Living Philosophers, Evanston, Ill., 1949. Essays by distinguished scientists on various aspects of Einstein's work.

9 BLACK HOLES

Black Horror scream'd, and all her goblin route
Diminish'd shrunk from the more withering scene!
— Coleridge (1772–1834), "To the author of 'the robbers' "

GRAVITATIONAL COLLAPSE

Confinement to the Black Hole . . . to be reserved for cases of Drunkeness, Riot, Violence, or Insolence to Superiors.
— British Army regulation (1844)

"INTO THE JAWS OF DEATH." Stars are luminous globes of gas in which the pull of gravity is balanced by internal pressure. Nuclear energy is released in the deep interior, and by a nice adjustment of the physical laws of the universe, this energy maintains the stars over long periods of time and is radiated from their surfaces in a manner that makes possible the chemistry of planetary life.

But for each star comes eventually a day of reckoning. Its central reservoir of hydrogen approaches exhaustion and the star begins to die. The tireless pull of gravity causes the central region to contract and rise to higher temperature, and as a consequence the outer region swells up and the star becomes a red giant. When it is a sunlike star it evolves into a white dwarf, with most of its matter compressed into a sphere roughly equal to the size of Earth. Many stars end as white dwarfs, slowly cooling, supported internally against the pull of gravity by incessant electron waves.

More massive stars do not give up the game so easily. They draw on slender reserves of nuclear energy, and their central regions continue to contract and rise to higher temperatures. They become luminous giant stars squandering energy at a prodigious rate. Soon the supply of nuclear energy is exhausted, and all that remains is gravitational energy, with its fatal price of continual contraction. In the final throes of death the cores of these stars rush inward, their exploding mantles hurl outward, and for a brief ecstatic moment they become blazing supernovas. Out of such cataclysms are born neutron stars in the form of rapidly rotating pulsars.

The most massive of all stars have imploding cores that cannot terminate in any known state of matter. Gravity is normally the weakest of nature's forces: It takes an entire Earth to cause a feather to fall. But in the imploding cores of massive stars gravity overwhelms all other known forces; nothing can withstand it and arrest the fall of matter "into the jaws of death." Gravity devours the star, and from this strange circumstance we trace the birth of black holes.

"INTO THE MOUTH OF HELL." If we follow the imploding core of a massive star, then in 1/10,000 of a second, or thereabouts, it attains virtually infinite density. It becomes a *singularity* – in some ways similar to the extreme early universe – about which we still understand very little. According to some arguments the density of the singular state is 10^{94} grams per cubic centimeter, or

10,000,000,000,000,000,000,000,000,
000,000,000,000,000,000,000,000,000,
000,000,000,000,000,000,000,000,000,
000,000,000,000,000

times the density of water. Particles, in whatever form they might exist under these ultimate conditions, are possibly a by-product of the quantum nature of spacetime, and spacetime itself is perhaps a chaos in which space and time are inextricably scrambled together and can never be distinguished. Conceivably, in the singular state, an orderly timelike sequence of events has vanished and there is no beginning or end.

But to a distant observer in the outside world the falling star never attains the singular state. The gravitational redshift gets progressively greater and the star appears to fall more and more slowly. As the star approaches a critical size the redshift rises to infinity. The star reddens, darkens quickly into blackness, and remains forever at the critical size, frozen in a permanent state of collapse. Nothing, not even light, can now escape to the outside world. Within the collapsing star nemesis lies only a fraction of a second away; to an outside observer the star is a black hole where time stands still. The ultimate fate of gravitational collapse is hence concealed from the outside world.

Consider two observers; or rather, because we shall treat them brutally, consider two particles: one particle inside the black hole and the other in the outside world. The outside particle says, "Why should I worry about the fate of the inside particle, seeing that in my space and time it is suspended in a static state and can never reach the singularity?" This raises an interesting question: Should the outside particle bother with the fate of the inside particle, when whatever is dreadful inside a black hole happens only in the infinite future of the outside particle?

The particle in a collapsing star sees the outside world blueshifted, and the blueshift attains an extreme value when the star reaches its critical size and becomes a black hole. Things in the outside world are seen speeded up, and at the critical size everything outside happens with extreme rapidity. The entire future history of the universe passes in a flash. Suppose that in tens of billions of years in our time the universe ceases to expand and then collapses back into a big bang. The inside particle sees all this happen almost instantaneously; the galaxies streak away and then streak back again, and the outside world soars in density until it matches the density of the black hole. Time inside the black hole and time in the outside world now tick away at the same rate, and holding hands, the inside and outside particles descend together into a cosmic singularity. The particle outside, which has hitherto congratulated itself on not falling into the black hole, now finds itself on equal footing with its lost and swallowed friend, and they plunge together to meet their doom.

Most massive stars are members of binary systems, and in Chapter 2 we noticed that stellar evolution in close binary systems is complicated by the exchange of matter. When a star in a close binary system swells into a red giant it spills matter onto its companion and with reduced mass then evolves into a neutron star or a black hole. The companion star in its turn swells up, returning matter to the collapsed star. The collapsed star as a result grows in mass, and if it is a neutron star, it will perhaps implode and become a black hole. The companion star also evolves and finally collapses into either a neutron star or a black hole. The most massive stars are 50 or more times the mass of the Sun, and it is therefore difficult to escape the conclusion that many stars, either isolated in a single state or as members in a binary state, have inevitably developed into black holes (see Figure 9.1).

Collapsed stars in binary systems are of considerable interest to astronomers. The gas that spills over onto a neutron star or is sucked into a black hole releases a large quantity of gravitational energy that is

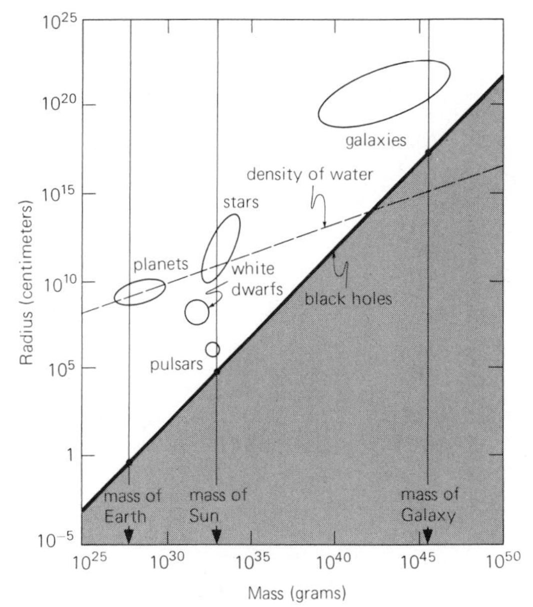

Figure 9.1. Mass and radius of black holes. When bodies contract they move downward and approach the black-hole line.

radiated away, usually in the form of X-rays. There are now numerous X-ray sources, many of them are known to be binary systems, and it is believed that in each case one of the companions is a collapsed star that is either a neutron star or a black hole. Neutron stars have masses less than about 3 solar masses, and collapsed stars of greater mass are probably black holes. One of the best-known candidates for a black hole is the collapsed star in the X-ray binary source known as Cygnus X–1. This was the first X-ray source discovered in the constellation Cygnus, and the collapsed star has a mass perhaps about 8 times that of the Sun.

CURVED SPACETIME OF BLACK HOLES

Deep into the darkness peering, long I stood there, wondering, fearing,
Doubting, dreaming dreams no mortal ever dared to dream before.
— Edgar Allan Poe (1809–49), "The Raven"

CURVATURE OF SPACETIME. As shown by John Mitchell and Pierre de Laplace, nearly 200 years ago, even the Newtonian laws warn us that something odd happens when the escape speed from the surface of a collapsed body equals the speed of light. From general relativity we know that spacetime is curved by gravitating bodies, and the curvature is a measure of the strength of gravity. As a body contracts, its surface gravity increases, and spacetime becomes more curved. At a critical size, known as the *Schwarzschild radius,* spacetime becomes so curved that in effect it encloses the body. It has become a black hole, wrapped in curved spacetime, and nothing, not even light, can leave it and escape to the outside world.

A stretched horizontal rubber sheet helps us to visualize what happens (see Figure 9.2). We consider a ball of constant mass

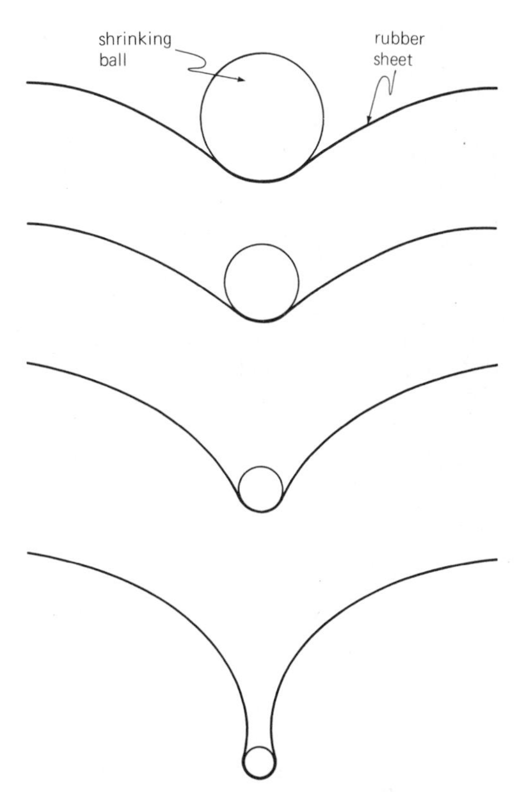

Figure 9.2. A rubber sheet is depressed in the center by a heavy ball. The ball is of constant weight and slowly shrinks in size. When the ball has shrunk to a small size, as shown, the sheet encapsulates the ball and forms a connecting link.

resting on the sheet and imagine that the ball slowly shrinks in size. At a distance from the ball the depression of the sheet is determined only by the mass of the ball, and because this is constant, the curvature of the sheet remains unchanged while the ball contracts. A similar state of affairs occurs in general relativity: At a distance the gravitational field, or curvature of spacetime, remains unchanged while a spherical body of constant mass contracts. The distant spacetime does not alter, and therefore no gravitational waves are radiated away from a collapsing spherical body. While we watch the ball on the rubber sheet get smaller, we notice how the sheet becomes more depressed and curved near the ball. Eventually, when the ball of constant mass is small enough, the sheet encapsulates the ball and forms a connecting neck. The sheet, now enveloping the ball, is like the curved spacetime of a black hole. The ball is not completely isolated but is connected through the neck with the rest of the sheet; black holes also are not entirely isolated but are connected with the outside world of spacetime.

The Schwarzschild radius of a collapsed body of a mass equal to that of the Sun is only 3 kilometers, and for a collapsed body of a mass equal to that of the Earth it is about 1 centimeter. This means that if the Sun collapsed to a sphere of 3 kilometers radius, or the Earth collapsed to a size slightly smaller than a Ping-Pong ball, then both would have infinite redshift and become black holes.

LIGHTRAYS. Rays of light leaving a luminous gravitating body are affected in two ways: They are redshifted and deflected. As the body contracts, its surface gravity increases, and the lightrays emitted are increasingly redshifted and deflected. Let R be the radius of a body whose ultimate Schwarzschild radius is R_s; the redshift of lightrays that escape to large distances is given by the equation

$$\text{gravitational redshift} = \frac{1}{\sqrt{(1 - R_s/R)}} - 1$$

Thus, when the radius is one-third greater than the Schwarzschild radius (i.e., $R = 4R_s/3$), the redshift is equal to 1. We saw in Chapter 8 that if λ is the emitted wavelength, and at great distances the wave-

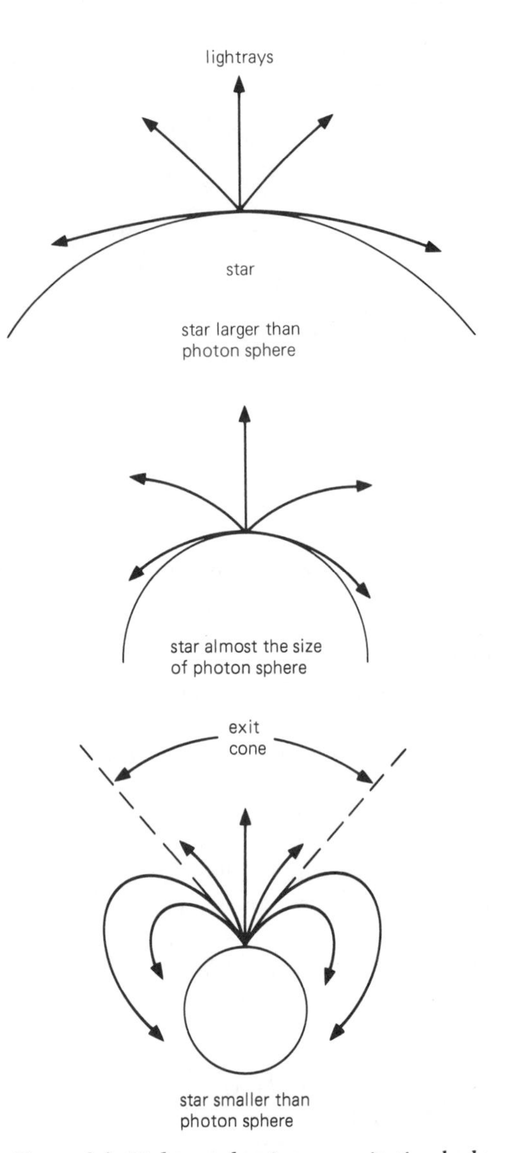

Figure 9.3. Lightrays leaving a gravitating body are curved as shown. As the body shrinks in size the rays become more curved. When the radius is less than 1.5 times the Schwarzschild radius, which is the radius of the photon sphere, the exit cone begins to close. Rays emitted within the exit cone escape, but those outside are trapped and fall back.

length increases to λ_0, then

$$\text{gravitational redshift} = \frac{\lambda_0 - \lambda}{\lambda}$$

and for a redshift of 1 an escaping ray has its wavelength increased twofold. If the radius R differs from R_s by only one-millionth, the redshift is 1000; and if the difference is one-trillionth, the redshift is 1 million; and so on to infinity.

Lightrays leaving the surface in a perpendicular direction are not deflected, as shown in Figure 9.3, whereas the rays moving in a direction tangential to the surface are deflected to the greatest extent. When the contracting body reaches a radius 1.5 times the Schwarzschild radius, all rays emitted tangential to the surface are curved into circular orbits. This is the radius of the *photon sphere;* an atom at this distance is able to "see the back of its head." With further contraction, the rays are more strongly deflected, and many now fall back to the surface. Only the rays emitted within an *exit cone* can escape, and this exit cone narrows as contraction continues. When the body reaches the Schwartzschild radius and becomes a black hole, the exit cone closes completely, so that no lightrays escape. Redshift and deflection conspire to ensure that no radiation escapes from a black hole.

The photon sphere has a radius 1.5 times the Schwarzschild radius and is the surface where light rays travel in circular orbits around a black hole. These orbits are unstable: If a circulating ray is disturbed slightly, it either spirals in and is captured or spirals out and eventually escapes at a radius $\sqrt{3}$ times that of the photon sphere (see Figure 9.4). The redshift of light leaking outward from the photon sphere is $\sqrt{3} - 1 = 0.73$. All lightrays approaching a black hole closer than $\sqrt{3}$ times the radius of the photon sphere spiral inward and are captured (see Figure 9.5).

Again we consider the surface of a

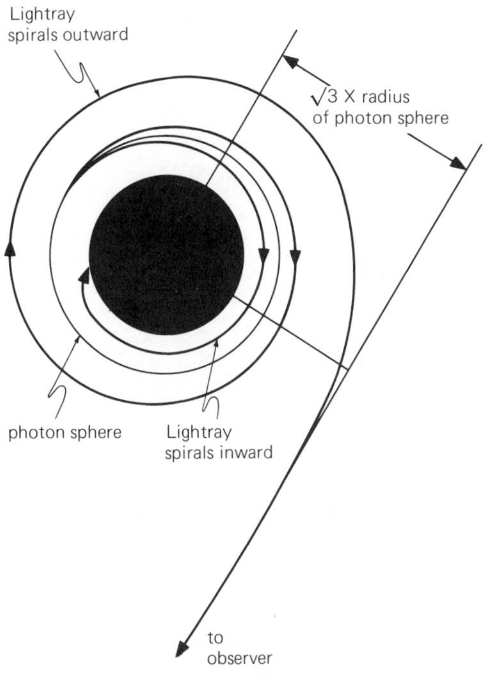

***Figure 9.4.** The photon sphere has a radius 1.5 times the Schwarzschild radius. Circular lightrays at the photon sphere either spiral in and are captured or spiral out and escape.*

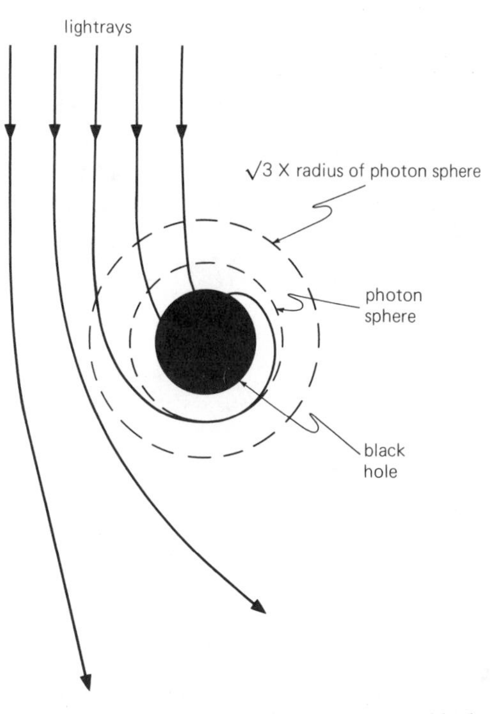

***Figure 9.5.** Deflection of lightrays by a black hole. Rays approaching closer than $\sqrt{3}$ times the radius of the photon sphere are captured.*

contracting spherical body and this time imagine that particles are shot out horizontally from a point on the surface. Usually we can find a velocity such that each particle will follow a circular orbit. Particles in circular orbits obey Kepler's law (P^2 is proportional to R^3) in general relativity, provided that the period P and the orbit radius R are measured in the space and time of a distant and stationary observer. Circular orbits, however, are not possible when the radius is less than three times the Schwarszchild value, and particles within this region spiral inward and are captured by the black hole.

LIGHTCONES AND THE EVENT HORIZON. At large distances from a black hole gravity is weak and spacetime is approximately flat and therefore the same as the spacetime used in special relativity. But close to a black hole spacetime is greatly deformed, and the intervals of space and time of an observer are not the same as those of the distant observer. When we say that a black hole has a certain radius, such as 3 or 30 kilometers, we are using not the space of a nearby observer but that of the distant observer. It is often convenient to discuss a black hole in terms of the normal space and time of the distant observer; let us therefore use our distant space and time and consider a particle falling toward a black hole.

At large distances from the black hole the particle has lightcones similar to those of the distant observer. As the particle approaches the black hole its lightcones are tilted and its future lightcone tips forward toward the black hole, as shown in Figure 9.6. This tilting of lightcones is caused by the curvature of spacetime. When the particle reaches the Schwarzschild surface its future lightcone is tipped so far forward that all light emitted by the particle falls into the black hole and no light can escape to the outer world. Its past lightcone is also tipped so far backward that light is received only from the outside world, and the particle therefore

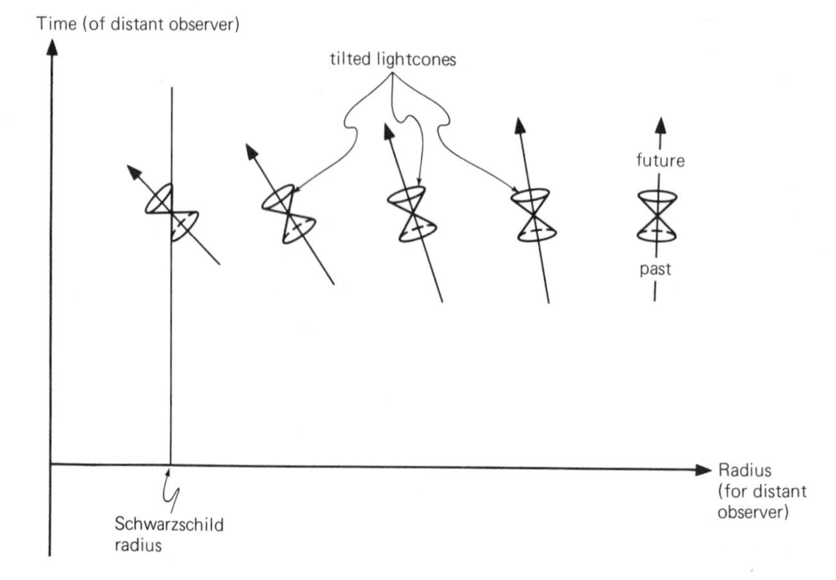

Figure 9.6. The effect of spacetime curvature near a black hole. Lightcones are tilted in such a way that the future-pointing lightcone tips toward the black hole and the past-pointing lightcone tips away from the black hole. At the surface of the black hole (the Schwarzschild surface), all rays emitted in the future direction fall into the black hole, and no rays from the past are received from the black hole. A person passing into a black hole therefore receives no information of what lies ahead.

has no advance warning that it is entering a black hole. It sees only the world it is leaving and not the fate that awaits it. Inside the Schwarzschild surface the lightcones are tilted even further; all light emitted by the particle moves ahead into the singularity, but the particle cannot see the singularity into which it is plummeting. The free-falling particle's own spacetime, in a small local region, is always flat and the same as that of ordinary special relativity, and therefore, the particle passes smoothly into the black hole without realizing that anything unusual has happened. If we are measuring its own intervals of time, it reaches the singularity, after entering the black hole, in a time equal to the Schwarzschild radius divided by the speed of light. To the distant observer, however, the particle takes an infinite time to enter the black hole.

Static points and *wavefront circles* help us to understand the effect of spacetime curvature (see Figure 9.7). A static point is fixed in space and does not move relative to the black hole and the distant observer. If we suppose that the static point emits a short burst of radiation, the wavefront circle shows us where the rays of light have reached a moment later. Far from the black hole spacetime is flat and the wavefront circle is centered on the static point. Near the black hole the circle is not centered on the point but is displaced toward the black hole. Curved spacetime displaces the wavefront circle, and we can imagine that gravity is dragging the lightrays into regions of stronger gravitation.

This dragging effect is the same as if space itself were flowing into the black hole and carrying the lightrays with it. At the Schwarzschild surface the static point is located on the wavefront circle, as if space were flowing inward at the speed of light. Lightrays moving outward at the Schwarzschild surface remain in the same place; they move locally at the speed of light and travel through space that is itself falling in at the speed of light. The surface of the black hole is the country of the Red Queen where one must move as fast as possible in order to remain on the same spot.

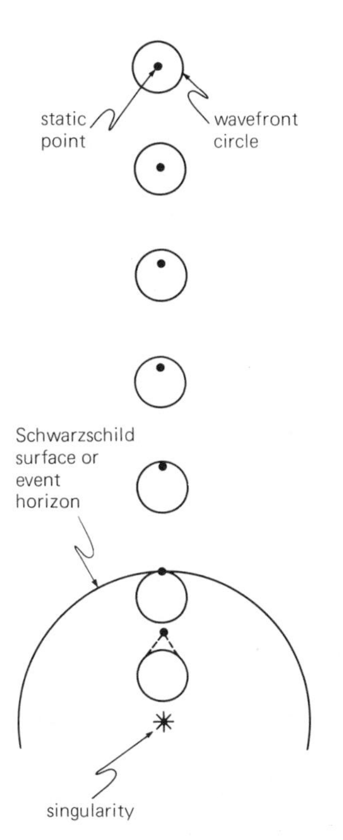

Figure 9.7. Static points and wavefront circles. A static point emits a pulse of light in all directions. A moment later the rays reach a surface shown as the wavefront circle. Far from the black hole the static point lies in the center of the wavefront circle. Near the black hole the rays are dragged inward, and the wavefront circle is displaced toward the black hole. At the Schwarzschild surface, the static point lies on the wavefront circle, and therefore no rays escape outward.

The surface of a black hole where space in effect falls inward at the speed of light is known as the *event horizon:* Events inside the horizon can never communicate with the events outside the horizon, because light signals cannot travel faster than the speed of light in local space (see Figure 9.8). A signal traveling outward at the speed of light remains static at the event horizon. Inside the horizon space flows inward faster than the speed of light, and outward-moving signals, traveling through space at the speed of light, are dragged in and cannot reach the horizon. Special relativity is still valid in

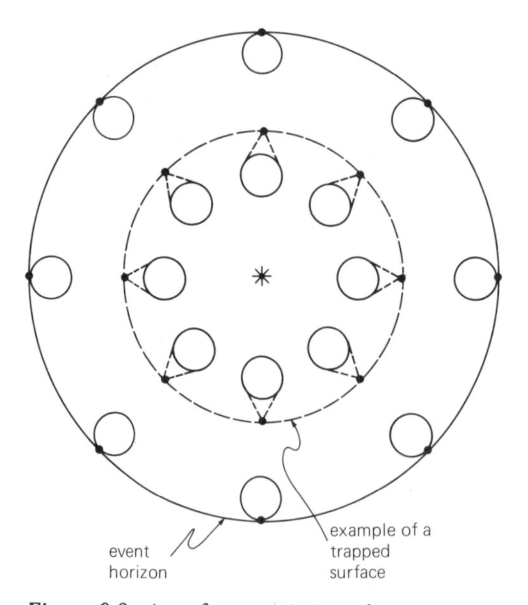

Figure 9.8. A surface consisting of static points in contact with their wavefront circles forms an event horizon of a nonrotating black hole. The event horizon is a one-way membrane: Light can move inward but not outward. Inside the event horizon all static points lie outside their wavefront circles. A closed surface consisting of static points outside their wavefront circles is called a trapped surface. *The event horizon is the outermost trapped surface. Roger Penrose and Stephen Hawking have shown that when there are trapped surfaces, collapse to a singularity is inevitable.*

small local regions; for instance, the local speed of light is the speed that we use, and nothing can move through space faster than light. But on the large scale, space is so deformed that we can regard it as flowing inward with a speed that has no limit, and small local regions – in which special relativity is valid – are carried along with it. If we were in a spaceship that fell into a black hole, we could never escape; no matter how strong the thrust of the rocket engines, we could not exceed the speed of light in local space and would therefore be carried to our doom.

PERFECT SYMMETRY. Black holes of the kind we have considered are perfectly spherical. This raises some interesting questions. For instance, if the collapsing body is not a perfect sphere, does this mean that the final black hole is not exactly spherical? The answer is no, for in the final moments of collapse the nonspherical irregularities in the gravitational field are radiated away as gravitational waves, leaving behind a spherical black hole. All long-range fields of force capable of having spherical symmetry, such as gravitational and electric fields, survive collapse; all fields of force that are not long range or perfectly radial, such as magnetic fields, are either sucked in or radiated away (see Figure 9.9),

It is sometimes asked how it is that, if nothing can escape, gravity can reach out beyond a black hole. One way of tackling this question is to think of gravity as radial lines of force: The collapsing body slides down these lines of force, and they are hence able to survive in the outside world. Another way is to think of the gravitation as a fossil field, forever frozen in the state it had at the last moments of collapse before the black hole was formed. But the best way of all is to realize that gravity is curved spacetime, and spacetime is continuous and cannot terminate at an edge. A black hole is not an

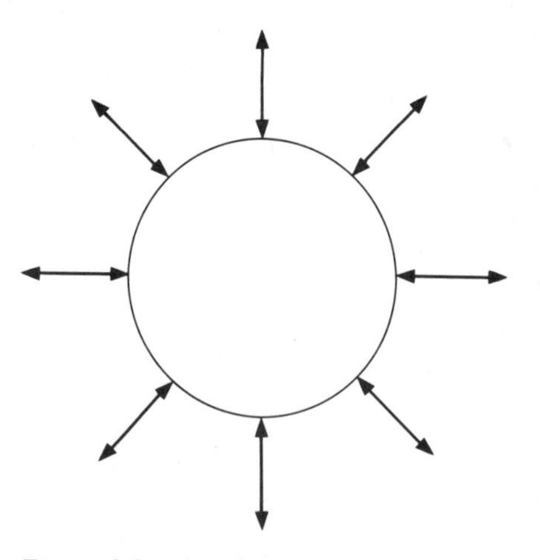

Figure 9.9. The "hairy theorem." Only "hair" that sticks out straight survives, and all else is sucked in or radiated away. Thus electric lines of force survive (hence a black hole preserves its electric charge), but magnetic lines of force are sucked in.

isolated universe of its own, but remains part of our universe.

ROTATING BLACK HOLES

ROTATION. Up till now we have considered black holes as having mass and nothing else. More generally, there are three basic quantities that determine a black hole: mass, rotation (or rather angular momentum), and electric charge. These are the only properties that survive when a collapsing body becomes a black hole; all other characteristics are either dragged in or radiated away.

Strong concentrations of electric charge are rare in astronomy, because they are easily neutralized, and most massive bodies tend to be electrically neutral. Consequently, we shall not explore the effects of electric charge in black holes. Rotation, on the other hand, is quite common in astronomy and cannot be ignored.

Almost all matter in the universe, whether concentrated in stars or dispersed as gas between stars, has too much rotation to collapse into black holes. The majority of stars would fly apart because of centrifugal force long before they had reached the size of black holes. But stars in their death throes have various tricks for getting rid of the rapid rotation of their contracting cores. The neutrinos that stream out of the core carry away some of the rotation, and the spinning core is also braked by the magnetic fields that link it to the overlying mantle. By such means the core is able to collapse and become a black hole, but there is little doubt that most black holes are born with rapid rotation – some may spin as fast as 10,000 revolutions per second.

The gas dispersed between the stars is generally in a swirling state of motion; so it cannot fall from a large distance directly into a black hole. Where gas is sufficiently dense it accumulates about a black hole and rotates as a disk-shaped cloud. Such a rotating cloud, called an *accretion disk* (see Figure 9.10), is supported against gravity by centrifugal force like the rings of Saturn. Because of Kepler's third law the inner parts of the disk rotate more rapidly than its outer parts, and friction within the disk slowly brakes the inner parts. As a result, much of the matter in the disk spirals inward until it is finally accreted by the black hole; at the same time, the accretion disk is continually replenished with gas falling in from large distances. The gas in galactic nuclei and in close binary systems is generally dense enough to form accretion disks. In these places, even if a black hole initially were not

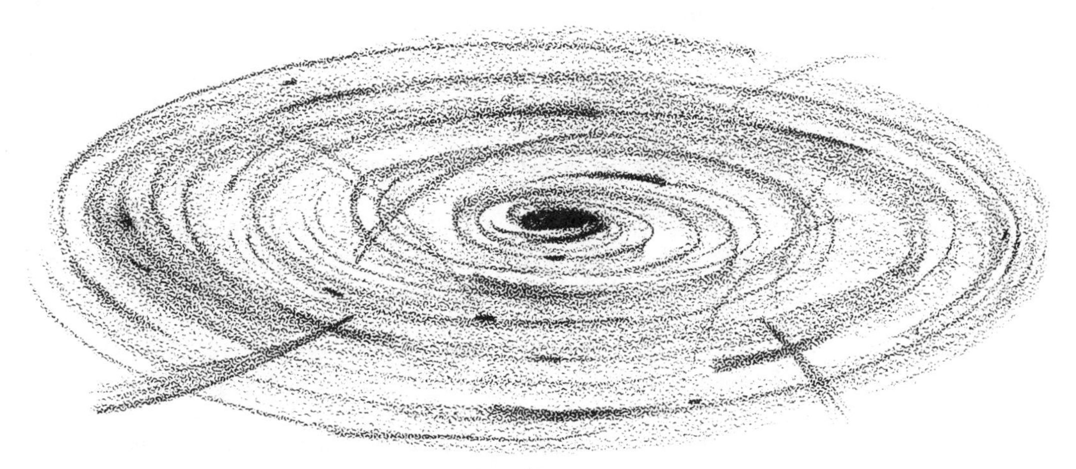

Figure 9.10. "All-devouring, all-destroying/Never finding full repast/Till I eat the world at last" (Jonathan Swift, "On Time"). An accretion disk encircling a black hole. Gas in the disk spirals into the black hole, which grows in mass.

rotating, it would acquire rotation from the swirling gas that is accreted.

KERR METRIC. The solution of Einstein's equation for the exterior spacetime of a rotating black hole was discovered in 1963 by Roy Kerr. The Kerr spacetime, or Kerr metric as it is called, is not quite so simple as the previous Schwarzschild spacetime. In the discussion of nonrotating black holes we saw that spacetime curvature causes the future lightcone to be tilted in the direction of the black hole. Now, in the neighborhood of a rotating black hole, we have the additional effect that spacetime curvature causes the future lightcone to tilt also in the direction of rotation.

We have imagined space falling into a nonrotating black hole, thus creating an event horizon at the Schwarzschild surface where space rushes in at the speed of light. We must now imagine that space also rotates like a whirlpool as it flows into a rotating black hole. Particles caught in this maelstrom are carried around as they are dragged inward.

We can visualize rather easily what happens by using static points and wavefront circles (see Figure 9.11). At large distances from a rotating black hole spacetime is flat and the static point is in the center of the wavefront circle. Near the black hole the lightrays emitted by the static point are dragged inward and around, and as a result the wavefront circle is displaced partly in the inward direction and partly in the direction of rotation. This creates two distinctly different surfaces about the black hole. The first surface, known as the static-limit surface (or just the *static surface*), is where space flows at the speed of light. At this surface the static points are at the edges of their wavefront circles. The second surface is the event horizon where space flows in the inward direction at the speed of light. At this surface all static points have their wavefront circles touching the horizon.

The outer static surface has the shape of a flattened spheroid, as shown in Figure 9.12, and has an equatorial radius equal to

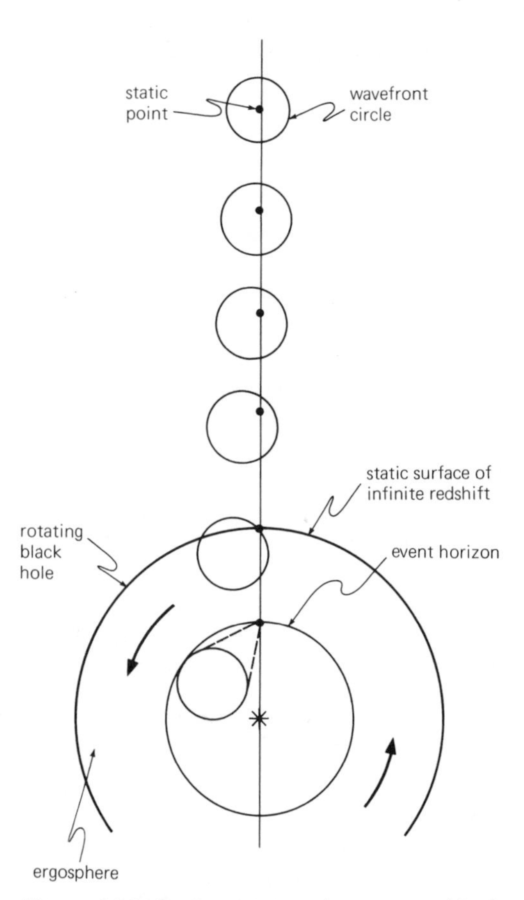

***Figure 9.11.** In the vicinity of a rotating black hole all wavefront circles are displaced inward and also in the direction of rotation. We can imagine space as a whirlpool that rotates as it flows into a rotating black hole. There are now two surfaces enclosing a black hole: the outer static surface and the inner event horizon. At the static surface all static points lie on their wavefront circles. At the event horizon all static points have their wavefront circles touching, but not overlapping, the horizon.*

the Schwarzschild radius. It is the surface where space flows at the speed of light, and a particle, stationary in this space, is dragged along at the speed of light relative to the distant observer. If the particle moves against the flow of space at the speed of light it will remain static, on the static surface (hence its name). The wavefront circle protrudes outside the static circle, and rays of light can therefore still escape to infinity. Inside the static surface space flows faster

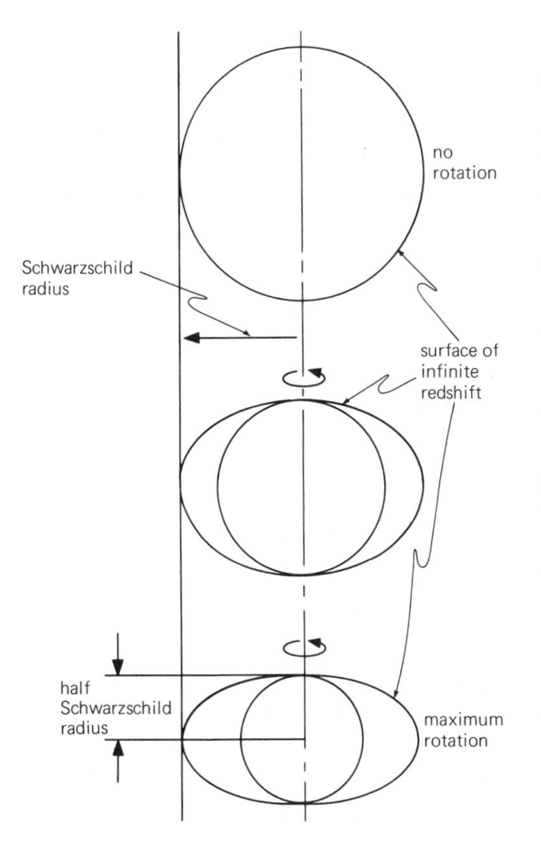

Figure 9.12. Black holes of the same mass with different amounts of rotation.

than the speed of light, so that any particle, no matter how fast it moves against the flow, is always dragged along and can never be static relative to the distant observer.

The inner surface, the event horizon, is spherical in shape. Across this surface space moves with an inward component of velocity that equals the speed of light. Lightrays traveling in any direction cannot escape to the outside world. Both surfaces, the static surface and the event horizon, are in contact at the rotation axis. In a nonrotating black hole both surfaces merge together and have the Schwarzschild radius. As rotation increases, the spherical horizon shrinks to a smaller radius, and at maximum rotation it has a radius that is half the Schwarzschild value. At maximum rotation the equator of the horizon rotates at the speed of light.

THE ERGOSPHERE. Between the event horizon and the static surface is the region known as the *ergosphere,* which consists of swirling space spiraling inward to the horizon (see Figure 9.13). Space flows faster than the speed of light in the ergosphere, but the inward component of velocity is less than the speed of light. All static points lie outside their wavefront circles, but some lightrays can still move outward and escape to the outside world. Roger Penrose was the first to show that it is possible, in principle, to gain energy in the ergosphere from the black hole. An object is shot into the ergosphere and made to explode into two components; one component moves against the rotation (in local space), and the other component moves forward with increased energy. The backward-moving component plunges into the event horizon, and the forward-moving component escapes across the static surface carrying energy that has come from the rotational energy of the black hole.

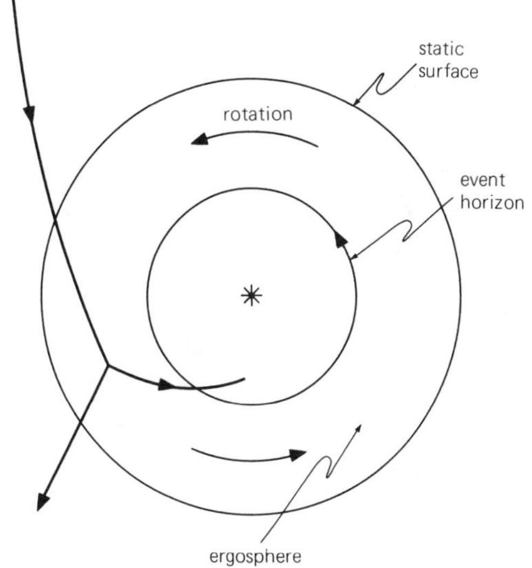

Figure 9.13. The ergosphere is the region between the outer static surface and the inner event horizon. It is possible for a body falling into the ergosphere to divide into two components in such a way that the outgoing component has more energy (or mass) than the initial infalling body. The gain in energy is at the expense of the rotational energy of the black hole.

Some of the mass of a black hole is due solely to its rotational energy. A black hole of maximum rotation has 29 percent of its mass contributed by the rotational energy, and this is the maximum amount of mass that can be extracted as energy by the Penrose method.

COSMIC CENSORSHIP. At maximum rotation the event horizon has a radius half the Schwarzschild value; it rotates at the speed of light, and space falls inward at the speed of light. Suppose that in some way it were possible to increase the rotation even further. In that case the horizon would vanish! The rotational velocity would increase, but the inward flow of space would decrease and would everywhere be less than the speed of light. Information could then escape, and the singularity in all its nakedness would become visible to the outside world. Naked singularities of matter crushed to almost infinite density, beyond the understanding of our known laws of physics, are so appalling a prospect that their indecency has been covered by what Penrose calls "cosmic censorship." The *cosmic censorship hypothesis* states that all singularities are cloaked from the view of the outside world by event horizons and that nature conspires in every way possible to avoid naked singularities. Perhaps universes exist in which naked singularities are common, but the eruption of energy would be so immense from a naked singularity that most likely life could not exist in these scorched universes. We exist because naked singularities do not exist, and the cosmic censorship hypothesis conforms with the anthropic principle.

SUPERHOLES AND MINIHOLES

VOYAGES TO BROBDINGNAG AND LILLIPUT. Once born, a black hole grows by accretion to Brobdingnagian size, and its growth comes to a halt only when the supply of matter is exhausted. In the nuclei of giant galaxies, where stars and gas abound, the possibilities are awesome. Black holes suck in the gaseous wreckage of ensnared stars; they even swallow each other and swell into superholes of thousands and even millions of solar masses. Gas, as it is drawn in, is compressed and heated, and radiates energy. The energy released is generally a small percentage, but may be as high as 40 percent, of the mass of the accreted gas. The conversion of mass into energy is hence much more efficient than in nuclear reactions. The energy is radiated over a wide spectrum and consists of infrared, optical, ultraviolet, and X-ray radiation.

It is estimated that superholes of hundreds of millions of solar masses grow in a time of a billion years in the matter-rich nuclei of giant galaxies and release energy continually at a prodigious rate. At present the "best-buy" theory of quasars is that they are accreting superholes, and it is possible that the powerful radio sources also gain their energy from superholes.

The tidal force produced by a black hole will tear apart a body, such as a planet or a star, when it is stronger than the gravity that holds the body together. As a rough-and-ready rule we can say that a body held together by its own gravity will be tidally disrupted by a black hole when its average density is less than the density of the black hole. The size of a black hole is proportional to its mass, and this means that its density is inversely proportional to the square of its mass (see Figure 9.14). A black hole of 1 solar mass has a density of 10^{16} grams per cubic centimeter; a black hole of 10^8 solar masses has therefore a density of 1 gram per cubic centimeter; and black holes of mass greater than 3 billion solar masses have a density less than that of air at sea level. A star similar to the Sun, with a density of approximately 1 gram per cubic centimeter, will be torn apart by tidal forces if it approaches a black hole of mass less than 10^8 solar masses. The gaseous remnants of the disrupted star will be sucked in and will release energy. Gas that is drawn inward and compressed is able to radiate energy efficiently, but once a swollen black hole begins to swallow stars whole, it is likely that

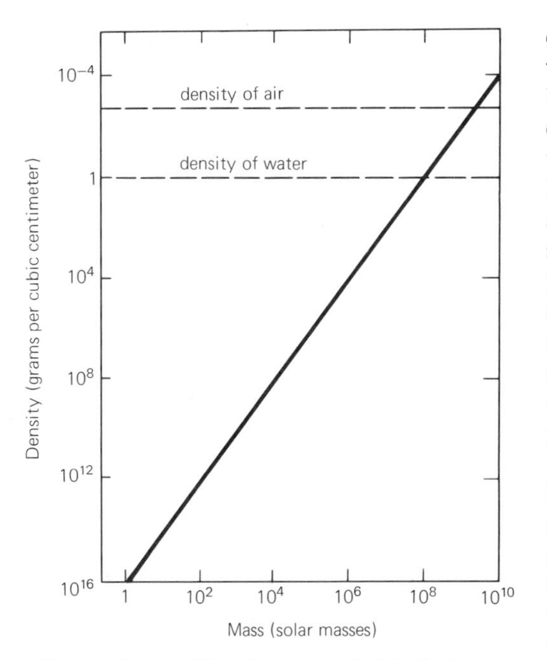

Figure 9.14. ***The density of black holes decreases as their mass increases. Thus a black hole of the density of water has a mass a hundred million times that of the Sun.***

its energy output will wane. Stars approaching black holes of mass much greater than 10^8 solar masses will not be torn apart; instead, they will dive straight in without releasing very much energy.

A spaceship could plunge into a superhole of a million or so solar masses without its human occupants experiencing discomfort from tidal forces. In the case of a quiescent superhole of much greater mass the occupants at first might not even know that they had passed the event horizon.

All black holes so far discussed have masses greater than that of the Sun, and it is unlikely that Lilliputian black holes of lesser mass are ever produced by gravitational collapse. A planet, such as the Earth, for example, can never collapse and become a black hole, because gravity in these smaller bodies is never strong enough to overwhelm the pressure of matter during collapse.

Yet it is conceivable that black holes of a wide range of masses were created in the early universe. During the first stages of expansion the density of the universe was very great, and it has been conjectured that black holes might have been born with densities comparable to the density of the universe. In this case, black holes would not have been caused by catastrophic collapse, and hence it would have been possible for small black holes to form. The formation of primordial black holes would depend on the condition of matter in the early universe, and also on the nature of its irregularities, and on both these subjects our knowledge is still insufficient to say for certain whether primordial black holes exist.

If they do exist, however, primordial black holes can have masses very much less than a solar mass. *Miniholes* the size of an atom have a mass of 10^{20} grams (100 trillion tons), and miniholes the size of a nucleon have a mass of 10^{15} grams (1 billion tons). If we divide the radius of a minihole by the speed of light we obtain very roughly the age of the universe at the time it was formed. An atomic-sized minihole is therefore created in a universe only 10^{-18} seconds old. The smallest of all miniholes, perhaps formed during the very earliest moments of the universe, are quantum black holes having the Planck mass of 10^{-5} grams and a size of 10^{-33} centimeters. These quantum miniholes have a mass 10^{20} times that of a nucleon mass and a size 10^{-20} that of a nucleon, and as we shall see, it is very unlikely that they now exist. Miniholes of atomic size or less do not easily accrete matter; for example, an atomic-sized minihole could pass easily through the Earth, and its mass of 100 trillion tons would increase by only 1 gram. Perhaps the universe contains primordial black holes, not only miniholes of atomic size or less, but also black holes of much greater mass. They are a real possibility and it would be surprising if none exist at all.

BLACK HOLE MAGIC

Some astronomers still scout the idea of black holes, perhaps because in recent years far-fetched notions have been advertised, verging on what might be called black-hole magic. Denied the magic of mythology, we must have it in science – hence science fiction. One such notion is the time-reversed black hole. In theory, a collapsing black hole can be reversed to make a *white hole,* in which everything rushes outward and abruptly releases immense energy. What happens when matter falls into a singularity? One answer, we are told, is that it emerges elsewhere in the universe as a white hole, at a different place and time, or even in another universe. Black holes in other universes, having equal rights, pop out in our universe as white holes. It has even been suggested that matter is siphoned from black holes back into the big bang, which is hot because the matter reappears as white holes. There is no evidence whatever that matter can be transported in this manner, and the whole concept of white holes is implausible. What falls down does not always fall up, as anyone knows who has seen a city bombed.

There is also an argument that black holes are bridges connecting widely separated regions of spacetime, perhaps used by technologically advanced civilizations as a transportation system for effortless travel to and fro in space and backward and forward in time (see Figure 9.15). Rotating black holes take the prize in modern sorcery and are said to be tunnels that connect our universe with the fairyland of an infinite number of other universes. Mathematical tricks of patching different spacetimes together are not necessarily good physics, nor do they necessarily create new physical universes.

The rule in totalitarian societies is that "everything which is not compulsory is forbidden." This rule is sometimes applied in science in the sense that everything which is not forbidden is compulsory. It is argued that if a thing is possible, then it must happen, and what is potential in nature must be actual. The totalitarian doctrine is sometimes of help in science, but it is flawed by the fact that we rarely know enough to determine what is possible and potential. There is still much that we have to learn about black holes before science-fiction speculations are constrained by the real world.

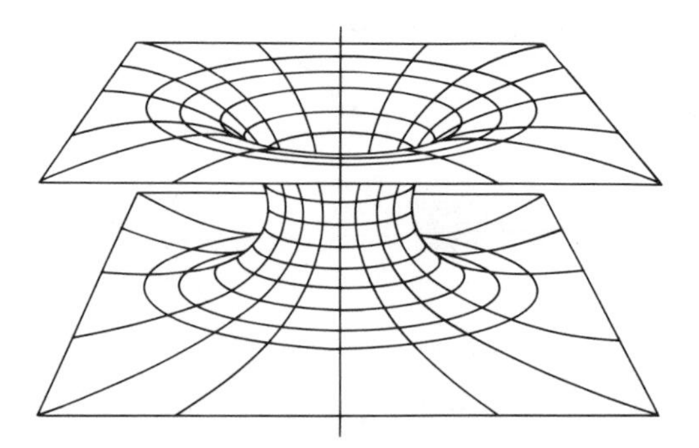

Figure 9.15. Figurative representation of a black hole bridging two universes, or separate regions of the same universe.

BLACK HOLES ARE HOT

Sir, I have found you an argument. I am not obliged to find you an understanding.
— Samuel Johnson (1709–84), in Boswell's *Life of Johnson*

In 1974, Stephen Hawking at Cambridge University showed that black holes emit thermal radiation. This remarkable discovery revealed that black holes are less black than previously supposed.

To understand why black holes radiate it is necessary to take a look at the nature of empty space. The vacuum, or empty space, is not so empty as one might at first think; it is actually a dense sea of virtual particles of every kind. Virtual particles, which are made from energy, exist for only fleeting moments. According to the uncertainty principle, energy can always be borrowed but must be paid back within a limited time. The greater the amount borrowed, the more quickly it must be repaid. For example, the energy needed to make two electrons can be borrowed for 10^{-21} seconds, and for two nucleons the time limit for repayment is 10^{-24} seconds. The whole of space is filled with virtual particles that incessantly appear and disappear. They do not exist long enough to produce gravitational fields, and yet they are responsible for many observed effects and significant changes in the structure of atoms.

Elementary particles are more than just energy. Among their various properties they possess spin and electric charge, which are conserved and cannot be borrowed. For this reason virtual particles are always accompanied by their antiparticles. A virtual electron, for instance, has negative electric charge and is accompanied by a virtual positron, which has positive charge and is the antielectron. The net charge borrowed is hence zero. The vacuum contains everywhere virtual electron pairs (electrons and positrons) whose combined spin and electric charge is zero. Particles and their antiparticles have the same mass, but their conserved quantities are always the opposite of each other. Thus, when a virtual particle and its antiparticle appear in the vacuum, only their combined energy has been borrowed, and all other quantities cancel out.

Photons, which are particles of radiation, are their own antiparticles: The antiparticle of a photon is another photon of opposite spin; both kinds are equally abundant and are emitted equally by luminous bodies. Luminous bodies of matter or of antimatter emit similar photons. The vacuum seethes with virtual photons that are usually associated with other virtual particles.

Whenever a virtual particle and its antiparticle can gain sufficient energy from somewhere, they become real particles and do not have to return to the limbo of the vacuum state (see Figure 9.16). Such energy can be gained in various ways. We shall consider two examples.

Consider an electric field set up between two parallel conducting plates at different potentials. By increasing the potential difference between the two plates we increase the strength of the electric field. In the space between the plates are hordes of virtual particle pairs, including electron pairs, each existing for only a brief period of time. During their short existences, a virtual electron is attracted toward the positive plate and a virtual positron is attracted toward the negative plate. Usually, in their short lifetimes, they move hardly at all

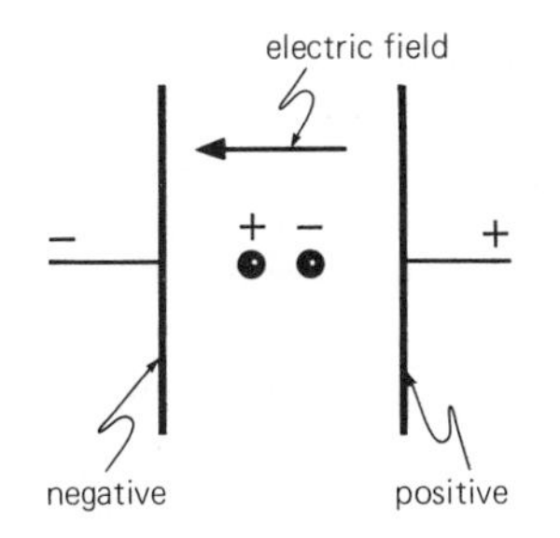

Figure 9.16. A virtual pair of electrons in the vacuum. In the brief time in which they exist, if they can be pulled apart with sufficient energy by an electric field or a tidal force, then they become real.

because the electric field is far too weak. But when the electric field is made intensely strong, approaching 10^{15} volts per centimeter, the oppositely charged electrons and positrons move apart a sufficient distance to gain the requisite energy to become real. Under these extreme circumstances the vacuum springs to life, and as energy is poured into it, a deluge of electrons, positrons, photons, and other particles is created.

Consider now virtual particles in a gravitational field. All are accelerated in the same direction, and the tidal force tends to tear virtual pairs apart. Black holes have the strongest of all tidal forces, and the smaller they are, the stronger are their tidal forces. When the tidal force is very strong, a virtual particle and its antiparticle are able to move apart a sufficient distance to gain the requisite energy to become real. Most particles created in this way in the vicinity of a black hole fall inward, and no energy is lost from the black hole; but some virtual pairs are torn apart with enough violence for a few particles to move outward and escape. Accompanying these particles are many photons also created from the vacuum state. Black holes therefore emit particles, including photons and neutrinos, and are continually losing energy.

Particles are wavelike and can tunnel through barriers that normally are impenetrable. This offers another way of looking at particle and photon emission from black holes. A virtual particle can tunnel through the event horizon and emerge as a wave (see Figure 9.17). When the tidal force is sufficiently strong to separate a particle from its antiparticle, the requisite energy is gained, and the emerging wave becomes a real particle that can escape.

Hawking discovered that the emission of particles and photons is thermal – just the same as that from hot bodies – and black holes are therefore hot. The temperature of a black hole is usually not very high:

$$\text{temperature} = 10^{-7}\frac{M_{\odot}}{M}\text{ degrees Kelvin}$$

but it increases as the mass decreases.

Black holes of solar mass or greater have extremely low temperatures. Even a black hole of radius 1 centimeter, of a mass equal to that of the Earth, has a temperature of only 1/30 of a degree. A black hole the size of a speck of dust — 1/1000 of a centimeter in radius, with a mass 1/10 that of the Moon – has a temperature of 30 degrees. But miniholes of atomic size have temperatures of 3 million degrees and emit intense radiation. Hawking's discovery has shown us that black holes are actually not black but can be intensely bright. They radiate energy and as a consequence their masses slowly decrease.

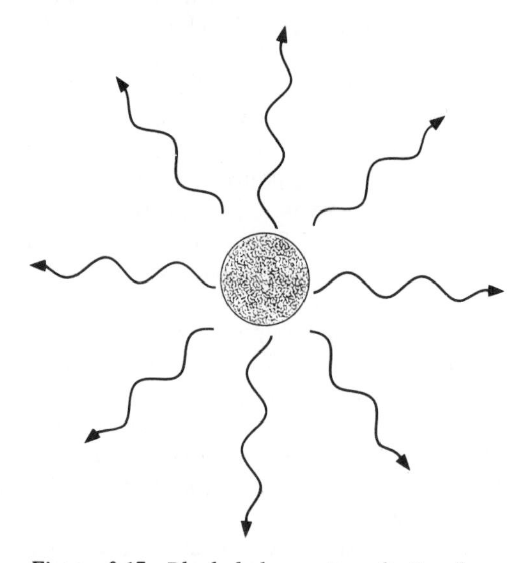

Figure 9.17. Black holes emit radiation by a quantum mechanical process. The wavelength is approximately equal to the size of the black hole.

With decreasing masses they get hotter and radiate faster (see Figure 9.18). Given sufficient time, all black holes will radiate away their entire masses and leave behind nothing but the radiation they have emitted: The equation giving this evaporation time is approximately

$$\text{evaporation time} = 10^{66}\,(M/M_{\odot})^{3}\text{ years}$$

In most cases it is an extremely long time: Even an intensely bright minihole of atomic size lasts for 10^{26} years; a minihole of nucleon size has a lifetime of 10^{10} years, which is roughly the age of the universe. This means that all primordial black holes of mass less than 10^{16} grams have burned themselves up and have vanished, and only those miniholes originally of greater mass now exist.

As a minihole dwindles in mass its temperature and luminosity rise, and in the last moment, it erupts in a crescendo of high-energy particles and photons. It is possible that in the future astronomers will discover these exploding miniholes.

A black hole has mass, rotation, and electric charge, and nothing else. When an object falls into a black hole, it contributes to these three properties, and everything else pertaining to the falling object is forever lost as far as the outside world is concerned. Each time an object falls into a black hole a vast amount of information is lost that can never be recovered.

A black hole might contain matter or antimatter, it might contain smaller black holes, or it might contain just radiation. In the course of time it loses mass that is radiated away mainly as photons. It also emits particles such as electrons and positrons, and always the same number of particles are emitted as antiparticles. From its emission it is impossible to tell what the black hole was originally made from. Given sufficient time, the black hole will evaporate away, and even in the final torrent of radiation it will emit equal numbers of particles and antiparticles. The emitted particles and antiparticles can then annihilate each other

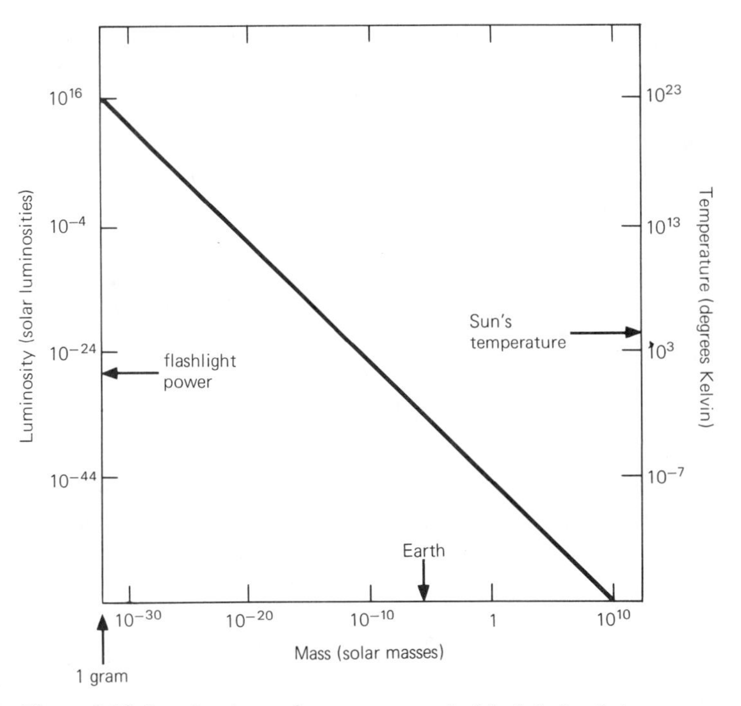

Figure 9.18. Luminosity and temperature of a black hole of given mass.

and produce further photons. It comes to this: A black hole is a machine for converting mass into radiation, and it has no memory of whether it was originally made of matter or of antimatter. It ignores the conservation of such things as baryon and lepton numbers (see Reflections section of this chapter).

In recent years it has become apparent that black holes are heat engines that obey the laws of thermodynamics. They have temperature and therefore entropy. Entropy is a measure of information lost; it is also a measure of thermal disorder; and in a closed system it either remains constant or increases, but never decreases. We can understand entropy if we think of energy in its various forms as forever cascading into less useful and accessible states: In such a case, entropy is increasing. A black hole is a sink of lost information; all the detailed structure of the things from which it was made, such as flowers, crystals, books, and other equally intricate objects, has vanished, and in return the black hole gives us photons of high entropy (see Figure 9.19).

The lost information in a black hole is proportional to the surface area of its event horizon. This area is hence a measure of its entropy. The area of the event horizon normally stays constant; although it increases in the case of accretion, it never decreases. When two black holes encounter each other and coalesce, they form a single black hole whose horizon area is greater than the sum of the horizon areas of the original black holes. A black hole can never split into two black holes because the total horizon area would decrease.

But we have seen that a black hole is hot and slowly radiates away its mass. The surface area of a black hole is proportional to its radius squared, and is therefore proportional to its mass squared. As its mass decreases, owing to radiation, its horizon area also decreases, and as a consequence the entropy gets less. This, however, does not violate the new law of black-hole thermodynamics. The entropy of the black hole decreases because the radiation emitted carries away entropy into space. The closed system in which entropy is constant or

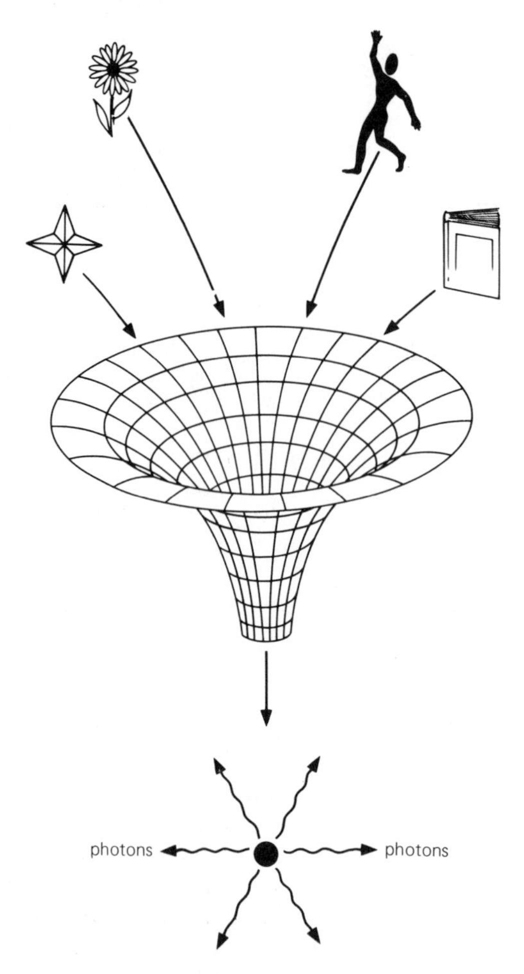

Figure 9.19. A black hole is a sink of lost information, a machine that creates entropy.

increases is not just the black hole by itself, but the universe, and in the universe the entropy of the black hole plus the entropy of the radiation emitted does not decrease. Black-hole thermodynamics, now securely founded on the theoretical discovery that black holes are hot, has opened up an intriguing new realm of physics.

REFLECTIONS

1 *Does an object falling into a black hole vanish from the universe?*

* *How can the presence of a black hole be detected?*

* *What is seen as one enters a black hole?*

* *Why is the event horizon a "one-way membrane"?*

* *How big would the Galaxy be if it were a black hole?*

* *Can a black hole be composed of smaller black holes?*

* *Is a minihole black or white?*

* *Consider a particle inside a black hole and a particle outside. Which reaches the singularity first in a collapsing universe?*

* *Use the anthropic principle to show why white holes and naked singularities are implausible.*

* *"The very hairs of your head are all numbered" (Matthew 10:30). Is this true of a friend who has fallen into a black hole?*

* *Discuss matter and antimatter.*

* *Why are black holes like heat engines? What is so strange about them?*

* *Is the universe a black hole? An essential feature of a black hole is that its interior spacetime is continuous with the spacetime of the outside world. It is* black *and is a* hole *as viewed from the outside world. The universe, however, is self-contained and is not continuous with an external world. Does the statement "the universe is a black hole" violate containment?*

2 *It has been suggested by some astronomers that the Galaxy contains billions of black holes. If this is true, a black hole one day may come our way, and the particles of our bodies in company with the wreckage of the Solar System will then be engulfed. Instead of continuing to exist for tens of billions of years, or an eternity, they will exist for only a fraction of a second. It is also possible that one day the universe will collapse and the particles of our bodies, in company with those in black holes, will fall into the oblivion of a big bang.*

Suppose that a person with a soul falls into a black hole. This person is eventually converted into photons. Where has the soul gone? Does this situation prove that souls are not contained in space and time?

* *The totalitarian doctrine was first enunciated by John T. Whittaker in "Italy's Seven Secrets,"* Saturday Evening Post, *December 23, 1939, p. 53. He wrote: "Cof-*

fee is forbidden, the use of motorcars banned and meat proscribed twice a week, until one says of Fascism, 'Everything which is not compulsory is forbidden.'" Show that the totalitarian principle, in the reversed form "what is not forbidden is compulsory," as applied to science, has much in common with the principle of plentitude.

3 *Even Newtonian laws warn us that something odd must happen when a star is very massive. This was realized in 1784 by John Mitchell, rector of Thornhill in Yorkshire, who was an innovative astronomer (see "John Mitchell and black holes," by Simon Schaffer, 1979). In order to escape from the Sun's surface a particle must have a speed of 1/500 of the speed of light. Mitchell argued that if a star had the same average density as that of the Sun, and a radius 500 times greater, then light would be unable to escape from the surface of the star. "All light emitted from such a body would be made to return to it, by its own power of gravity," he said. William Herschel was impressed with Mitchell's theory and used it to interpret some of his own observations (but erroneously, as we now know). The search for evidence of black holes is 200 years old!*

Mitchell used Newton's idea that light-rays consist of particles, and he assumed that they obey the Newtonian equations of motion. The escape speed v from the surface of a body of mass M and radius R is given by

$$v^2 = \frac{2GM}{R}$$

and when v is equal to the speed of light c, or the radius is

$$R_s = \frac{2GM}{c^2}$$

then the light is extinguished because it is unable to escape.

Pierre Simon, Marquis de Laplace, in Exposition of the System of the World, *published at the end of the eighteenth century, made a similar prediction "that the attractive force of a heavenly body could be so large that light could not flow out of it," and it is possible that he was aware of Mitchell's earlier work. Laplace's discussion is translated into English in* The Large Scale Structure of the Universe, *by Stephen Hawking and George Ellis. It has become common in recent years to cite the work of Laplace, and Mitchell's earlier work has been overlooked.*

* *Karl Schwarzschild in 1916, shortly after Einstein had published the final version of general relativity, solved the Einstein equation for the exterior spacetime of a spherical nonrotating body. This solution shows that there is an infinite redshift when a body of mass M contracts to a radius R_s, now known as the Schwarzschild radius, given by the relation $R_s = 2GM/c2$. It is interesting that general relativity gives the same result as Newtonian theory, although for different reasons. When the mass of a spherical body is measured in units of the Sun's mass, the Schwarzschild radius is simply*

$$R_s = 3\frac{M}{M_\odot}\text{ kilometers}$$

If the Sun were to contract to a radius of 3 kilometers it would become a black hole. The Earth's mass is $3 \times 10^{-6} M_\odot$, and if the Earth were to contract to a radius of approximately 1 centimeter, it also would become a black hole.

Subsequent developments were as follows:

In 1930, Subrahmanyan Chandrasekhar showed that white dwarfs of mass greater than $1.4\ M_\odot$ cannot be supported against gravity by electron pressure and must collapse.

In 1934, Walter Baade and Fritz Zwicky advanced the concept of neutron stars. They proposed that these dense objects are born in catastrophic stellar events, which they termed supernovas. *This was only 2 years after the discovery of the neutron by James Chadwick.*

Robert Oppenheimer and George Volkoff in 1939 used general relativity to investigate the structure of neutron stars. In the same year, Oppenheimer and Hartland

Snyder studied the collapse of a spherical body from the point of view of inside and outside observers.

In 1963, Roy Kerr discovered the equivalent of the Schwarzschild spacetime for rotating black holes. The discovery of quasars in 1963 stimulated renewed interest in gravitational theory, and in the same year, Fred Hoyle and William Fowler proposed that the energy released is gravitational in origin and comes from supermassive objects in the nuclei of galaxies. In 1964, Edwin Salpeter of Cornell University proposed that the supermassive objects are black holes. Pulsars were discovered in 1967 and Thomas Gold in 1968 proposed that pulsars are rotating neutron stars. The term black hole *was first used by John Wheeler in 1968 in an article entitled "Our universe: the known and the unknown." Many theoretical discoveries have since been made concerning the nature of black holes, of which the most outstanding is Stephen Hawking's demonstration that black holes emit thermal radiation.*

4 *The word* antimatter *was first coined in 1898 by the scientist Arthur Schuster in a letter to* Nature, *the science journal, entitled "Potential matter: a holiday dream." He discussed the possible properties of antimatter, speculated on the existence of antistars, and said: "Astronomy, the oldest and yet most juvenile of the sciences, may still have some surprises in store. May antimatter be commended to its care!"*

Paul Dirac in 1930 developed a relativistic equation in quantum theory that indicated the existence of antielectrons of positive electric charge and antiprotons of negative charge. Only two years later the positron was discovered, and the antiproton was found in 1955. Numerous particles and their antiparticles have since been discovered with high-energy accelerators. Particles and antiparticles are distinguished by conserved quantities of opposite sign, such as electric charge, and baryon and lepton numbers. Baryons, *such as nucleons, are heavy particles (their name derives from the Greek* bary, *meaning heavy). Each has a* baryon number *of +1. Antiprotons and antineutrons each have a baryon number of −1. A nucleon and its antinucleon have a combined baryon number of zero, and can therefore annihilate and produce photons that have no baryon number.* Leptons, *such as electrons and neutrinos, are light particles (their name derives from the Greek* lepto, *meaning small or light). Each has a* lepton number *of +1; their antiparticles, the positrons and antineutrinos, each have a lepton number of −1. An electron and a positron have a combined lepton number equal to zero and can therefore annihilate, and their energy goes into photons that have no lepton number. Baryon and lepton numbers are conserved in all particle interactions. This is why it is impossible to annihilate matter by itself or convert it into antimatter. But black holes apparently ignore these powerful laws of conservation. Once an object has passed beyond the event horizon we have lost all information on whether it was made of matter or antimatter.*

Matter is made of ordinary particles and antimatter of antiparticles. Everything in our part of the universe – the Earth, Solar System, and the Galaxy – is apparently made of matter only. A substantial amount of antimatter in the Galaxy, if it existed, would be betrayed by its violent and recognizable interactions with matter. We see no sign of a copious emission of high-energy photons, known as gamma rays, of the expected energy. What about other places in the universe? Antistars in antigalaxies emit the same kind of photons as stars in galaxies, and we cannot distinguish them by their radiation. Most galaxies are members of clusters, however, and if antigalaxies are mixed with galaxies we should be able to see unmistakable signs of their dislike for each other. There are indications of violence, it is true, but not of the kind that emits gamma rays of distinctive energies. The gas dispersed between galaxies in rich clusters does not contain an admixture of antigas because we again do not see the expected radiation that would result from mutual annihilation. The evidence so far suggests that antimatter is rare in the universe. According to recent grand unified theories the universe is now made only of matter because in the very early universe the dominant hyperweak force favored matter slightly more than antimatter.

5 *The thermal radiation emitted from a hot body has a characteristic wavelength, which is the wavelength corresponding to the point at which the spectrum has its maximum intensity. For radiation of temperature T this wavelength, measured in centimeters, is equal to 0.3/T. A black hole has a size, also measured in centimeters, of* $3 \times 10^5 M/M_\odot$. *Because a black hole and the thermal radiation it emits both have characteristic scales of length, it would not be too surprising to find that they are approximately equal. By equating the two we obtain a result that is not greatly different from Hawking's more exact calculation. (The agreement is better if we equate the wavelength to the circumference of the black hole.)*

FURTHER READING

Davies, P. C. W. *Space and Time in the Modern Universe.* Cambridge University Press, Cambridge, 1977.

Gursky, H., and Heuvel, E. van den. "X-ray emitting double stars." *Scientific American,* March 1975.

Hawking, S. "The quantum mechanics of black holes." *Scientific American,* January 1977.

Penrose, R. "Black holes." *Scientific American,* May 1972.

Peters, P. C. "Black holes: new horizons in gravitational theory." *American Scientist,* September–October 1974.

Ruffini, R., and Wheeler, J. A. "Introducing the black hole." *Physics Today,* January 1971.

Sciama, D. W. "The ether transmogrified." *New Scientist,* February 2, 1978.

Smarr, L. L., and Press, W. H. "Our elastic spacetime: black holes and gravitational waves." *American Scientist,* January–February 1978.

Sullivan, W. "A hole in the sky." *New York Times Magazine,* July 14, 1974.

Sullivan, W. *Black Holes.* Anchor Press, Doubleday, Garden City, N.Y., 1979.

Thorne, K. S. "The search for black holes." *Scientific American,* December 1974.

Wald, R. "Particle creation near black holes." *American Scientist,* September–October 1977.

Wheeler, J. A. "Our universe: the known and the unknown." *American Scholar 37,* 248 (1968) and *American Scientist,* 1968. First mention of the term *black hole.*

Wheeler, J. C. "After the supernova, what?" *American Scientist,* January–February 1973.

SOURCES

Gold, T. "Multiple universes." *Nature 242,* 24 (March 2, 1973).

Hawking, S. W., and Ellis, G. F. R. *The Large Scale Structure of Space–Time.* Cambridge University Press, Cambridge, 1973, pp. 365–8.

Misner, C. W., Thorne, K. S., and Wheeler, J. A. *Gravitation.* W. H. Freeman, San Francisco, 1973.

Oppenheimer, J. R., and Snyder, H. "On continued gravitational contraction." *Physical Review 56,* 455 (1939).

Penrose, R. "Gravitational collapse: the role of general relativity." *Revista Nuovo Cimento,* numero special 1, 252 (1969).

Schaffer, S. "John Mitchell and black holes." *Journal for the History of Astronomy 10,* 42 (1979).

Schuster, A. "Potential matter: a holiday dream." *Nature 58,* 367 (August 18, 1898).

Sciama, D. W. "Black holes and their thermodynamics." *Vistas in Astronomy 19,* 385 (1976).

EXPANSION

For the history that I require and design, special care is to be taken that it be of wide range and made to the measure of the universe. For the world is not to be narrowed till it will go into the understanding (which has been done hitherto), but the understanding is to be expanded and opened till it can take in the image of the world.
— Bacon (1561–1626), *Novum Organum*

THE GREAT DISCOVERY

DOPPLER SHIFTS. From a historical viewpoint the *Doppler formula,* discovered in 1842 by the Austrian scientist Christian Doppler, paved the way to the discovery of the expanding universe. Nowadays we do not use the Doppler formula in cosmology, and we shall therefore examine it only briefly and defer to Chapter 11 a more careful study of cosmic redshifts.

A spectrum of the light from a luminous source consists of bright and dark narrow regions: These are the emission lines and absorption lines produced by excited atoms in a hot gas. When a luminous source, such as a candle or a star, moves away from an observer, the wavelengths of the radiation received are increased and the spectral lines are shifted toward the redder end of the spectrum. A receding luminous source therefore has *redshift*. The source may move away from the observer, or the observer may move away from the source, and in either case the separating distance increases and there is an observed redshift. When the luminous source moves toward the observer, the wavelengths of the radiation received are decreased and the spectral lines are shifted toward the bluer end of the spectrum. An approaching luminous source therefore has *blueshift*. The source may move toward the observer or the observer may move toward the source; in either case the separating distance decreases and there is an observed blueshift.

The shift in the observed wavelength can be expressed in terms of the relative velocity by means of the Doppler formula (see Figure 10.1). Let v be the relative velocity of a source moving away, and let λ represent an emitted wavelength and λ_0 the observed wavelength. According to the Doppler formula, we have

$$\frac{\lambda_0}{\lambda} = \frac{\text{observed wavelength}}{\text{emitted wavelength}} = 1 + \frac{v}{c} \qquad (10.1)$$

where c is the velocity of light. The emitted wavelength of a spectral line is usually known from measurements made previously with atoms in the laboratory, and a shift in the observed wavelength of the same spectral line betrays the relative velocity of the source. The Doppler formula given here applies only when v is very much less than c. For higher velocities we must use the exact formula (still referred to as the Doppler formula) derived from special relativity, which we shall give in the next chapter. Notice that when the source moves toward the observer with relative velocity v the sign of v is changed. The redshift of any source is defined as the fractional increase in wavelength:

$$\text{redshift} = \frac{\lambda_0 - \lambda}{\lambda} \qquad (10.2)$$

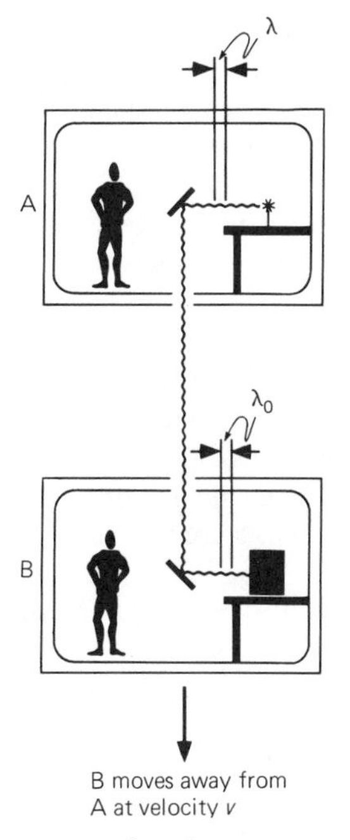

Figure 10.1. Observers A and B are in separate laboratories that move apart at velocity v. Both study the light emitted by atoms in their laboratories and find that emission and absorption lines have identical wavelengths in the two laboratories. Let λ be the wavelength of one of these spectral lines. A sends light to B, who finds that the wavelength λ of the transmitted light is received at wavelength λ_0. The Doppler formula says that the fractional increase in wavelength is equal to v/c, where c is the velocity of light.

From equations (10.1) and (10.2) we get

$$\text{Doppler redshift} = \frac{v}{c} \tag{10.3}$$

CELESTIAL SPEED CHAMPIONS. Vesto Slipher, an astronomer at the Lowell Observatory at Flagstaff, Arizona, began in 1912 to measure the shift in the spectral lines of light received from spiral nebulae. By 1923, as a result of Slipher's painstaking measurements, it was known that of the 41 galaxies studied, 36 had redshifts, and the remaining 5, including the Andromeda Nebula, had blueshifts. Slipher's results were strange and surprising. For if galaxies have random motions, one would expect that those moving away with redshifts would be approximately equal in number to those approaching with blueshifts. His observations indicated that most galaxies were moving away.

Edwin Hubble was the measurer of the universe. He developed to a fine art distance-measuring techniques that had been pioneered by Harlow Shapley, and in 1924 he began to determine the distances of galaxies (see Figure 10.2). We have already recounted how Hubble classified the galaxies by their appearance and was the first to show that the Andromeda Nebula is a galaxy beyond the Milky Way. The expansion law of the universe was firmly established by Hubble in 1929, but he did not discover the expansion of the universe, as is often claimed.

Howard Robertson, the American cosmologist, in 1928 had already used Slipher's redshifts and Hubble's published distances

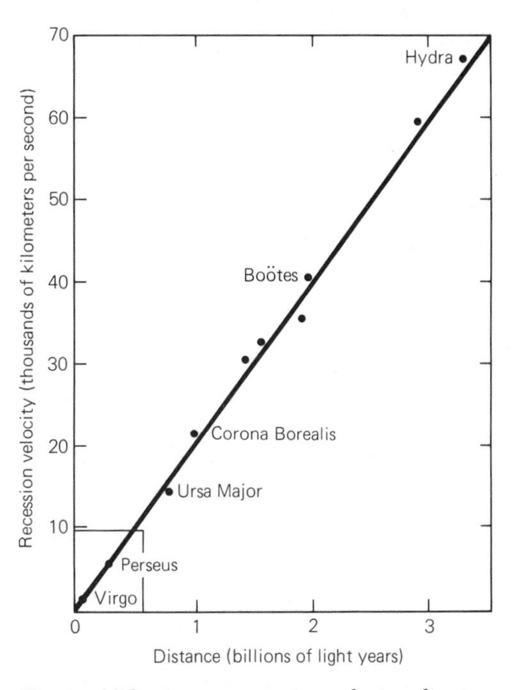

Figure 10.2. A representative velocity–distance curve showing the expansion of the universe. The lower left corner is the region surveyed by Hubble up to 1929.

to show that they agreed with the theoretical prediction now known as the *velocity–distance law*:

velocity of galaxies = constant × distance

Robertson was a theoretician and his "rough verification," as he called it, was tucked away in a theoretical paper published in a physics journal that possibly was not widely read by astronomers. In the following years, Hubble extended his distance measurements and, with redshifts determined by Milton Humason, founded the velocity–distance law on a secure basis. By 1955, using the 200-inch telescope of Mount Palomar, Humason, Nicholas Mayall, and Allan Sandage had observed and analyzed the data of more than 800 galaxies with redshifts up to 0.2. According to the velocity–distance law the more distant a galaxy, the faster it moves away. Thus, when distance is doubled, the recession velocity is doubled. The galaxies have their own random velocities superimposed on the cosmic recession, and additionally, the Sun moves around the center of the Galaxy; this is why the nearest galaxies are sometimes seen to have blueshifts.

The velocity–distance law of the universe,

recession velocity = constant × distance (10.4)

was arrived at in two ways. Theoreticians assumed a uniform (homogeneous and isotropic) universe and showed that if there is expansion, then there is a velocity–distance law that is exact for all distances. But, as we shall shortly see, all three of the terms *velocity, constant,* and *distance* require careful interpretation. Observers discovered a *redshift–distance law:*

redshift = constant × distance (10.5)

and by assuming that redshifts were caused by the Doppler effect they deduced from equation (10.3) the velocity–distance law of equation (10.4). Yet a velocity–distance law and a redshift–distance law are not identical, and one of the hurdles that must be leaped in modern cosmology is the realization that extragalactic redshifts are not the result of the Doppler effect. The laws fortunately become identical when the velocities of galaxies are small compared with the velocity of light. The velocity–distance law, or the Hubble law, as it is often called, is in fact the theoretician's law that holds true for all distances.

DISCOVERERS OF THE EXPANDING UNIVERSE. The discovery of the expanding universe did not come as an abrupt revelation. To unfold the historical record we must anticipate developments that will become clearer in later chapters.

The first intimation of an expanding universe came in 1917 in the work of the Dutch astronomer Willem de Sitter. He predicted "a systematic displacement of spectral lines toward the red." Without doubt, Slipher's redshift measurements and de Sitter's studies in cosmology initiated the idea of an expanding universe.

In 1917, Einstein and de Sitter proposed two different kinds of universe, both based on Einstein's theory of general relativity, which had been published in its final form the previous year. Einstein's universe was uniform: It contained uniformly distributed matter and had uniformly curved spherical space. A main feature of the Einstein universe was its static nature; it was unchanging, neither expanding nor collapsing. We must remember that at that time the astronomical universe was believed to be static on the cosmic scale. To conform with this belief, and to enable his universe to maintain a static state, Einstein introduced a *cosmological constant* into the theory of general relativity. The cosmological constant is equivalent to a repulsive force that opposes the force of gravity. By adjusting the value of the cosmological constant it is possible to make it counterbalance the gravity resulting from a uniform distribution of matter. We must mention that the cosmological constant (denoted by Λ and therefore sometimes called the *lambda constant*) was introduced in a rather ad hoc way, and

Einstein sought to justify it by appeal to Mach's principle. In his static universe the local mass density was related directly to the cosmological constant, and Einstein argued that the bootstrapping together of local and global things was in accord with Mach's philosophy.

But in the same year came the completely different de Sitter universe. It was an isotropic universe that contained absolutely no matter. It incorporated the cosmological constant and was also assumed to be static. This strange alternative universe, derived from the same theory of general relativity, showed clearly that the cosmological constant does not imply a unique universe, as Einstein had supposed. The empty de Sitter universe might have been ignored as a mere curiosity were it not for one very interesting property: When particles were sprinkled in it they behaved as if they were moving away from each other. This finding was thought to have some bearing on the redshift results obtained by Slipher, and for a long time it was referred to as the "de Sitter effect."

The German astronomer Carl Wirtz, inspired by Slipher's redshift measurements and the de Sitter effect, proposed in 1922 a velocity–distance relation. He used the apparent diameters of galaxies as distance indicators – the larger their distances, the smaller their average apparent diameters – and found that recession velocity increased with distance. It would seem, then, that Wirtz was the first to propound a velocity–distance law.

The apparent static nature of the de Sitter universe was a mathematical fiction. This universe was able to masquerade as static simply because it contained no matter that could exhibit its actual dynamic condition. Robertson later showed that a simple readjustment in the distinction between space and time had the effect of making the de Sitter universe homogeneous as well as isotropic. In this form it was spatially flat and infinite and was also expanding. An apt distinction was then made: The Einstein universe was "matter without motion," and the de Sitter universe was "motion without matter."

By this time more general investigations, those undertaken by Alexander Friedmann (a Russian scientist) in 1922 and 1924 and by Georges Lemaître (a Belgian cosmologist) in 1927, had become known and had opened the door to a variety of universes, all homogeneous and isotropic, either expanding or collapsing, and all containing matter.

The velocity–distance law is usually expressed in the form

$$\text{recession velocity} = H \times \text{distance} \tag{10.6}$$

where H is referred to as the *Hubble term*. Both Robertson in 1928 and Hubble in 1929 found for H a value of 150 kilometers a second per million light years; in other words, the recession velocity increased by 150 kilometers a second for every million light years' increase in distance.

In 1952, Walter Baade discovered the two stellar populations. As we have already seen, his discovery showed that the distances of galaxies had been greatly underestimated. A second revision in distances came in 1958 when Sandage discovered that what had previously been supposed to be bright stars in more distant galaxies were very luminous regions of hot gas, and these more remote galaxies were therefore at even greater distances. Revisions and other adjustments in distance estimates have increased the scale of the universe and reduced the original value of H by a factor between 5 and 10. Modern estimates place the value of the Hubble term somewhere between 15 and 30 kilometers a second per million light years. A main theme in the history of twentieth-century cosmology has been the progressive and often bewildering decline in the value of the Hubble term as determined by astronomers. The reason for this is the extraordinary, almost unbelievable, difficulty of measuring the distances of remote extragalactic systems.

We shall assume that the Hubble term has a value of 20 kilometers a second per million light years. Hence at a distance of 5 billion light years the recession velocity is 100,000 kilometers a second, or one-third the velocity of light.

EXPANDING RUBBER SHEET UNIVERSE

The time has come to introduce ERSU – short for "Expanding Rubber Sheet Universe" – an imaginary universe with which we shall perform several thought experiments to illustrate the nature of expanding space.

A two-dimensional model has one serious drawback: It consists of an expanding surface within our three-dimensional space. The universe does not expand within space, either three-dimensional or higher-dimensional space, but instead consists of expanding space. We must therefore occasionally think of ourselves as living in the surface as two-dimensional creatures and consider the expanding surface as representing space itself. We live in the surface as ordinary observers; when we stand back, however, and survey the surface as it is at any instant, we are then like the cosmic explorer who is able to move from place to place instantaneously. Because there is no cosmic edge we must suppose that the surface is indefinitely large. Instead of an expanding rubber sheet we could imagine a spherical balloon that is slowly inflated, as suggested by Eddington. But when it comes to performing experiments a flat surface has several advantages.

EXPERIMENT 1: DILATION, SHEAR, AND ROTATION. We start by supposing that the surface exhibits every kind of motion possible: In some places it expands and in others it contracts, at varying rates. On the surface we draw a large number of triangles and notice that in the course of time the triangles change in size and shape. There are three basic kinds of motion apparent: *dilation, shear,* and *rotation,* and most triangles will have complex motions that combine all three.

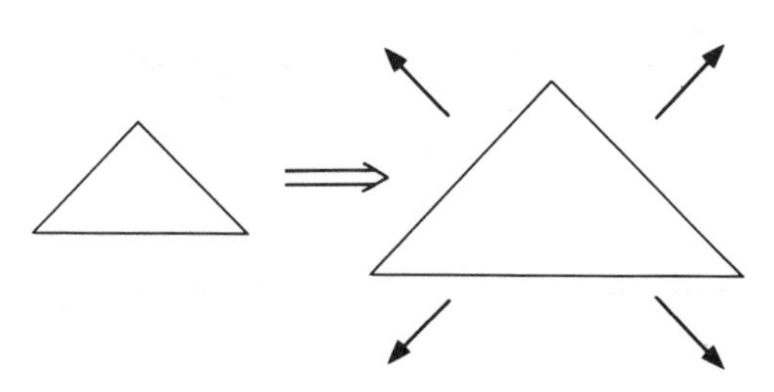

Figure 10.3. Dilation only. Triangles change their size but not their shape.

Those triangles that dilate (or contract) only have shape-preserving motion, and in the regions they occupy the expansion (or contraction) is isotropic (see Figure 10.3). Those triangles with shear only have shape-changing motion, and in the regions they occupy there is no dilation or rotation (see Figure 10.4). And those triangles that rotate only have shape-preserving motion, and in the regions they occupy there is no dilation or shear (see Figure 10.5). In general, complex motions combine dilation, shear, and rotation, and when all three vary from place to place, we have inhomogeneous motion (see Figure 10.6).

EXPERIMENT 2: HOMOGENEITY. Again we draw a large number of triangles on the surface, but this time we impose the condition that they all change in the same way. The surface is consequently homogeneous and has everywhere the same amount of

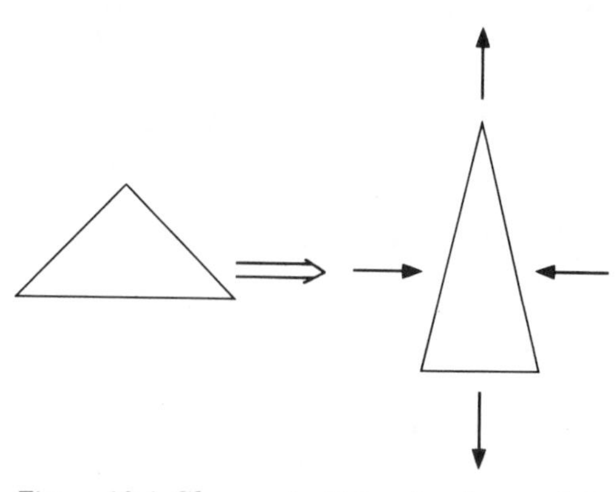

Figure 10.4. Shear only. Triangles change their shape but not their area.

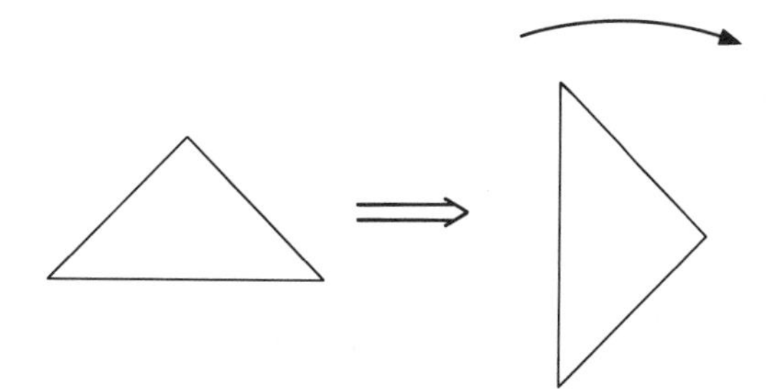

Figure 10.5. Rotation only. Triangles preserve their shape and size.

dilation, shear, and rotation. The amounts of dilation, shear, and rotation can be separately adjusted to provide a wide variety of homogeneous motions. Homogeneity of the universe was discussed in Chapter 4, where we concluded that the constituents and the laws of nature are everywhere alike. We have now extended the meaning of homogeneity to include cosmic motion.

Hitherto we have not mentioned that homogeneity also means that all clocks in the universe – apart from timekeeping variations owing to local irregularities – agree on their intervals of time. The explorer, whom we must once again summon, goes around the universe and adjusts all clocks to show a common time. On subsequent tours the explorer finds the clocks running in synchronism and showing the same time. This common time in the universe is known as *cosmic time.* All irregularities in timekeeping, departures from cosmic time, are due to local velocities and gravitational fields in the astronomical systems where the clocks are installed. Because we are disregarding irregularities in space, we shall also disregard irregularities in time.

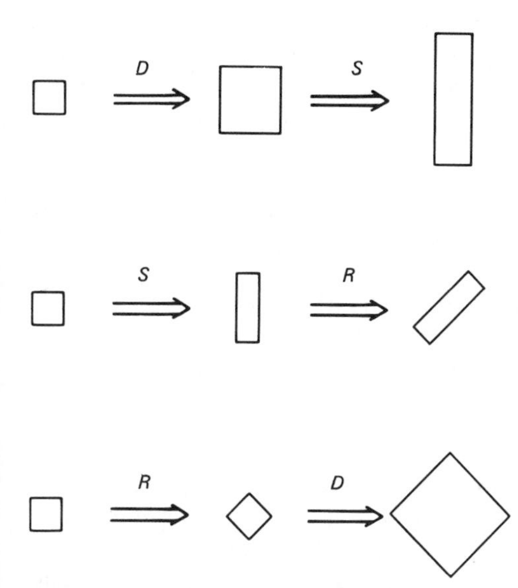

Figure 10.6. Combinations of dilation (D), shear (S), and rotation (R).

A homogeneously expanding universe that has dilation, but no shear and rotation, is shape-preserving and expands equally in all directions. This is isotropic expansion and each triangle retains its shape and is enlarged in size. This is a homogeneous (all places are alike) and an isotropic (all directions are alike) universe and displays the simplest of all dynamic states.

In the next example, the surface stretches in such a way that triangles change their shapes but preserve their areas. The dilation is zero and the motion consists of pure shear. Circles become ellipses and squares become oblongs, because the motion is now anisotropic.

Finally, we consider pure rotation, in which dilation and shear are zero and triangles preserve their shapes and areas. Because the motion is homogeneous, the rotation is not about one point but about all points in the surface. Remember, as observers we are in the surface and must not think of three-dimensional space. Rotation of our three-dimensional universe of space is not easy to imagine. It is the same everywhere and consists of anisotropic motion about one of the three perpendicular directions. Such rotation can be detected by shooting a particle toward a distant point and noting that its trajectory appears to curve away from the target. In other words, the compass of inertia does not rotate with the universe.

EXPERIMENT 3: HOMOGENEITY AND ISOTROPY. When the expansion is homogeneous and isotropic, triangles drawn on the surface dilate everywhere in the same way, do not change their shape, and do not rotate. For brevity, we shall refer to homogeneous and isotropic expansion as uniform expansion. The word *uniform* is used here in its most general sense: Figures change everywhere in space in the same way and preserve their form (or shape) in time. In the following experiments we shall assume that the surface, representing space, expands uniformly.

EXPERIMENT 4: DRAWN CIRCLES ARE NOT "GALAXIES." With a piece of chalk we draw a circle on the surface of the sheet and declare that it represents a galaxy (see Figure 10.7). As the sheet expands we observe that the "galaxy" gets bigger. This result is quite misleading: A real galaxy is held together by its own gravity and is not free to expand with the universe. Similarly, if the chalked circle is labeled "atom," "Earth," "Solar System," or almost anything, the result is misleading because most systems are held together in equilibrium and are not free to partake in cosmic expansion. If we call the chalked circle "cluster of galaxies" the result can also be wrong because most clusters are bound together and are not free to expand. Some clusters consisting of a few widely separated galaxies are weakly bound and tend to expand, but most do not, particularly those consisting of many galaxies. Superclusters are vast systems of numerous clusters that are weakly bound, and they expand almost freely with the universe.

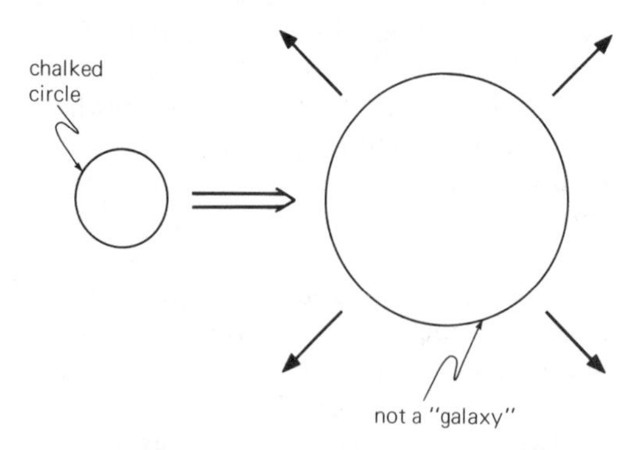

Figure 10.7. A circle chalked on the surface of the rubber sheet expands and cannot therefore represent a galaxy.

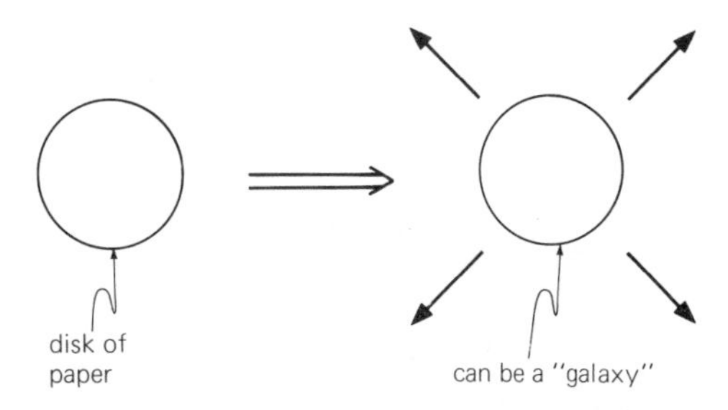

Figure 10.8. A disk of paper retains its size when placed on the surface and can therefore represent a galaxy of fixed size in an expanding universe.

This experiment teaches us a useful lesson. We are able to detect expansion because our measuring instruments, and the astronomical systems in which we are located, tend to have fixed sizes and do not expand. If everything were like the chalk circle, free to expand, then clearly there would be no way of detecting expansion. It is an amusing thought that perhaps the universe is not expanding but static, and we do not know this because atoms, we ourselves, and our laboratories and observatories are all shrinking. With tongue in cheek, Eddington in 1932 proposed that the theory of the "expanding universe" might also be called the theory of the "shrinking atom." He said: "We walk the stage of life, performers of a drama for the benefit of the cosmic spectator. As the scenes proceed he notices that the actors are growing smaller and the action quicker. When the last act opens the curtain rises on midget actors rushing through their parts at frantic speed. Smaller and smaller. Faster and faster. One last microscopic blurr of intense agitation. And then nothing."

EXPERIMENT 5: DISKS OF PAPER ARE "GALAXIES." We place on the surface of the rubber sheet a disk of paper to represent a galaxy or any other bound system (see Figure 10.8). As the sheet expands the disk stays constant in size. This result is not misleading, and we have hence found a way

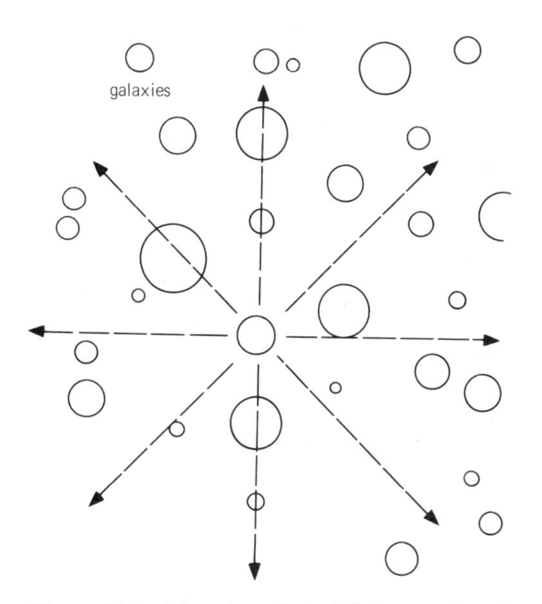

Figure 10.9. A large number of disks are placed on the expanding surface. About any one disk the other disks recede isotropically.

of correctly representing a galaxy or a cluster of galaxies of fixed size.

EXPERIMENT 6: WORLD MAP AND WORLD PICTURE. We sprinkle uniformly over the surface a large number of disks of various sizes to represent the galaxies (see Figure 10.9). Strictly speaking, the disks should denote the largest bound systems, the clusters of galaxies, but for convenience we shall continue to refer to them as "galaxies." Each galaxy is surrounded by receding galaxies, and its inhabitants might therefore think that they occupy a cosmic center from which everything is flying away. But because the universe is uniform, with no cosmic center and no edge, this impression of being at the center is shared with all the inhabitants of other galaxies.

We, the experimenters, can stand back and look down on the surface and see that it is uniform. The disks are sprinkled more or less uniformly and the expansion is everywhere the same. We are able to see things as they are everywhere in space at a single instant of cosmic time. We see what Edward Milne called the *world map*, with everything laid out in space and visible for immediate comparison (see Figure 10.10). This is also what is seen by the mobile cosmic explorer, who rushes around and notes that all places are alike.

The observers in a particular galaxy look out in space and back in time and therefore cannot see the way things are everywhere in space at the moment of observation. They cannot see the world map, but see things distributed on their backward lightcone, and the universe they observe is what Milne called the *world picture*. These observers notice that other galaxies are scattered about them isotropically and also that the galaxies are moving away isotropically. They therefore assert that all directions are alike. They then invoke the location principle and come to the conclusion that probably all places are alike. Their immobility in the cosmic sense has forced them to become philosophers as well as scientists.

EXPERIMENT 7: VELOCITY–DISTANCE EXPANSION LAW. Our next experiment shows that the surface obeys the velocity–distance law, known also as the Hubble law (see Figure

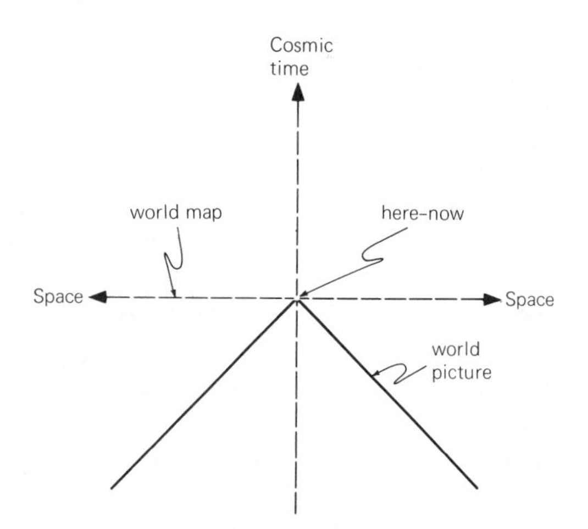

Figure 10.10. A spacetime diagram showing the "world map" of uniform space as seen by the cosmic explorer, and the "world picture" on the backward lightcone as seen by an ordinary observer.

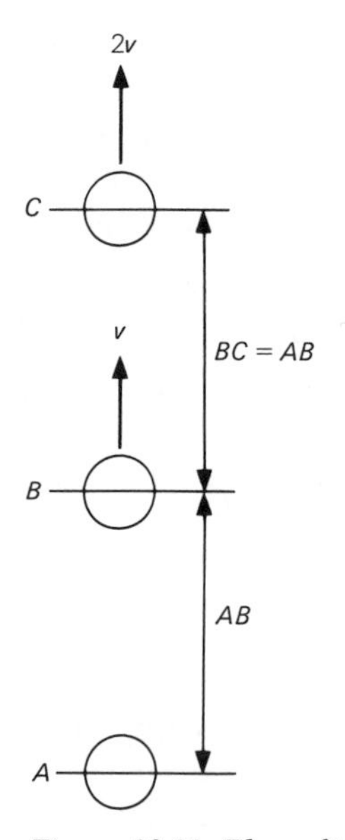

Figure 10.11. Three disks A, B, and C, with the distance between A and B equal to the distance between B and C. The velocity at which disk B recedes from A is the same as that at which disk C recedes from B. C therefore recedes from A at a velocity twice that at which B recedes from A. Hence the velocity of recession is proportional to distance.

10.11). We choose any disk and label it A. A second disk, labeled B, at a certain distance moves away from A at a certain velocity. A third disk, labeled C, in the same direction as B and at twice the distance, moves away from A at twice the velocity. It has to. The expansion is homogeneous, and therefore C moves away from B at the same velocity as that at which B moves away from A. In a particular direction the velocity relative to A is always proportional to distance. Hence we can say

recession velocity = constant × distance

This simple law is a direct consequence of homogeneity and is independent of the location of A. The recession velocity relative to any disk obeys the same law. If the expansion is anisotropic (faster in one direction than in another), the "constant" will have different values in different directions. But for isotropic expansion, which is what we are considering, the constant has the same value in all directions. The law we have derived by this simple experiment is the velocity–distance law, and the constant is the Hubble term denoted by the symbol H in equation (10.6).

We now must pause and take note that *recession velocity, Hubble term,* and *distance* require careful consideration, for each is open to misinterpretation. The velocity–distance law is obviously true for the cosmic explorer – and also for us who are experimenting with ERSU – and is moderately simple and straightforward. The case is not the same, however, for the unfortunate observers in their galaxies.

The expression *recession velocity* needs careful handling. On the surface of ERSU the disks are stationary; they move apart because the surface is expanding. The disks do not move on the surface, but are at rest and are carried apart by the expansion of the surface. Similarly, the galaxies are stationary and yet recede from each other because intergalactic space is expanding. The galaxies are not hurling through space; they are at rest in space and are carried apart by the expansion of space. Recession velocity is therefore not an ordinary velocity in the usual sense and is quite unlike the velocities encountered on Earth, in the Solar System, or in the Galaxy. For this reason we must be careful in cosmology when using the word *velocity,* and to avoid confusion we shall use *recession velocity* or just *recession* to indicate motion owing to the expansion of space.

The Hubble term is the same everywhere in space at any instant, but is usually not constant in time. The expansion may have been faster in the past, in which case the value of H would have been greater; if the expansion was slower, the value of H was similarly smaller. The observers look out from a galaxy to great distances and therefore look back through time and see H having different values at different distances. To them the velocity–distance law is true only out to short distances; at larger distances the law breaks down because H changes with distance. The velocity–

distance law is true in the world map seen by the cosmic explorer, but is not true in the world picture seen by the ordinary observer.

The measurement of distances is no great problem to the cosmic explorer, and whenever in doubt, the explorer can always use a tape measure. But the stay-at-home observers, who look out in space and back in time, use roughhewn scales of distance and have a terrible problem trying to determine how far away some object is. Their difficulty is twofold. First, everything seen is distributed on the sky, and distances are not apparent but must be estimated. Second, distances in an expanding universe change with time (see Figure 10.12). We can speak of the distance of a galaxy at the time when it emitted the light we now see, or of its distance now. In an expanding universe the distance a galaxy now has is greater than the distance it had when it emitted the light we see. In the velocity–distance law we must use the distances that galaxies now have. All distances must therefore be adjusted before they can be used in the velocity–distance law.

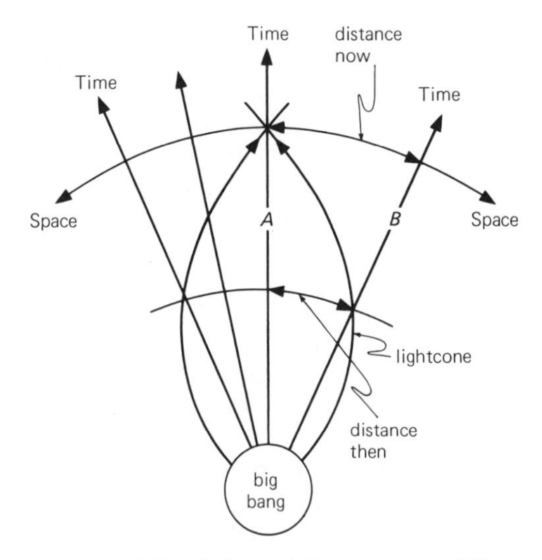

Figure 10.12. Galaxies shown as worldlines diverging from the big bang. An observer on galaxy A looks out into space and back in time and sees galaxy B in the past. There are hence two distances: the "distance now" between galaxies A and B, and the "distance then" between the two galaxies at the time B emitted the light that is now seen. The "distance now" is the distance used in the velocity–distance relation.

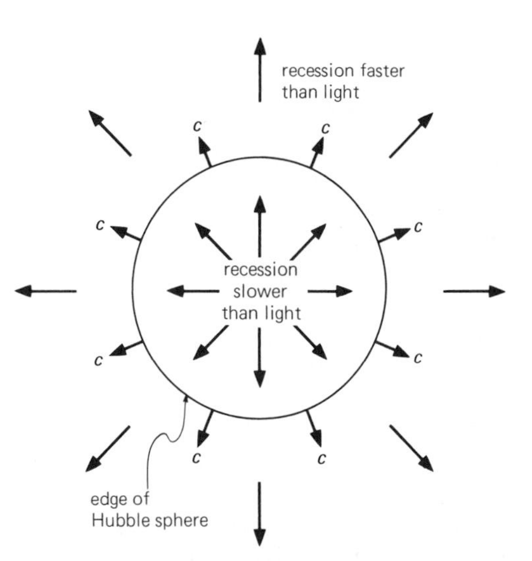

Figure 10.13. At the edge of the Hubble sphere the recession velocity of the galaxies equals the velocity of light. Inside the Hubble sphere all galaxies recede slower than the velocity of light, and outside the Hubble sphere all galaxies recede faster than the velocity of light. The observable universe is approximately the size of the Hubble sphere.

EXPERIMENT 8: HUBBLE SPHERE. The recession velocity increases with distance and at a certain distance equals the velocity of light (see Figure 10.13). This distance is c/H and is known as the *Hubble length*. If the Hubble term H is 20 kilometers a second per million light years, the Hubble length is 15 billion light years.

Let us draw about any disk a large circle whose radius represents the Hubble length. The circle on the surface encloses the *Hubble sphere*. Within the Hubble sphere the recession is less than the velocity of light, and outside the Hubble sphere the recession is greater than the velocity of light. Each disk has its Hubble sphere.

Observers on all galaxies find that the recession velocity is equal to the velocity of light at a distance equal to the Hubble length. The universe has no edge and cannot end abruptly; and an edge, even if it existed, could not be at the same distance from all observers everywhere. There is no edge, and therefore galaxies further away than the Hubble length recede faster than the velocity of light. This is what puzzles most people when they first study cosmology.

In special relativity we learn that nothing moves through space faster than light. How is it then possible for recession to be faster than light? The answer is that galaxies are not moving through space at all but are wafted apart by the expansion of intergalactic space. No galaxy can move through space faster than light, and in its local space it obeys always the rules of special relativity. But recession velocity is a result of the expansion of space and is not like motion through space that obeys the laws of special relativity. Recession velocity is without limit, and in an infinite universe a galaxy at infinite distance has infinite recession velocity. People who find it difficult to understand that recession is without limit usually make the mistake of thinking that the receding galaxies are projectiles shooting away through space. This is an incorrect view; the correct picture consists of galaxies at rest in expanding space.

EXPERIMENT 9: STEADY STATE EXPANDING UNIVERSE. The steady state universe is easily simulated with ERSU. As the surface expands and the disks move apart, we continually sprinkle fresh disks on the surface so that their average separating distance always remains the same. The surface thus presents the same appearance and nothing changes with time. Expansion also must never change, and the Hubble sphere has therefore constant radius. Instead of drawing a large circle to represent the Hubble sphere about any disk, we should place a large hoop of fixed size on the surface.

EXPERIMENT 10: COMOVING AND PECULIAR MOTIONS. Disks stationary on the expanding surface and galaxies stationary in expanding space are said to be *comoving*. They comove with expansion. Clocks on comoving galaxies all measure cosmic time. We can imagine that in a homogeneous universe the cosmic explorer has set all clocks on comoving objects to read the same time. In subsequent tours of the universe the explorer finds that these clocks continue to agree. Comoving objects have their worldlines perpendicular to *cosmic space* – that space which has uniform curvature and is uniformly expanding.

In all experiments so far we have supposed that the disks are comoving on the expanding sheet. But galaxies are not exactly comoving: They wander around relative to their neighbors and have independent motion in their local space. Independent motion of this kind is known as *peculiar motion.* Owing to this peculiar motion the worldlines of galaxies are slightly crinkled.

We can suppose in ERSU that all disks have independent motion and jitter around slowly on the surface. Observers on any one disk now see other disks with their peculiar velocities superposed on their recession velocities. The observers cannot tell, when looking at any other disk, how much of its motion is recession and how much is peculiar. But peculiar motions are usually random, and the observed velocities of many disks at a certain distance can be averaged to find the recession velocity at that distance. Observers must also take into account the peculiar motion of their own disk, and this can in principle be done by averaging the observed velocities of all surrounding disks at a particular distance.

It sounds easy, but in the real universe the determination of peculiar velocities is exceedingly difficult. Fortunately, recession dominates at large distances and peculiar motions can then often be neglected. Some nearby galaxies are approaching us and others are moving away; further out, a few are approaching and most are receding; and further out still, none are approaching and all are receding. Typical peculiar velocities of galaxies are 300 kilometers a second; therefore, beyond 15 million light years, the recession dominates. Some galaxies, particularly those in rich clusters, have peculiar velocities of 1000 or more kilometers a second, and in their case recession dominates beyond 50 million light years. Our own peculiar motion in the universe must also be taken into account; it consists of the Earth moving about the Sun (30 kilometers a second with annual variation in direction), the motion of the Sun owing to the rotation of the Galaxy (300 kilometers a second), the motion of the Galaxy in the Local Group, and the peculiar motion of the Local Group itself. Determining the peculiar velocity of the Sun in the universe is not at all easy, but nowadays it is believed that eventually the anisotropy of the 3-degree cosmic radiation will provide this information.

The implications of the experiments performed so far are startling. In Newtonian theory we are taught there is no such thing as absolute rest and all velocities are purely relative. But in cosmology a comoving body is in a state of absolute rest that can in principle be verified. Also, all peculiar velocities have absolute values that can be determined by comparison with a local comoving frame of reference. In special relativity we are taught that there is no preferred way of decomposing spacetime into space and time, and that a body cannot move faster than light. Yet in cosmology spacetime separates naturally into uniformly curved and expanding space and cosmic time, and comoving objects obey a velocity–distance law in which recession can exceed the velocity of light without limit.

What we have been taught in special relativity applies only to motion in the laboratory, the Solar System, and the Galaxy, and not to that in the universe as a whole. All local velocities are peculiar in a cosmic sense and cannot exceed the velocity of light. Recession velocity, however, is not a local phenomenon; it is the result of the expansion of space and does not conform to special relativity. In summary, we can say that motion in an expanding universe is compounded from recession and peculiar velocities; recession velocities are due to the expansion of space and are without limit; and peculiar velocities are due to motion through space and conform to special relativity.

EXPERIMENT 11: COMOVING COORDINATES. With a piece of chalk we draw on the expanding surface a network of lines. These intersecting lines form what is called a *comoving coordinate system* (see Figure 10.14). The coordinate lines are fixed on the surface and expand with it. Whether the lines are straight or curved is unimportant; what matters is that comoving bodies are fixed in relation to these lines, and their positions are specified by means of comoving coordinates. Distances between comoving bodies, measured in comoving coordinates,

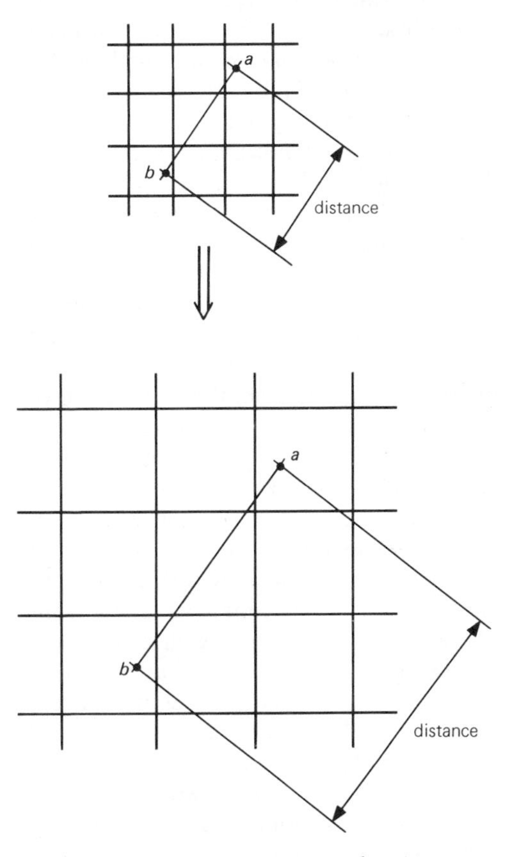

Figure 10.14. A comoving coordinate system consists of a network of lines drawn on the expanding surface. All coordinate distances, such as that between points a and b, remain constant in value.

remain constant, and these constant distances are called comoving *coordinate distances.*

A comoving coordinate system enables us to distinguish between peculiar and recession velocities, because the peculiar motion of the disks can be detected and measured relative to the system of comoving coordinates. In the real universe imaginary comoving coordinates are of little help to observers, but they are a great conceptual aid to cosmologists, who require a way of distinguishing between peculiar motion and expansion.

EXPERIMENT 12: EMPTY UNIVERSE. Let us remove all disks from the surface except the one on which our observers are located. These observers have now no way of determining their peculiar motion, and further, they cannot tell if the surface is expanding, static, or contracting. Confronted with such a situation they will probably recall Mach's famous words: "When, accordingly, we say, that a body preserves unchanged its direction and velocity *in space,* our assertion is nothing more or less than an abbreviated reference to the *entire universe.*" By the entire universe, Mach meant the "remote heavenly bodies" that we have, figuratively speaking, removed from the sheet. Even the ubiquitous explorer is puzzled and is tempted to believe that empty space is meaningless and that the physical properties of space are totally dependent on the presence of matter. Then we learn the trick of scattering around a few tiny comoving particles, or – what amounts to the same thing – of chalking on the sheet a grid of comoving coordinates. This illustrates what happened originally with the empty universe proposed by de Sitter: Nobody knew that it was expanding until it was sprinkled with test particles and equipped with a system of comoving coordinates and found to have dynamic properties even though it contained no matter.

EXPERIMENT 13: SMOOTHED-OUT UNIVERSE. The expanding sheet model of the universe has been covered with disks and looks rather like the ground scattered with confetti after a wedding. The distribution is clumpy on small scales and not obviously uniform except when perceived over large areas. The accidental irregularities of this distribution are distracting when we wish only to study the general principles of how the model behaves. In the last experiment we collected up all the disks; let us take these disks, grind them into powder, and then carefully and uniformly distribute the powder over the expanding surface so that there is no irregularity whatever. The surface is now covered with a uniform comoving layer of dust that consists of the same amount of matter as before. We have a model of a "smoothed-out" universe that we shall refer to as an *idealized universe,* in which all galaxies are smoothed out into a continuous fluid that is everywhere uniform.

Ideally smooth universes have many uses. It is of interest, for instance, to know what happens to light when it propagates in a universe free of all irregularity. Corrections for irregularities can always be made afterward. Most cosmologists take the view that an idealized universe is a convenient fiction useful for making easy calculations – a sort of cosmic undergarment onto which the ostentatious details of the real world are later tacked.

An idealized universe is also useful for studying the origin of galaxies. These vast celestial systems have not always existed, and prior to their formation it is possible that the universe was much less irregular than at present. The idealized universe may therefore resemble the way things were once upon a time. How the original unstructured universe evolved into its present highly structured state is a major unsolved riddle in cosmology.

Idealized universes, perfectly homogeneous and isotropic, are known as Robertson-Walker models after the two scientists (Howard Robertson and Arthur Walker) who demonstrated that uniform universes have a spacetime that separates into curved space and cosmic time common to all comoving observers. In the observed universe all regions, in an average sense, are probably alike; in an idealized universe we can truly say that all regions, no matter how small, are alike. By studying the expansion of a small region of an idealized universe we automatically learn how all other small regions expand, and by piecing this information together we then know how the whole universe expands. The behavior of large regions, even the universe itself, is mirrored in the behavior of small regions. By smoothing the universe into an idealized state we are, in a sense, able to study it in a nutshell. We shall exploit this aspect of cosmology in Chapter 13.

MEASURE OF THE UNIVERSE

THE UNIVERSAL SCALING FACTOR. Distances between galaxies, or clusters of galaxies, increase in an expanding universe, whereas distances within galaxies and even clusters of galaxies do not increase. This difference, important in astronomy and other sciences, is a distracting nuisance in cosmology. The solution is simple: Because we are interested only in cosmic phenomena, we abolish all astronomical systems and use instead an idealized universe. We are now free to consider how distances vary in time without reservations about distances not being smaller than the size of galaxies or clusters of galaxies. In a smoothed universe distances can be as small or as large as we please.

Over an interval of time all distances between comoving points increase by the same factor. If one distance increases by 1 percent, then all other distances increase by 1 percent. A comoving triangle, for example, has its three sides scaled by the same factor and the dilated triangle retains its original shape. There exists a universal *scaling factor,* usually denoted by the symbol R, which increases with time in an expanding universe and which at any instant has the same value everywhere in space (see Figure 10.15). Distances between comoving points increase in proportion to R; areas of two-dimensional figures increase in proportion to R^2; and volumes of three-dimensional figures increase in proportion to R^3.

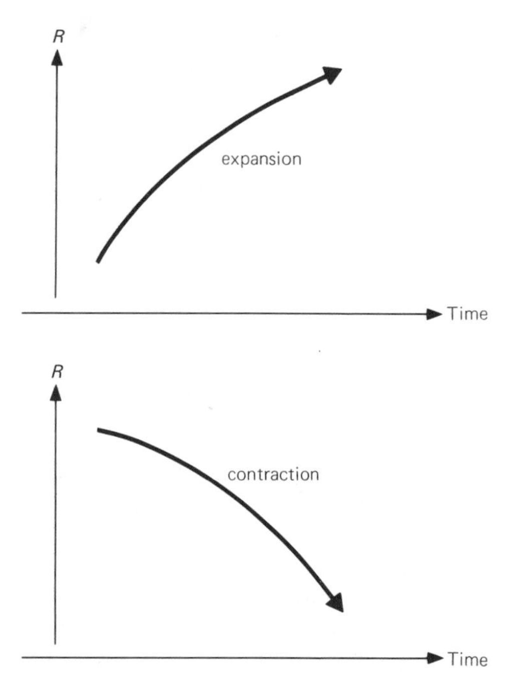

Figure 10.15. The scaling factor R *increases with time in an expanding universe and decreases with time in a collapsing universe.*

COORDINATE DISTANCES AND THE SCALING FACTOR. The scaling factor has important uses. We can best show this by beginning with comoving coordinates. A comoving coordinate system is fixed in expanding space, and all comoving points are separated by coordinate distances that stay constant. By using such a coordinate system we can say that the actual distance is equal to the coordinate distance multiplied by the scaling factor, or

$$\text{distance} = R \times \text{coordinate distance} \qquad (10.7)$$

Instead of saying *actual distance* we shall usually just say *distance.* The coordinate distance stays constant, and the distance itself increases in the same manner as R.

Instead of a flat surface with a grid of lines chalked on it, as in ERSU, we could use a rubber balloon with latitude and longitude coordinates drawn on its surface. Latitude and longitude are comoving coordinates, and a point on the surface of an expanding balloon does not change its position in terms of these coordinates. In this case we can think of the scaling factor as the radius of the balloon's surface. The distance between two points on the surface is the constant coordinate distance measured in latitude and longitude multiplied by the radius R. Some years ago the scaling factor was often referred to as the "radius of the universe," and this is why to this day it is denoted by the symbol R. The phrase *radius of the universe* can be misleading, however, because some universes are flat; it is better therefore to use an expression such as *scaling factor*.

THE VELOCITY–DISTANCE LAW. The scaling factor increases with cosmic time in an expanding universe. But how fast does it increase? Some thought on this matter soon makes it clear that the rate of increase of R must have something to do with the Hubble term. Consider a comoving body: It is at a fixed coordinate distance from us, and its actual distance is given by the equation

$$\text{distance} = R \times \text{coordinate distance}$$

As R increases, the distance increases also and the body recedes. The faster R increases, the faster the body recedes.

The recession velocity of a comoving body is just the rate at which its distance is increasing. This must therefore equal the rate of increase of R multiplied by the constant coordinate distance; that is,

$$\text{recession velocity} = \text{rate of increase of } R \times \text{coordinate distance}$$

For convenience we use Newton's notation and let $\dot{R}$ stand for the rate of increase of R; therefore

$$\text{recession velocity} = \dot{R} \times \text{coordinate distance}$$

We again use the expression distance = $R \times$ coordinate distance, and get

$$\text{recession velocity} = \frac{\dot{R}}{R} \times \text{distance} \qquad (10.8)$$

This is the velocity–distance law:

$$\text{recession velocity} = H \times \text{distance}$$

where the Hubble term is given by the expression

$$H = \frac{\dot{R}}{R} \qquad (10.9)$$

This important derivation of the velocity–distance law reveals that the law is the consequence of uniform expansion.

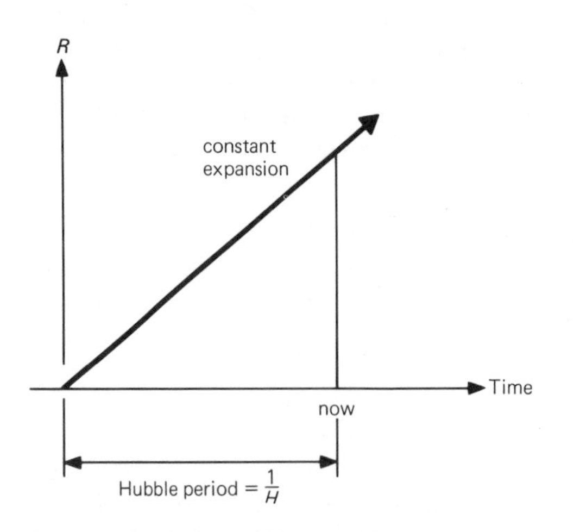

Figure 10.16. A Hubble period is the age the universe would have reached if it had expanded at a constant rate.

HUBBLE PERIOD AND THE AGE OF THE UNIVERSE. A *Hubble period* is the age the universe would have reached if it had expanded at a constant rate equal to the present rate of expansion (see Figure 10.16). This age is given by

$$\text{Hubble period} = \frac{1}{H} \qquad (10.10)$$

and is approximately 15 billion years. A Hubble period is sometimes referred to as the *expansion time*.

In almost all universes studied by cosmologists the scaling factor does not increase at a constant rate but either accelerates or decelerates. In accelerating universes R increases more rapidly as time goes by, so that the actual age is always longer than a Hubble period and such universes are older than 15 billion years. In decelerating universes R increases more slowly as time goes by; the age of such universes is always shorter than a Hubble period and hence less than 15 billion years.

At present we cannot tell precisely the

age of the universe, and a Hubble period is useful as a rough indication of age. We have to be on our guard, because in some universes a Hubble period may be a thoroughly misleading indicator of age. The de Sitter, steady state, and "hesitation" universes are examples.

In the steady state universe nothing ever changes in a cosmic sense, and therefore the Hubble term is constant (see Figure 10.17). Because $\dot{R} = HR$, and H is constant, $\dot{R}$ increases as R gets larger, and it is an accelerating universe. In this particular case the age of the universe is infinite.

HUBBLE SPHERE AND THE OBSERVABLE UNIVERSE. On multiplying the Hubble period by the velocity of light we obtain the *Hubble length,* given by

$$\text{Hubble length} = c \times \text{Hubble period} = \frac{c}{H} \quad (10.11)$$

which is approximately 15 billion light years. A Hubble length serves as a cosmic yardstick, and when we speak of distances of cosmic magnitude we have in mind distances comparable with the Hubble length. It is not only a sort of cosmic yardstick but also the distance at which the recession velocity is equal to the velocity of light. We have previously seen that the Hubble sphere has a radius equal to the Hubble length and that each point in space is the center of a Hubble sphere.

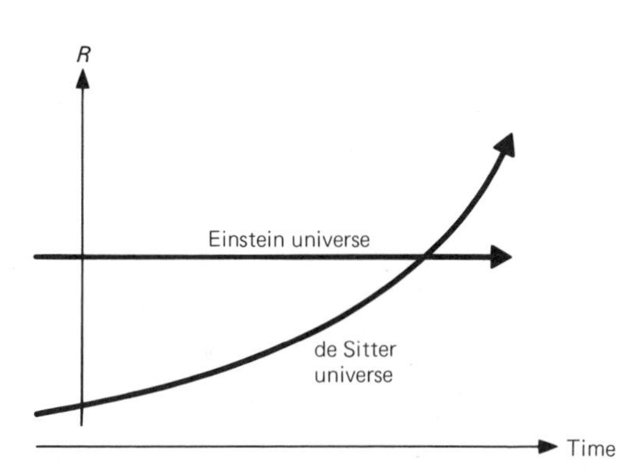

Figure 10.17. The Einstein universe and the de Sitter universe. In the de Sitter universe, as in the steady state universe, the Hubble term H is constant and the scaling factor R increases exponentially with time, as shown.

The *observable universe* is that part of the universe about an observer which can be seen; it will be discussed in more detail in Chapter 19, which is devoted to the subject of cosmic horizons. Broadly speaking, the observable universe has a size comparable with that of the Hubble sphere. We can understand why this is so with the aid of the following argument: The age of the universe is approximately a Hubble period and the maximum distance light can have traveled is approximately a Hubble length. We do not see things outside the observable universe because their light has not had sufficient time to reach us. In most expanding universes the Hubble sphere also expands. When the universe was 1 million years old the Hubble sphere had a radius of roughly 1 million light years, and when it was 1 year old the Hubble sphere had a radius of roughly 1 light year. In almost all expanding universes the Hubble sphere expands more rapidly than the universe itself; the edge of the Hubble sphere recedes faster than the galaxies, and in the course of time we see more and more galaxies that were previously unobservable.

THE DECELERATION TERM. We have seen that distances increase according to the law

$$\text{distance} = R \times \text{coordinate distance}$$

and the recession velocity is the rate of increase of distance:

$$\text{recession velocity} = \dot{R} \times \text{coordinate distance}$$

Acceleration is just the rate of increase of velocity, and if we use the symbol $\ddot{R}$ to denote the rate of increase of $\dot{R}$, we have

$$\text{recession acceleration} = \ddot{R} \times \text{coordinate distance}$$

because, as before, the coordinate distance remains constant. We now use our first relation, distance $= R \times$ coordinate distance, and find

$$\text{recession acceleration} = \frac{\ddot{R}}{R} \times \text{distance} \quad (10.12)$$

The term $\ddot{R}/R$, sometimes indicated by the symbol h, is the acceleration term. The acceleration term is not often used; instead, in common use nowadays is the *deceleration term*, indicated by the symbol q and defined as $-h/H^2$, or alternatively,

$$q = -\frac{\ddot{R}}{RH^2} \tag{10.13}$$

The deceleration term, like the Hubble term, can change with time but is everywhere the same in space at any instant.

When the rate of expansion never changes, and hence $\dot{R}$ is constant, the deceleration term is zero (as in Figure 10.16). When the Hubble term is constant, the deceleration term is also constant, as in the de Sitter and steady state universes where q has the fixed value of -1. In most universes the deceleration term changes with time (see Figure 10.18).

When the deceleration term is positive, there is deceleration (slowing down of expansion), and when it is negative, there is acceleration (speeding up of expansion). It can be seen from the curves in Figure 10.19 that in a deceleration universe, where q is positive, the age of the universe is shorter than a Hubble period; and in an accelerating universe, where q is negative, the age of the universe is longer than a Hubble period.

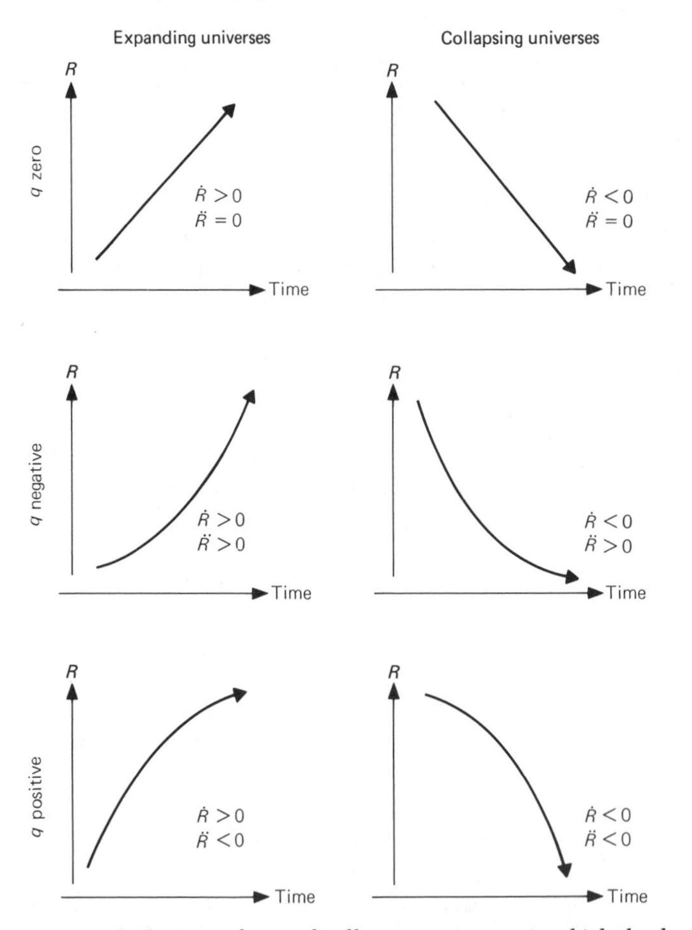

Figure 10.18. Expanding and collapsing universes in which the deceleration term q is zero, negative, and positive. These diagrams indicate how the scaling factor R changes with time.

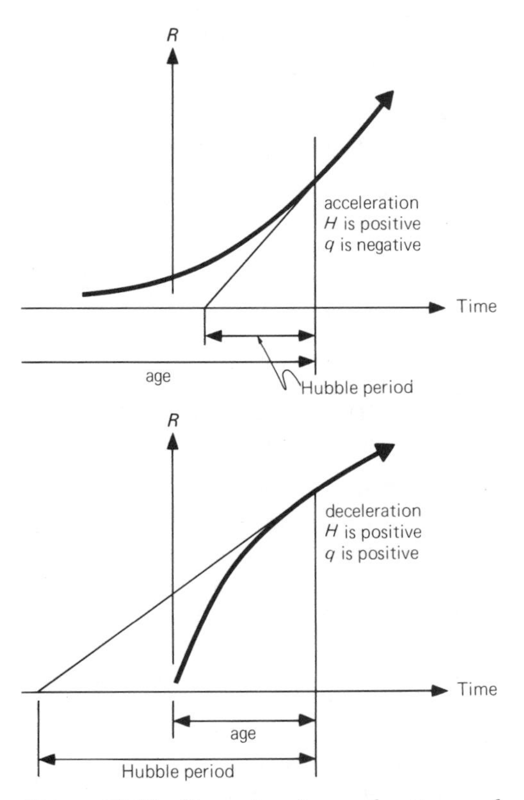

Figure 10.19. Expansion in accelerating and decelerating universes. Note that the accelerating universe has an age longer than a Hubble period, and the decelerating universe has an age shorter than a Hubble period.

DISTANCES, AREAS, VOLUMES, DENSITIES. Usually we are not interested in the actual value of the scaling factor R. Changing R by multiplying it by 2, or any other fixed number, does not alter the Hubble and deceleration terms. We can always adjust comoving coordinates so that distances are R × coordinate distances. Often we are interested only in the way R changes with time and in a comparison of its values at different stages in cosmic expansion.

Let R_0 be the value of the scaling factor at the present time, and let L_0 be the present distance between two comoving points:

$$L_0 = R_0 \times \text{coordinate distance}$$

At any other time the scaling factor is R, and the distance between the comoving points of constant coordinate distance is L, where

$$L = R \times \text{coordinate distance}$$

Hence, from these two relations, we find

$$L = L_0 \left(\frac{R}{R_0}\right) \tag{10.14}$$

This result is obvious, for if the universe doubles its size, then the scaling factor is increased twofold and all distances between comoving points are doubled. Pursuing similar arguments, we can say that if an area A_0 is comoving, then at any other time the area is given by

$$A = A_0 \left(\frac{R}{R_0}\right)^2 \tag{10.15}$$

Thus we see how areas change with expansion. The case is similar with volumes: If V_0 is a comoving volume, then at any other time the volume is expressed by

$$V = V_0 \left(\frac{R}{R_0}\right)^3 \tag{10.16}$$

We see that if distances double in size, then comoving areas increase fourfold, and comoving volumes increase eightfold.

Let us consider an expanding volume and suppose that it contains N particles. We assume that no particles are created or destroyed, so that N is a fixed number. The density of particles, call it n, is the number in a unit of volume, such as a cubic centimeter, and is $n = N/V$. The present density n_0 is the number N divided by the present volume: $n_0 = N/V_0$. We already know how volumes vary, and hence we now know how densities vary:

$$n = n_0 \left(\frac{R_0}{R}\right)^3 \tag{10.17}$$

With this result it is possible to find the density in the past or the future from the present density. The average density of matter in the universe is about 1 hydrogen atom per cubic meter. Back in the past when the scaling factor was 1 percent of its present value, the density was a million times greater and equal to 1 hydrogen atom per cubic centimeter. This is a typical value for the density of galaxies, and from it we infer that galaxies had not yet formed when the universe was smaller than 1 percent of its present size. Incidentally, when we say that the universe "changes in size" we imply not that it is necessarily finite but only that all comoving distances have changed.

BANG AND WHIMPER UNIVERSES

In the expanding universe the scaling factor increases, and by measuring the Hubble term, we determine how fast the scaling factor is increasing at present. The rate of increase is perhaps slowing down, and the expansion is therefore decelerating. From observations it is very difficult to determine precisely the value of the deceleration term. The available evidence at present suggests that the deceleration term is only slightly greater than zero. Its estimated value has changed frequently, and even current guesses must be viewed with reservations. Fortunately, there are many subjects in cosmology that can be discussed in general terms without exact knowledge of how the scaling factor varies with cosmic time. The way in which the scaling factor can vary provides us with a means of classifying a wide variety of universes.

Our universe is expanding and was hence more dense in the past that it is at present. The universe also evolves and presents a slightly different face each billion years. Probably it was extremely dense prior to the birth of galaxies and also extremely hot. The 3-degree cosmic radiation is widely regarded as providing evidence of a dense and hot early universe. Universes that start or end at high density, or pass through a high-density phase, are called *big bang* universes. The descriptive name *big bang* was first coined by Fred Hoyle in 1950. Universes that begin or die "not with a bang but a whimper" we shall call *whimper* universes.

We can classify all universes in the following simple way. A big bang occurs whenever the scaling factor R is either zero or extremely small, and a whimper is a long-drawn-out state that occurs when R is large and without limit. Hence we have

big bang: R approaches 0, density approaches ∞
whimper: R approaches, ∞, density approaches 0

where the symbol ∞ denotes infinity. Because a universe can begin either as a bang or as a whimper, and can end either as a bang or as a whimper, there are altogether four possibilities (see Figure 10.20):

bang–bang:	has finite lifetime
bang–whimper:	has infinite lifetime
whimper–bang:	has infinite lifetime
whimper–whimper:	has infinite lifetime

Of these four kinds of universes, we note that only one has a finite lifetime.

Because we live in an expanding universe we can rule out the whimper–bang kind that continually contracts. The whimper–whimper kind of universe is also unlikely, because the turnaround, or bounce, must occur at a density great enough to create the cosmic radiation that now has a temperature of 3 degrees, and a dense turnaround state of such a nature is indistinguishable from a big bang. We are left with two possible kinds of universes: the bang–bang kind that expands from a big bang and then collapses back to a big bang, with us at present existing in the expanding phase; and the bang–whimper kind that expands continually for an infinite period of time. We shall see later, in the case of Friedmann universes, that the bang–bang kind not only have finite lifetimes but also

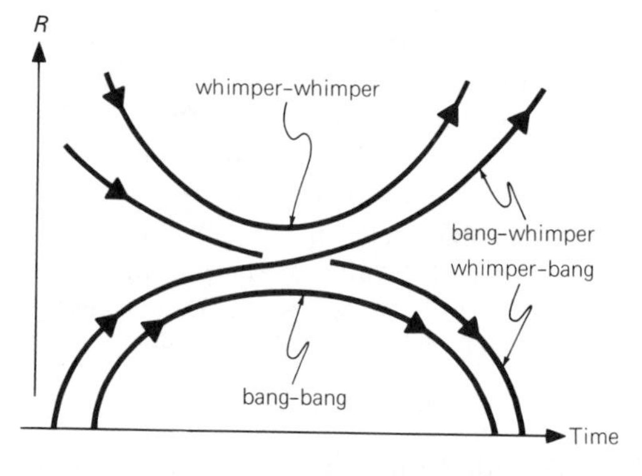

Figure 10.20. The four kinds of universes in the bang–whimper classification scheme.

contain space of finite extent (spherical space), whereas the bang–whimper kind not only have infinite lifetimes but also contain space of infinite extent (flat and hyperbolic spaces).

REFLECTIONS

1 *Show that the observable universe increases its volume each second by an amount roughly equal to the volume of the Galaxy.*

* *Explain why, in a universe 1 year old, we cannot see further than a distance of roughly 1 light year.*

* *What is wrong with the concept of a big bang occurring at a point in space?*

* *Explain the difference between recession velocity and ordinary velocity and comment on the following statement, taken from a textbook on astronomy: "There are mathematical models of the universe that have galaxies . . . going even faster than the velocity of light. Of course the laws of relativity forbid this, and such models are only of academic interest."*

* *Take a length of elastic, place markers on it, and slowly stretch it (see Figure 10.21). Show that the markers move apart according to the velocity–distance law. The markers illustrate the nature of comoving coordinates (or of other bodies, such as galaxies, that comove with expansion). Show how distances are related to constant coordinate distances by means of the scaling factor. Show that the way in which the elastic is stretched – by slow or fast expansion and contraction – demonstrates the change in time of the scaling factor. By increasing and decreasing the rate of expansion, show how accelerating and decelerating universes behave.*

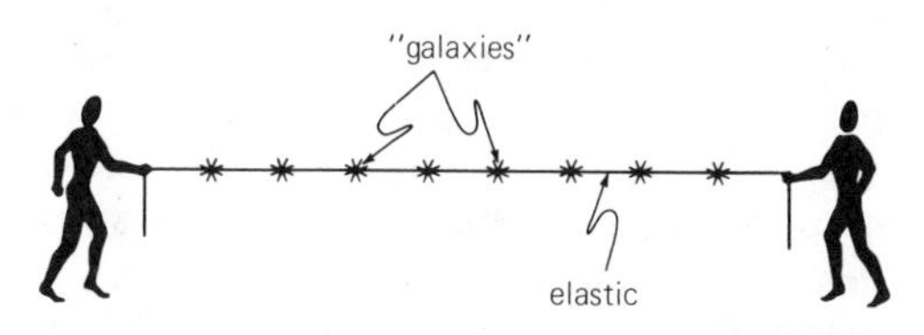

Figure 10.21. A length of elastic with attached markers denoting galaxies. The markers act as comoving coordinates and can be used to demonstrate the velocity–distance relation.

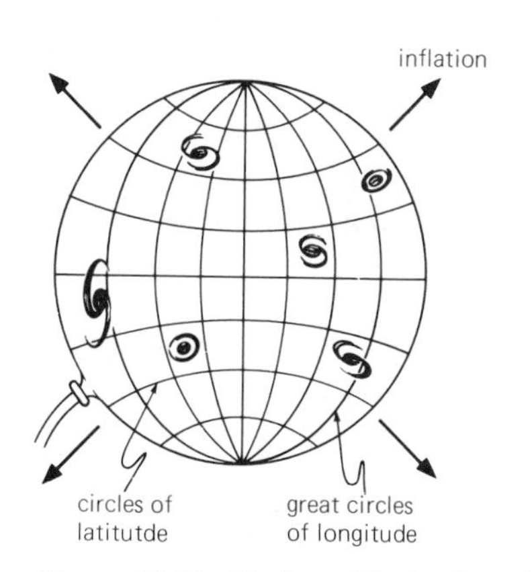

Figure 10.22. Circles of latitude and longitude on the surface of a rubber balloon are comoving coordinates.

* *Inflate a spherical balloon and with a felt pen draw on it circles of latitude and longitude, and also a few galaxies (see Figure 10.22). Notice, as the balloon is inflated and deflated, that the circles of latitude and longitude form a comoving coordinate system, and that the galaxies, although stationary on the surface, have recession velocities relative to each other. The expanding balloon experiment was suggested by Arthur Eddington in* The Expanding Universe *(1933): "For a model of the universe let us represent spherical space by a rubber balloon. Our three dimensions of length, breadth, and thickness ought all to lie in the skin of the balloon; but there is room for only two, so the model will have to sacrifice one of them. That does not matter very seriously. Imagine the galaxies to be embedded in the rubber. Now let the balloon be steadily inflated. That's the expanding universe."*

* *Draw diagrams showing how the scaling factor R may vary with time. Invent universes, such as the static and steady state type, that do not fit into the simple bang and whimper classification.*

* *Why do the bang–whimper and whimper–bang universes have infinite lifetimes?*

* *Let the smoothed-out density of the universe be equivalent to 1 hydrogen atom per cubic meter. Now gather this matter together into marbles and find their separating distance. Gather the matter together into stars similar to the Sun and find their separating distance.*
* *Who discovered the expansion of the universe?*

2 ***DREYER NEBULA NO. 584 INCONCEIVABLY DISTANT***
Dr. Slipher Says the Celestial Speed Champion Is 'Many Millions of Light Years' Away.
By Dr. Vesto Melvin Slipher.
Assistant Director of the Lowell Observatory,
Flagstaff, Ariz.

FLAGSTAFF, Ariz., Jan. 17. – The Lowell Observatory some years ago undertook to determine the velocity of the spiral nebulae – a thing that had not been previously attempted or thought possible. The undertaking soon revealed the quite unexpected fact that spiral nebulae are far the most swiftly moving objects known in the heavens. A recent observation has shown that the nebula in the constellation Cetus, number 584 in Dreyer's catalogues, is one of very exceptional interest.

Like most spiral nebulae, this one is extremely faint, and to observe its velocity requires an exceedingly long photographic exposure with the most powerful instrumental equipment. This photograph was exposed from the end of December to the middle of January in order to give the weak light of the nebula's spectrum time to impress itself upon the plate. It is necessary to disperse the nebular light into a spectrum in order to observe the spectral lines, and to measure the amount they are shifted out of their normal positions, for it is this displacement of the nebula's lines that discloses and determines the velocity with which the nebula is itself moving.

The lines in its spectrum are greatly shifted, showing that the nebula is flying away from our region of space with a marvelous velocity of 1,100 miles per second.

This nebula belongs to the spiral family, which includes the great majority of the nebulae. They are the most distant of all celestial bodies, and must be enormously large.

If the above swiftly moving nebula be assumed to have left the region of the sun at the beginning of the earth, it is easily computed, assuming the geologists' recent estimate of the earth's age, that the nebula now must be many millions of light years distant.

*The velocity of this nebula thus suggests a further increase to the estimated size of the spiral nebulae themselves as well as to their distances, and also further swells the dimensions of the known universe (*New York Times, *January 19, 1921, p. 6).*

3 *"The unanimity with which the galaxies are running away looks almost as though they had a pointed aversion to us. We wonder why we should be shunned as though our system were a plague spot in the universe . . . But the theory of the expanding universe is in some respects so preposterous that we naturally hesitate to commit ourselves to it. It contains elements apparently so incredible that I feel almost an indignation that anyone should believe in it – except myself . . . For the reader resolved to eschew theory and admit only definite observational facts,* all *astronomical books are banned.* There are no purely observational facts about the heavenly bodies. *Astronomical measurements are, without exception, measurements of phenomena occurring in a terrestrial observatory or station; it is only by theory that they are translated into knowledge of a universe outside" (Arthur Eddington,* The Expanding Universe, *1933).*

4 *According to general relativity, matter deforms spacetime in the vicinity of stars and black holes. We must remember, however, that these are static situations, and that matter can also affect the dynamics of spacetime. In cosmology we have homogeneous universes in which space* *is either flat or curved. The reason why space can contain a uniform distribution of matter and yet be flat, as well as curved, is that matter affects the dynamical behavior of space. In the presence of matter, flat and static space is impossible; curved and static space is possible, and flat and expanding space is also possible.*

5 *Hermann Weyl (1885–1955), a German scientist and philosopher, was a pioneer in general relativity theory and quantum mechanics. In 1923, Weyl supposed that the galaxies are diverging worldlines and the galaxies are stationary in space that is perpendicular to the worldlines (see Figure 10.23). In this space the galaxies have a common time (cosmic time). This was the beginning of comoving coordinates, and* Weyl's principle *was developed by Howard Robertson and Arthur Walker in 1935 into the homogeneous and isotropic spacetimes now known as the Robertson–Walker metrics of cosmic time and uniformly curved space. In the preface of his book* Space, Time, Matter, *Weyl wrote in 1919 (after World War I), "To gaze up from the ruins of the oppressive towards the stars is to recognize the indestructible world of laws, to strengthen faith in reason, to realize the 'harmonia mundi' [harmony of the worlds] that transforms all phenomena, and that never has been, nor will be, disturbed."*

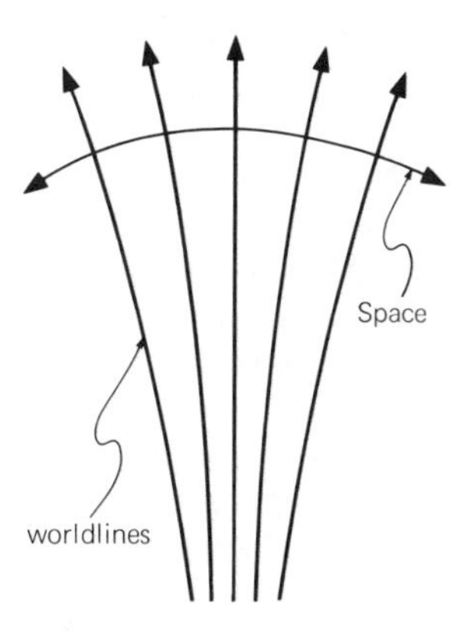

Figure 10.23. The Weyl principle. Diverging worldlines are stationary in a space that is perpendicular to the worldlines. By means of this principle, Hermann Weyl argued in 1923 that galaxies would recede from each other according to the velocity–distance law.

6 *Can we prove that the universe is homogeneous? Remember, we cannot observe the world map that is supposedly homogeneous; we see only a world picture that slices back through space and time. From this world picture we must project and construct a world map that shows what the universe is like everywhere at present; and so we must carefully allow for evolution and expansion. We must therefore assume that the laws of nature are everywhere the same and that things evolve everywhere as in our own neighborhood. But this presupposes an underlying homogeneity. We cannot prove homogeneity with arguments that assume homogeneity already exists. At best, we can show that the observed world picture is consistent with a homogeneous world map; we can never prove that the world map is necessarily homogeneous. It is possible for an infinite number of isotropic universes to mimic at some instant our world picture and have world maps that are not homogeneous. Yet each of those mischievous universes requires that we have special location at a cosmic center. On philosophical grounds and by appeal to the location principle we dismiss them as unlikely. We believe that the universe is homogeneous because special location is improbable.*

It is a curious consequence of homogeneity that on the average everybody in the universe thinks alike. Variety in thought and outlook is to be found within a galaxy and not by exploring the uttermost depths of the universe.

7 *Anything, such as a human being or the Earth itself, is made of a finite number of particles. A finite number of particles can be arranged into only a finite number of different configurations. If the particles are rearranged an infinite number of times, then each configuration of finite probability will occur an infinite number of times. Consider now a uniform universe containing space (flat or hyperbolic) of infinite extent.*

The observable part extends out to approximately a Hubble length of 15 billion light years. The unobservable universe that lies beyond extends endlessly. Trillions of trillions of Hubble lengths are nothing compared with infinity. And if the universe is homogeneous, then out there must exist an infinite number of identical Solar Systems, having identical Earths, having identical human populations, living identical lives. All things of finite probability are repeated an infinite number of times in an infinite universe. This is the principle of plenitude revived with a vengeance. The probability argument of the location principle has led to the improbable absurdity of infinite numbers of identical things. Whichever way we turn we are the victims of a cosmic hoax; either we are at a cosmic center and isotropy hoaxes us into a belief that the universe is homogeneous, or we live in an infinite universe and we are hoaxed into the belief that each of us is unique. The escape from this dilemma is the finite uniform universe of spherical space. See "Life in the infinite universe," by G. F. R. Ellis and G. B. Brundrit (1974).

While on this theme we should also consider the eternal steady state universe in which everything goes on forever in the same way. It is a universe of infinite and uniform space in which everything is eternally the same. Out there in space are an infinite number of Harrisons at this instant writing this book. Furthermore, every configuration of finite probability has been repeated in the past and will be repeated in the future an infinite number of times. In the past and in the future lies the astonishing picture of an infinite sequence of Harrisons writing this identical book, each thinking about the "monotonous cosmological principle" implicit in the steady state theory. I have felt repelled by the steady state theory of the universe almost from its inception because of its eternal sameness and its implication that nothing is unique in the past, present, or future.

When confronted with spatially infinite universes and eternal steady state universes we feel impelled to ask, What's the point of it all when once is often more than enough? Perhaps you can understand why some cosmologists, if only for philosophical reasons, favor universes that are finite in both space and time.

8 *The words* big bang *were first used by Fred Hoyle in his series of BBC radio talks on astronomy, published in* The Nature of the Universe, *1950. "This big bang idea seemed to me to be unsatisfactory even before examination showed that it leads to serious difficulties. For when we look at our own Galaxy there is not the smallest sign that such an explosion ever occurred. This might not be such a cogent argument against the explosion school of thought if our Galaxy had turned out to be much younger than the whole Universe. But this is not so. On the contrary, in some of these theories the Universe comes out to be younger than our astrophysical estimates of the age of our own Galaxy . . . On philosophical grounds too I cannot see any good reason for preferring the big bang idea. Indeed it seems to me in the philosophical sense to be a distinctly unsatisfactory notion, since it puts the basic assumption out of sight where it can never be challenged by direct appeal to observation." Discuss this quotation.*

* *The word* whimper, *used to denote a universe that does not begin or end with a big bang, is suggested by T. S. Eliot's "The hollow men":*

This is the way the world ends
Not with a bang but a whimper.

George Ellis has used whimper *to indicate universes that do not terminate in a uniform big bang. Our alternative use of the word is not likely to confuse the reader, and is perhaps more in accord with the spirit of Eliot's poem. Instead of* bang *and* whimper *we could use* fire *and* ice, *as suggested by Robert Frost in "Fire and ice":*

Some say the world will end in fire,
Some say in ice.
From what I've tasted of desire
I hold with those who favor fire.
But if it had to perish twice,
I think I know enough of hate
To say that for destruction ice

Is also great
And would suffice.

* *In classical theory a* singularity *occurs when the density is infinitely great. A bang type of universe has a singularity when the scaling factor R is zero. What happens at infinite density is not known, and for physical reasons (see Chapters 9 and 18) it is probable that a singular state of this nature is unattainable. But extreme densities are possible in black holes and collapsed universes, and these congested states are still referred to as singularities.*

* *Many people have disliked the notion of a big bang. In* The Expanding Universe, *Eddington said, "Since I cannot avoid introducing this question of a beginning, it has seemed to me that the most satisfactory theory would be one which made the beginning* not too unaesthetically abrupt." *He also wrote in 1931 in "The expansion of the universe," "Philosophically, the notion of a beginning of the present order of Nature is repugnant to me." Eddington's aversion to a big bang was shared by others, including the advocates of the steady state universe.*

Cosmic birth and death were common notions in mythology, and the modern fear of such ideas may stem from the way that people now live. No longer are we members of large family communities, surrounded by numerous relatives, young and old, among whom birth and death are common incidents. We live instead in small families, isolated from one another, and birth and death are remote incidents that frequently occur out of sight in hospitals. Eddington, who was outspoken in his dislike of cosmic birth and rejected catastrophic cosmic death, was a bachelor who lived with his sister. When a cosmologist states a preference for a certain type of universe, perhaps we should not bother to examine too carefully the scientific arguments offered, but instead examine the philosophical and psychological reasons!

FURTHER READING

Davies, P. C. W. *The Runaway Universe.* J. M. Dent, London, 1978.

Eddington, A. S. *The Expanding Universe.* Cambridge University Press, Cambridge, 1933. Reprint. University of Michigan Press, Ann Arbor Paperback, Ann Arbor, 1958.

Gamow, G. *The Creation of the Universe.* Viking Press, New York, 1952.

North, J. D. *The Measure of the Universe: A History of Modern Cosmology.* Oxford University Press, Clarendon Press, Oxford, 1965.

Sandage, A. R. "Cosmology: a search for two numbers." *Physics Today,* February 1970.

Sciama, D. W. *Modern Cosmology.* Cambridge University Press, Cambridge, 1971.

Whitrow, G. T. *The Structure and Evolution of the Universe.* Hutchinson, London, 1959.

SOURCES

Eddington, A. S. "The expansion of the universe." *Monthly Notices of the Royal Astronomical Society 91,* 412 (1931).

Eddington, A. S. *The Expanding Universe.* Cambridge University Press, Cambridge, 1933. Reprint. University of Michigan Press, Ann Arbor Paperback, Ann Arbor, 1958.

Ellis, G. F. R., and Brundrit, G. B. "Life in the infinite universe." *Quarterly Journal of the Royal Astronomical Society 20,* 37 (1974).

Hoyle, F. *The Nature of the Universe.* Blackwell, Oxford, 1950. Rev. ed. Penguin Books, London, 1960.

Hubble, E. *The Observational Approach to Cosmology.* Oxford University Press, Clarendon Press, Oxford, 1937.

Lemaître, G. "A homogeneous universe of constant mass and increasing radius accounting for the radial velocity of extra-galactic nebulae." *Monthly Notices of the Royal Astronomical Society 91,* 483 (1931).

McCrea, W. H. "Cosmology after half a century." *Science 160,* 1295 (1968).

McCrea, W. H. "Willem de Sitter, 1872–1934." *Journal of the British Astronomical Association 82,* 178 (1972).

Metz, W. D. "The decline of the Hubble constant: a new age for the universe." *Science 178,* 600 (1972).

Murdoch, H. S. "Recession velocities greater than light." *Quarterly Journal of the Royal Astronomical Society 18,* 242 (1977).

Robertson, H. P. "On relativistic cosmology." *Philosophical Magazine 5,* 835 (1928).

Robertson, H. P. "The expanding universe." *Science 76,* 221 (1932).

Sandage, A. R. "Observational cosmology." *Observatory 88,* 91 (1968).

Sandage, A. R. "Distances to galaxies: the Hubble constant, the Friedmann time and the edge of the world." *Quarterly Journal of the Royal Astronomical Society 13,* 282 (1972).

Sitter, W. de. *Kosmos.* Harvard University Press, Cambridge, Mass., 1932.

Tolman, R. C. "The age of the universe." *Review of Modern Physics 21,* 374 (1949).

Weyl, H. *Space, Time, Matter.* Methuen, London, 1922. Reprint. Dover Publications, New York, 1950.

REDSHIFTS

O ruddier than the cherry,
O sweeter than the berry,
O nymph more bright
Than moonshine night,
Like kidlings blithe and merry.
— John Gay (1685–1732), *Acis and Galatea*

COSMIC REDSHIFTS

WAVELENGTH STRETCHING. In the previous chapter we saw that the Doppler formula played a historical role in the discovery of the expanding universe. Distant galaxies have redshifted light, and this redshift was interpreted, in accordance with the Doppler formula, to mean that the galaxies were rushing away from us. Georges Lemaître and Howard Robertson in the late 1920s discovered a new interpretation – the *expansion-redshift effect* – and it then became evident that the observed redshifts of extragalactic systems are not manifestations of the Doppler effect. The expansion-redshift effect is simple and extremely easy to understand.

We suppose that all galaxies are comoving and that their light is received by observers who are also comoving. Light leaves a galaxy, which is stationary in its own local region of space, and is received by observers who are also stationary in their own local region of space. Between the galaxy and the observers light travels through vast regions of expanding space. What happens is immediately obvious: All wavelengths of the light are stretched by the expansion of space (see Figure 11.1). It is as simple as that.

A lightray is emitted and, after having traveled across expanding space, is received by an observer. If, while it travels, all comoving distances are doubled, then all wavelengths of the lightray are also doubled. Waves are stretched by expanding space, and their expansion is proportional to the increase in the scaling factor. Let λ represent a wavelength of an emitted wave of light and λ_0 represent the wavelength of the wave when it is received by an observer. If R is the value of the scaling factor at the time of emission and R_0 is the value at the time of reception, then

$$\frac{\lambda_0}{\lambda} = \frac{R_0}{R} \tag{11.1}$$

and wavelengths increase in just the same way as comoving distances in an expanding universe. This applies to radiation of all kinds and is of fundamental importance in cosmology.

EXPANSION REDSHIFTS. A spectrum shows how the intensity of radiation varies with wavelength and contains recognizable features in the form of spectral lines (see Figure 11.2). These lines, usually produced in the source, are the result of electrons within atoms jumping up and down between energy levels and emitting and absorbing radiation of specific wavelengths. From measurements made in the laboratory we know the *standard wavelengths* of spectral lines of different atoms in various states of excitation. Because the universe is homogeneous –

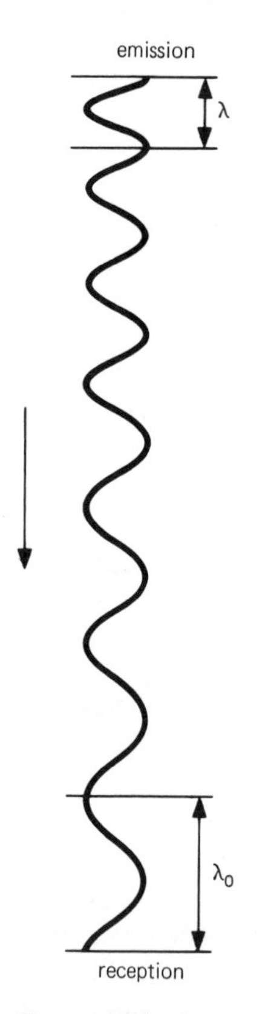

Figure 11.1. A wave of radiation is stretched as it travels through expanding space.

or so we assume – atoms everywhere are the same and the wavelengths of radiation they emit are the same as those observed in the laboratory. Hence, when a recognizable spectral line from a distant galaxy is observed at wavelength λ_0, we know that far away and long ago it originated at wavelength λ. From the ratio λ_0/λ we are given the ratio R_0/R, and we know immediately how much the universe has expanded since the time of emission. This is the miracle of cosmic redshifts: They measure the expansion of the universe.

The redshift, as we have said earlier, is the increase of wavelength, $\lambda_0 - \lambda$, divided by the original wavelength λ:

$$\text{redshift} = \frac{\lambda_0 - \lambda}{\lambda} \tag{11.2}$$

It is convenient to use the symbol z to denote redshift of any kind, and we have

$$1 + z = \frac{\lambda_0}{\lambda} \tag{11.3}$$

from equation (11.2). The *expansion redshift*, caused by the expansion of the universe, is therefore

$$1 + z = \frac{R_0}{R} \tag{11.4}$$

from the previous equations (11.1) and (11.3) (see Figure 11.3). All wavelengths of radiation received from a particular source have the same expansion redshift – not only optical waves but also the short-wavelength X-rays and long-wavelength radio waves. If the universe doubles in size – which means that R_0/R is equal to 2 – during the time

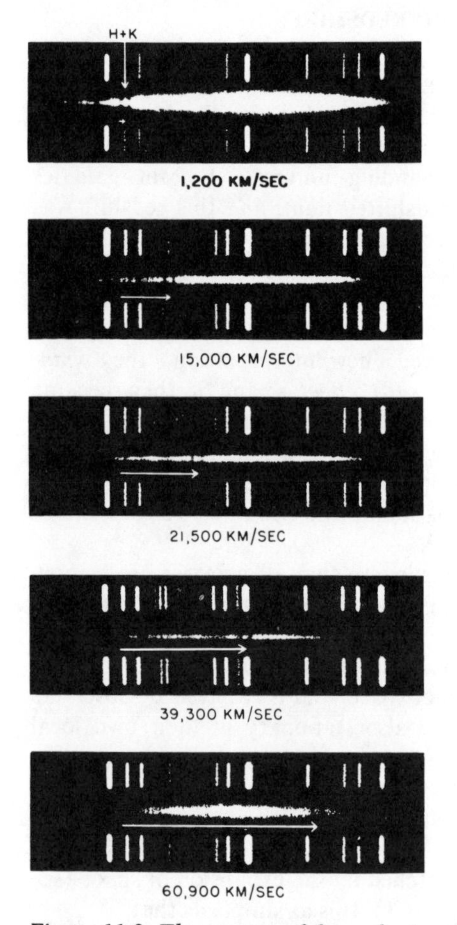

Figure 11.2. The spectra of five galaxies, showing displacement of the H and K lines of calcium. The recession velocities are estimated with the classical Doppler formula.

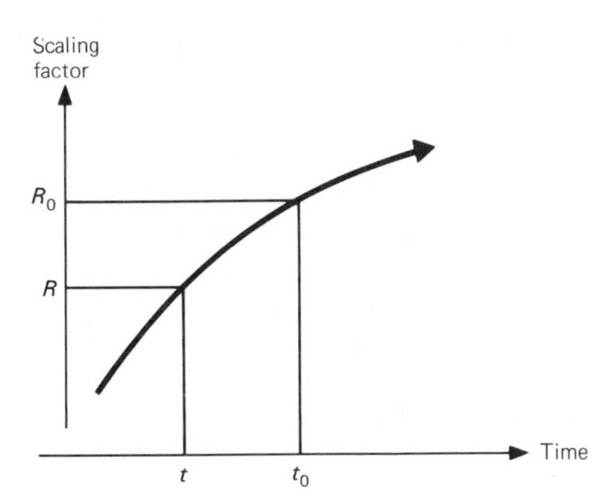

Figure 11.3. In an expanding universe the scaling factor increases with time. Radiation is emitted from a source at time t when the scaling factor has a value R, and is received at time t_0 when the scaling factor has a value R_0. The redshift z of the received radiation is given by $1 + z = R_0/R$.

between emission and reception of radiation, then all wavelengths, short and long, are increased twofold and the redshift is hence $z = 1$. If the universe trebles in size (i.e., $R_0/R = 3$), then all wavelengths are increased threefold and $z = 2$.

In an expanding universe the cosmic redshift has any value from zero to infinity. This is because in equation (11.4) the scaling factor R can range from zero (the beginning of the universe) to R_0, which denotes the present. Small redshifts indicate that light has not traveled for very long and large redshifts that it has traveled for immense periods of time. Redshifts of optically observed galaxies have measured values up to about 0.5, whereas the measured redshifts of the luminous quasars extend to values greater than 3. In a collapsing universe the cosmic redshift has any value from -1 to 0. This is because R_0 is now less than R, and at the time of reception the universe is smaller in size than at the time of emission. Negative redshifts are blueshifts, and large cosmic blueshifts occur only in collapsing universes. If a collapsing universe halves its size during the time of emission and reception of radiation, then $R_0/R = 0.5$, and the redshift is -0.5 (or the blueshift is 0.5).

Simply by determining the redshift of a distant galaxy or quasar we know immediately the ratio of the scaling factors for the epochs of emission and reception of its radiation. We know, in other words, how much the universe has expanded since the time its light was emitted.

RED MEANS SLOW. Wavelength multiplied by frequency equals the speed of light for radiation traveling through space:

$$\lambda \times f = c \tag{11.5}$$

where f stands for frequency at wavelength λ. The light speed c is constant and always the same when measured locally in the region of space through which radiation is passing. In expanding space the wavelengths are stretched and therefore the frequencies of vibration are reduced. From the way wavelengths vary we find from equations (11.4) and (11.5) that

$$1 + z = \frac{f}{f_0} \tag{11.6}$$

where f is the emitted frequency and f_0 is the received frequency.

It must be understood that the slowing down of vibrations applies to all frequencies, not only the ultraviolet, optical, infrared, and radio, but also the slow variations in the intensity of radiation that may have periods of hours, days, or even years. Some quasars fluctuate in brightness with periods of days and even years; these are the observed periods that have been increased by the passage of radiation through expanding space. If a quasar of redshift $z = 1$ is seen to vary in brightness with a period of 1 month, the original period of variation at the quasar was 2 weeks.

Things appear to happen more slowly the further we probe into the depths of outer space. This strange law can be understood more easily in the following way. Suppose that a distant galaxy sends out sharp pulses of radiation at intervals of 1 second. In a universe that is not expanding, but is static like the Newtonian universe, we will receive these pulses at 1-second intervals. In this case the distance between adjacent pulses, as they journey through space, will remain constant and equal to 1 light second. But in an expanding universe the distance between the pulses steadily increases. The distance is initially 1 light second; thereafter the distance between the pulses is continually stretched and the received pulses are separated not by 1 light second, but by $1 + z$ light seconds. The pulses are therefore received at intervals of $1 + z$ seconds, as shown in Figure 11.4. Clocks at great distances appear to us to run slow, and their intervals of time are increased by an amount $1 + z$. Whatever happens at great distances, whether atomic vibrations or a waxing and waning of light, is seen by us as slower and more sluggish than it actually is. When something happens at a redshift of $z = 1$ in 1 second, it is seen by us to happen in 2 seconds.

At the far frontier of the observable universe, where the expansion redshift rises to infinity, nothing ever seems to change; everything is apparently in a frozen state of immobility (see Figure 11.5). Infinite redshifts of course are not observed, and the maximum encountered so far – that for the cosmic background radiation – is approximately 1000. If we should ever succeed in detecting neutrinos from the big bang then we shall be able to look back to redshifts equal to about 10 billion. At $z = 1000$ a second is seen by us to last 17 minutes, and at $z = 10$ billion a second is seen by us to last 300 years.

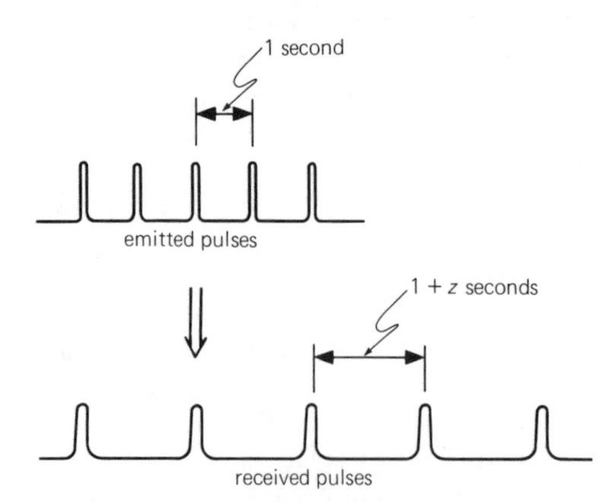

Figure 11.4. Pulses of radiation are emitted from a source at 1-second intervals. After traveling through expanding space the pulses are more widely separated; they arrive at intervals of 1 + z seconds. This explains why redshift makes everything appear to happen more slowly.

THE USEFULNESS OF REDSHIFTS. In the previous chapter it was mentioned that the actual value of the scaling factor R is not always important, and that usually we require only a comparison of its values for different epochs. Whenever the ratio R_0/R occurs we can substitute $1 + z$. Thus for comoving length L, area A, and volume V we are able to write:

$$\frac{L}{L_0} = \frac{1}{1 + z}$$

$$\frac{A}{A_0} = \frac{1}{(1 + z)^2}$$

$$\frac{V}{V_0} = \frac{1}{(1 + z)^3}$$

A comoving length L at the time of emission has now a length L_0 equal to $L(1 + z)$; a comoving area A at the time of emission has now an area A_0 equal to $A(1 + z)^2$; and a comoving volume V at the time of emission has now a volume V_0 equal to $V(1 + z)^3$. For instance, if $z = 1$, then at the time of emission comoving lengths were one-half their present value, comoving areas were one-quarter their present value, and comoving volumes were one-eighth their present value.

The change in density of the universe provides an important example of the relation between expansion redshift and the scaling factor. At any time the number n of things (atoms or galaxies) in a unit of volume is equal to the present density n_0 multiplied by $(R_0/R)^3$, and therefore the density at the time of emission at redshift z is given by

$$\frac{n}{n_0} = (1 + z)^3 \qquad (11.7)$$

For $z = 1$ the density at the epoch of emission was 8 times greater than the present density. For a redshift $z = 100$ the density was 1 million times greater than now, or approximately 1 hydrogen atom per cubic centimeter, which is typical of the average density of matter in galaxies. At a redshift of 100 it is therefore unlikely that

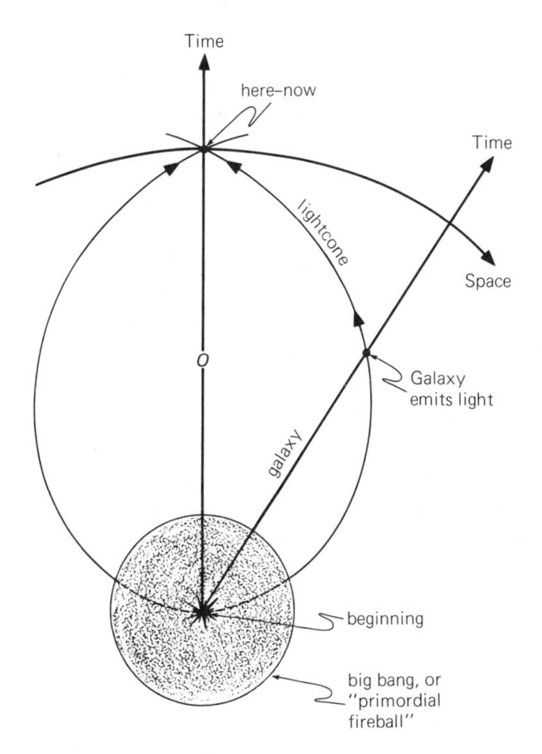

Figure 11.5 We (represented by worldline O) look out in space and back in time, and all information comes to us on our backward lightcone. The furthest we can look back is the primordial fireball from which comes the 3-degree cosmic radiation. If we could look back even further, to the beginning of the universe, the scaling factor R at the time of emission would be zero and the redshift would be infinite.

galaxies in their present form had come into existence. We cannot normally look out in space and back in time to such large redshifts because luminous galaxies and quasars did not then exist. The 3-degree cosmic radiation that fills the universe is an exception, and it is believed that it has been redshifted by approximately 1000 since it last interacted with matter, at which time the universe was hot and had a density a billion times greater than at present.

THE THREE REDSHIFTS

GRAVITATIONAL, DOPPLER, AND EXPANSION REDSHIFTS. The gravitational redshift is due to the change in the strength of gravity and occurs mostly in the vicinity of massive bodies. If a spherical body has a mass M and radius R, then

$$z = \frac{1}{\sqrt{(1 - R/R_s)}} - 1 \text{ (gravitational redshift)} \tag{11.8}$$

where R_s is the Schwarzschild radius and equals $2GM/c^2$. We discussed this kind of redshift previously, in Chapters 8 and 9, and will not pursue it further at the moment.

The Doppler redshift is due to relative motion in space. When a luminous body moves away from an observer, in the laboratory, the Solar System, or the Galaxy, it moves through space, and the radiation received from it is redshifted. If v represents the velocity of a luminous body moving away, then

$$z = \frac{v}{c} \qquad \text{(Doppler redshift)} \tag{11.9}$$

This is the *classical Doppler formula* that applies only when v is small compared with the light velocity c. The exact formula, or the *special relativity Doppler formula,* which must be used when v is not small, is

$$z = \left(\frac{c + v}{c - v}\right)^{1/2} - 1 \quad \text{(Doppler redshift)} \tag{11.10}$$

Most of the peculiar velocities encountered in astronomy, such as those of stars in galaxies and of galaxies in clusters, are small in comparison with the velocity of light and thus allow us to use the original classical Doppler formula. When z is not small, we find from the special relativity Doppler formula of equation (11.10) that

$$\frac{v}{c} = \frac{z^2 + 2z}{z^2 + 2z + 2} \tag{11.11}$$

Thus from the classical Doppler formula (of equation 11.9) we have $v = 0.1c$ when $z = 0.1$; and from the special relativity Doppler formula of equation (11.11) we have $v = 0.6c$ when $z = 1$, $v = 0.8c$ when $z = 2$, and $v = c$ when z equals infinity.

The expansion redshift is due to the expansion of space in an expanding universe. Comoving bodies, stationary in expanding space, receive radiation from each other that is redshifted. The radiation propagates through the expanding space, and all wavelengths are continually stretched. This redshift is determined by the amount of expansion according to the law

$$z = \frac{R_0}{R} - 1 \qquad \text{(expansion redshift)} \qquad (11.12)$$

where R is the value of the scaling factor at the time of emission and R_0 is the value at the time of reception. Once the expansion redshift of a distant galaxy has been determined, the ratio R_0/R tells us how much the universe has expanded during the time in which the light from the galaxy has been traveling toward us. The quasar 3C 273 has a redshift of 0.16, which means that the universe is now 1.16 times larger than when this quasar emitted the light we now see. The universe has therefore increased in size by 16 percent since 3C 273 emitted its light. Quasar 3C 48 has a redshift of 0.37; accordingly, the universe has expanded by 37 percent since its light was emitted. The rule is quite simple: Multiply the expansion redshift by 100 to get the percentage of increase in the size of the universe since the radiation was emitted.

DIFFERENCE BETWEEN DOPPLER AND EXPANSION REDSHIFTS. Astronomers in the early 1920s thought that Doppler redshifts were the same as expansion redshifts. Rather curiously, the habit of referring to expansion redshifts as Doppler redshifts has survived and is now widespread. Cosmologists feel compelled to use this inexact terminology in popular literature (otherwise, few people would know what they were talking about), and astronomers even catalogue the recession velocities of galaxies by simply multiplying each redshift by the velocity of light. If the observed redshift is 0.05, for example, the galaxy is catalogued with a recession velocity v of 15,000 kilometers per second. Professionals know what they are doing and therefore avoid the pitfalls that by a misuse of words they have unfortunately prepared for others. The truth is that expansion redshifts are totally different from Doppler redshifts, and the velocities catalogued by astronomers are not the recession velocities used in the velocity–distance law.

Despite the widespread confusion between expansion and Doppler redshifts, the difference is quite marked and easily understood. Doppler redshifts are the result of relative motion of bodies moving through space: They depend on the relative velocity of the emitter and receiver at the instants of emission and reception; they are the result of peculiar velocities and not recession velocities; and they are governed by the rules of special relativity. Expansion redshifts are produced by the expansion of space between bodies that are stationary in space: They depend on the increase of distance between the emitter and the receiver during the time of propagation; they are the result of recession velocities and not peculiar velocities; and they are not governed by the rules of special relativity.

It is interesting to note that the expansion redshift is independent of the way the universe expands. This is perhaps not very surprising; after all, when a length of elastic is stretched by a certain amount, it is unimportant how the stretching is done, whether slowly, quickly, or in a series of jerks. In the end it is always stretched by a stated amount. The time taken to expand from a given value R of the scaling factor to the present value R_0 and the way in which the expansion occurs have nothing to do with the expansion redshift.

An extragalactic source of light that has a small expansion redshift is not, cosmologically speaking, very far away. The scaling factor difference $R_0 - R$ is small compared with R itself, and the light travel time from the source is small compared with the Hubble period. For sources that are near, the redshift

$$z = \frac{R_0 - R}{R}$$

becomes approximately

$$z = H \times \text{light travel time}$$

The distance to these nearby sources is also approximately the light travel time multiplied by the velocity of light, and hence

$$z = H \times \frac{\text{distance}}{c}$$

If we now use the velocity–distance law ($v = H \times$ distance), we find for the recession velocity

$$z = \frac{v}{c}$$

which is the classical Doppler formula. This shows us that when comoving objects are at short distances (short compared with the cosmic yardstick of the Hubble length), the expansion-redshift law and the velocity–distance law together give an expression that is the same as the classical Doppler formula.

Originally, Slipher, Humason, and others used the Doppler formula to interpret the meaning of extragalactic redshifts. They were correct in their conclusion that the universe is expanding because the galaxies then observed by astronomers are nearby and have small recession velocities. This is the origin of the belief that Doppler redshifts are actually expansion redshifts. But as we have seen, they are coincidental only for nearby galaxies of small redshift. The Doppler formula is convenient and can be used when expansion redshifts are less than about 0.1. At $z = 0.1$ the recession velocity (equal to cz) is 30,000 kilometers per second, and the distance (equal to $z \times$ Hubble length) is 1.5 billion light years for a Hubble length of 15 billion light years.

THE REDSHIFT BALLET. We have supposed that light is emitted by extragalactic comoving sources and is received by comoving observers; consequently, the redshift is purely a cosmic effect resulting from the expansion of the universe. In reality the situation is slightly more complicated. Light is emitted from the surfaces of stars in a galaxy and has therefore an initial gravitational redshift. It also has an initial Doppler shift owing to the peculiar velocity of the galaxy. After this radiation traverses expanding space and acquires an expansion redshift, it is then Doppler shifted by the peculiar motion of the observer, and receives a final gravitational redshift contribution from the system in which the observer is located. The total redshift observed is thus contaminated with gravitational and Doppler contributions at both ends, and one of the astronomer's many skills consists of knowing how to average out or otherwise discount these irrelevant contributions. In most cases the gravitational effects are exceedingly small. Light entering our Galaxy and striking the Earth's surface receives a gravitational redshift of -0.001, and this is negligible. The Doppler shifts of distant sources are usually small in comparison with their expansion redshifts and either are negligible or can be averaged out over many sources. Our own peculiar velocity of the Solar System moving in the Galaxy, and of the Galaxy moving at approximately 500 kilometers per second, perhaps in the direction of the constellation of Leo, produces a Doppler shift that must always be taken into account. The Three Redshift Graces, *Gravi-*

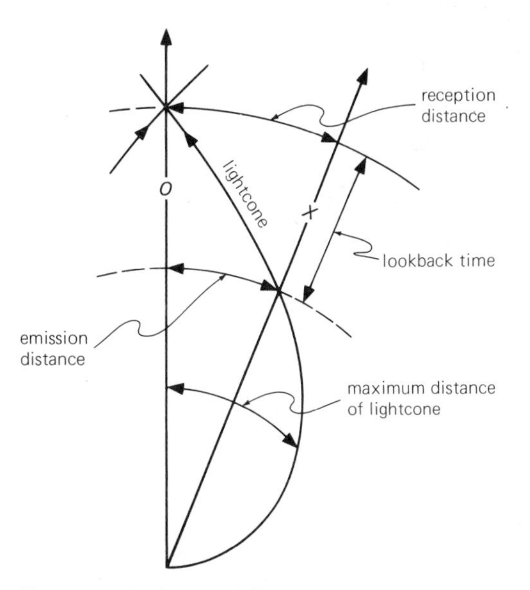

Figure 11.6. Our worldline is O and the worldline of a distant galaxy is X. Our backward lightcone intersects X at the time of emission of the light we now see. The reception (or present) distance and the emission (or past) distance are shown for the galaxy. The lookback time, from the present to the time of emission, is also shown.

ty, Doppler, and *Expansion,* are usually together, and rarely do we find one alone by herself.

DISTANCES AND RECESSION VELOCITIES. Once the redshift of a distant galaxy has been found, there are several things we would like to know, such as the distance of the galaxy and its recession velocity at the present epoch, the time the radiation has taken to reach us (the *lookback time* to the galaxy), and the age of the universe at the epoch of emission (see Figure 11.6). This information, regrettably, cannot be obtained directly from the observed redshift. If we had all this information we would know the kind of universe in which we live. Because we do not know for certain what our kind of universe is, we must make calculations for different universes, and each universe gives different answers.

A redshift measurement tells us directly how much the universe has expanded since the epoch of emission; to determine distances, recession velocities, and the lookback time we must also know the geometry of space and how the scaling factor changes with time. But these things are not known for certain at present, and therefore we cannot translate redshifts unambiguously into distances, recession velocities, and other things of interest.

A popular universe, because of its extreme simplicity, is the bang–whimper universe proposed by Einstein and de Sitter in 1931. This *Einstein–de Sitter universe* has Euclidean (flat) space and a deceleration term of constant value $q = \frac{1}{2}$. Nothing could be simpler. For any redshift in the Einstein–de Sitter universe the corresponding distance, recession velocity, lookback time, and age of the universe at the epoch of emission are shown in Figures 11.7, 11.8, and 11.9. Our universe is possibly not of the

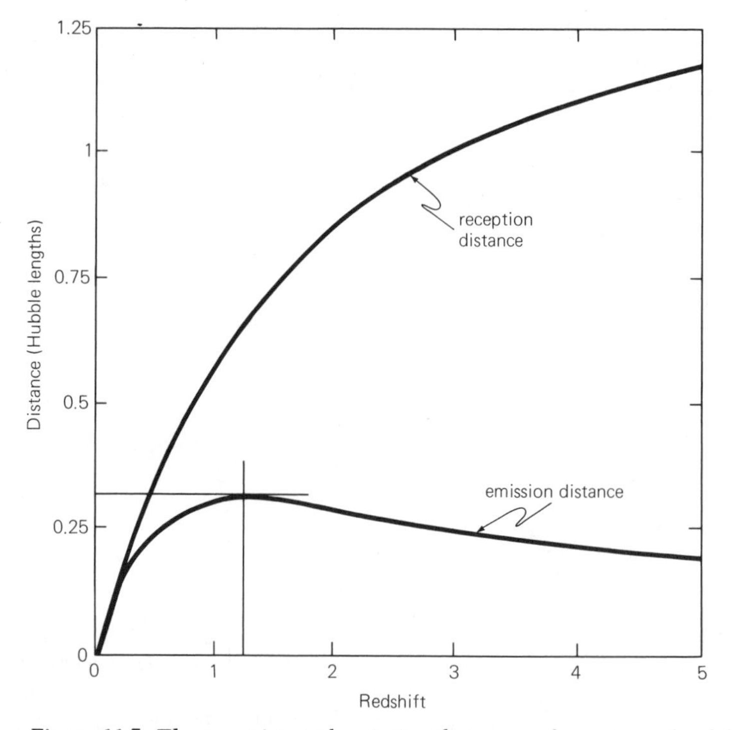

Figure 11.7. The reception and emission distances of a source of redshift z in the Einstein–de Sitter universe. A Hubble length is 15 billion light years for H = 20 kilometers a second per million light years.

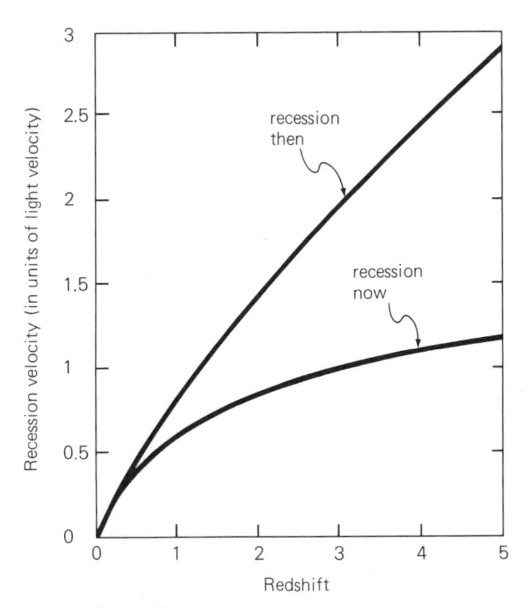

Figure 11.8. The present recession velocity at the time of reception, and the past recession velocity at the time of emission, of a source at redshift z in an Einstein–de Sitter universe.

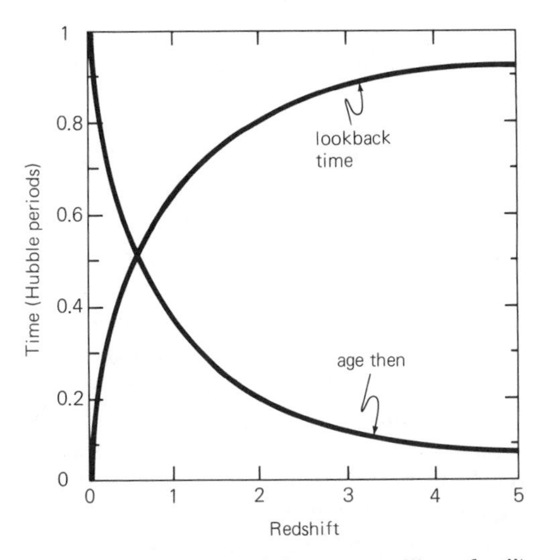

Figure 11.9. The age of the universe ("age then") at the time a source of redshift z emitted the light that we now see, and also the lookback time, in an Einstein–de Sitter universe. A Hubble period is 15 billion years for H = 20 kilometers a second per million light years, and the present age is ⅔ of a Hubble period.

Einstein–de Sitter type, but the results shown are far more meaningful than the erroneous results derived from the special relativity Doppler formula.

A COSMOLOGICAL PITFALL

O Thou who didst with pitfall and with gin
Beset the road I was to wander in.
— Edward FitzGerald (1809–83), *The Rubáiyát of Omar Khayyâm*

It is the custom when a galaxy or a quasar is discussed in the popular literature to give its recession velocity and distance. The velocity and distance quoted are usually not stated within the context of a particular cosmological model nor hedged with reservations concerning the validity of the model, for fear of puzzling the audience and losing its interest. Instead, the Doppler formula is used. The redshift is multiplied by the light velocity to give the recession velocity: Thus if $z = 0.2$, the recession velocity is said to be one-fifth the velocity of light, or 60,000 kilometers per second. The velocity–distance law then states that the distance is one-fifth the Hubble length, or 3 billion light years, and light has traveled for 3 billion years. The classical Doppler formula is very convenient for making quick estimates, but the results obtained in this way are incorrect. All that can be said is that the universe has expanded 20 percent since the emission of light. More specific information must be obtained within the framework of a particular cosmological model, such as the Einstein–de Sitter universe.

Suppose that the redshift of a quasar is larger than unity. The classical Doppler formula cannot be used for large velocities, and the special relativity Doppler formula is used instead. For $z = 1$ the velocity is said to be $0.6c$, or 180,000 kilometers per second; and the distance is 0.6 times the Hubble length, or 9 billion light years. For $z = 2$ the velocity given is $0.8c$, or 240,000 kilometers per second; and the distance is 0.8 times the Hubble length, or 12 billion light years.

These statements regrettably are also incorrect: All that can be said is that the universe has expanded 100 percent for $z = 1$, and 200 percent for $z = 2$, since the time of emission.

Although the Doppler formula enables us easily and quickly to make statements about recession velocities and distances, for redshifts greater than about 0.1 these statements are meaningless in cosmology; their only virtue is that they capture the interest of the public. The damage done, however, is that students are led to believe that this is the correct method. The Doppler formula applies only to bodies that move through space, not to bodies that are comoving in expanding space. By using this formula we erroneously depict a universe in which galaxies hurl through space, with redshifts that are the result of velocities limited by the velocity of light. The recession velocity is thus reduced to the status of an ordinary velocity – such as that of an automobile or a rocket – and is made subject to the rules of special relativity. It is easy to see that this leads to a violation of the principles of containment and location. For if the recession velocity cannot exceed the velocity of light, then the velocity–distance law terminates abruptly at the edge of the Hubble sphere. This creates a cosmic edge with us at the cosmic center. By failing to distinguish between recession velocity and ordinary velocity, and by failing to distinguish between expansion redshift and Doppler redshift, a confused student is presented with a situation that is tantamount to proof that the edge of the universe is at a distance of 15 billion light years and that we occupy the center of a bounded universe.

In cosmology we have two very simple and beautiful laws: the velocity–distance law

$$\text{recession velocity} = H \times \text{distance}$$

and the expansion-redshift law

$$1 + z = \frac{R_0}{R}$$

But these laws cannot be combined to give recession velocities and distances in terms of redshifts, except when redshifts are small or when a particular model of known geometry is used, such as the Einstein–de Sitter universe, in which it is known how the scaling factor changes with time. The custom of referring to the expansion redshifts as Doppler redshifts ranks among the most curious aspects of the whole redshift story. To say the least, it is a regrettable practice that creates confusion and ensnares the unwary.

REDSHIFT CURIOSITIES

FATIGUED LIGHT. The expansion interpretation of cosmic redshifts, delightfully simple, has been challenged several times. Fritz Zwicky, a Swiss-American astronomer who studied galaxies and pioneered the study of supernovas, advanced in 1929 the idea that light loses energy progressively while traveling across large distances of extragalactic space. According to this *tired light* hypothesis, resurrected several times since Zwicky first proposed it, the vibrations of light are steadily slowed down over long periods of time. Various suggestions have been made to explain why light might suffer from fatigue while traveling in the universe, but so far none has been successful. Either their authors have failed to look fully into the consequences (for example, if interaction with intergalactic gas were the cause, then scattering would occur, point sources of light would become blurred, and we would not see quasars as starlike objects), or they have failed to realize that a hypothesized unknown law is rarely an attractive substitute for a known law.

According to the tired light hypothesis, redshift is the result of fatigue and not expansion, and hence the universe is static and not expanding. It is a quaint idea, unlikely to be true because it must explain why the universe is static and why the redshift of an extragalactic source is the same for a wide range of wavelengths. It must also explain the 3-degree cosmic radiation. In the distant past the cosmic radiation had a higher temperature, and the tired light hypothesis confronts us with the startling prospect of a big bang within a static universe. One is left wondering where all the energy has gone. (A more subtle question is where all the entropy has gone. In an expanding universe the entropy of the cosmic radiation remains constant; but in a static universe, in which radiation suffers from fatigue and is reddened by old age, the entropy declines, and no one has yet been able to say where it goes. See Chapter 13.)

DISCORDANT REDSHIFTS. In more recent years further controversy has arisen and the expansion theory of redshifts has been challenged and brought to combat by formidable knights such as Halton Arp, Geoffrey and Margaret Burbidge, and Fred Hoyle. In Chapter 4 it was mentioned that Gerald de Vaucouleurs has questioned the simplistic doctrine that the universe is homogeneous almost up to our galactic doorstep. Meanwhile, Arp has mounted an attack on the sweeping hypothesis that all extragalactic redshifts are primarily due to expansion.

Many galaxies and quasars, perhaps most, have redshifts caused by expansion, but there are others, argues Arp, whose redshifts appear to be the result of unknown causes. Galaxies within a group are all at practically the same distance from us and should therefore have almost equal redshifts. But Arp claims that there are chains and groups of galaxies whose members have widely different redshifts (see Figure 11.10). Some galaxies appear to have

Figure 11.10. This group of galaxies is known as Stephan's Quintet. The largest galaxy, lower left of center, has a recession velocity 5000 kilometers a second less than that of the other companion galaxies. Perhaps it is a foreground galaxy and not a true member of the group. (Association of Universities for Research in Astronomy, Inc., The Kitt Peak National Observatory.)

companion galaxies connected by bridges and filaments of luminous material. The members of these systems would seem to be at the same distance; yet the smaller companions sometimes have larger, and occasionally much larger, redshifts than their parent galaxies. Arp believes that these companions are young and have been ejected from their parent galaxies, and he suggests that possibly all young galactic systems have large intrinsic redshifts of an unknown origin. Arp's claim to have discovered discordant redshifts is contested by more conservative astronomers who argue that the apparent physical connections between galaxies of different redshifts is accidental, the result of seeing one galaxy superposed on more distant galaxies.

Arp's arguments have one important virtue. The possibility of discordant redshifts prompts us to scrutinize more carefully the nature of the three redshifts and the role they play in cosmology. When we leap to defend conventional wisdom we should remember that it cannot be proved true but only be proved false, and science is lost without those few people who are bold enough to interrogate its treasured doctrines.

ARE QUASARS NEAR OR FAR? Soon after the discovery of quasars an attack on the cosmological interpretation of redshifts came from a different quarter. We recall that Maarten Schmidt in 1963 discovered a redshift of 0.16 for the bright starlike object 3C 273, previously known to be a radio source. The word *quasar* is now used to denote all starlike objects of large redshift regardless of whether they are radio sources. Undoubtedly quasars are still among the most puzzling of all known celestial objects. According to the cosmological hypothesis their large observed redshifts are due to the expansion of the universe. This interpretation places the quasars at vast distances and therefore indicates that they are extraordinarily powerful sources of radiation. The light from some quasars varies rapidly, in a matter of only days, a time so brief as to suggest that they are extremely compact bodies not much larger than the Solar System. This makes it rather difficult to understand their source of energy. Perhaps, then, their redshifts are not entirely due to expansion?

Jessie Greenstein and Maarten Schmidt showed that it is unlikely that quasars have large gravitational redshifts. Many of the emission lines come from gas surrounding the quasars and are formed by radiation originating at different depths within the gas. If the redshifts were mainly gravitational the emission lines would not be sharp, but would be spread over a continuous range of redshifts.

An entirely different interpretation, proposed by James Terrell and advocated by Geoffrey Burbidge and Fred Hoyle, is that the redshifts are due almost entirely to the Doppler effect. According to this idea the quasars are nearby extragalactic bodies that have been expelled at high velocities by violent explosions in the Galaxy and neighboring galaxies. This *local hypothesis,* as it is called, reduces the distances of quasars from billions to millions of light years and alleviates the problem of explaining their immense output of energy. It does not, however, explain the powerful radio sources that are still at cosmologically large distances.

If quasars are flying out of galaxies, as suggested by Terrell, then some will move away and some will move toward us. But there are no quasars known to have blueshifts. They are all moving away, and this is possible only if they have all originated locally. The local hypothesis faces two difficulties. It is estimated that millions of quasars are observable with the largest telescopes, and this number is far too great to attribute to a local origin. The other difficulty is that large numbers of quasars must also originate in other galaxies similar to our own. In that case the night sky would be much brighter than is actually observed. The local hypothesis has been weakened even further by the discovery of a few quasars with redshifts similar to those of galaxies in the same region of the sky, which suggests that these quasars at least are members of distant clusters of galaxies and are not merely nearby bodies. Also, many quasars are radio sources, and there is little doubt that most radio sources are at cosmologically large distances.

The *cosmological hypothesis* assumes that quasars are at large distances and attributes their redshifts to the expansion of the universe. It is a widely accepted hypothesis and at present seems moderately secure. The expansion redshift interpretation will undoubtedly be continually attacked and – who knows? – might in the future be overthrown by the onslaught of a new and more profound theory.

MULTIPLE REDSHIFTS. The spectra of quasars often contain bright emission lines and also dark absorption lines. The bright lines are produced by atoms emitting radiation in hot gaseous regions, the dark lines by atoms absorbing radiation in cooler gaseous regions that are presumably between us and the quasar. In many cases the emission and absorption lines have similar redshifts. Yet the spectra of some quasars are complex and not easy to understand, containing absorption lines that appear to have several different redshifts. These multiple *absorption redshifts* observed in a single spectrum are often very much less than the *emission redshift* that is presumably the true expansion redshift of the quasar. A typical case is the quasar PHL 957: It has an emission redshift of 2.69, and its spectrum contains numerous absorption lines grouped into eight absorption redshifts having values between 2.0 and 2.7. Such multiple-redshift spectra have so far not been satisfactorily explained. One possible explanation is that the quasar expels streams or shells of absorbing gas at high velocity, and the absorptions at different redshifts occur in clouds of different velocities in the vicinity of the quasar. Another possibility is that the absorption occurs in the extended halos of galaxies nearer to us than the quasar, through which the light from the quasar passes. Conceivably there is some truth in both explanations, although we are still a long way from understanding quasars and their complicated spectra.

REFLECTIONS

1 *In a city at night one can hear frequently the wail of sirens approaching and then receding, and the Doppler effect is often quite pronounced (see Figure 11.11). Have you noticed how at first sirens are always high-pitched and then, as they fade away into the distance, become low-pitched? The opposite effect – first a low pitch that later fades away into a high pitch – never occurs. Why is this?*

2 *"On the other hand, the plausible and, in a sense, familiar conception of a universe extending indefinitely in space and time, a universe vastly greater than the observable region, seems to imply that red-shifts are not primarily velocity-shifts" (Edwin Hubble,* Observational Approach to Cosmology, *1937). By "velocity-shifts" Hubble meant Doppler shifts.*

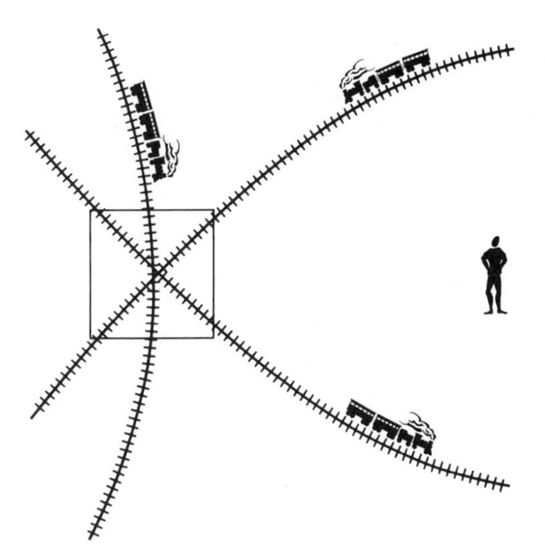

Figure 11.11. The "train-whistle" effect. Trains rush through a railway station from all directions blowing their whistles. People who live nearby always hear the pitch of the whistles steadily decreasing from high to low.

* *"The red shift is something that happens to the light on its journey, along with the expansion of space, not really a Doppler effect" (Erwin Schrödinger,* Expanding Universes, *1957).*

* *"Note that the cosmological red shift is really an* expansion *effect rather than a* velocity *effect" (Wolfgang Rindler,* Essential Relativity, *1969).*

3 *Bend a piece of stiff wire into a wave shape. Now slowly stretch the wire and observe how the wavelength increases. This experiment illustrates the stretching of waves of radiation as they travel through expanding space.*

* *Draw redshift–velocity and redshift–distance curves using the classical Doppler formula. What is wrong with this interpretation?*

* *What are the difficulties created by use of the special relativity Doppler formula?*

* *Use Figures 11.7, 11.8, and 11.9 to derive as much information as possible about the galaxies and quasars listed in Table 11.1.*

4 *Consider the following symmetrical arrangement (see Figure 11.12). Two widely separated comoving galaxies X and Y emit signals of identical wavelength at the same instant of cosmic time. X receives the signal from Y and Y receives the signal from X at the same instant and at the same wavelength. This state of perfect symmetry exists because of the homogeneity of the universe, which applies not only to expansion, but also to the laws of nature and the fundamental constants (such as the mass and electric charge of the electron) that determine the structure of atoms and the wavelengths of the radiation they emit and absorb.*

If by mischance the laws of nature and the fundamental constants are not the same everywhere, then symmetry between X and Y is lost and we are in grave difficulty. We are moderately confident, for the reasons that led us to believe in homogeneity, that symmetry does exist. The universe is isotropic; hence the laws and constants of nature are the same in all directions; and if we are not at a cosmic center, it follows that the laws and constants must be the same everywhere in space.

Perhaps the laws and constants are the same everywhere in space but change in time? In this case we still have homogeneity, and symmetry between X and Y is preserved, but the interpretation of cosmological redshifts as an expansion effect is then most likely in error. But a theory that postulates a fundamental change in the structure of atoms with time must be well contrived; a multiplicity of wavelengths, short and long, produced by different physical processes, must all give the same redshift. It has been found, for example, that 21-centimeter wavelength radiation from hydrogen atoms in distant galaxies

Table 11.1. *Sample redshifts*

Object	Redshift	Remarks
Virgo	0.016	cluster
BL Lacertae	0.07	radio galaxy with optically bright nucleus
Boötes	0.13	cluster
Ton 256	0.13	radio-quiet quasar
3C 273[a]	0.16	first radio source to be identified as a quasar
PKS 2251+11[b]	0.32	quasar, member of a cluster
3C 48	0.37	quasar
3C 295	0.46	radio galaxy
3C 9	2.01	quasar
PHL 957[c]	2.69	quasar, multiple absorption redshifts: 2.67, 2.55, 2.54, 2.31, 2.23
4C 05.34[d]	2.88	quasar, multiple absorption redshifts from 2.87 to 1.75

[a]3C = Third Cambridge survey (England).
[b]PKS = Parkes radio survey (Australia).
[c]PHL = Palomar, Haro, Luyten (USA).
[d]4C = Fourth Cambridge survey (England).

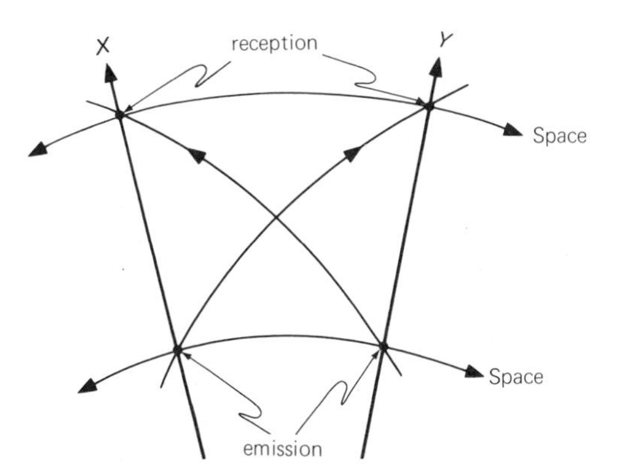

Figure 11.12. Galaxies X and Y have worldlines that diverge in expanding space. Each sends a signal to the other at the same instant of cosmic time. Because of symmetry the signals arrive at the same instant of cosmic time and also have identical redshifts.

has the same redshift as optical wavelengths almost a million times shorter. The emission mechanisms in these two cases are entirely different and involve different combinations of the fundamental constants. The expansion interpretation of cosmological redshifts is therefore moderately secure and will not be easily overthrown in favor of a "shrinking atom" theory.

5 *Consider a luminous source of large redshift. Imagine that we can arrange a number of observers strung out in a line between ourselves and the distant source. Radiation leaves the source and travels a short separating distance, and the first observer in line calculates the Doppler redshift (from $cz = H \times$ distance), by using the separating distance and the Hubble term of that particular epoch. The second observer in line then calculates the Doppler redshift of the radiation arriving from the first observer, using their separating distance and the new value of the Hubble term. The third observer in line then does the same thing. This procedure is repeated step by step until the radiation finally reaches us, when we make the same calculation, using the Hubble term of the present epoch. We now add up all the small successive Doppler redshifts obtained in this way and find that the total is equal to the expansion redshift of the source. The expansion redshift can thus be considered as a sum of incremental Doppler redshifts. We cannot, however, regard the expansion redshift as directly equivalent to a large single Doppler redshift.*

6 *Perhaps you are not convinced that there is a difference between expansion and Doppler redshifts? Let us then demonstrate the difference in the following imaginary experiments.*

Consider first the Doppler redshift. Two bodies, call them X and Y, are separated by a fixed distance in the Galaxy – or the laboratory (see Figure 11.13). Let X emit a pulse of radiation toward Y. After the pulse has left X, and while it is traveling toward Y, let the distance between X and Y increase. Before the pulse arrives at Y let the separating distance again become fixed. In this case the two bodies X and Y have a

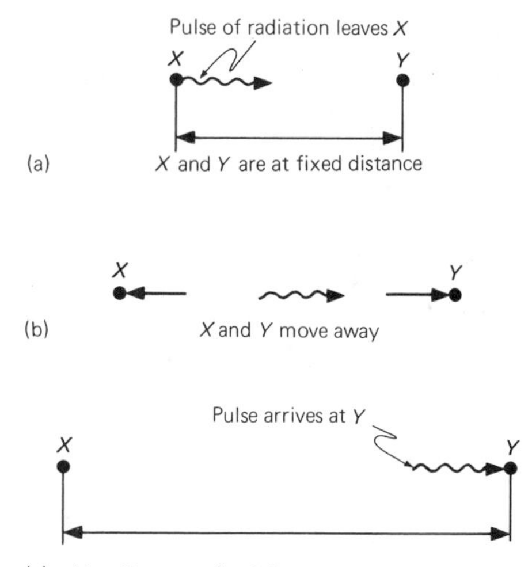

Figure 11.13. (a) First, X and Y are separated by a fixed distance in the laboratory or elsewhere in the Galaxy, and X emits a pulse of radiation toward Y. (b) While the pulse is traveling, X and Y move apart and come to rest at a wider fixed distance. (c) The pulse then arrives at Y while the separating distance is again constant. In this case the pulse is emitted by X and received by Y at the same wavelength. There is no Doppler redshift because X and Y have a relative velocity of zero at the instants of emission and reception of the pulse.

relative velocity of zero at the instants when the pulse is emitted and received. Hence, Y receives the pulse of radiation at the wavelength at which it was emitted by X and the Doppler redshift is zero.

Consider now the expansion redshift in a similar situation (see Figure 11.14). Two comoving bodies X and Y are stationary in expanding space. We suppose, in this imaginary experiment, that the universe is initially static and that the distance between X and Y is therefore constant and fixed, and the recession velocity is zero. The body X now emits a pulse of radiation toward Y, and while this pulse is traveling, we suppose that the universe ceases to be static and starts to expand. Before the pulse arrives at Y the universe stops expanding and again becomes static, and the distance between X and Y is once more a fixed amount with a recession velocity of zero. When the pulse eventually arrives at Y its wavelength has been increased by the expansion of space through which it has traveled. If the scaling factor has the value R_1 when the universe is static the first time, and the value R_2 when the universe is static the second time, the wavelength has increased by the amount R_2/R_1, and the redshift is given by

$$1 + z = \frac{R_2}{R_1}$$

Even though the distance between X and Y is constant at the instants of emission and reception, and their relative recession velocity is zero, an expansion redshift has nonetheless occurred. Yet in the previous experiment the Doppler redshift vanished under similar conditions. These two experiments show that the Doppler redshift depends on motion through space at the instants of emission and reception, whereas the expansion redshift depends on the expansion of space between the instants of emission and reception.

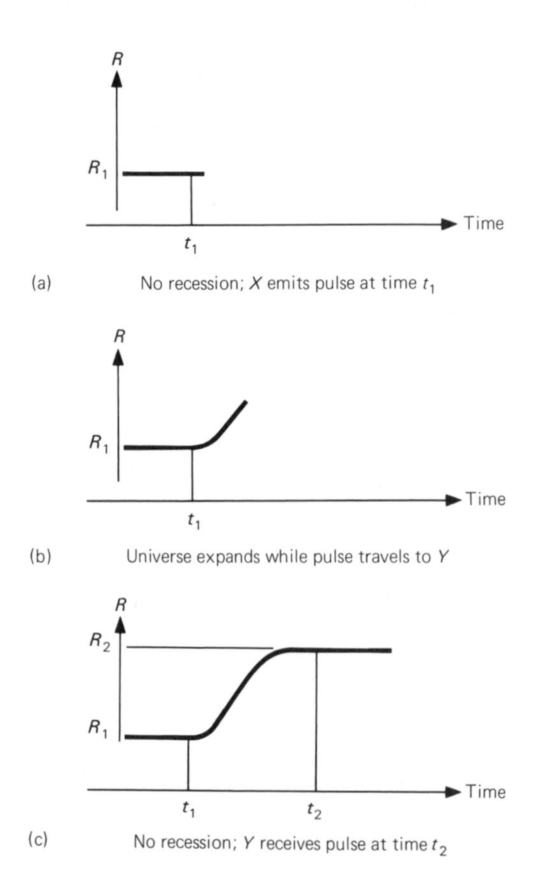

***Figure 11.14.** (a) A universe is first static, and its scaling factor has a value R_1. (b) The universe then expands and (c) later again becomes static with a value of R_2 for the scaling factor. Now consider a pulse of radiation emitted by a comoving body X while the universe is static the first time, and received by a comoving body Y while the universe is static the second time. At the instants of emission and reception the recession velocity is zero. Yet the pulse of radiation is received by Y with redshift z, given by $1 + z = R_2/R_1$. This proves that the Doppler and expansion redshifts are not identical.*

7 *In an expanding universe distant galaxies are seen redshifted, and in a collapsing universe distant galaxies are seen blueshifted. Imagine that our universe ceased to expand and began to collapse. What would we see? Galaxies near us would be seen approaching with blueshifts, but distant galaxies, because we look back to a time when the universe was still expanding, would be seen receding with redshifts. The boundary between blueshifts and redshifts would recede with time, and more and more galaxies would be seen with blueshifts. In the last moments of a bang–bang universe, just before we plunged into the big bang, everything would be blueshifted.*

8 *We do not know the kind of universe in which we live, but probably it is of the big bang type. There are multitudes of different types of big bang universes, and we do not necessarily live in the simple Einstein–de Sitter universe that consists of flat space (see Chapters 14 and 15). When we construct redshift–distance and redshift–velocity curves for big bang universes of various types, their differences are not large: The Einstein–de Sitter curves shown in the text are reasonable approximations. These are the curves to use when quoting distances and recession velocities of galaxies and quasars of large redshift. They are at least much more reliable than the meaningless Doppler curves.*

Examine the Einstein–de Sitter curves; they reveal several surprising facts. Notice that there are two distances and two recession velocities, and in each case one must always state which of the two is being used.

*Notice that the present distances (*reception distances*) of the galaxies increase steadily with redshift, and at infinite redshift the reception distance attains a maximum value equal to twice the Hubble length and is therefore approximately 30 billion light years. This maximum reception distance is the* particle horizon *(see Chapter 19). What is the recession velocity at the particle horizon?*

Notice that the distances of the galaxies, at the time they emitted the light we now see (these are the emission distances*), do not continually increase with redshift (see Figure 11.15). They attain a maximum value at a redshift of 1.25, and at greater redshifts the emission distances get less. What is the maximum emission distance at $z = 1.25$, and what is the recession velocity at this redshift at the time of emission?*

Two quasars have redshifts of 1 and 3, respectively; which of the two was closer at the time of emission? If you are to be close to an object, it must have either very small or very large redshift: If the redshift is small you are close to it now; if the redshift is large you were close to it long ago. Can you understand how it is possible for bodies at large redshift to be nearer to us at the time of their emission than bodies of lesser redshift?

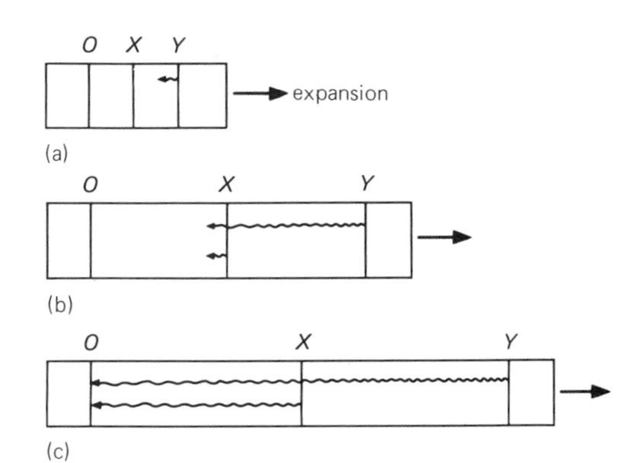

Figure 11.15. On a piece of elastic let O represent our position, and X and Y the positions of two galaxies. If signals from X and Y are to reach us at the same instant, then Y must emit before X. In (A) Y emits a signal. In (b) X emits its signal at a later instant, when it is further away than Y was when Y emitted its signal. In (c) both signals arrive simultaneously at O. Y's signal has the greater redshift (it has been stretched more), despite the fact that Y is closer than X at the time of its emission. This strange situation occurs only when redshifts are large.

9 *The following relations apply to an extragalactic body of redshift z in the simple Einstein–de Sitter universe:*

$$\text{reception distance} = 2\left(1 - \frac{1}{\sqrt{(1+z)}}\right)$$

$$\text{emission distance} = \frac{2}{1+z}\left(1 - \frac{1}{\sqrt{(1+z)}}\right)$$

$$\text{recession velocity now} = 2\left(1 - \frac{1}{\sqrt{(1+z)}}\right)$$

$$\text{recession velocity then} = 2[\sqrt{(1+z)} - 1]$$

$$\text{age of universe now} = \frac{2}{3}$$

$$\text{age of universe then} = \frac{2}{3(1+z)^{3/2}}$$

$$\text{lookback time} = \frac{2}{3}\left(1 - \frac{1}{(1+z)^{3/2}}\right)$$

The distances are given in terms of the Hubble length c/H, equal to 15 billion light years for $H = 20$ kilometers a second per million light years. The recession velocities are in terms of the velocity of light c, equal *to 300,000 kilometers a second. The recession velocity now is at the reception distance, and the recession velocity then is at the emission distance. The age now, the age then, and the lookback time are in terms of the Hubble period of 1/H, equal to 15 billion years. The lookback time is equal to the age now (at the time of reception) minus the age then (at the time of emission).*

SOURCES

Arp, H. "Observational paradoxes in extragalactic astronomy." *Science 174,* 1189 (1971).

Einstein, A., and Sitter, W. de. "On the relation between the expansion and the mean density of the universe." *Proceedings of the National Academy of Sciences 18,* 213 (1932).

Field, G. B., Arp, H., and Bahcall, J. *The Redshift Controversy.* Benjamin, Reading, Mass., 1973.

Harrison, E. R. "Radiation in homogeneous and isotropic models of the universe." *Vistas in Astronomy 20,* 341 (1977).

Hubble, E. *The Observational Approach to Cosmology.* Oxford University Press, Clarendon Press, Oxford, 1937.

McCrea, W. "Cosmology today." *American Scientist,* September–October 1970.

McVittie, G. C. "Distance and large redshifts." *Quarterly Journal of the Royal Astronomical Society 15,* 246 (1974).

Rindler, W. *Essential Relativity: Special, General, and Cosmological.* Van Nostrand Reinhold, New York, 1969. See p. 239.

Sandage, A. R. "The red-shift." *Scientific American,* September 1956.

Sandage, A. R. "Travel time for light from distant galaxies related to the Riemannian curvature of the universe." *Science 134,* 1434 (1961).

Schrödinger, E. *Expanding Universes.* Cambridge University Press, Cambridge, 1957. See p. 62.

Terrell, J. "Quasi-stellar objects: possible local origin." *Science 154,* 1281 (1966).

DARKNESS AT NIGHT

If this is true, and if they are suns having the same nature as our sun, why do not these suns collectively outdistance our sun in brilliance?
— Kepler (1571–1630), *Conversation with the Starry Messenger*

THE GREAT PARADOX

AN INFERNO OF STARS. There is a simple and important experiment in cosmology that almost everybody can perform: It consists of gazing at the night sky and noting its state of darkness. When we ask why the sky is dark at night the response usually is that the Sun is shining on the other side of the Earth and starlight is much weaker than sunlight. It takes a genius to realize that the relative weakness of starlight is of great cosmological significance, and such a person was Johannes Kepler, imperial mathematician to the emperor of the Holy Roman Empire.

When we stand in a forest of trees our distant view in any horizontal direction is obstructed by a background of tree trunks. On looking away from Earth at night we see in all directions a "forest" of stars (see Figure 12.1). And if the stars stretch away endlessly, like trees in a celestial forest, then surely our distant view is also obstructed by stars? Every line of sight must ultimately intercept the surface of a star.

In a universe of infinite extent, populated everywhere with shining stars, the entire sky should be covered by stars with no dark spaces in between. Hence, when all stars are bright like the Sun, the entire sky at every point should blaze with a brilliance equal to the Sun's disk. In the midst of this inferno of intense light, life should cease in seconds, the atmosphere and oceans should boil away in minutes, and the Earth should turn to vapor in hours.

Yet the sky at night is dark. What then is wrong with the forest analogy?

KEPLER TERRIFIED BY INFINITY. Kepler believed in the Copernican heliocentric theory and was excited by Galileo's astronomical observations with the newly discovered telescopes. He believed also, in company with Copernicus and in keeping with Aristotelian cosmology, that the starry universe is finite and bounded with a cosmic edge. Thomas Digges in 1576, however, had seized the Copernican system, torn away its outer edge, and transformed it into an infinite universe of fixed stars. The concept of an infinite universe, outlined in the epic poem *The Nature of the Universe* by Lucretius (discovered in 1417) and advanced in Digges's popular work *A Perfit Description of the Caelestiall Orbes,* was enthusiastically championed by Giordano Bruno and William Gilbert. But Kepler was terrified by such a monstrous idea and vehemently rejected it whenever possible.

In 1610, Kepler received a copy of Galileo's small book *The Starry Message.* After only a few days he dashed off a letter to Galileo, and a month later this letter was published as a short book entitled *Conversation with the Starry Messenger*. In this book

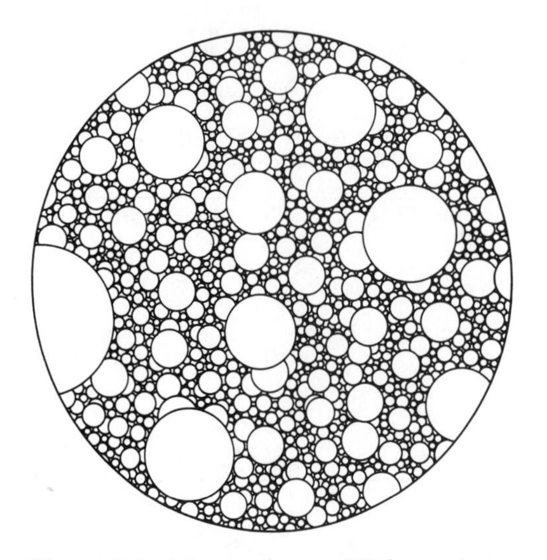

Figure 12.1. A forest of stars. (With permission from E. R. Harrison, American Journal of Physics 45, *120, 1977.)*

can be found Kepler's most potent argument against the concept of an infinite universe. "You do not hesitate to declare," he said, "that there are visible over 10,000 stars. The more there are, and the more crowded they are, the stronger becomes my argument against the infinity of the universe." For if the universe stretched away endlessly, with stars like the Sun swarming everywhere, then the whole "celestial vault would be as luminous as the sun"; hence it was clear to Kepler that "this world of ours does not belong to an undifferentiated swarm of countless others."

According to Kepler the universe was not like an endless forest of trees; instead, it was like a finite clump of trees in which we looked out between the tree trunks to a dark enclosing wall. Kepler did not use the forest analogy. He realized, nonetheless, that in an infinitely large universe the stars would collectively outshine the Sun and flood the heavens with light far more intense than is observed. This is the origin of the famous *dark night sky paradox*.

Only 34 years after the first edition of Digges's *Perfit Description,* the infinite stellar universe encountered in Kepler's *Conversation* its most devastating criticism. The choice was clear: either a cosmic edge and a dark night sky, or no cosmic edge and a blazing sky. Over the centuries since astronomers have sought for a way out of the impasse into which Kepler had led cosmology.

"I HAVE HEARD URGED . . . "

THE INFINITE NEWTONIAN UNIVERSE. Kepler's *Conversation with the Starry Messenger* was widely read, and many no doubt pondered on the problem of the infinite universe and the darkness of the night sky. The problem became particularly acute, tantamount to a paradox, with the rise of the infinite Newtonian universe. More than likely, Newton was aware of the paradox, but he was apparently more concerned with a similar problem in the theory of gravity. The emitted light and the gravitational pull of a star both decrease as the inverse square of its distance. We should not only receive large quantities of light from numerous distant stars but also be pulled by large gravitational forces. Newton resolved the gravity problem by assuming that the infinite universe is homogeneous – the same everywhere – and that all forces pulling in opposite directions are therefore equal and cancel each other.

The light received from distant stars in opposite directions does not cancel out, but instead adds up, and in this case theory is brought into direct conflict with observation. The first person to discuss the problem of the dark night sky within the context of the Newtonian universe, so far as we know, was Edmund Halley. In 1720 he published two short papers on the subject and wrote: "Another Argument I have heard urged, that if the number of Fixt Stars were more than finite, the whole superficies of their apparent Sphere would be luminous." Where he had heard the argument, we do not know, but presumably it came directly or indirectly from the work of Kepler.

Halley's solution to the problem was that "the more remote Stars, and those far short of the remotest, vanish even in the nicest Telescopes, by reason of their extreme minuteness; so that, tho' it were true, that some such Stars are in such a place, yet their Beams, aided by any help yet known, are not sufficient to move our Sense; after the same manner as a small Telescopical fixt Star is by no means perceivable to the naked Eye." Halley was wrong; he tried to explain the darkness of the night sky by arguing that light from distant stars, even their collective light, was too faint to be detected by the eye. We know that the light emitted by a single atom is too weak to be perceived by the eye, but the collective light from many atoms is easily seen; yet by Halley's argument, the collective light would also be imperceptible. According to the forest analogy, Halley assumed that distant trees were invisible, so that the background of fused tree trunks vanished and we saw only a surrounding clump of nearby trees.

Newton, as president of the Royal Society, was in the chair when Halley read his paper to the society, but the Journal Book of the society contains no record of any comment made by Newton. Why did Newton remain silent about a subject on which he had perhaps often thought? Michael Hoskin, in a Christmas lecture, has made the suggestion that Newton, who was then nearly eighty years old, was possibly asleep.

"OLBERS' PARADOX." Only a few years later, in 1744, the darkness of the night sky was discussed by a young Swiss astronomer, Jean-Philippe Loys de Cheseaux. He attributed the darkness to absorption of starlight by a fluid distributed throughout interstellar space. In 1823, Heinrich Olbers, a distinguished German astronomer, presented a similar argument and also said that starlight was absorbed while traveling in space. The corresponding analogy is that of a foggy forest in which distant trees are obscured from view and only the nearest trees are seen clearly.

The solution proposed by both Cheseaux and Olbers fails, as was shown by John Herschel in 1848, because the absorbing gas would heat up and eventually emit as much radiation as it received.

The puzzling darkness of the night sky, with its direct conflict between theory and observation, is now widely but incorrectly known as "Olbers' paradox." The paradox has been discussed in recent decades by numerous authors who were unaware of the earlier work by Kepler, Halley, and Cheseaux, and who believed that it had been discovered by Olbers. Stanley Jaki, who has written a book entitled *The Paradox of Olbers' Paradox,* says: "This constitutes, in effect, the most paradoxical aspect of Olbers' paradox. In this sense, Olbers' paradox is not Olbers', nor is it Halley's. It is the paradox of the unscientific habits of scientific workers and writers. For it is no small matter that some scientists can be shockingly careless when it comes to the presentation of a detail of scientific history." Olbers was not the first to stumble on the paradox; indeed, it is probable that neither he nor Cheseaux, any more than Halley, conceived the paradox independently. It is possible that the idea, floating around in nebulous form since the time of Kepler, and occasionally referred to in conversations and correspondence, had become common knowledge and only emerged in publications whenever sharpened into a form that permitted a resolution.

BRIGHT-SKY UNIVERSE

We adopt for the time being the pre-twentieth-century Newtonian picture of an infinite universe populated uniformly with sunlike stars, in an attempt to understand the paradox without the complication of galaxies, expansion, and other twentieth-century discoveries.

CONCENTRIC SHELLS OF STARS. Let us occupy any position in space and add up the contributions of light received from all stars. Nearby stars are few in number, and yet each gives a large contribution; distant stars are numerous, and yet each gives only a small contribution. Institutions that send out requests for help are aware that small sums from numerous contributors add up to a large income. We should not be surprised therefore if the light from the distant numerous stars also turns out to be very large. We follow Cheseaux and consider a series of imaginary spheres, of increasing radius, each with its center at the point in space we occupy. Let the radius of each successive sphere increase by a fixed amount, as shown in Figure 12.2, so that the spaces between the spherical surfaces are shells of equal thickness. The volume of each shell increases as the square of distance. Because stars are uniformly distributed, their number in each shell must also increase as the square of distance.

But the light received from any single star is inversely proportional to the square of its distance. Consequently, when the number of stars in a shell is multiplied by the amount of light received from each of these stars, we obtain a quantity of light that is fixed and independent of the distance of the shell. All shells contribute equal quantities of light. It follows that the total amount of light reaching us is the quantity of light from one shell multiplied by the number of shells.

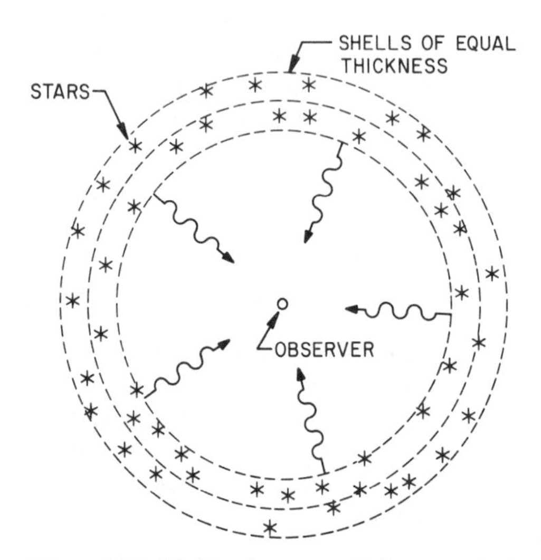

Figure 12.2. Shells of stars at all distances (only two shells are shown) with the observer at the center. The amount of light reaching the observer is the same from each shell. (With permission from E. R. Harrison, American Journal of Physics 45, *121, 1977.)*

In a universe of stars stretching away endlessly there is an infinite number of shells. Each shell contributes a finite quantity of light and therefore, according to this argument, at our chosen point in space there is an infinite amount of light! Our chosen point can be anywhere – hence at all points in space light is infinitely intense. This conclusion is of course absurd, and the error in the argument can be spotted almost immediately.

THE LOOKOUT LIMIT. When standing within a forest we do not see all the trees of the forest. Tree trunks obstruct our view of more distant trees, and a line of sight extends to a background consisting of a fusion of trees that lies not very far away (see Figure 12.3). Let A be the area in a forest that contains on

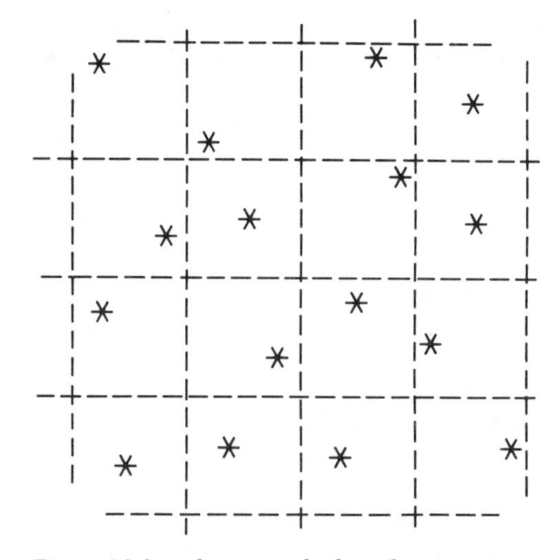

Figure 12.3. A forest in which each tree occupies an average area A. If each tree has a width w the lookout limit in the forest is A/w. In the Newtonian universe each star occupies an average volume V. If the cross section of each star is a (the area of its disk), the lookout limit in the universe is V/a. (With permission from E. R. Harrison, American Journal of Physics 45, *121, 1977.).*

the average one tree, and let w be the typical width of a tree trunk at eye level. The distance seen in the forest is the *lookout limit* and is given by

$$\text{lookout limit} = \frac{A}{w} \qquad (12.1)$$

and the total number of trees visible from any point is therefore found to be

$$\text{number of visible trees} = \frac{\pi A}{w^2} \qquad (12.2)$$

As an example, if the average distance between trees is 10 meters, the area A occupied by a single tree is 100 square meters. If the diameter w of a tree trunk at eye level is typically half a meter, the lookout limit in the forest is 200 meters, and the number of visible trees is 1257.

Stars, like tree trunks, have a certain size and tend therefore to obstruct our view of more distant stars. A line of sight extends to a continuous background of stars that lies at finite distance. The lookout limit in the universe, as in a forest, is easily calculated. Let V be the volume of space that contains on the average one star, and let a be the typical cross-sectional area of a star (equal to π times its radius squared). The expression for the distance seen in the universe is then

$$\text{lookout limit} = \frac{V}{a} \qquad (12.3)$$

and the total number of stars visible from any point in space is given by the equation

$$\text{number of stars visible} = \frac{4\pi V^2}{3a^3} \qquad (12.4)$$

Although the number of stars is infinite in a universe of infinite extent, only a finite number can be seen from any single point, and these stars cover the sky and prevent us from seeing the rest.

We now realize that the light reaching us comes only from stars within the lookout limit, and all light from stars further away is intercepted by the nearer stars and never reaches us. We must therefore add up only the contributions of light from successive shells out to a distance equal to the lookout limit. The light that reaches us is consequently of finite and not infinite intensity.

A BRIGHT-SKY UNIVERSE. We have found that a line of sight in every direction terminates at the surface of a star, and the sky is covered with bright stars with no dark spaces between them. It is as if we were enclosed within a spherical surface, of radius equal to the lookout limit, that has a temperature equal to the surface temperature of the stars. We live, it would seem, in a furnace that has incandescent walls. Wherever we stand in space we are surrounded by an unbroken wall of stars. Everything is bathed in a flood of intense light, and the temperature everywhere is the same as that at the surfaces of stars. Because sunlike stars have a surface temperature of about 6000 degrees Kelvin, this is the temperature everywhere in space.

This seemingly logical argument, leading to a conclusion in utter disagreement with reality, constitutes the dark night sky paradox.

Absorption of starlight by gas in interstellar space, as suggested by Cheseaux and Olbers, cannot avert a bright sky. It is useless putting gas into a furnace in the hope that it will keep the objects inside cool, because the gas will quickly heat up to the same temperature as the furnace. Whatever is put in a bright-sky universe to shield us from the blinding rays of trillions and trillions of stars will almost immediately heat up and become part of its incandescent brilliance.

THE COSMIC-EDGE RESOLUTION. There is available an obvious resolution of the night-sky paradox. All we need do is restore the cosmic edge of antiquity and place it at a distance much less than the lookout limit. A finite and bounded universe, either Aristotelian or Copernican, contains a finite number of stars, and if its radius is less than the lookout limit, then it has a sky that is dark at night. This was Kepler's resolution of the problem; but a spatially bounded universe is nowadays totally unacceptable because space cannot terminate abruptly at a wall-like cosmic edge.

Wall-like edges went out of fashion with the rise of the Newtonian universe, but cliff-like edges retained a declining popularity right up until the early years of this century. An island universe – or a Stoic cosmos floating in an infinite void of empty space – was frequently proposed as a resolution of the paradox: We stand, as it were, within a clump of trees and look out through the trees to a vast treeless plain. Harlow Shapley, as recently as 1917, said: "Either the extent of the star-populated space is finite or 'the heavens would be a blazing glory of light' . . . since the heavens are not a blazing glory, and since space absorption is of little moment throughout the distances concerned in our galactic system, it follows that the defined stellar system is finite." An island or Stoic universe is now also out of fashion, without a shred of observational evidence to support it.

SPHERICAL SPACE. Traditional arguments have led us to the conclusion that an unbounded and infinite universe has a bright sky. What about an unbounded universe that is of finite size?

Many times it has been suggested that the night sky is dark because the universe is finite and unbounded and has therefore spherical space. At first glance it seems plausible that the sky will be dark in a finite universe if the lookout limit is much greater than the size of the universe itself. This proposed resolution is in some ways similar to that adopted by Kepler but avoids using the objectionable cosmic edge. Unfortunately, it does not work.

The surface of a ball is unbounded and yet of finite area and can be taken as representing a spatially finite and unbounded universe. We saw earlier, in a universe of positive curvature having spherical geometry, that when we travel in any direction we eventually arrive back at our starting point – in the same way as an ant crawling on the surface of a globe. Light rays circumnavigate such a universe and continue to go around and around until absorbed by some obstruction.

Let us suppose that the spherical surface of a planet is covered with a forest of trees and we stand within this finite but endless forest. We must imagine, in this analogy, that all lightrays are bent and travel parallel to the surface of the planet at eye level. (Think of the photon sphere.) On looking out horizontally we look always through an endless forest of trees. When the lookout limit is more than half the circumference of the planet we see the fronts and backs of some trees by looking in opposite directions; and when the lookout limit is more than the circumference we see all trees repeated more than once. The forest is obviously endless and therefore we must always see a fused background of trees at the lookout limit. If, for example, the lookout limit is 100 times the circumference of the planet, we see in any direction trees repeated 100 times, and they all form a continuous background.

The same thing happens in a universe of finite and unbounded space. We see a continuous background of stars created by circumnavigations of lightrays, and the sky is covered with stars as in an infinite universe. A finite but unbounded universe therefore fails to resolve the paradox.

THE HIERARCHICAL RESOLUTION. Hierarchical resolutions of the paradox were first proposed by John Herschel and Richard Proctor in the nineteenth century. This approach was adopted by Fournier d'Albe and Carl Charlier and much publicized early in this century. The Kantian idea of a hierarchy of clusters of increasing size was adapted by Charlier and modified by Fournier d'Albe, who put forward the quaint notion that the visible universe is only one of a series of universes nested inside each other like Chinese boxes. In either case it was shown that a bright sky could be averted. Each cluster of larger size is of lower density, and if we arrange for the density of the clusters to decrease sufficiently rapidly with size, the lookout limit becomes infinitely great and the sky remains dark.

A hierarchical resolution is not very satisfactory. On all scales the universe will be anisotropic, contrary to optical observations and the isotropy of the 3-degree cosmic radiation, and on the largest scale a Kantian universe has a cosmic center and is not homogeneous. Furthermore, a hierarchical resolution of the paradox is completely unnecessary, as we shall now see.

THE PARADOX RESOLVED

A MORE REALISTIC UNIVERSE. So far we have considered an infinite Newtonian universe populated uniformly with sunlike stars. Is the paradox applicable in a more realistic universe?

We have seen that absorption by dust and gas is of no help. The gathering together of different kinds of stars into galaxies, and of galaxies into clusters, is also of no help. This clustering merely alters the value of the lookout distance while the sky continues to blaze with light. When various kinds of trees are clumped together into groups in a forest, our line of sight in all horizontal directions is still obstructed by distant tree trunks. Also we have seen that a finite but unbounded space fails to avert a bright sky; in this case a line of sight stretches around and around the universe until it eventually intercepts the surface of a star. Expansion has not been considered, and this will be taken up shortly. But first, we must turn to more important matters.

ENERGY CONSIDERATIONS. The conclusion that the sky everywhere should be as bright as the Sun derives from pre-twentieth-century science. That something is seriously wrong with this conclusion is shown by the following argument, which considers the subject from the point of view of energy.

The average density of matter of all kinds in the universe is about equal to the mass of 1 hydrogen atom per cubic meter. Mass and energy are equivalent, and we shall imagine that in some way all matter in the universe is annihilated and converted directly into thermal radiation. Calculation then shows that the thermal radiation everywhere has a temperature of only 20 degrees Kelvin. This is considerably less than the surface temperature of stars, and we are forced to conclude that the universe does not contain enough energy to create a bright sky.

A bright-sky universe at a temperature of 6000 degrees Kelvin contains radiation whose mass is 10 billion times greater than that of all the stars together. The mass density of radiation in our universe, if it had a bright sky, would be vastly greater than that of matter and our universe would be *radiation dominated*. Such a situation occurred long ago in the radiation era of the early universe but is quite impossible nowadays. The brilliance of the early universe cannot be restored because there is now not enough energy.

If the universe contained 10 billion times more matter in the form of stars, and if all this matter were annihilated and converted totally into thermal radiation, the temperature everywhere would equal that at the surface of the Sun. But stars do not convert their masses into radiation with 100 percent efficiency. Sunlike stars burn hydrogen into helium and during their luminous lifetimes convert only about 0.1 percent of their masses into starlight. Instead of increasing the number of stars by a factor of 10^{10} (10 billion), we need to use a factor of 10^{13} (10 trillion) to create a bright-sky universe. Bright skies can in principle exist, but only in universes at least 10 trillion times more dense than our own.

By using energy arguments we have shown that in our universe a bright sky cannot exist. The traditional arguments that deduce a bright sky are therefore wrong, and the paradox collapses in the face of twentieth-century science. Let us try to track down what is actually wrong with the traditional argument used since the seventeenth century.

LOOKBACK LIMIT IS GREATER THAN THE LUMINOUS LIFETIME. We continue to suppose, merely for convenience, that all stars are similar to the Sun. With an average cosmic density of 1 hydrogen atom per cubic meter, and with this matter all lumped into stars, we find that the lookout limit in the universe is 10^{23} (100 billion trillion) light years. Most of the starlight contributing to a bright sky comes therefore from immensely remote regions of the universe. The number of visible stars covering the entire sky has the fantastic value of 10^{60}, or a trillion trillion trillion trillion trillion. These large numbers – of the lookout limit and the visible stars – look suspiciously improbable and suggest that there is something fishy about the so-called Olbers' paradox. They provide the essential clue in an investigation that will abolish the paradox.

Light travels at finite velocity and when we look out in space we also look back in time. A lookout limit in space of 10^{23} light years is equivalent to a lookback limit in time of 10^{23} years (see Figure 12.4). This means that the most distant stars contributing to a bright sky were shining 10^{23} years ago when they emitted the light that we now see. In a homogeneous universe the distant stars are similar to nearby stars, and because the nearby stars are still shining, it follows that the distant stars are also still shining. Hence, according to the traditional argument, the distant stars have been shining continuously for a hundred billion trillion years. But this is utterly impossible.

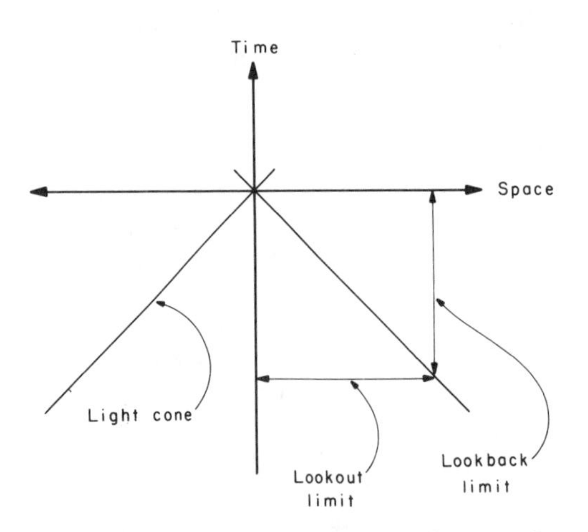

Figure 12.4. Spacetime diagram showing the lookout limit in space and the lookback limit in time.

Sunlike stars shine for a few billion years; a rough-and-ready luminous lifetime for such stars is 10^{10} years. This typical luminous lifetime is exceedingly short compared with the lookback limit in time. A bright sky was obtained earlier by adding up the light contributions from successive shells of stars out to the lookout limit in space. We now realize that beyond a distance of 10^{10} light years we are looking back to a time before the stars became luminous (see Figure 12.5). The stars at distances greater than 10^{10} light years are now shining, just the same as nearby stars, but their light has not yet reached us.

All shells of visible stars contribute equal quantities of light, and the total amount of light reaching us from stars out to a distance of 10^{10} light years is therefore only $10^{10}/10^{23} = 10^{-13}$ (1 ten-trillionth) of the amount required to create a bright sky. The number of stars visible is hence only 10^{21} and not the 10^{60} that is needed to cover the entire sky.

This then is why the sky is dark at night in the infinite and static Newtonian universe; it is because the luminous lifetime of stars is very much less than the lookback limit in time. The paradox is thus resolved within the historical framework in which it was conceived. According to the forest analogy, we stand within a clump of trees, ringed with successive zones of progressively younger trees, and we look out beyond the furthest seedlings to a treeless plain.

It is now seen that a hierarchical distribution of clustered stars is quite unnecessary, for it merely increases the lookout limit, which is already quite large enough to ensure a dark night sky.

Suppose that luminous stars are not all created at the same moment – what then happens? For instance, imagine that only 10 percent of all uniformly distributed stars are luminous at any one moment. The lookout limit is now 10 times greater and the corresponding lookback limit in time is 10^{24} years. When these stars begin to die after 10^{10} years, the next 10 percent become luminous,

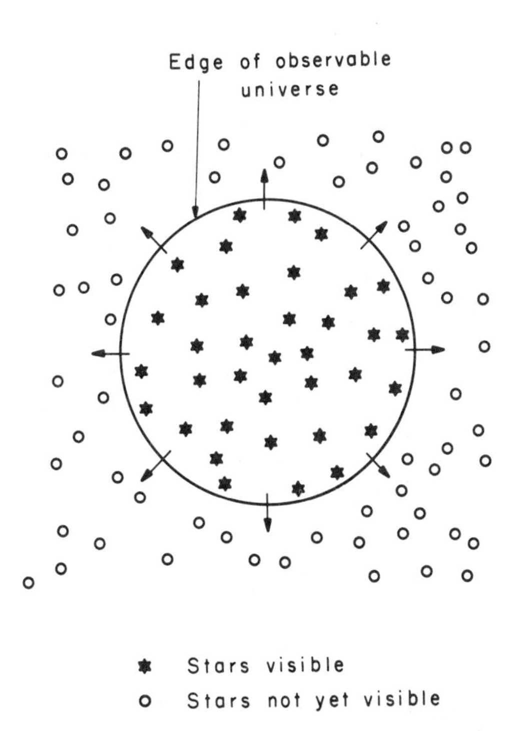

Figure 12.5. Why the sky at night is dark in the static Newtonian universe. We look out and see luminous stars surrounding us out to a distance of roughly 10 billion light years. At greater distances we look back to a time before the stars were luminous. Although the stars are stationary, the outer boundary of the sphere of visible stars in a static universe expands at the velocity of light. If we wait long enough the stars around us will begin to die out. Thereafter we shall be surrounded by an expanding dark sphere of burned-out stars; beyond the dark sphere will lie an expanding shell of luminous stars that has a constant thickness of 10 billion light years; and beyond this shell of visible stars will lie a dark universe of unborn stars.

and then the next, and so on, thus giving an overall luminous lifetime for all stars of 10^{11} years. The total amount of light is still only $10^{11}/10^{24} = 10^{-13}$ of that required for a bright sky. Switching stars on sequentially over many generations fails to increase the brightness of the night sky.

BRIGHT-SKY UNIVERSES. It is not difficult to design hypothetical universes with bright skies. All that is needed is a lookback limit less than the luminous lifetime of stars. If the luminous lifetime of the stars is kept the same, we must increase the number of stars until they are sufficient to cover the entire sky. With a lookback limit of 10^{10} years, equal to the luminous lifetime, we require 10^{13} times as many stars as in our universe, and the sky is then covered with 10^{34} stars.

A bright sky can also be created by abandoning homogeneity. We arrange that all stars are luminous on an observer's backward lightcone. The more distant a star, the earlier it is switched on. Starlight converges on the observer, growing in intensity, and creates in a region about the observer an incandescently bright sky. In this way it is possible to manufacture a bright sky in a particular region even when the lookout limit is quite large. The universe is isotropic about the observer, but not homogeneous, and the observer in this case, roasted in the glare of focused starlight, is not in the least privileged by occupying the cosmic center.

THE POETIC VISION. For three and a half centuries astronomers proposed various resolutions of the dark night sky paradox. They were all wrong. Only one person, the American poet and writer Edgar Allan Poe, proposed the correct resolution. In his imaginative essay *Eureka,* published in 1848 two years before he died at the age of 40, he wrote: "Were the succession of stars endless, then the background of the sky would present us a uniform luminosity, like that displayed by the Galaxy – since there could be absolutely no point, in all that background, at which would not exist a star. The only mode, therefore, in which, under such a state of affairs, we could comprehend the *voids* which our telescopes find in innumerable directions, would be by supposing the distance of the invisible background so immense that no ray from it has yet been able to reach us at all." Poe's explanation of the darkness at night, the reason why the sky is not covered with stars, is that the lookout limit (the "invisible background") is too far away for light to have yet reached us. In other words, the lookback limit is greater than the time that stars have been shining.

When I first read Poe's words I was astounded: How could a poet, at best an amateur scientist, have perceived the right explanation 140 years ago when in our colleges the wrong explanation (see the next section) is still being taught? Poe, however, did not believe in the infinite Newtonian universe, and he continued: "That this *may* be so, who shall venture to deny? I maintain, simply, that we have not even the shadow of a reason for believing that it *is* so." He discarded the correct resolution in favor of a finite universe, and in effect, reverted to Kepler's original solution. Even so, his prescient vision on this and other subjects discussed in *Eureka* is remarkable.

EXPANSION AND DARKNESS

EXPANSION DOES NOT RESOLVE THE PARADOX. A belief popular in recent years asserts that "Olbers' paradox" is resolved by the expansion of the universe. Starlight from distant regions of the universe is weakened by the expansion redshift, and frequently it has been said that the redshift effect explains why the sky is dark at night. According to this argument the act of gazing at the night sky and noting its state of darkness provides sufficient proof that the universe is expanding. If the universe were static, it has been repeatedly said, the sky would blaze with light and we would be scorched to death. Only the expansion redshift saves the universe and keeps it cool and habitable.

Little thought is needed to realize that something is seriously wrong with the redshift resolution of the paradox. John Herschel and Richard Proctor, when they proposed that hierarchy resolved the paradox, assumed that the traditional argument was correct and that the sky should be ablaze with light. Similarly, those who have advocated that redshift resolves the paradox also assume that the traditional argument is correct. But we have seen that a uniform and static Newtonian universe has a dark night sky, and the traditional argument is therefore incorrect; hence there is no paradox. Hierarchy and redshift cannot resolve a paradox that no longer exists; they merely make the night sky slightly darker than what it is in a uniform and static universe. The redshift resolution, if it were correct, would mean the sky was covered with 10^{60} stars that we could not see because their collective light had been weakened by the expansion redshift. Yet we have found that it is impossible for the sky to be covered by so many stars because of their relatively short luminous lifetimes. The sky at night is dark because it is not covered with stars; it is certainly not dark because of the expansion redshift.

THE COSMIC BOX. We turn now for a moment to a more powerful method of resolving the paradox (see Figure 12.6). This method anticipates the discussion of the next chapter and will be considered here only briefly. Each star can be thought of as occupying an

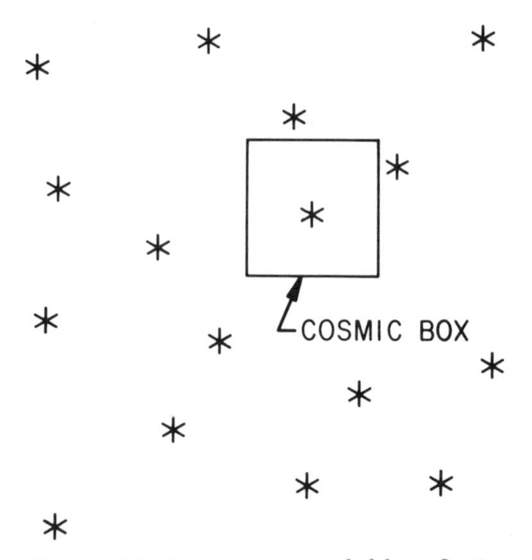

Figure 12.6. A star is surrounded by reflecting walls that form a box of the same volume as the average volume V occupied by each star. The conditions for filling this cosmic box with radiation from a single star are the same as the conditions for filling the whole of space with radiation from all stars. (With permission from E. R. Harrison, American Journal of Physics 45, *123, 1977.)*

average volume V of the universe. We imagine a single average star surrounded by perfectly reflecting walls that form a box of volume V. Light emitted by this star, instead of streaming away into endless space, bounces from wall to wall and is trapped inside the box. It is intuitionally obvious that the radiation inside this *cosmic box* is always the same as the radiation outside the box where there is a uniform distribution of stars and each star occupies an average volume V. The star inside the box retains its radiation in its vicinity, and the stars outside the box mingle their radiation, but otherwise there is no difference.

A bright sky, as visualized in the traditional argument, implies that the temperature everywhere is the same as at the surfaces of stars. Space is therefore filled with radiation up to the level where it equals the intensity at the surfaces of stars. The time required to fill all space with radiation from all stars up to this level is equal to the time required by a single star to fill the cosmic box with its own radiation. This *fill-up time* is easily calculated. A ray of light from the star in the box travels to and fro between the reflecting walls and is eventually intercepted and absorbed by the star. The average time traveled by a ray of light, from the moment of emission to the moment of absorption by the star, is the fill-up time of the box. After that, the star absorbs as much radiation as it emits, and the box is filled with radiation in equilibrium with the star.

Now the average distance traveled in the cosmic box by rays of light between emission and absorption by the star is nothing more than the lookout limit that was derived earlier. To understand this, it might help the reader to imagine one tree surrounded by mirrors. Reflected in the mirrors are multiple images of the tree, which form a forest with a background at the lookout limit. The average time traveled by rays of light is therefore the lookback limit. From this it follows that the time taken to fill the box with radiation is equal to the lookback limit:

fill-up time = lookback limit

The lookback limit is the fill-up time not only of the cosmic box but also of the whole universe.

The single star within the box must shine for 10^{23} years in order to fill the box with radiation; but it is capable of shining for only a very small fraction of this time, and the box always remains almost empty of radiation. The same law applies to the universe as a whole: The radiation level remains low because the luminous lifetime of stars is short compared with the fill-up time.

AN EXPANDING COSMIC BOX. We consider now an expanding universe and imagine an average star inside a cosmic box that expands with the universe. The box has a comoving volume V, and V is always an average volume occupied by a star. Lightrays in the box are reflected and Doppler redshifted repeatedly by the expanding walls. Repeated small Doppler redshifts are the same as the expansion redshift, as we saw in the previous chapter, and the radiation inside the box at any instant is exactly the same as the radiation outside the box in the expanding universe with its uniform population of stars.

If the sky at night is dark because the universe is expanding, then the radiation in the box is feeble because the box is expanding. Take two boxes, one expanding and the other static, and let both contain identical stars that have been shining for identical periods of time. Calculation shows that radiation in the expanding box is not much weaker than radiation in the static box at the instant when they have equal volumes. Expansion reduces the intensity of radiation by a factor that generally is not less than one-half. The effect of expansion is therefore not the cause of a dark night sky, because in a bright-sky universe the light must be multiplied not by a factor of one-half, but by a factor of one ten-trillionth.

It is easy to design hypothetical universes that expand and have either dark or bright skies, and the designing of such universes helps us to understand our own universe better. The general condition for a dark sky is that whichever is the shorter, the luminous lifetime of the stars or the Hubble period, must be shorter than the fill-up time. Correspondingly, the general condition for a bright sky is that whichever is shorter, the luminous lifetime or the Hubble period, must be longer than the fill-up time of the universe.

We can have expanding dark-sky universes, such as our own, in which the luminous lifetime and the Hubble period are both shorter than the fill-up time. We can have expanding dark-sky universes in which expansion is the explanation of their darkness: They have long luminous lifetimes, longer than the fill-up time, and the Hubble period is shorter than the fill-up time. And there are dark-sky universes in which the Hubble period is longer than the fill-up time and the luminous lifetime is shorter than the fill-up time; these latter universes resemble the Newtonian universe for which the Hubble period is infinitely long.

The Hubble period is roughly the age of a big bang universe. Expansion, but not its redshift effect, plays therefore an important role, in the sense that it determines how long radiation can have been pouring into the universe. If the lookback limit, equal to the fill-up time, is larger than the Hubble period, then we cannot look out to a sky covered with stars, because the universe is too young and light has not had sufficient time to traverse the immense distance of the lookout limit.

When we gaze at the night sky we cannot say that its state of darkness proves that the universe is expanding. All we can say is that either the age of the universe or the luminous age of the stars is shorter than the fill-up time.

STARLIGHT IS TOO FEEBLE TO FILL THE DARK UNIVERSE. The darkness of our night sky is due not to absorption of starlight, nor to hierarchical clustering of stars, nor to the finiteness of the universe, nor to the expansion redshift, nor to any of the other proposed resolutions of the last three and a half centuries. The explanation is actually quite simple: The universe does not contain enough energy to create a bright sky. To cover the entire sky with stars we must look out in space to vast distances and look back in time over vast periods that are much greater than the age of stars and the age of the universe. Another way of saying the same thing is to observe that the spaces between the stars are too large for stars to fill these spaces with bright-sky radiation in the time available. Why is the sky dark at night? Strange to say, it needs twentieth-century science to give the delightfully simple answer: Starlight is too feeble to fill the dark universe.

REFLECTIONS

1 *"I am always surprised when a young man tells me he wants to work at cosmology; I think of cosmology as something that happens to one, not something one can choose" (William McCrea, presidential address to the Royal Astronomical Society [February 1963]).*

2 *What is the average distance traveled by an arrow in a forest before it strikes a tree? When we speak of a background of trees at the lookout limit it must be understood that the lookout limit is an average sort of distance: Some visible trees are nearer and some are further away; their average distance is the lookout limit.*

* *Stand amidst a large crowd of people who are taller than you and look around. How many people do you see? Now walk in a straight line; how far on the average can you go before colliding with somebody? In this case you must add the width of your own body to the width of a person in the crowd.*

* *Decide what resolutions of the dark night sky paradox the following analogies apply to: a clump of trees surrounded by a high wall; a clump of trees on a small island; a clump of trees in which the nearest are the oldest and the furthest are seedlings; a forest-filled Sphereland; a foggy*

forest; an endless forest consisting of clumps of trees in small copses, which form woods that are parts of larger woods, and so on; a forest in which the foreground trees are white, the middleground trees are red, and the background trees are black.

3 *In his book* The New Star, *Kepler wrote in 1606, concerning the infinite universe: "This very cogitation carries with it I don't know what secret, hidden horror; indeed one finds oneself wandering in this immensity to which are denied limits and centre and therefore also all determinate places" (quoted from Alexandre Koyré,* From the Closed World to the Inifinite Universe, *1958).*

"Suppose that we took 1000 fixed stars, none of them larger than 1′ (yet the majority in the catalogues are larger). If these were all merged in a single round surface, they would equal (and even surpass) the diameter of the sun. If the little disks of 10,000 stars are fused into one, how much will their visible size exceed the apparent disk of the sun? If this is true, and if they are suns having the same nature as our sun, why do not these suns collectively outdistance our sun in brilliance? . . . Hence it is quite clear that . . . this world of ours does not belong to an undifferentiated swarm of countless others" (Kepler, 1610). See Edward Rosen, Kepler's Conversation with Galileo's Sidereal Messenger *(1965). In his haste, Kepler mistranslated Galileo's* Starry Message *into* Starry Messenger, *and according to Rosen, "He thereby unintentionally supplied a powerful weapon to the deadliest enemies of Galileo, whom he would never have deliberately injured in the slightest way."*

Koyré in From the Closed World to the Infinite Universe *translates Kepler's words as follows: "The explanation of this fact is easy: whereas the planets shine by the reflected light of the sun, the fixed stars shine by their own, like the sun. But if so, are they not really suns as Bruno has asserted? By no means. The very number of the new stars discovered by Galileo proves that the fixed stars, generally speaking, are much smaller than the sun, and that there is in the whole world not a single one which in dimensions, as well as luminosity, can be equal to our sun. Indeed, if our sun were not incommensurably brighter than the fixed stars, or these so much less bright than it, the celestial vault would be as luminous as the sun."*

4 *"The enormous difference which we find between this conclusion and actual experience shows* either *that the sphere of the fixed stars is not only not infinite but that it is actually much smaller than the finite extent I have supposed for it,* or *that the power of light diminishes in greater proportion than the inverse square of the distances. This latter supposition is plausible enough, it requires only that the heavens are filled with some fluid capable of intercepting light, however slightly" (J. P. Loys de Cheseaux,* Treatise on Comets, *1744).*

* *"But because the celestial vault has not, in all its points, the lustre of the sun, must we reject the infinity of the stellar system? Must we restrict this system to a confined portion of limitless space? By no means. In the reasoning, by means of which we arrive at the inference of the infinite number of the stars, we have supposed that space was absolutely transparent, or that the light composed of parallel rays was not impaired, as it removed to a distance from the bodies from which it emanated. Now, not only is this absolute transparency of space not demonstrated, but, moreover, it is altogether improbable" (Heinrich Olbers,* On the Transparency of Space, *1823). In this work, Olbers does not mention Cheseaux but refers to Halley's papers.*

* *"Light, it is true, is easily disposed of. Once absorbed, it is extinct forever, and will trouble us no more. But with radiant heat the case is otherwise. This, though absorbed, remains still effective in heating the absorbing medium, which must either increase in temperature the process continuing,* ad infinitum, *or in its turn becoming radiant, give out from every point at every instant as much heat as it receives" (John Herschel, in the* Edinburgh Review,

1848). Herman Bondi, in Cosmology *(1960), writes: "What happens to the energy absorbed by the gas? It clearly must heat the gas until it reaches such a temperature that it radiates as much as it receives, and hence it will not reduce the average density of radiation."*

* *"It is easy to imagine a constitution of a universe literally infinite which would allow of any amount of such directions of penetration as* not *to encounter a star. Granting that it consists of systems subdivided according to the law that every higher order of bodies in it should be immensely more distant from the centre than those of the next inferior order – this would happen" (John Herschel, in a letter written in 1869 and quoted by Richard Proctor in* Other Worlds Than Ours, *1870).*

* *"It is worth noticing that . . . if we adopt the belief in an infinite succession of orders of systems; that is, first satellite-systems, then planetary-systems, then star-systems, then systems of star-systems, then systems of systems of star-systems, and so on to infinity; . . . we no longer have as a conclusion that the whole heavens should be lighted up with stellar (that is solar) splendor; even though, in this view of the subject, there are in reality an infinite number of stars, just as in the view according to which the sidereal system extends without interruption to infinity" (Richard Proctor,* Other Worlds Than Ours, *1870).*

* *"The reason for the cosmological significance of such a simple fact as the darkness of the night sky is that this is one of the phenomena that depend critically on circumstances far away" (Herman Bondi,* Cosmology, *1960).*

5 *Assume that the average density of matter in the universe is equal to 1 hydrogen atom per cubic meter and that this matter is contained in uniformly distributed sunlike stars. Calculate the average volume occupied by each star, the lookout limit, the fill-up time, and the number of stars required to cover the sky.*

* *If there are universes with bright skies and dark skies, what role does the anthropic principle play?*

* *Stars, as we know, exist in a dark-sky universe, but they surely cannot also exist in a bright-sky universe that is flooded with radiation of intensity equal to that at the surfaces of stars. We can discuss the conditions necessary for a dark-night sky, but not the conditions necessary for a bright-night sky. Perhaps by a "bright sky" we should mean a sky not so bright that stars cannot exist.*

6 *Edmund Halley (1656–1742) was one of Newton's few and enduring friends. He is now best known for Halley's Comet, which he observed in 1682 and predicted would return in 1758. It has since returned again in 1835 and 1910, and it will be back in 1986. Halley was elected a fellow of the Royal Society at the age of 22 and became astronomer royal in 1720. The two short papers Halley published in 1720 concerning the darkness of the night sky were "Of the infinity of the sphere of fix'd stars" and "Of the number, order, and light of the fix'd stars." An account of his argument can be found in the Journal Book of the Royal Society: "The other objection against an infinite number of stars is from the small quantity of light which they all give whereas were there an infinite number it would seem to be much more. To this Dr Halley replies that light is not divisible in infinitum and consequently when the stars are at very remote distances their light diminishes in a greater proportion than according to the common rule and at last becomes entirely insensible even to the largest telescopes" (quoted from Michael Hoskin's Christmas lecture "Dark skies and fixed stars," 1973).*

7 *The phrase* Olbers' paradox *was introduced by Herman Bondi in his book* Cosmology. *A typical dictionary definition of* paradox *is "any person, thing, or situation exhibiting an apparent contradictory nature." From the point of view of historical priority the name should perhaps be* Kepler's paradox. *But darkness at night was not in the least paradoxical to Kepler, and he thought that the night sky was dark for the obvious reason that all stars constituted a finite and bounded universe. Nor was it* *paradoxical to Halley, Cheseaux, Olbers, or even Bondi. Most people from the time of Kepler who have shown an interest in the subject have proposed their own resolutions. It would be incongruous to refer to the darkness of the night sky as* Mr. X's paradox, *when in fact Mr. X proposed an explanation and did not himself find the subject paradoxical: As a titillating topic for astronomical speculation, it is of the nature of a paradox; when, however, Mr. X solves the problem to his complete satisfaction, for him at least it ceases to be a paradox. It seems therefore that* dark night sky paradox *is an acceptable title for the subject if we wish to retain the word* paradox.

8 *The condition for a dark night sky in a hierarchical universe was calculated by Carl Charlier. In this kind of universe there are stars, systems of stars, systems of systems of stars, and so on, indefinitely, with each successive system having a larger radius than the previous system. Charlier showed that the lookout limit is infinitely great, and the sky is not covered with stars, when the total number of stars in each successive system increases more slowly than the radius of each system. Let us call systems of stars (galaxies) the 1st level, systems of systems of stars (clusters of galaxies) the 2nd level, and so on; and let us consider systems of the ith level, where i is a number in the range 1 to infinity. A system of the ith level has radius* R_i *and contains* N_i *systems of the next lower level of radius* R_{i-1}*. Charlier showed that the night sky is dark at all points when* R_i *is always greater than* R_{i-1} *multiplied by* $\sqrt{N_i}$*. If a cluster contains 900 galaxies, its radius must exceed 30 times that of a galaxy; similarly, if a supercluster contains 900 clusters, its radius must exceed 30 times that of a cluster; and so on, for clusters of higher and higher level. The unrealistic assumption in this analysis is that stars are reservoirs of unlimited energy and can shine for eternity. With stars of finite energy and limited luminous lifetime, hierarchy is unnecessary for the attainment of a dark night sky in our universe.*

9 *Herman Bondi writes in* Cosmology: *"Since . . . the rigour of the deductive argument employed appears to be unimpeachable, we must conclude that some of Olbers' assumptions are wrong. The assumptions may be restated here as:*

(i) The average density of stars and their average luminosity do not vary throughout space.

(ii) The same quantities do not vary with time.

(iii) There are no large systematic movements of the stars.

(iv) Space is Euclidean.

(v) The known laws of physics apply." He suggests that the paradox can be resolved if assumption (iii) is dropped: "If distant stars are receding rapidly the light emitted by them will appear reddened on reception and hence will have lost part of its energy. If the recessional velocity of distant stars is great enough the loss of energy may be sufficient to reduce the radiation density to the observed level." Calculation shows, however, that redshift by itself is not sufficient. The assumption that must be dropped is (ii), because stars cannot shine for a time equal to the lookback limit.

In elementary texts we find statements of the following kind: In pre-twentieth-century cosmology it was not known that the universe is expanding. The traditional argument is in error because it fails to take into account the redshift that weakens the light received from distant stars. Most of the light entering the eye at night comes from extremely distant stars of very large redshift. The sky is covered with stars, but only a small number can be seen and the rest are invisible owing to their large redshift. By going out at night and noting the darkness of the sky we perform the most elementary and most important experiment in cosmology: We prove that the universe is expanding.

If this argument were correct we could finish on a wistful note by asking, "Why didn't Olbers realize that the darkness of the night sky is proof that the universe is expanding? He missed the chance of a lifetime by overlooking this obvious fact.

Surely from a scientific point of view the paradox is that nobody had the wit to perceive that the darkness of the night sky is direct proof that the universe is expanding."

10 *Let us imagine that we have constructed on Earth a large box with perfectly reflecting walls. Inside this box we place a source of light such as the filament of a flashlight bulb. The filament has a cross-sectional area of about 1 square millimeter, and we shall assume that the box is a large cube with sides measuring 1 kilometer. When the filament is switched on, the emitted rays of light travel on the average a distance $V/a = 10^{17}$ centimeters, or 0.1 light years, before returning to the filament. After 0.1 years, or roughly 5 weeks, the filament absorbs as much radiation as it emits and the box is filled with radiation. But suppose that the filament is connected to a supply of limited electrical energy, such as an ordinary flashlight cell, which is capable of maintaining a bright filament for only 10 hours. The luminous lifetime of the filament is now much less than the fill-up time of the box, and the radiation level remains low for the same reason the sky remains dark at night.*

11 *Darkness at night can be understood with the aid of a water-tank analogy (see Figure 12.7). Water pours into a tank from a faucet, and the time required to fill the tank is the "fill-up" time. The time during which the faucet is open is the "luminous lifetime." The condition for the tank not to fill and the water level to remain low is simply that the luminous lifetime be less than the fill-up time.*

Now suppose that the tank is expanding. This tends to keep the water level low and the sufficient condition for the tank not to overflow is the same as for a static tank. For if the tank does not fill when it is static it most certainly will not fill when it is expanding. We can be more precise. Let

t^* = *luminous lifetime (time faucet is turned on)*

τ = *fill-up time (time to fill tank at volume V)*

T = *Hubble period (roughly the time taken by the tank to expand to volume V)*

When the luminous lifetime t^* is smaller than τ (as in our universe), the sky is dark independent of whether the universe is expanding or static. When T and τ are both small compared with t^*, and the expansion is rapid enough that T is less than τ, the sky is dark because of expansion. The water level is kept low either by expanding the volume of the tank rapidly or by not allowing water to flow into the tank long enough to fill it.

Stated in slightly more mathematical language, we can say that the condition for a dark night sky is that

$$\frac{1}{\tau} \text{ must be less than } \frac{a}{t^*} + \frac{b}{T}$$

where a and b are numbers, close to unity, that depend on the nature of the universe.

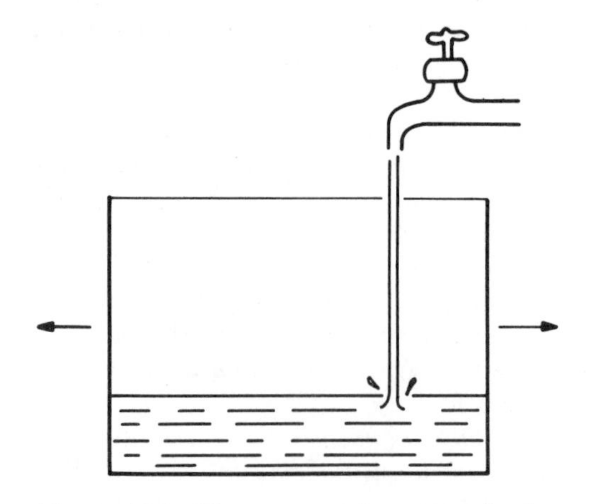

Figure 12.7. Water pours into a tank that expands. The conditions for the water level remaining low in the tank are similar to the conditions for the sky being dark at night.

FURTHER READING

Harrison, E. R. "The dark night sky paradox." *American Journal of Physics 45,* 119 (February 1977). This gives a treatment similar to that in the present chapter. See also "The paradox of the dark night sky." *Mercury,* July–August 1980.

Hoskin, M. "Dark skies and fixed stars: a Christmas lecture." *Journal of the British Astronomical Association 83,* 4 (1973).

Rosen, E. *Kepler's Conversation with Galileo's*

Sidereal Messenger. Johnson Reprint Corp., New York, 1965.

SOURCES

Beaver, H. *The Science Fiction of Edgar Allan Poe.* Penguin Books, New York, 1976. See Poe's "Eureka: an essay on the material and spiritual universe," first published in 1848.

Bondi, H. *Cosmology.* Cambridge University Press, Cambridge, 1960.

Clayton, D. D. *The Dark Night Sky: A Personal Adventure in Cosmology.* Quadrangle, New York, 1975. With an introduction by Fred Hoyle.

Harrison, E. R. "Olbers' paradox and the background radiation density in an isotropic homogeneous universe." *Monthly Notices of the Royal Astronomical Society 131,* 1 (1965).

Harrison, E. R. "Why the sky is dark at night." *Physics Today,* February 1974.

Hoskin, M. "Dark skies and fixed stars: a Christmas lecture." *Journal of the British Astronomical Association 83,* 4 (1973).

Jaki, S. L. *The Paradox of Olbers' Paradox.* Herder, New York, 1969. This contains much historical material, but does not credit Kepler with the discovery of the cosmological significance of a dark night sky. Further, the advocated resolution (a finite universe) is erroneous. The "paradox" of the paradox, according to Jaki, is that scientists have failed to consult historical sources, and have also failed to realize that the night sky is dark because the universe is "finite in space."

Koyré, A. *From the Closed World to the Infinite Universe.* Johns Hopkins Press, Baltimore, 1957. Reprint. Harper Torchbooks, New York, 1958.

Ronan, C. *Edmund Halley: Genius in Eclipse.* Doubleday, Garden City, N.Y., 1969.

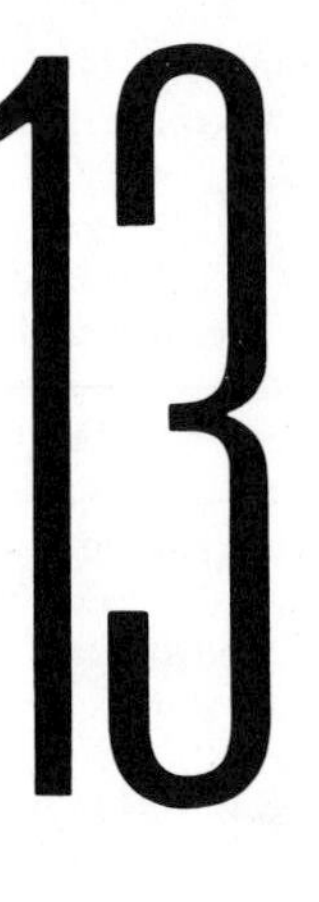

THE UNIVERSE IN A NUTSHELL

I could be bounded in a nutshell and count myself king of infinite space, were it not that I have bad dreams.
— Shakespeare, *Hamlet*

COSMIC BOX

REFLECTING WALLS. We look out in space and back in time, and the things seen at large distances are similar to the things that existed in this part of the universe long ago. The scenery billions of light years away is the same as the scenery here billions of years ago. With a time machine that could travel back into the past we would have less need of large telescopes that strain to reach the limits of the observable universe.

This consideration prompts the following thought: Because everywhere things are the same, why not confine our attention to a single region, study its history, and ignore the rest of the universe? What happens in a sample region over a long period of time is the same as what happens elsewhere. The life history of one elephant is similar to that of all elephants.

This argument has a drawback. A sample region is influenced by other regions, which therefore cannot be ignored. Light, for instance, comes from great distances and influences what happens in the sample region. If we are to pay undivided attention to a sample region we must in some way allow for the influence of all other regions. But things at great distances that influence the sample region are no different from the things already in the sample region that existed long ago. Can we therefore contrive a way in which things in the sample region of long ago are substituted for the things at great distances?

Cosmologists, because they cannot potter around botanizing and experimenting like most other scientists, have become adept at performing imaginary experiments. The problem of isolating a region of the universe, and making it self-influencing, offers no difficulty. The trick is as follows.

Our own region of the universe, or any other sample region, is isolated by surrounding it with imaginary reflecting walls (see Figure 13.1). It is enclosed within a *cosmic box*. All lightrays emitted by things inside the box are mirrored to and fro by the reflecting walls and are not allowed to escape. Objects inside the box are now influenced by the light emitted long ago by local objects, and on the average this light is the same as that which normally would have come from distant objects. We have thus succeeded in isolating a sample region in such a way that its present condition is influenced by its own past conditions, which were identical with those elsewhere. The imaginary reflecting walls of the cosmic box must be perfect in every sense: They must transmit nothing, absorb nothing, and reflect everything, such as light, gravitational waves, particles, neutrinos, and whatever else moves from place to place in the universe.

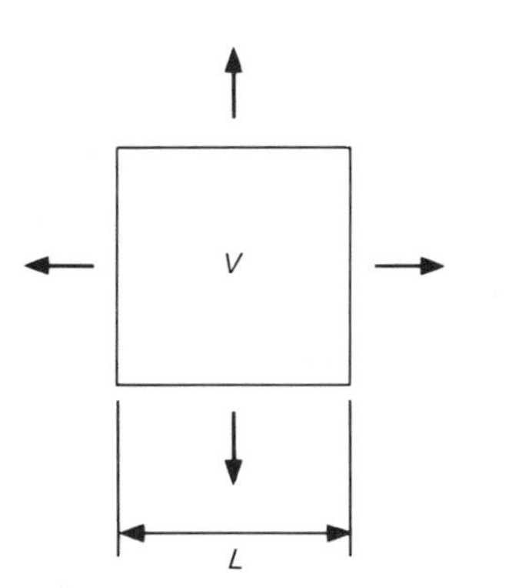

Figure 13.1. The expanding cosmic box of volume V. The box is shown as a cube and L is the side length.

PARTITIONS DO NOT AFFECT THE UNIVERSE. A *partitioned universe* helps to clarify our ideas on this subject. Imaginary partitions, comoving and perfectly reflecting, are used to divide the universe into numerous separate cells (see Figure 13.2). Each cell encloses a representative sample and is sufficiently large to contain galaxies and clusters of galaxies. Each cell is larger than the largest scale of irregularity in the universe, and the contents of all cells are therefore in identical states.

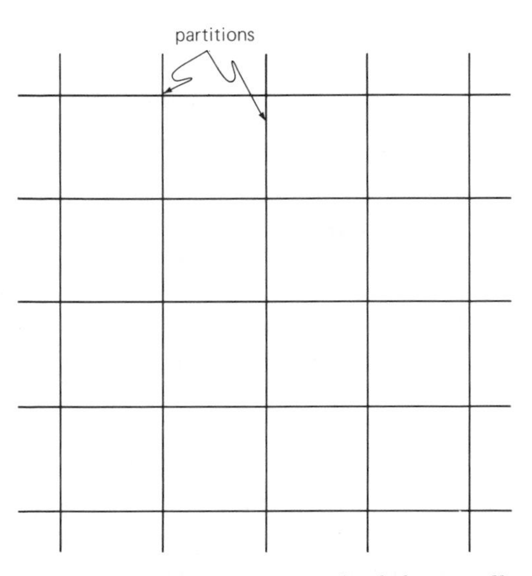

Figure 13.2. The universe is divided into cells with imaginary and perfectly reflecting partitions. The contents of all cells are in identical states and we need study only what happens in a single cell.

A partitioned universe behaves in exactly the same way as a universe without partitions. The partitions, we assume, have no mass, and their insertion therefore does not alter the dynamical behavior of the universe. Inside each cell everything performs in the same way as when there were no partitions. The contents of all cells are in identical states and are also in the same state as previously, when there were no partitions. Lightrays that normally would have come from distant galaxies come instead from local galaxies and travel similar distances by multiple reflections. What normally would have passed out of a cell is reflected back and is the same as what normally would enter a cell. Hence partitioning has no effect whatever on the behavior of the universe.

We have shown that comoving and reflecting partitions do not change the nature of the universe. Let us now remove all partitions and leave only those walls that enclose a single cell. It is clear that what happens inside this cell, or cosmic box, is similar in every way to what happens everywhere outside. Observers inside and observers outside the cosmic box perceive essentially the same scenery. The cosmic box is the universe in a nutshell, and an inside observer can truly say, with Hamlet, "I am king of infinite space." The bad dreams in this case pertain to the nightmare thought that we live in a "mirrored universe" created by a cosmic jester!

COSMIC BOX IN AN IDEALIZED UNIVERSE. The comoving walls of the cosmic box move apart at a velocity given by the velocity–distance law. If the box is a cube with sides of length L, then L is proportional to R, where R is the scaling factor that is everywhere the same in space. Opposite walls move apart at a relative velocity HL, where H is the Hubble term. When L is 100 million light years, this relative velocity is 2000 kilometers per second, if H is 20 kilometers a second per million light years.

Irregularities, we must admit, are a

distraction and serve little purpose in the present discussion. For simplicity we shall therefore ignore all irregularities and assume that the universe is ideally smooth. Effects caused by irregularities can always be considered at some other time if they should turn out to be important. In an idealized universe the cosmic box can be small, and its contents will still remain in a representative state.

When L is as little as 1 light year the walls move apart with a relative velocity of 2 centimeters per second; when L is 1 million kilometers the velocity is 6 centimeters per year; when L is 1 kilometer the velocity is 6 centimeters per million years; and so on. The relative velocity of the expanding walls may be as small as we please while we still retain in the box an average sample of an idealized universe. The box must naturally be larger than the things under investigation; if we study radiation, for example, the box must be large enough to contain the longest wavelengths of interest. A star, of course, cannot be put into a box having a volume of only a cubic centimeter, or a cubic kilometer; but matter in such small boxes can nonetheless be luminous in the same average way as that in which matter is luminous in the unsmoothed universe. When the smoothing process is more than one is willing to tolerate, then L may be made sufficiently large to embrace the objects of interest. This was done in the previous chapter with an average star inside an average volume V. The important point is that L must be kept small in comparison with the Hubble length of 15 billion light years.

SOME ADVANTAGES OF THE COSMIC BOX. The advantages of the cosmic box are numerous; some we shall mention now and others will become apparent later in this chapter.

In a small box we are free to use Euclidean geometry. Space may be curved, with either spherical or hyperbolic geometry, and yet in a small region it is virtually flat and Euclidean. It is a great convenience to study cosmic phenomena in a box without the bother of taking into account the curvature of space. Whatever happens inside a cosmic box, and is unaffected by the presence of confining walls, is independent of the large-scale geometry of space. The box therefore helps us to decide when space curvature is important. We recall that in the last chapter we found that the darkness of the night sky is independent of whether the universe is open or closed, and this is why we were able to use the cosmic box.

The walls have an expansion velocity that is small compared with the velocity of light. Inside the box there is hence no need to distinguish between Doppler and expansion redshifts. Light bounces to and fro repeatedly, and because the walls are moving apart slowly, the sum of small Doppler redshifts is the same as the expansion redshift. There is also no need to distinguish between peculiar and recession velocities, and everything inside the box can be regarded as moving through space as if in the laboratory.

Inside a relatively small cosmic box we use ordinary everyday physics and are able to determine easily the consequences of expansion. The physics of what happens in a slowly expanding box are well understood, and this knowledge is now available for immediate use in cosmology. A gas that is uniformly distributed in the universe serves as an example. It behaves exactly the same as a sample of gas in an expanding box, and because we know how gases behave in expanding boxes, we also know how they behave when uniformly distributed in an expanding universe.

The cosmic box is the universe in a nutshell; it works because the universe is homogeneous, and we need study only what happens in a small, isolated, self-influencing region to know what happens everywhere.

PARTICLES AND WAVES

FREELY MOVING PARTICLES. Let us consider a particle moving freely in an expanding universe. It moves freely in the sense that it is not trapped in a bound system such as a galaxy. It has peculiar motion as seen by a local comoving observer who is stationary in expanding space. Strange to say, this freely moving particle slows down in the course of time and ultimately becomes stationary. It is an astonishing fact that all freely moving particles (including galaxies) slowly lose their peculiar motion and ultimately become stationary and comoving in expanding space.

We shall try to understand what happens by considering a moving particle inside an expanding box (see Figure 13.3). The cosmic box must be small enough so that the walls move apart with a velocity much less than the velocity of the particle, and the particle therefore repeatedly rebounds from the expanding walls.

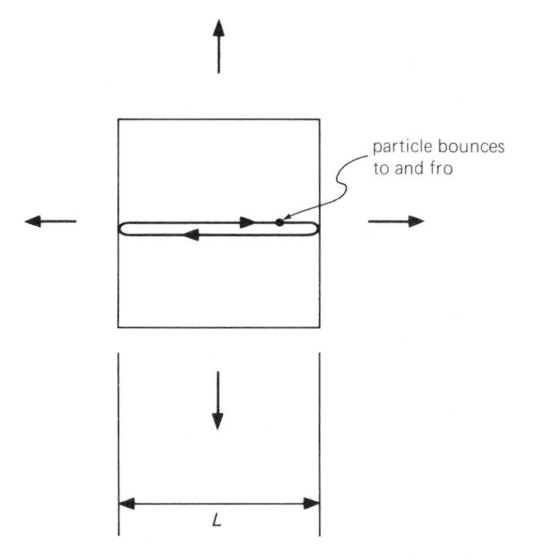

Figure 13.3. A particle bounces to and fro in an expanding box and slowly loses energy, in exactly the way that a particle moving freely in an expanding universe does, and all peculiar motions are decelerated and ultimately reduced to zero.

A particle moving freely in the universe travels in a straight line. The expanding regions through which it passes, however, are always identical at each instant to the region inside the cosmic box. The speed of the particle in the box is the same as the speed of the particle moving freely in a straight line. The particle in the box continually changes its direction but nothing else. The straight-line trajectory is folded up inside the box, and the rebounding particle has the same speed and energy at each instant as if the reflecting walls did not exist.

It is well known that a particle bouncing around in an expanding box loses energy. Each time it strikes a wall it rebounds with slightly reduced energy. It moves continually in different directions, and on the average it is not important what particular direction it has at any instant. For simplicity we suppose that the particle moves in a direction perpendicular to two opposite walls. The walls are perfect reflectors and therefore, *relative to a wall,* the particle rebounds with the same speed as that with which it strikes the wall. During the collision the direction of motion is reversed, but the speed relative to the wall remains unchanged. The wall is receding, however, and the particle therefore returns to the center of the box with slightly reduced speed. Each time the particle strikes a receding wall it returns with reduced speed.

It can be shown that a particle of mass m and speed v, moving within an expanding box, obeys the law that mv is proportional to $1/L$. The product mv is the momentum, and this means that the momentum is reduced as L expands. The length L expands in the same way as the scaling factor R, and the momentum therefore obeys the important law

$$\text{momentum is proportional to } \frac{1}{R} \qquad (13.1)$$

This law holds not only for particles in an expanding box but also for particles moving freely in an expanding universe. The general relativity equation of motion of a freely moving particle in the curved space of an expanding universe gives exactly the same result. This illustrates how the cosmic box not only helps us to understand what is happening but also allows us to employ very simple methods to derive important results.

When the speed is much less than the speed of light the mass m is constant and

$$v \text{ is proportional to } \frac{1}{R} \qquad (13.2)$$

or, in terms of redshift,

$$v = v_0(1 + z) \qquad (13.3)$$

where v_0 is the present speed. The energy varies as the square of the speed and it follows that

$$\text{energy is proportional to } \frac{1}{R^2} \qquad (13.4)$$

In terms of the redshift this gives

$$\text{energy} = \text{energy now} \times (1 + z)^2 \qquad (13.5)$$

These results apply to all freely moving bodies, and a galaxy not in a cluster of galaxies loses energy in this fashion.

Consider now a particle with a speed close to the speed of light – a relativistic particle of high energy such as a cosmic ray particle. In this case the speed is almost constant, the mass is proportional to the energy of motion, and we get from equation (13.1)

$$\text{energy is proportional to } \frac{1}{R} \qquad (13.6)$$

In terms of the redshift this gives

$$\text{energy} = \text{energy now} \times (1 + z) \qquad (13.7)$$

At a redshift of 1, when the universe was half its present size, the energy of an ordinary freely moving particle was four times its present value, and the energy of a relativistic particle was twice its present value.

LIGHT WAVES. The cosmic box also helps us to understand why light loses energy in an expanding universe. Each time a ray of light is reflected from a receding wall it is slightly redshifted because of the Doppler effect. In 1913, Max Planck showed that the cumulative result of repeated small Doppler redshifts in an expanding box obeys the law that wavelengths are proportional to L. All wavelengths are stretched as L increases. Because L is proportional to the scaling factor R, this yields

$$\text{wavelength is proportional to } R$$

and therefore, in terms of redshift,

$$\text{wavelength} = \text{wavelength now} \times \frac{1}{1 + z}$$

Lightrays bouncing to and fro inside an expanding box and lightrays traveling in expanding space have their wavelengths stretched according to the same law. The cumulative Doppler redshift is independent of how the box expands, just as the expansion redshift is independent of how the universe expands, and in both cases the redshift depends only on the amount of the expansion.

Another way of understanding the behavior of radiation in an expanding box is to consider a resonant cavity containing radio waves. These *standing waves* have zero amplitude at the walls where they are reflected, as shown in Figure 13.4, and as the cavity expands all wavelengths increase in step with the size of the cavity.

Yet another way is to consider light – and all electromagnetic radiation – as composed

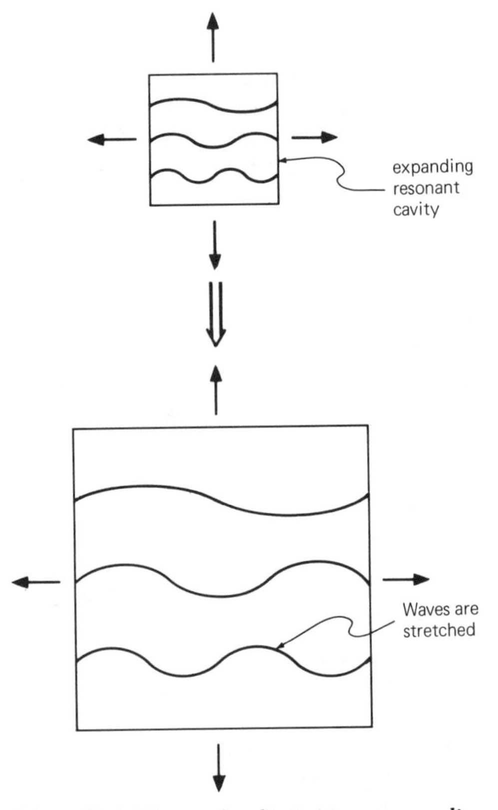

Figure 13.4. Waves of radiation in an expanding box; as the box expands, the waves are stretched.

of photons. These are particles that travel at light speed and have energy proportional to their frequency. A photon may be viewed as a relativistic particle, and its energy must therefore change as $1/R$. Because frequency is also inversely proportional to wavelength, it follows that the wavelength increases with the scaling factor R. This argument applies to all particles that move at the speed of light and can hence be used for neutrinos. In an expanding universe neutrinos lose their energy in the same way as photons.

THERMODYNAMICS AND COSMOLOGY

TEMPERATURE. The full power of the cosmic box is realized when we turn to the science of thermodynamics. In thermodynamics we study the properties of heat – applied to things in general – under various circumstances, in isolated systems of varying volume. How does a uniformly distributed gas behave in an expanding universe? By isolating some of it in a cosmic box we are able immediately to harness thermodynamics in the service of cosmology.

A system consisting of particles in thermal equilibrium has everywhere the same temperature. The particles move in all directions, and have various speeds as shown in Figure 13.5, and their average energy is everywhere the same. Temperature is in fact a measure of thermal energy. Radiation in thermal equilibrium, sometimes referred to as blackbody radiation, consists of photons of various energies, as shown in Figure 13.6, and their average energy is also the same everywhere.

We have already seen that individual particles, moving freely, lose their energy when enclosed in an expanding box. Exactly the same thing happens to a gas comprising many particles. The particles of the gas collide with one another, but between collisions they move freely and lose energy, and on the average collisions do not change their

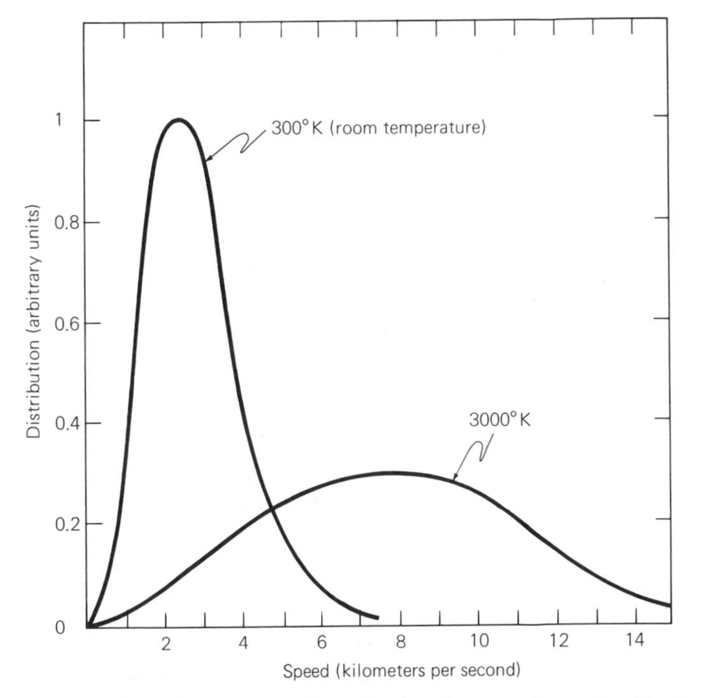

Figure 13.5. These curves show the distribution of speeds of hydrogen atoms in gases of the same density and different temperatures.

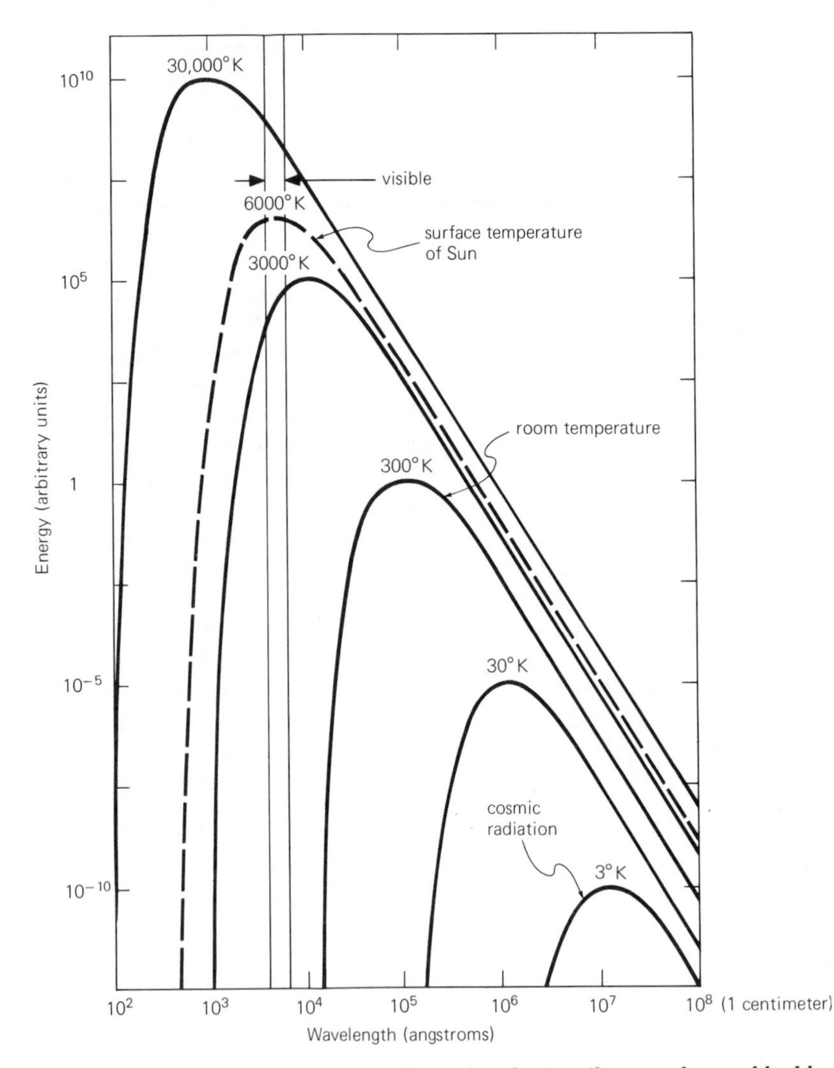

Figure 13.6. These curves are for thermal radiation (known also as blackbody radiation) and show how the energy is distributed at different wavelengths for various temperatures. The average separating distance of the photons is approximately the wavelength at the peak energy.

energy. The temperature of a gas therefore varies with expansion in the same way as the energy of a single particle, and hence

$$\text{gas temperature is proportional to } \frac{1}{R^2} \qquad (13.8)$$

If T denotes temperature, and T_0 is the present temperature, then

$$T = T_0(1 + z)^2 \qquad (13.9)$$

The temperature diminishes with expansion and the gas remains in thermal equilibrium.

A gas can also consist of photons: Such a gas in thermal equilibrium is known as thermal radiation or blackbody radiation. The temperature of thermal radiation is a measure of the average energy of the photons and, because we know how photon energy varies with expansion, we have

$$T = T_0(1 + z) \qquad (13.10)$$

The 3-degree cosmic radiation is thermal radiation uniformly distributed in the universe. It has remained thermal during expansion and at a redshift of 100 it had a temperature of 300 degrees Kelvin, which is slightly more than room temperature. At a redshift of 1000 it was incandescent and had a temperature of 3000 degrees; and this was roughly when it last interacted with all the matter in the universe.

FIRST LAW OF THERMODYNAMICS. The walls of the cosmic box are nonabsorbing and perfectly reflecting; hence nothing passes through the walls, including heat and all other forms of energy (see Figure 13.7). No energy passes into or out of the box, and such an isolated system, which neither gains nor loses energy through its walls, is said to be *adiabatic*. The universe as a whole is adiabatic; and each representative part is also adiabatic, because the energy that leaves is replaced by energy of the same kind, coming from other regions.

The *first law of thermodynamics* can be broken down into two parts. The first part states that energy entering a closed volume increases the energy already inside the closed volume. Because a cosmic box is adiabatic, and no energy enters or leaves, we can ignore this first part. The second part states that the energy in an expanding volume is reduced because of the work performed by the pressure. A classic example is the steam engine: Steam in a cylinder exerts pressure on the piston, and as the steam expands and pushes out the piston, the steam performs work, loses energy, and consequently gets cooler (see Figure 13.8). Our cosmic box has pressure on the inside walls, and the contained energy therefore decreases with expansion. This tells us that energy in the expanding universe is decreasing.

Let V be the volume of the cosmic box; let E denote the total energy inside; and let p stand for pressure. For adiabatic changes, the first law of thermodynamics states that a small increase in energy (call it dE), plus the pressure times a small increase in volume (call it dV), is equal to zero:

$$dE + pdV = 0 \tag{13.11}$$

We see that as V increases (dV is positive), the energy E decreases (because dE is negative). The greater the pressure, the greater the extent to which the energy is decreased

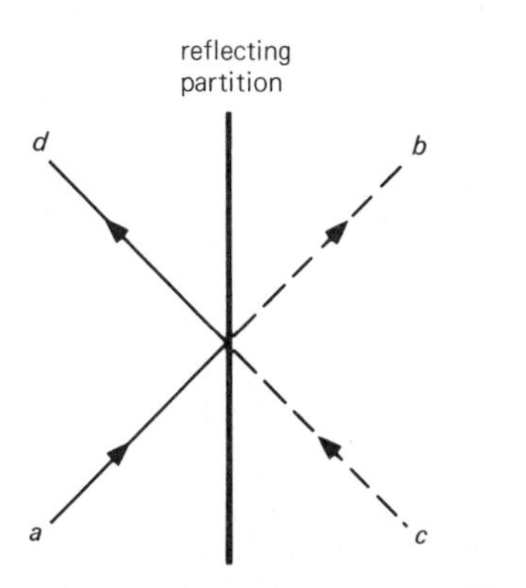

Figure 13.7. Perfectly reflecting partitions change nothing in a uniform universe. A particle that would normally have traveled from a to b is deflected by a partition and travels instead to d. In a uniform universe there exists on the average a similar particle of the same energy that would normally have traveled from c to d, and the partition deflects this second particle into direction b. Such deflections occur for all particles incident on both sides of the partiion, and by "detailed balancing" arguments of this kind we see why reflecting partitions do not affect a uniform universe.

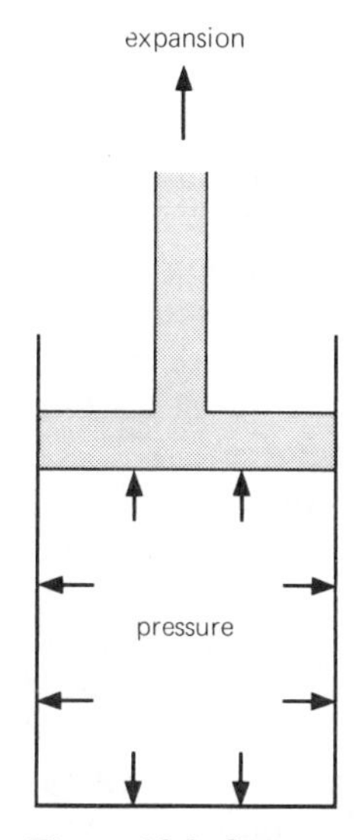

Figure 13.8. Steam in a cylinder pushes against a piston. As the piston withdraws, and the volume expands, the heat energy of the steam is converted into mechanical energy.

by expansion. The first law, as expressed by equation (13.11), is of fundamental importance in cosmology.

SECOND LAW OF THERMODYNAMICS. The *second law of thermodynamics* states that entropy either remains constant or increases in an isolated system. Perfect systems can be devised in which entropy remains constant, but usually they are idealizations. The most perfect system realized by nature is the universe itself, and even this is not totally perfect: Its entropy is increasing.

Energy forever cascades into less useful and effective forms, and this is just another way of saying that entropy increases. In practice one cannot use the heat discarded by the house next door to heat one's own house adequately, because energy is lost in the process to the environment and the discarded heat has reduced temperature.

The cosmic box is adiabatic – no energy enters or leaves – and we must ask how it is possible for entropy to increase in such an isolated system. This isolated system, which is a representative part of the universe, contains stars. In these stars hydrogen burns to helium and heavier elements, and nuclear energy is transformed into other forms of energy. The nuclear energy is released at a temperature of millions of degrees and is radiated away from the surfaces of stars at a temperature of only thousands of degrees. Stars obtain their energy at high temperature and discard it at low temperature and are therefore entropy-generating machines. Stars pour out radiation into space, and the amount of starlight in space is a measure of the entropy they have generated.

The entropy of the universe is mainly in the radiation that fills space. Everything emits radiation, even black holes, and therefore entropy is increasing. Usually, to measure entropy in cosmology, all we need do is count photons: The total number of photons in the cosmic box is a reasonably good measure of its total entropy.

I cannot use my next-door neighbor's discarded energy to heat my house effectively because some energy escapes into the

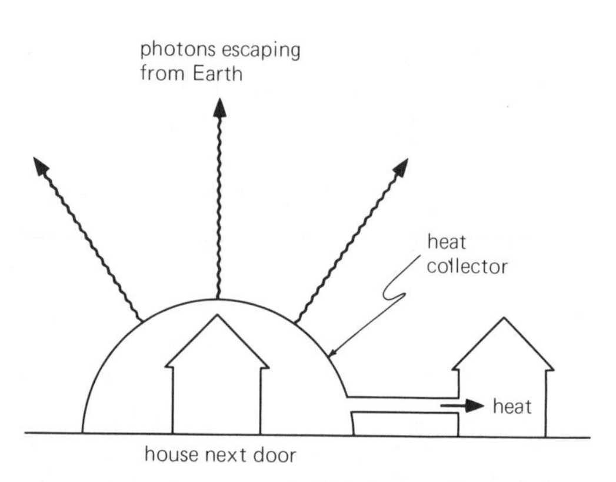

Figure 13.9. My inability to exploit my neighbor's discarded energy is registered in the universe by an increased number of photons.

atmosphere, which then radiates it away into space as infrared photons. My inability to exploit fully my neighbor's discarded energy is registered in the universe by an increased number of photons (see Figure 13.9).

Now comes the surprise. The number of photons in the 3-degree cosmic radiation is at least 10,000 times greater than the number of all other photons that have been emitted by the stars. Most of the entropy of the universe is already in the 3-degree cosmic radiation, and whatever stars do, or whatever anything else does, is not going to affect the total entropy very much. Some decades ago, when scientists discussed the universe, they would foretell with bated breath the eventual heat death of the universe, noting how everything would fade and die, and entropy would rise inexorably and attain its ultimate value. We now realize that the heat death has already occurred; it happened long ago, and we live in a universe that has very nearly attained its maximum entropy.

The number of photons per cubic centimeter of thermal radiation at temperature T is given by

$$n_{\text{photons}} = 20 \times T^3 \qquad (13.12)$$

and for the 3-degree cosmic radiation (T is closer to 2.7 degrees) this is 400 photons per cubic centimeter. If you go out of doors, in daytime or nighttime, and hold the palm of your hand out for just 1 second, a thousand trillion (or 10^{15}) photons of the cosmic radiation will strike it in this short interval of time. These photons continually bombard everything in the universe, and their number is so colossal that this loss is utterly negligible.

The number of photons in the universe is a measure of its entropy. In the cosmic box the total number of photons is n_{photons} multiplied by the volume V. According to equation (13.12), n_{photons} varies as T^3, and therefore varies as $1/R^3$; but V itself varies as R^3: Hence the entropy of thermal radiation in the box remains constant during expansion. This is just another way of saying that the total number of photons in the box remains constant. Actually, their number is slowly increased by emission from the stars and other things, but this increase is so small that for most purposes it can be ignored. We conclude that the entropy of the universe is immense and almost constant.

A convenient way of measuring entropy is to count the number of photons for each nucleon. On the average there is approximately 1 nucleon per cubic meter, and because there are 400 photons per cubic centimeter, this means there are between 10^8 and 10^9 photons to each nucleon. The number of photons per nucleon, almost a billion, is the *specific entropy* of the universe. The specific entropy does not alter much during expansion, and either the universe was created with a large reservoir of entropy or the entropy was generated in some unknown way during the earliest moments of the big bang. An advantage of the big bang type of universe is that we are able to see how the specific entropy determines the nature of the universe. If the specific entropy were much smaller, almost all hydrogen would have been converted into helium in the big bang, and if it were much larger, the universe would have been too hot for the formation of galaxies. In the first case stars would not be luminous over long periods of time; in the second case stars would probably not exist; and in either case the universe would not contain life.

WHERE HAS ALL THE ENERGY GONE?

Radiation, freely moving particles, and also gases lose energy in an expanding universe. Where does this energy go? We take for granted that light is redshifted and usually do not concern ourselves about where its energy has gone. But the deceleration of freely moving particles is something more tangible, and the question of where the vanishing energy goes can no longer be evaded.

It is easy to see why energy is lost in an expanding box. Lightrays and particles push on the walls; as the box expands their pressure performs work, and the energy of the rays and particles therefore decreases. Many engines make use of this principle. Steam engines and internal combustion engines are familiar examples of how pressure on a piston produces mechanical energy at the expense of the energy within the cylinder.

The cosmic box contains a representative sample of the universe, and what is inside is always in the same state as what is outside. Rays of light and particles push on the inside of the walls, and similar rays and particles push back with equal force on the outside of the walls. The rays and particles inside lose energy because they push against the receding walls and they therefore also push against the rays and particles outside. But those outside do not gain the energy lost by those inside; and, vice versa, those inside do not gain the energy lost by those outside. To try to understand what is happening we return to the partitioned universe. In any expanding cell there is a progressive loss of energy, which cannot be gained by adjacent cells because they are also expanding and losing energy. The pressures inside the different cells work against each other, and consequently energy is lost everywhere and reappears nowhere as useful work.

The universe is not in the least like a steam engine and we must not jump to the conclusion that pressure is the cause of expansion. Pressure has nothing whatever to do with why the universe expands; the universe could just as easily contract, and in the future it may pass from its present state of expansion into a state of contraction. If the universe possessed a cosmic edge, the situation would be different; the pressure at the edge could then do work, and we would have a universe similar to a steam engine. But the universe has no edge, and the pressure everywhere is therefore impotent and unable to produce mechanical energy. The conclusion, whether we like it or not, is obvious: Energy in the universe is not conserved.

Most of the things with which we are familiar are surrounded by edges, or gradients that merge one discrete object into another. Energy is transported across these edges, or flows down these gradients, and whenever it is lost in one place it reappears elsewhere, either in the same form or in some other form, such as mechanical energy. A fraction of the energy we use escapes from Earth and is written off as an inevitable loss that nonetheless balances the budget. What escapes into outer space increases the cosmic entropy; but its energy will slowly vanish in the expanding universe.

Science clings tenaciously to concepts of conservation, the most fundamental of which is the *conservation-of-energy principle.* Whenever scientists have found that energy has vanished in a mysterious fashion they have searched diligently for its reappearance in some other form, and such searches have led to the discovery of new and hitherto unrecognized forms of energy. The discovery of the neutrino is a classic example. Radioactive nuclei decay into lower-energy states and in the 1920s it was known that the electrons emitted by the decaying nuclei failed to carry away all the energy released. Wolfgang Pauli, who had discovered the exclusion principle (not more than two electrons can have the same kind of wave in an atom), suggested in 1931 that another particle is emitted, having no electric charge, that carries away the missing energy. Enrico Fermi shortly afterward dubbed the hypothetical particle the *neutrino,* and it was detected experimentally in 1956.

The conservation-of-energy principle serves us well in all sciences except cosmology. In regions that do not partake of the expansion of the universe, that are dense compared with the average density of the universe, we can trace the cascade and interplay of energy in its multitudinous forms and claim that it is conserved. But in the universe as a whole it is not conserved. The total energy decreases in an expanding universe and increases in a collapsing universe. To the questions where the energy goes in an expanding universe and where it comes from in a collapsing universe the answer is – nowhere, because in this one case energy is not conserved.

REFLECTIONS

1 *The cosmic box helps us to understand in a simple way how things change in an expanding universe. We do not worry about whether space is curved because in a small enough region space can always be considered flat. We do not have to distinguish between peculiar motion through space and comoving motion in expanding space, because the recession velocity of the walls is extremely small compared with the velocity of light, and the two forms of motion have become indistinguishable. Nor do we have to distinguish between Doppler and expansion redshifts. Finally, we are able to draw on the familiar laws of science that govern the behavior of self-influencing things in isolated and expanding boxes.*

The convenience of the cosmic box is demonstrated in the following way. Let Albert be an observer inside the cosmic box and Bertha an observer outside. Bertha outside tries to estimate the radiation she receives from all sources in the universe. She adds together the contributions from individual sources everywhere, taking into account their distances in curved space, their redshifts, and the absorption of their rays of light while they are traveling through curved space, and she has to be particularly careful about the possibility of rays circumnavigating in a closed universe.

Imagine her surprise when, having done this tedious calculation, she finds that the result is independent of the curved geometry of space. Albert inside finds out how much light has been emitted by the local sources, makes an allowance for the loss owing to expansion and absorption, and by a much simpler method obtains exactly the same result as Bertha. He is not in the least surprised that his result is independent of the curved geometry of space. Bertha has the tricky problem of writing down an integration formula, whereas Albert merely writes down a differential equation of the kind that is quite common in physics.

2 *When a tennis player is running forward and strikes a ball, the ball crosses the court faster than when the tennis player is running backward. Why is this?*

* *Resolve the dark night sky paradox with the aid of a cosmic box.*

* *Stars emit neutrinos as well as photons. Why is the universe not flooded with neutrinos to enormous density?*

* *The 3-degree photons are continually being absorbed by matter in the universe. How long would it take to deplete the cosmic radiation in this way? (If you take an average volume V and put inside it an average star, the answer is the fill-up time discussed in the last chapter.)*

* *Most of the photons in the universe belong to the 3-degree cosmic radiation that has survived from the big bang. Their total number is almost constant, and hence the entropy of the universe is almost constant. Their thermal energy, once large, now is small. Where has this energy gone? Can we say in cosmology that the second law of thermodynamics provides us with a better principle of conservation than the more familiar principle of the conservation of energy?*

* *In what way is the anthropic principle related to the value of the specific entropy?*

* *Consider the following puzzle: In cosmology we look for those things that will ultimately unify the universe. But the partitioned universe shows that when regions are isolated from one another in cells they continue to behave as before, as if nothing had changed. What then unifies the universe? Whatever it is, surely it is something that cannot be isolated in a cosmic box?*

3 *Imagine that our part of the universe, extending hundreds of millions of light years, is enclosed within a box whose comoving walls are perfect mirrors. We see our part of the universe mirrored repeatedly and apparently stretching away endlessly. Each time light is reflected off the expanding walls it is slightly redshifted, and the* mirrored universe *that stretches away has a progressively increasing redshift in the same way as the real universe. The mirrored universe faithfully depicts conditions as they were in the past, and when we look out in space we also look back in time as in the real universe.*

Why should a creator make an entire universe when a cosmic box suffices to reproduce all known diversity? Perhaps, for reasons of cosmogenic economy, we live in a mirrored universe, and what we believe to be a vast universe stretching away without apparent limit is actually nothing but multiple reflections that depict our own region in the past! The mirrored universe restores the wall-like Keplerian cosmic edge, and we must reject it on the grounds that space cannot terminate at a boundary. Apparently, cosmic plentitude exists because cosmic edges cannot.

4 *The big bang is often referred to as an explosion. This can be very misleading. An explosion occurs at a point in space, whereas the big bang embraces the whole of space. In an explosion the gas is driven outward by a pressure gradient, and there is a large difference of pressure between the center of the explosion and the edge of the expanding gas. In the universe there are no gradients of pressure because the pressure is everywhere the same. There is no center and there is no edge. The expression* exploding big bang, *although vivid, conveys the wrong idea and should be avoided. G. C. McVittie, a British cosmologist, wrote in 1974: "Furthermore, a 'bang' suggests that sound waves are emitted and a noise is heard. But again the equations defining the model universe show that no such sound* *waves are produced. In semi-popular expositions of cosmology terms such as 'the big bang hypothesis' or 'the big bang theory' are to be found. If these expressions have any meaning at all, they must be disguised ways of referring to the singular state found in the model universes of general relativity. For all these reasons it is unfortunate that the term 'big bang,' so casually introduced by Hoyle, has acquired the vogue which it has achieved."*

5 *"Once Nernst, the great physiochemist, pondering on the great quandary of the physical sciences that all energy forms of nature seem to be converted eventually into heat and the universe is aging more and more without visible signs of rejuvenation, concocted an ingenius scheme which allowed a reconversion of heat into matter and thus made a periodic universe possible in which the beginning and end of time was eliminated. In his enthusiasm he called up Einstein (with whom he had the best of scientific and the worst of personal relations; the personalities of these two great men of science were utterly clashing) and explained to him how he envisaged the evolution of the world over billions of years, asking his opinion about the theory. Einstein's comment was: 'I was not present'" (Cornelius Lanczos,* Albert Einstein and the Cosmic World Order, *1965). See "The last static steady state universe" in Chapter 15 of this text.*

* *"Physics tells the same story as astronomy. For, independently of all astronomical considerations, the general physical principle known as the second law of thermodynamics predicts that there can be but one end to the universe – a 'heat death' in which the total energy of the universe is uniformly distributed, and all the substance of the universe is at the same temperature. This temperature will be so low as to make life impossible. It matters little by what particular road this final state is reached; all roads lead to Rome, and the end of the journey cannot be other than universal death . . . Thus, unless this whole branch of science is wrong, nature permits herself, quite literally, only two alternatives, progress and death: the only standing still she permits is in the stillness of the grave.*

"Some scientists, although not, I think, very many, would dissent from this last view. While they do not dispute that the present stars are melting away into radiation, they maintain that, somewhere in the remote depths of space, this radiation may be reconsolidating itself again into matter. A new heaven and a new earth may, they suggest, be in process of being built, not out of the ashes of the old, but out of the radiation set free by the combustion of the old. In this way they advocate what may be described as a cyclic universe; while it dies in one place the products of its death are busy producing new life in others.

"This concept of a cyclic universe is entirely at variance with the well-established principle of the second law of thermodynamics, which teaches that entropy must forever increase, and that cyclic universes are impossible in the same way, and for much the same reason, as perpetual motion machines" (James Jeans, The Mysterious Universe, *1930). See "The last static steady state universe" in Chapter 15 of this text.*

6 *The first law of thermodynamics for the cosmic box is*

$$dE + pdV = 0$$

as mentioned in the text. Often we are interested only in energy density, such as the energy in a cubic centimeter. Let $\epsilon = E/V$ be the energy density. The first law then becomes

$$Vd\epsilon + (\epsilon + p)\, dV = 0 \qquad (13.13)$$

In an ordinary gas the pressure p is extremely small compared with the energy density ϵ of matter; the total energy E is therefore unchanged during expansion, and the mass of the gas in the cosmic box stays virtually constant. This is what happens in our everyday experiences: Matter has a mass that is constant because the pressures exerted on it are small compared with its energy density.

Thermal radiation has a pressure p that is equal to one-third its energy density ϵ. Because the volume V of the cosmic box increases as R^3 we find that ϵR^4 is constant during expansion, and therefore

$$\epsilon = \epsilon_0(R_0/R)^4 = \epsilon_0(1 + z)^4$$

where ϵ_0 is the present radiation density. To obtain mass we divide energy by c^2, where c is the speed of light. Let ρ stand for mass density – such as grams per cubic centimeter – and let ρ_r be the mass density of radiation. The mass density of the radiation in an expanding universe is given by the energy density equation, and we have

$$\rho_r = \rho_{r0}(1 + z)^4 \tag{13.14}$$

where ρ_{r0} is the present density. The total mass is density multiplied by volume, and this mass decreases with expansion. If we have in the laboratory a box containing radiation, then as the box expands, its weight will get less. In our everyday experiences we are not familiar with things that alter their mass in this way.

Let ρ_m stand for the mass density of ordinary matter. We know that

$$\rho_m = \rho_{m0}(1 + z)^3 \tag{13.15}$$

where ρ_{m0} is the present density of matter in the universe. The ratio of the radiation and matter densities is

$$\frac{\rho_r}{\rho_m} = \frac{\rho_{r0}}{\rho_{m0}} \times (1 + z) \tag{13.16}$$

and we see that as we go back into the past the radiation density increases faster than the density of matter. The present density of the 3-degree radiation is about 1/3000 of the average density of matter in the universe. Back in the past at a redshift of 3000 the two densities were therefore equal and the temperature was almost 10,000 degrees. Earlier still, at higher temperatures, the radiation was more dense than matter and the universe was radiation dominated. *The universe nowadays is* matter dominated.

FURTHER READING

Harrison, E. R. "The dark night sky paradox." *American Journal of Physics 45,* 119 (February 1977). Discussion of the cosmic box.

SOURCES

Davidson, W. "Local thermodynamics and the universe." *Nature 206,* 249 (April 17, 1965).

Dutta, M. "A hundred years of entropy." *Physics Today,* January 1968.

Harrison, E. R. "Olbers' paradox." *Nature 204,* 271 (October 17, 1964). Introduction of the cosmic-box method of treating radiation in the universe.

Harrison, E. R. "Why the sky is dark at night." *Physics Today,* February 1974.

Jeans, J. *The Mysterious Universe.* Cambridge University Press, Cambridge, 1930. Reprint. Macmillan, London, 1937.

Klein, M. J. "Thermodynamics and quanta in Planck's work." *Physics Today,* November 1966.

Klein, M. J. "Maxwell, his demon, and the second law of thermodynamics." *American Scientist,* January–February 1970.

Lanczos, C. *Albert Einstein and the Cosmic World Order.* Interscience Publishers, New York, 1965.

McVittie, G. C. "Distance and large redshifts." *Quarterly Journal of the Royal Astronomical Society 15,* 246 (1974).

Plank, M. *The Theory of Heat Radiation.* Dover Publications, New York, 1959.

Tolman R. C. *Relativity Thermodynamics and Cosmology.* Oxford University Press, Clarendon Press, Oxford 1934. See chap. 10, pt. 2.

14 NEWTONIAN COSMOLOGY

And the Coyners are now thinned in this City, yet by their flight from hence and by the turning of clippers to coyners, they seem more numerous in other places where I cannot reach them.
— Newton (1642–1727), *Letter to the Treasury*

STATIC NEWTONIAN UNIVERSE

Until this century it was generally thought that the universe was neither expanding nor contracting but was naturally static (see Figure 14.1). Faith in a static order had been carried over from the Aristotelian universe and remained unshaken even when the theory of gravity was discovered in the seventeenth century. Richard Bentley, a young clergyman who had prepared a collection of his sermons under the title *A Confutation of Atheism,* asked Newton for his comments before their publication. Newton, in one of four famous letters to Bentley, said that in an infinite universe it would be impossible for all matter to fall together into a single large mass; "some of it would convene into one mass and some into another, so as to make an infinite number of great masses, scattered at great distances from one to another throughout all that infinite space."

The theory of gravity reinforced the new belief that the universe is centerless and edgeless, and furthermore made secure the idea of an infinite universe. For, if the universe were finite and bounded by a cosmic edge, it would have a center of gravity; according to Newton, the attraction between its parts would then cause it, contrary to observation, to "fall down into the middle of the whole space, and there compose one great spherical mass," whereas in an infinite universe, without center and edge, there is no preferred direction in which matter might move and so fall into the "middle." Each piece of matter is pulled from all directions with equal forces and stays where it is put. As Newton said later, in the second edition of the *Principia,* "The fixed stars, being equally spread out in all points of the heavens, cancel out their mutual pulls by opposite attractions." The Newtonian universe therefore was stable on the cosmic scale, but unstable in finite regions where local gravity would cause irregularities of matter to condense into heavenly bodies.

WARRING COSMIC FORCES. The Newtonian universe was plagued by two paradoxes that have much in common: the dark night sky paradox and the gravity paradox. We have considered the first; let us take a look here at the gravity paradox.

Again, we occupy any point in a universe that is uniformly populated with stars, and we consider concentric shells of stars, with each shell of equal thickness. In any shell the number of stars is proportional to the square of the radius of the shell, and each star in the shell exerts a gravitational pull that is inversely proportional to the square of its distance. The combined gravitational effect of the stars in the shell is therefore indepen-

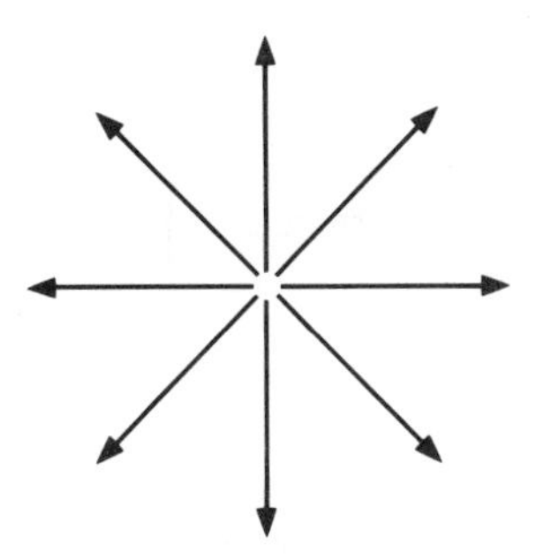

Figure 14.1. The war of cosmic forces. It seemed obvious to everyone that the infinite Newtonian universe was necessarily static. The universe had no center or edge and the whole of space was a container not in the least affected by all the contained matter. Distant matter pulled equally in all directions, and at each point the contending cosmic forces canceled each other out.

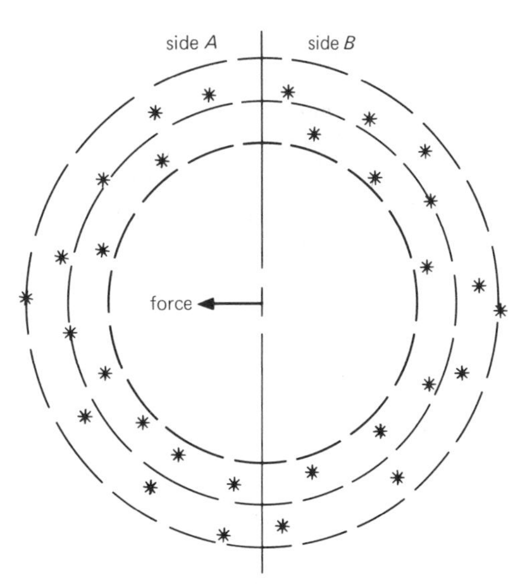

Figure 14.2. Shells of equal thickness that contain stars. Imagine that on side A the stars have on the average one more atom than the stars on side B. Each shell will exert a slight force as shown. In an infinite universe there are an infinite number of shells, and the net force exerted is also infinitely large, even though the force owing to each shell is very small. For the cosmic forces to be in equilibrium, exact isotropy must exist at all points, as was argued by Richard Bentley.

dent of the radius of the shell. The universe is isotropic about the chosen point, and the stars in any one direction have an attraction that opposes the attraction of stars in the opposite direction. We now add up the shells and find that, as their number increases, the opposing forces get stronger, and when the shells are added up to an infinite distance the opposing forces become infinitely great. The implications are disturbing, to say the least, and this situation has frequently been described as paradoxical.

Previously, when discussing the light emitted by stars, we added up shells to the lookout limit. But gravity is not obstructed by stars in the same way as light, and the unimpeded gravitational forces therefore mount up and attain infinite magnitude in a universe of infinite extent. Hence there must be exact isotropy about each point in space. Suppose that the universe is not exactly isotropic, and on one side of the sky each shell contains slightly more matter than on the other side (see Figure 14.2). To be specific, we imagine that on one side of the sky each star contains on the average one atom more than stars on the other side. A single shell now exerts a tiny residual force in one direction, but an infinite number of shells produces an infinite residual force in that direction. It is evident that exact isotropy must exist; otherwise the warring cosmic forces will not balance, and their difference will be infinitely great.

RESOLUTION OF THE GRAVITY PARADOX. In 1872 the German astronomer Johann Zöllner proposed a resolution of the gravity paradox that was a landmark in the history of cosmology. Inspired by Riemann's work on curved space, he suggested that space was curved and finite, so that the total amount of matter in the universe was finite. At each point, in any direction, the gravitational force from all matter in the universe was therefore finite in magnitude. This was a remarkable anticipation of the Einstein universe of 1917. Zöllner did not realize, of course, that we cannot put gravitational fields into curved space, for curved space is already equivalent to gravity, and gravity cannot be used twice.

Toward the end of the last century, Carl Neumann, a German physicist, and Hugo Seeliger, a German astronomer, sought to break the deadlock of infinite cosmic forces by simply abolishing them. They proposed that gravity falls off at large distances faster than the inverse-square law. Gravity was permitted to operate at full strength in local regions but was rendered impotent on the cosmic scale. They introduced in effect a repulsive force that canceled cosmic gravity, and thereby anticipated the *cosmological constant* that was used later by Einstein to create a static universe with general relativity.

Early this century, Carl Charlier, a Swedish astronomer, resuscitated the hierarchical universe in order to resolve the deadlock by again abolishing the cosmic forces. In a hierarchical universe the density of matter becomes progressively less when averaged over larger and larger regions, and it is possible to arrange the hierarchy so that in the limit, on the cosmic scale, the average density of the universe approaches zero. With this arrangement gravity operates effectively in the smaller systems, but gets weaker in the larger systems, and eventually vanishes on the largest of all scales. Charlier was concerned also with the dark night sky paradox, and he showed that in a hierarchical universe in which the cosmic forces vanish, the night sky is also dark because it is not covered with stars. The hierarchical universe slayed two dragons: It ended the war of cosmic forces and created a dark night sky – or so it seemed at the time.

It was often thought, perhaps not surprisingly, that the static Newtonian universe was also in a steady state – everything remained forever the same – and it was taken for granted that the stars could shine endlessly. Streams of light from the stars would therefore inevitably fill space with an ocean of radiation. The realization that matter is a reservoir of limited energy has put an end to the quaint notion that stars have interminably long luminous lifetimes. We saw in Chapter 12 that the night sky is dark and not covered with stars simply because the lookout limit is much greater than the luminous lifetime of stars. The resolution of the gravity paradox follows similar principles.

In Newtonian theory gravity acts simultaneously at all distances, and when a particle changes its position all other particles in the universe respond immediately. The theory of general relativity has taught us that gravity is the dynamic curvature of spacetime that propagates at the same speed as light and cannot therefore act instantaneously at a distance. When the dark night sky paradox was first propounded it was not known that light travels at finite speed; similarly, while the gravity paradox held sway for more than two centuries, it was not known that gravity also travels at finite speed.

The gravitational forces that pull in all directions at every point are produced by matter at great distances, and these forces, we now realize, have taken long periods of time to reach us. In a static universe of infinite age the forces have traveled from infinite distances and have attained infinite magnitude. But in a universe of finite age the forces have traveled from finite distances and are therefore only of finite magnitude. The war of cosmic forces in a universe only 10 billion years old abates and becomes a tussle between gentle forces that are a trillion times weaker than the gravitational pull of the Earth that we experience at its surface. The resolution of the gravity paradox, within the framework of a Newtonian-like universe, is that gravity travels at finite speed and the universe is of finite age. Otherwise, in a universe of infinite age, the slightest irregularities on vast scales will cause matter to move vast distances in the unlimited time available, and the universe will not be in the least uniform.

A hierarchical arrangement of matter, beyond the distance that light travels during the luminous lifetime of stars, does nothing toward resolving the dark night sky paradox. Additionally, a hierarchical arrangement of matter, beyond the distance that gravity travels during the age of the universe, does nothing toward resolving the gravity paradox. Both paradoxes are resolved by a finite propagation speed, which eliminates any need for a hierarchical arrangement.

EXPANDING COSMIC BALL

GENERAL RELATIVITY LEADS THE WAY. The universe is static according to Newtonian laws: Contending cosmic forces pull equally in all directions at every point, and consequently nothing dramatic happens. But according to general relativity the universe is dynamic and is made static only by contrived conditions, and even then is unstable. It was general relativity and not Newtonian theory that opened the door to the realization that the universe is nonstatic. In 1917, Willem de Sitter of Holland approached the door; in 1922, Alexander Friedmann of Russia was the first to enter; in 1927, Georges Lemaître in Belgium also passed through the door; and he was shortly followed by Arthur Eddington in England and Howard Robertson and Richard Tolman in the United States. Before considering the marks that these and other scientists left on cosmology, we must leap ahead to the year 1934.

In that year an unexpected thing happened. The British cosmologists Edward Milne and William McCrea showed that the dynamic laws controlling the universe, which previously had been derived from general relativity by dint of mathematical labor and great skill, could be obtained directly from simple Newtonian theory. This discovery has created a problem in cosmology, and learned discussions are still not conclusive concerning its solution. Why is it that Newtonian theory can now be manipulated so that a static universe is no longer mandatory? Why is it that Newtonian theory, when applied to a uniform universe, yields exactly the same result as general relativity? In all its applications, Newtonian theory is only approximately true, and yet in this most unlikely of all instances it yields the correct answer. We shall return to this problem at the end of the chapter.

AN EXPANDING BALL. Let us use Newtonian ideas, in the manner shown by Milne and McCrea, to construct a dynamic picture of the universe. This picture is often referred to as *Newtonian cosmology*.

We start with a comoving box of matter, but this time it is spherical in shape. We take a step that would have dismayed Newton, who, as master of the mint, would probably have regarded us as little better than coin clippers (see the Reflections section of this chapter). We forget about the matter outside the "cosmic ball" and in effect abolish it. We are left with a ball of matter expanding in empty Euclidean space. The ball has an edge and also a center of gravity, and the gravitational force at each point is easily calculated. We suppose the ball to be of uniform density and to consist of particles in free fall in the gravitational field produced by the ball.

To help us understand what happens we first suppose that the ball is static, of fixed radius; then we imagine a particle leaving the surface and traveling away (see Figures 14.3 and 14.4). If the ball has radius r and mass M, the escape speed v_{escape} is given by

$$(v_{escape})^2 = \frac{2GM}{r}$$

The escape speed is the minimum speed needed by the particle to escape and reach infinity. Any particle moving with the

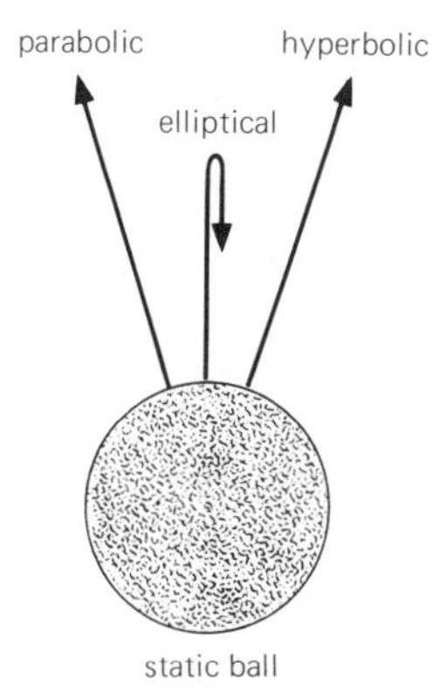

Figure 14.3. Particles are ejected radially from the surface of a self-gravitating ball. According to their initial velocity they follow orbits that are parabolic, elliptical, or hyperbolic.

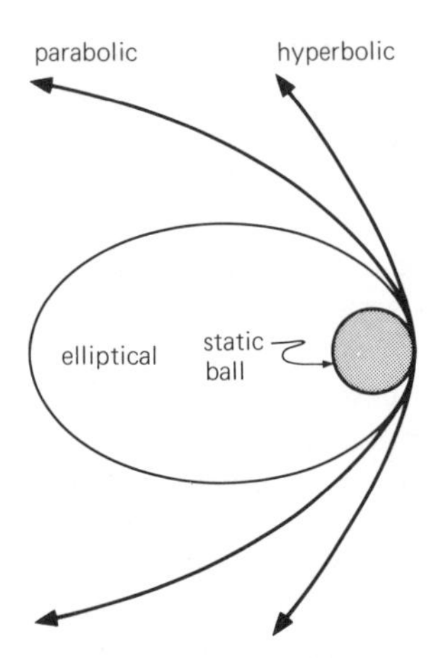

Figure 14.4. Particles are ejected tangentially from the surface of a self-gravitating ball. Here the shapes of the different kinds of orbits are clear.

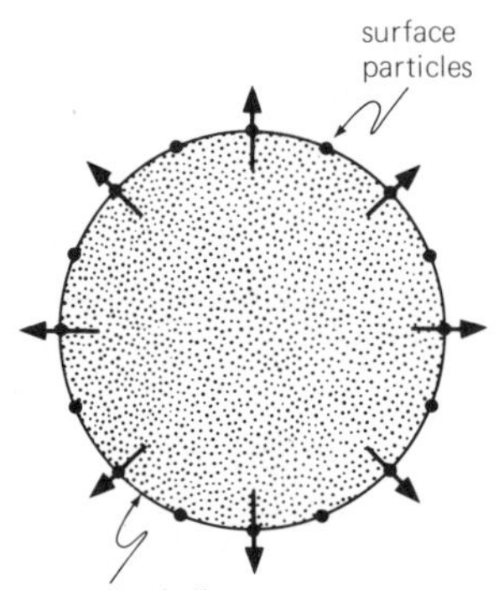

Figure 14.5. If the ball expands its surface follows the particles, which have the same orbits as in the case of the static ball shown in Figure 14.3. This is the "cosmic ball," and all particles in the ball are in free fall in the gravitational field of the ball.

escape speed follows a parabolic orbit. When the particle has less than the escape speed it cannot escape and must eventually turn around and fall back; in this case it follows an elliptical orbit. When the speed is greater than the escape speed the particle follows a hyperbolic orbit. If v is the speed of the particle when it leaves the surface, we have the following conditions:

v equal to v_{escape}:	orbit is a parabola
v less than v_{escape}:	orbit is an ellipse
v greater than v_{escape}:	orbit is a hyperbola

In our model the particle moves radially outward and the orbits are straight lines. In the following we shall regard all velocities as radial in direction.

The strength of the gravitational field of the ball, at any distance, depends on the mass M of the ball and not its radius. We can therefore imagine the ball itself expanding, keeping up with the particle, and the motion of the particle will not change as a result (see Figure 14.5). In other words, the particle we have considered can be a surface particle that remains at the surface of the expanding ball. The ball expands like a cloud of gas, and its particles all follow similar kinds of orbits. When the surface velocity is less than the escape velocity the ball first expands and then later collapses, and when the surface velocity is equal to or greater than the escape velocity the ball expands continually and never collapses.

The total energy, kinetic and gravitational, is constant for each particle:

kinetic energy + gravitational energy
= total energy

For convenience we shall continue to consider only surface particles. For each unit of mass, the kinetic energy is $\frac{1}{2}v^2$ and the gravitational energy is $-GM/r$, where r is the radius of the expanding ball. We therefore have

$$v^2 = \frac{2GM}{r} + \text{constant} \qquad (14.1)$$

where the "constant" is twice the total energy of a unit of mass. When v is equal to the escape velocity the total energy is zero and the constant is therefore zero; when v is less than the escape velocity the total energy is negative (the constant is negative), and the ball lacks sufficient kinetic energy to expand to infinity and therefore collapses back after reaching a maximum radius; and when v is greater than the escape velocity the total energy is positive (the constant is positive), and the ball has more than sufficient energy to expand to infinite size. The behavior of the expanding ball depends therefore on the constant in equation (14.1).

If the constant is zero, all particles follow parabolic orbits and have just enough kinetic energy to reach infinity. If the constant is negative, all particles follow elliptical orbits, without enough kinetic energy to escape, and the ball is gravitationally bound. If the constant is positive, all particles follow hyperbolic orbits; they have more than enough kinetic energy to escape, and the ball is not gravitationally bound.

THE EXPANDING BALL REPRESENTS THE UNIVERSE. We have gained some grasp of the way an expanding ball behaves under the influence of its own gravity. The next step is one of those inspirational flights of the mind that leave us gasping. We suppose that the ball represents the universe. Let us explore the consequences and defer until later a justification for such an extraordinary assumption.

With the aid of the scaling factor R and the velocity–distance law, the simple equation (14.1) transforms into

$$\dot{R}^2 = \frac{8\pi G\rho R^2}{3} - k \qquad (14.2)$$

as shown in the Reflections at the end of this chapter. We have already met $\dot{R}$: It is the rate at which R increases with time. The density (such as grams per cubic centimeter) is denoted by ρ, and in place of the constant of equation (14.1) we have a new constant k that is everywhere the same. All particles of the expanding ball obey equation (14.2). The scaling factor tells us how distances vary with time, and usually we are concerned only with ratios such as R_0/R. That is, we are interested in comparing the values of R at different epochs, and the actual value of R is itself not of primary importance. We now make use of this freedom and adjust the value of R by multiplying it by a constant, so that k has the simplest value possible. When the particle orbits are of the elliptical kind, k has a positive value, and multiplying R by a suitable constant permits us in this case to make k equal to 1. When the particle orbits are of the hyperbolic kind, k has a negative value, and multiplying R again by a suitable constant lets us in this case make k equal to -1. For the values

$k = 0, 1, -1$

the corresponding orbits are parabolas, ellipses, and hyperbolas.

Equation (14.2), which has been obtained by quite simple methods, shows how the ball expands as a dynamic system, and it is interesting to note that only the scaling factor, the density, and the constant k appear in the equation. The radius r has dropped out: Therefore balls of large mass (large r) and small mass (small r) behave in precisely the same way and are governed by the same equation. Even more remarkable is the fact that our equation is identical with that obtained from general relativity. In honor of Alexander Friedmann, who first derived such an equation, it is known as the *Friedmann equation.*

We have used Newtonian theory to derive the Friedmann equation. In our treatment of an expanding ball, a treatment nowadays referred to as Newtonian cosmology, space is flat and static, particles move through this space, and k determines the nature of their orbits. We can take either of two views: We can say that the cosmic ball represents a sample of the universe or that it is indefinitely large and represents the whole universe. Milne and McCrea took the latter view and supposed that the ball represents adequately the entire universe. We shall return to this controversial subject shortly.

FRIEDMANN UNIVERSES. In modern cosmology, or *relativity cosmology,* as it is called, the curvature of space is $K = k/R^2$, where k is the curvature constant. In the relativity and Newtonian pictures k has always one of the three values 0, 1, -1. The meaning of these different values in the two pictures is shown in Table 14.1. In brief, when $k = 0$, orbits are parabolas in the Newtonian picture, and expanding space is flat in the relativity picture; when $k = 1$, orbits are ellipses, and expanding space is spherical; and when $k = -1$, orbits are hyperbolas, and expanding space is hyperbolic. In the

Table 14.1. *Comparison of Newtonian and relativistic cosmologies*

Curvature constant k	Newtonian	Relativistic
1	elliptical orbits	spherical space
0	parabolic orbits	flat space
-1	hyperbolic orbits	hyperbolic space

Newtonian picture k distinguishes between orbits, whereas in the relativity picture it distinguishes between geometries; and orbits and geometries have names that are not entirely dissimilar. The orbits and geometries also have similar topology: Parabolic orbits and flat space (which was once occasionally called parabolic space) are both open; elliptical orbits and spherical space (which was sometimes known as elliptical space) are both closed; and hyperbolic orbits and hyperbolic space are both open.

The way the scaling factor changes with time tells us how the universe expands. Because there are many possible ways in which R can vary, there are potentially many universes, or models of the Universe, and one aim of cosmology is to find the model that best fits the observations. In the present treatment there are basically three kinds of universe that correspond to the three values of the curvature constant. These Friedmann universes are as follows:

The first is that in which $k = 0$ (see Figure 14.6). This kind of universe has flat expanding space and is infinite and unbounded. It expands continually, is of the bang–whimper type, and endures for an infinite period of time in the future. It is the simplest of all known universes, but was not considered by either Friedmann or Lemaître, and was first proposed by Einstein and de Sitter in 1932. It is known as either the Friedmann universe of zero curvature or the Einstein–de Sitter universe. In the Newtonian picture it corresponds to a ball that expands continually, and its free-falling particles follow parabolic orbits and have velocities equal to their escape velocity.

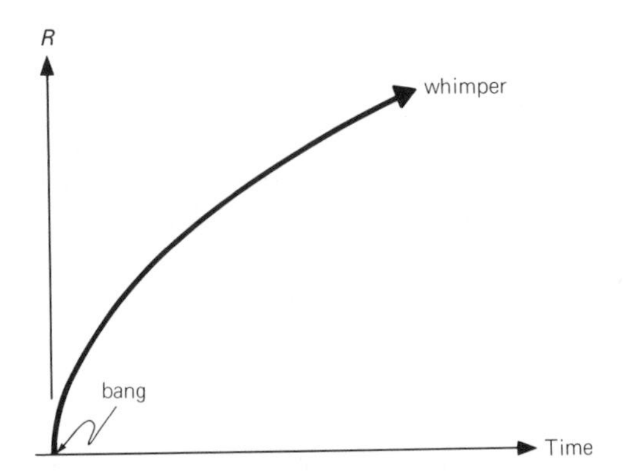

Figure 14.6. Friedmann universe of zero curvature ($k = 0$) and zero cosmological constant, also known as the Einstein–de Sitter universe.

The second is that in which $k = 1$ (see Figure 14.7). This kind of universe has spherical expanding space and is finite and unbounded. It expands to a maximum size and then collapses, is of the bang-bang type, and endures for only a finite period of time. This universe was discovered by Alexander Friedmann in 1922 and rediscovered by Georges Lemaître in 1927. In the Newtonian picture it corresponds to a ball that expands and then collapses, and its free-falling particles follow elliptical orbits and have velocities less than their escape velocity.

The third is that in which $k = -1$ (see Figure 14.8). This kind of universe has hyperbolic expanding space and is infinite

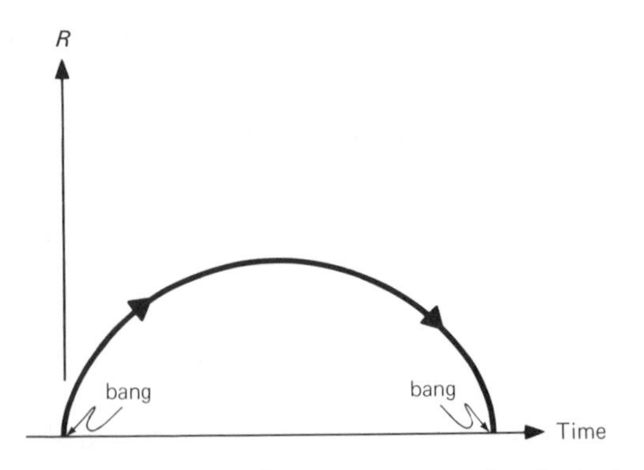

Figure 14.7. Friedmann universe of spherical space ($k = 1$) and zero cosmological constant.

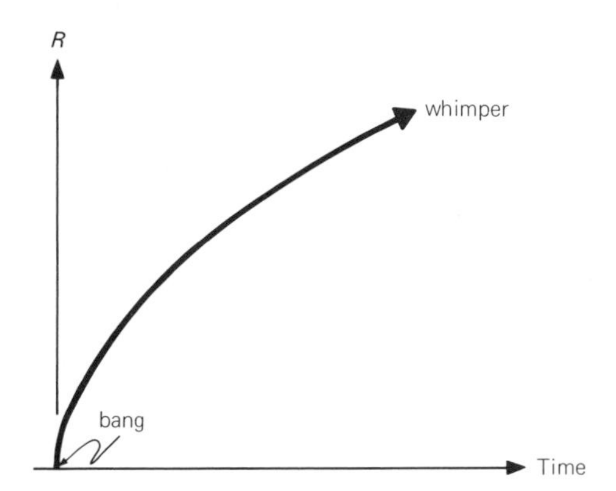

Figure 14.8. Friedmann universe of hyperbolic space ($k = -1$) and zero cosmological constant.

and unbounded. It expands continually, is of the bang–whimper type, and endures for an infinite period of time in the future. This universe was discovered by Friedmann in 1924 and was investigated in 1932 by the German cosmologist Otto Heckmann. In the Newtonian picture it corresponds to a ball that expands continually, and its free-falling particles follow hyperbolic orbits and have velocities greater than their escape velocity.

WHY DOES NEWTONIAN COSMOLOGY GIVE THE RIGHT ANSWER?

All are but parts of one stupendous whole.
— Alexander Pope, *An Essay on Man*

In the Newtonian picture our cosmic ball expands within flat and ordinary space, whereas in the general relativity picture the universe consists of expanding space. These two pictures, despite their marked difference, are governed by the Friedmann equation.

We have already noticed that the radius of the cosmic ball is apparently unimportant. A small sphere the size of a golf ball and a large spherical cloud millions of light years in radius behave in the same way when they have similar densities and curvature constants. (In the Newtonian picture k is still called the curvature constant.) It is tempting therefore to go the whole way and imagine that the cosmic ball is of unlimited size and represents an unbounded universe. This was the step taken by Milne and McCrea when they first introduced Newtonian cosmology.

Later, in 1954, David Layzer at Harvard University revived the argument that gravitational forces are impotent in a universe without center and edge. He showed that the deadlock of cosmic forces in an infinite universe cannot be resolved in the simple way adopted by Milne and McCrea. One possible way of evading the dilemma, which has since been suggested, is to regard the universe in the Newtonian picture as finite but nonetheless vast in size. The universe has then a center of gravity, but its edge is so far away that for most purposes its existence need not be of concern. There are several objections to this way of explaining why Newtonian cosmology gives the same answer as relativity cosmology. The first is the distasteful fact that the universe is supposedly surrounded by an edge and is embedded in infinite space that is empty beyond the edge. With such a "clipped" cosmos we are back again to the Stoic island universe. Another objection is that matter moves through space in the Newtonian picture, and therefore at large distances the expansion velocity must exceed the velocity of light. Motion through space faster than light flatly contradicts relativity theory. An indefinitely large cloud, with its edge expanding faster than the velocity of light, leads to contradiction and not to reconciliation between the Newtonian and relativity pictures of the universe. Furthermore, we are left mystified about why they have dynamic laws in perfect agreement.

The other way of understanding Newtonian cosmology is to regard the cosmic ball as representing only a small part of the expanding universe. The Friedmann equation is for an idealized universe, entirely free of irregularity, and in an idealized universe we are at liberty to study a representative part that is as small as we please. The sample may have a radius as small as 1 light year, or 1 kilometer, 1 centimeter, or even less. Smallness is very important, because there are two requirements to be satisfied if Newtonian theory is to give the same answer as general relativity. The first is that the expansion velocity must be small compared with the velocity of light. The expanding ball obeys the velocity–distance law, and when it is small its expansion velocity is also small. The second is that gravity must be weak. The square of the escape velocity is a measure of the strength of gravity, and when the ball is small the escape velocity is small compared with the velocity of light, and gravity is therefore weak. Newtonian and relativity theories must come into exact agreement when the ball is very small. It seems that when Newtonian theory is used in cosmology we must always assume that the region under consideration is extremely small compared with the Hubble length. This is also the condition that justifies the assumption that space is flat in the sample region.

This way of understanding Newtonian cosmology, of regarding the cosmic ball as a tiny sample of an idealized universe, is more satisfying than the clipped Stoic cosmos that violates relativity theory. From our experience with the cosmic box we know that in small regions space is essentially flat, and also there is no need to distinguish between motion through space and comoving motion in expanding space. Newtonian cosmology is yet another example of how a cosmic box helps us to understand the universe. Caution, however, must be exercised; when the universe is not homogeneous, and even when it is homogeneous but not isotropic, Newtonian cosmology may give results not in agreement with relativity cosmology.

The Newtonian picture helps us to understand the relativity picture of the universe. Elementary accounts of cosmology were confusing prior to the discovery of Newtonian cosmology. Readers gained the impression that the universe is expanding because spacetime is endowed with a mysterious power. The Newtonian picture now makes it clear that the universe does not expand because spacetime insists that it must; it expands for the same reason the cosmic ball expands: Both are in a state determined by their initial conditions, and both were initially launched into a state of expansion. They could both just as easily be in a state of contraction, and the equations that govern their dynamics show no preference for expansion rather than contraction.

REFLECTIONS

1 *"It seems to me, that if the matter of our sun and planets and all the matter of the universe, were evenly scattered throughout all the heavens, and every particle had an innate gravity towards all the rest, and the whole space throughout which this matter was scattered was but finite, the matter on the outside of this space would, by its gravity tend towards all the matter on the inside, and, by consequence, fall down into the middle of the whole space, and there compose one great spherical mass. But if the matter was evenly disposed throughout an infinite space it could never convene into one mass; but some of it would convene into one mass and some into another, so as to make an infinite number of great masses, scattered at great distances from one to another throughout all that infinite space" (Isaac Newton, letter to Richard Bentley, 1692).*

* *"But to this we reply: that unless the very mathematical center of gravity of every system be placed and fixed in the very mathematical center of the attractive power of all the rest, they cannot be evenly attracted on all sides, but must preponderate in some way or other. Now he that considers, what a mathematical center is, and that quantity is infinitely divisible, will never be persuaded, that such an universal equilibrium arising from the coincidence of infinite centers can naturally be acquired or maintained" (Richard Bentley,* A Confutation of Atheism, *1693).*

2 *Why was the Newtonian universe static?*

* *Discuss the gravity paradox.*

* *If the redshift of the nebulae had been discovered in the nineteenth century, how would the expansion of the universe have been explained?*

* *How should we interpret Newtonian cosmology?*

* *Discuss the three Friedmann universes.*

3 *Now that Milne and McCrea have shown that Newtonian theory can be used in cosmology, the whole subject looks deceptively simple – so simple that Newton is sometimes faulted for not having predicted the expanding universe. (Olbers is also blamed for not making this prediction by those who believe that the night sky is dark because of expansion.) I think Newton was wiser than the people who make such remarks. The Newtonian laws are incapable of breaking the deadlock of cosmic forces in an infinite universe. These same laws, however, spring to life, and the force at each point is no longer indeterminate, in the cosmic-ball picture. We then stumble on equations that with hindsight, gained from general relativity, are recognized to be correct. But the cosmic-ball picture is not the Newtonian universe, and Newton would never have adopted it and reversed history by restoring the cosmic center and edge. Newton and his illustrious successors were not in the least willing to consider a clipped universe. It is general relativity that justifies and makes possible Newtonian cosmology in which the cosmic ball is an isolated but representative sample of an unbounded universe. These comments should dispel any lingering thought that Newton could have predicted an expanding universe.*

* *Coin clippers were people who snipped pieces of metal off the edges of coins in the days when coins had their face value and were not made of base metal. Coin clippers and counterfeiters (or "coyners," as they were known in Newton's day) were punished with extreme severity when caught. As master of the mint, Newton had the task of protecting the coinage from the depredations of clippers and coiners. He probably would have regarded modern Newtonian cosmologists as little better than clippers.*

* *It has been suggested that in Newtonian cosmology we can regard the universe as a Stoic cosmos surrounded by infinite and empty space. In such an arrangement a large number of inhabitants will live near the edge and observe that all places are not alike. When the cosmos is large, however, the number of inhabitants living near the edge is a small fraction of the total, and most of them will therefore observe homogeneity. The larger the cosmos the smaller the fraction of inhabitants who know that the universe is not totally homogeneous. Is this a satisfactory argument for retaining the cosmic edge? "However large the system there must be observers within sight of the boundary. In this respect the theory seems to have more loose ends than are justified, even in a theory whose principal function is merely one of suggestion" (J. D. North,* The Measure of the Universe, *1965).*

4 *Anything that travels faster than light, such as Newtonian gravity, can be made to do impossible things and is therefore ruled out by modern physics. Let us imagine that in the fictional world of* Star Wars *everybody has faster-than-light guns. To make the point clear, we shall suppose that the guns shoot projectiles at infinite speed. Albert (A) and Bertha (B), in separate spaceships, are fleeing side-by-side at very high speed from enemy X. The enemy fires and destroys A. In X's space the act of firing and the destruction of A are simultaneous events. But they are not simultaneous in the space of A and B, and A's destruction occurs before X fires. Bertha sees Albert destroyed, and fires back, and is thus able to destroy the enemy X before X has fired. In this way she saves Albert from destruction. We have here a situation where an effect, after it has occurred, is canceled by the elimination of the cause. This violates what physicists call causality. The absurd situation we have described becomes farcical if there is a second enemy Y who is in the neighborhood of X. Y sees X destroyed, and* *fires back and eliminates B; A sees B destroyed, and fires back and eliminates Y; X sees Y destroyed, and fires back and eliminates A; and the struggle goes on, with each side creating effects and eliminating their causes.*

With faster-than-light travel, one can journey forward into the future. Suppose that we travel instantaneously (in Earth's space) to a distant star. On arrival, we turn around, and accelerate to a speed close to that of light. Our space is now different from that of Earth, and by entering into faster-than-light travel, we shall arrive back at Earth somewhere in the future. We can also travel back into the past; in this case, while at the distant star, we accelerate away from Earth to a speed close to that of light, and then return instantaneously.

5 *It was the custom once to refer to Euclidean geometry as parabolic to distinguish it from elliptical and hyperbolic geometries. Elliptical and spherical spaces have similar geometry, but are different in this sense: The surface of a sphere is analogous to spherical space, whereas the surface of a hemisphere is analogous to elliptical space (see Figure 14.9). The antipodal hemisphere is regarded as the same as one's own hemisphere in elliptical space, and when a receding object reaches halfway to the antipode, it disappears and reappears on the other side of the universe, and is seen approaching.*

6 *"It is therefore clear that from the direct data of observation we can derive neither the sign nor the value of the curvature, and the question arises whether it is possible to represent the observed facts without introducing a curvature at all" (Albert Einstein and Willem de Sitter, "On the relation between the expansion and the mean density of the universe," 1932). It was in this paper that the Einstein–de Sitter universe of zero curvature was first proposed.*

7 *The Friedmann equation (14.2) is derived from equation (14.1) in the following way. Let the radius of the cosmic ball at some instant be r_0, when the scaling factor has a value R_0. At any other instant the radius is given by*

$$r = r_0\left(\frac{R}{R_0}\right)$$

and is proportional to R. The velocity of the surface is expressed as

$$v = r_0\left(\frac{\dot{R}}{R_0}\right)$$

where $\dot{R}$ is the rate at which R increases with time. The mass of the ball is equal to its volume multiplied by the density ρ, and hence $M = 4\pi r^3\rho/3$. The energy equation (14.1) can now be written in the form

$$\dot{R}^2 = \frac{8\pi G\rho R^2}{3} + \frac{R_0^2}{r_0^2} \times \text{constant}$$

It is perhaps not immediately obvious, but the second term on the right, containing the constant, has the same value for all particles in the ball, because the total energy of any particle is proportional to the square of its distance r_0 from the center of the ball. We therefore write:

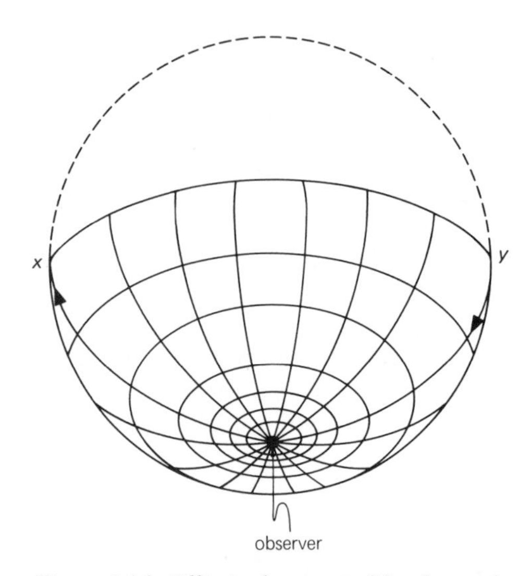

Figure 14.9. Elliptical space is like the surface of a hemisphere, and the antipode does not exist. Points x and y, halfway to the antipode in opposite directions, are equivalent. A particle approaching x reappears at y as an approaching particle. Elliptical space is nowadays regarded as a historical curiosity.

$$\frac{R_0^2}{r_0^2} \times \text{constant} = -k$$

where k is a new constant and has the same value for all particles in the ball. The negative sign is included for conventional reasons.

The equation for the expanding ball now takes the rather simple form

$$\dot{R}^2 = \frac{8\pi G\rho R^2}{3} - k$$

of equation (14.2), and because the mass M is constant during expansion we must remember that ρR^3 does not change. This basic equation is obeyed by all particles in the ball. By adjusting or "normalizing" the value of R, we can arrange for k to have the simplest values possible, as shown in the text. We then have k equal to 0, 1, or −1, and it becomes the curvature constant.

Let us be a little more thorough and start from the equation of motion which says that the acceleration of a particle is equal to the gravitational field. The acceleration is given by the expression

$$\text{acceleration} = r_0\left(\frac{\ddot{R}}{R_0}\right)$$

where r_0 is the distance of the particle from the center when the scaling factor has the value R_0, and $\ddot{R}$ is just the rate of change of $\dot{R}$, which is the rate of change of R itself. The gravitational field that is produced by the mass M interior to the position r of the particle is expressed by

$$\text{gravity} = -\frac{GM}{r^2} = -\frac{4\pi G\rho r}{3}$$

where ρ is the density of matter. On equating the acceleration and the gravitational field, and using $r = r_0(R/R_0)$, we find

$$\ddot{R} = -\frac{4\pi G\rho R}{3} \tag{14.3}$$

which is the acceleration equation. This can be integrated to give

$$\dot{R}^2 = \frac{8\pi G\rho R^2}{3} - k \tag{14.4}$$

where ρR^3 is unchanging, and R is adjusted so that the constant k has the value 0, 1, or −1, as discussed before. Equations (14.3) and (14.4) are the Friedmann equations when the cosmological constant is ignored.

The gravitational field exerts an attractive force that is proportional to the distance r, as we have seen. Any other force proportional to the radius r will have an effect similar to that of gravity. Einstein introduced a new force, equal to $\Lambda r/3$, where Λ is the cosmological constant. *When Λ is positive the new force opposes gravity and is equivalent to a cosmic repulsion, and when Λ is negative the force reinforces gravity and is equivalent to a cosmic attraction. We now take the cosmological constant into account and repeat the procedure used above, obtaining*

$$\ddot{R} = -\frac{4\pi G\rho R}{3} + \frac{\Lambda R}{3} \tag{14.5}$$

$$\dot{R}^2 = \frac{8\pi G\rho R^2}{3} + \frac{\Lambda R^2}{3} - k \tag{14.6}$$

Equations (14.5) and (14.6) were first used by Alexander Friedmann and later discovered independently by Abbé Georges Lemaître; they are known generally as either the Friedmann or the Friedmann–Lemaître equations. Although Einstein introduced the cosmological constant he did not derive these cosmological equations for an expanding universe. Einstein in 1923 in fact criticized Friedmann's discovery, and it is noteworthy that neither Friedmann nor Einstein thought of relating this dynamic picture to the receding nebulae that were then known to astronomers. In 1927, Lemaître said, "We have found a solution such that . . . the receding velocities of extragalactic nebulae are a cosmical effect of the expansion of the universe."

8 *All regions of a uniform universe expand in the same way – "all are but parts of one stupendous whole" – and when we know how any one small sample expands, we also know how the whole universe expands. Newtonian cosmology tells us how a relatively small region behaves, and we then know how all regions behave. But we have yet to explain why a small sample of the universe, when transplanted to an empty*

Euclidean space, continues to expand by itself as if it were still part of the universe. This is not an easy matter to understand.

Newtonian and relativity theories of gravity are founded on differential equations; in the former theory it is the gravity potential and in the latter theory it is the metric coefficients that change with time and vary from place to place. The Newtonian gravity potential is just the negative value of the binding energy per unit of mass, and for a body such as a star it is everywhere of finite value. But for the whole Newtonian universe it is everywhere of infinite magnitude because of the infinite distribution of matter. This is another way of stating the gravity paradox: The potential is infinite and the cosmic forces are therefore indeterminate. On the other hand the metric coefficients determine the curvature, and unlike the gravity potential, they are finite at every point. According to Newtonian theory gravity is an active force within passive space and time, and according to general relativity gravity is the dynamic activity of space and time. In the Newtonian universe the cosmic forces produced by distant matter are indeterminate (or cancel each other out) and nothing ever happens, but in the relativity universe the curvature at each point is specific and has definite dynamic properties.

It comes to this: In a small region of a uniform universe the metric coefficients can be simplified and made equivalent to the Newtonian gravity potential, provided that the small region is regarded as isolated and embedded in empty Euclidean space. The Einstein equation, when applied to the small region, is the same as the Newtonian equation when that region is isolated as if it were a cosmic ball.

FURTHER READING

Bondi, H. *Cosmology*. Cambridge University Press, Cambridge, 1960.

Callan, C., Dicke, R. H., and Peebles, P. J. E. "Cosmology and Newtonian mechanics." *American Journal of Physics 33,* 105 (February 1965).

Hoskin, M. A. "Newton, providence and the universe of stars." *Journal for the History of Astronomy 8,* 77 (1977).

Newton, I. *Four Letters to Richard Bentley*. See M. K. Munitz, ed. *Theories of the Universe: From Babylonian Myth to Modern Science.* Free Press, Glencoe, Ill., 1957.

Tinsley, B. M. "The cosmological constant and cosmological change." *Physics Today,* June 1977.

SOURCES

Alexander, H. G., ed. *The Leibniz–Clarke Correspondence.* Philosophical Library, New York, 1956.

Einstein, A., and Sitter, W. de. "On the relation between the expansion and the mean density of the universe." *Proceedings of the National Academy of Sciences 18,* 213 (1932).

Friedmann, A. "On the curvature of space." Trans. B. Doyle, in *A Source Book in Astronomy and Astrophysics, 1900–1975.* Eds. K. R. Lang, and O. Gingerich. Harvard University Press, Cambridge, Mass., 1979.

Landsberg, P. T., and Evans, D. A. *Mathematical Cosmology: An Introduction.* Oxford University Press, Clarendon Press, Oxford, 1977. A technical treatment at the undergraduate level using Newtonian theory.

Layzer, D. "The significance of Newtonian cosmology." *Astronomical Journal 59,* 168 (August 1954).

Lemaître, G. "A homogeneous universe of constant mass and increasing radius accounting for the radial velocity of extra-galactic nebulae." *Monthly Notices of the Royal Astronomical Society 91,* 483 (1931). Published in 1927 and translated into English in 1931.

McCrea, W. H. "On the significance of Newtonian cosmology." *Astronomical Journal 60,* 271 (1955).

McCrea, W., and Milne, E. "Newtonian universes and the curvature of space." *Quarterly Journal of Mathematics 5,* 73 (1934).

Milne, E. "A Newtonian expanding universe." *Quarterly Journal of Mathematics 5,* 64 (1934).

North, J. D. *The Measure of the Universe: A History of Modern Cosmology.* Oxford University Press, Clarendon Press, Oxford, 1965.

Peebles, P. J. E. *Physical Cosmology.* Princeton University Press, Princeton, N.J., 1971.

15 THE MANY UNIVERSES

Hereafter, when they come to model Heav'n
And calculate the stars: how they will wield
The mighty frame: how build, unbuild, contrive
To save appearances . . .
— Milton (1608–74), *Paradise Lost*

STATIC UNIVERSES

THE EINSTEIN UNIVERSE. By great contriving, Einstein in 1917 created a static universe with the theory of general relativity. In this universe, as in all those we shall discuss, all places are alike and matter is distributed with uniform density. Space and time in the new theory of general relativity had at last been awakened from the dead and had become active participants in the world at large. Einstein, believing that the universe was static, had to tranquilize their dynamic urgency by inventing a new counteracting agency.

In his 1917 paper "Cosmological considerations on the general theory of relativity," Einstein said, "I shall conduct the reader over the road that I have myself traveled, rather a rough and winding road, because otherwise I cannot hope that he will take much interest in the result at the end of the journey. The conclusion that I shall arrive at is that the field equations of gravitation which I have championed hitherto still need a slight modification." The modification referred to was the introduction of the cosmological constant Λ. When this new constant Λ is positive it acts as a force of repulsion that opposes gravity; it reduces the dynamic effect of gravity but not the curvature of space. A universe made static at one moment is not necessarily static at earlier and later moments. To ensure that the universe remains static, in a state of equilibrium, the curvature of space must be positive. The static Einstein universe therefore has spherical space: It is closed and finite, and contains a mysterious Λ force that offsets the attraction of gravity.

When distances are measured in light travel time, the radius of curvature of space in the Einstein universe is the scaling factor R (see Figure 15.1). The distance around the universe, or the circumnavigation time of light, is $2\pi R$, and the antipode of an observer, or point on the opposite side of the universe, is at a distance πR. Such a universe, when ideally smooth, acts as a giant optical lens. A body moving away appears at first to get smaller in the usual way; when halfway to the antipode, however, it ceases to get smaller, and thereafter as it recedes further it appears to get larger. All objects in the antipodal region are seen imaged as if they were close by in the local region (see Figure 15.2). Persons at the antipode see us as if we were close to them and we see them as if they were close to us. Because light circumnavigates the cosmic globe we also see ourselves from behind.

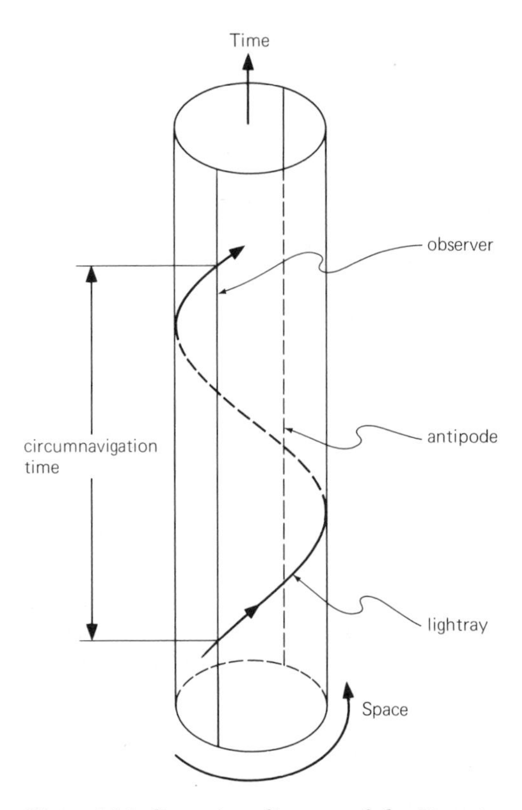

Figure 15.1. Spacetime diagram of the Einstein static universe, showing only one of the three dimensions of space. As a lightray moves in space it advances also in time and describes a helical path on the surface of a cylinder of radius R.

The time light takes to travel once around the Einstein universe is given by

$$\text{circumnavigation time} = 2\pi R = (\pi/G\rho)^{1/2} \quad (15.1)$$

where G is the universal constant of gravity and ρ is the average density of matter. When the density is measured in grams per cubic centimeter the circumnavigation time in hours is $2/\sqrt{\rho}$. Water has a density of 1 gram per cubic centimeter, and in a universe filled with transparent water the time taken by light to circulate once is 2 hours. In this "hydrocosmos," which is smaller than the Solar System and has a radius of curvature R of 20 light minutes, antipodal objects are seen as they were 1 hour ago and observers see themselves as they were 2 hours ago.

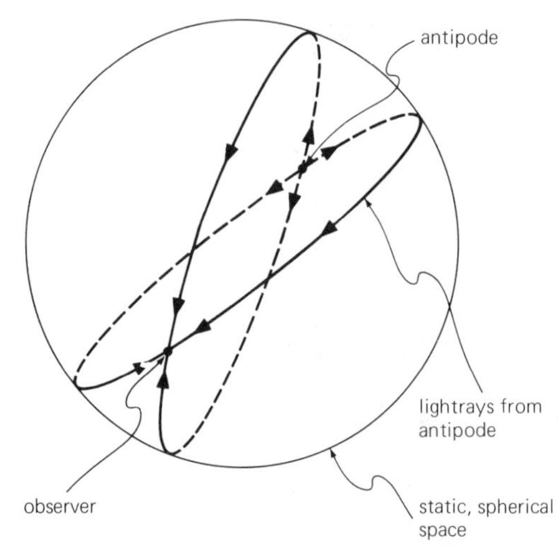

Figure 15.2. The Einstein static universe of spherical space. Lightrays from the antipode first diverge and then converge, and we see objects at the antipode as if they were close to us.

Light keeps on circulating and the inhabitants are continually reminded of what they were doing 2 hours ago, 4 hours ago, 6 hours ago and so on. These Einsteinian creatures see their past antics in graphic detail and are deprived of the convenience of short memories. The circumnavigation time in an Einstein universe filled with a gas of density equal to that of our atmosphere is slightly more than 60 hours; and when the density is considerably less, the inhabitants observe only the ghostly performances of their ancestors.

The distance around spherical space of radius R is $2\pi R$, but the volume of spherical space is $2\pi^2R^3$ and not the more familiar $4\pi R^3/3$. The total mass M of the universe is this volume multiplied by the density; thus, $M = 2\pi^2R^3\rho$, and we find

$$R = \frac{2GM}{\pi c^2} \quad (15.2)$$

When the radius is measured in light years and the mass is measured in solar masses, this gives

$$R = 5 \times 10^{-14}M$$

and in a universe of 10^{10} light years' radius

the mass is 2×10^{23} times that of the Sun, or a trillion times that of the Galaxy. The radius of a black hole is $2GM/c^2$, and the radius of an Einstein universe of the same mass is therefore $1/\pi$ times smaller. Although it is sometimes said that closed universes are black holes, this is misleading because there is no external space and there are no external observers to say, "Look at that universe over there."

The Einstein universe is unstable, as was shown by Arthur Eddington in 1930, and its inhabitants must tread on tiptoe and speak in hushed voices (see Figure 15.3). It stands on the edge of a razor: If nudged very slightly one way, gravity begins to dominate and it collapses into a big bang in a time approximately equal to a circumnavigation time; if nudged very slightly the other way, the repulsive Λ force begins to dominate and it inflates and develops into a whimper universe. If somebody lights a match it will start collapsing, and if some radiation is absorbed it will start expanding.

THE LAST STATIC STEADY STATE UNIVERSE. Static universes are motionless in the cosmic sense: They neither expand nor contract, although the objects they contain may evolve and have peculiar motions. Steady state universes, on the other hand, never change their appearance, and what happens now has always happened and will always happen. A steady state universe, in which the endless vistas of the future perpetuate those of the past, obeys what some of its inhabitants out of sheer boredom might call the "monotonous cosmological principle."

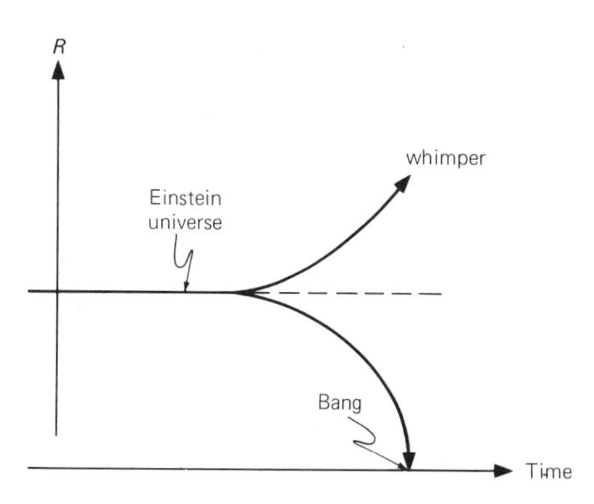

Figure 15.3. The Einstein universe is unstable; when disturbed, it either collapses or expands.

A static universe in a steady state is cosmically motionless, and the contents, on the average, never appear to change. Universes of this kind are not uncommon in the history of cosmology. They seem to be most popular in great empires that have gone into decline, in which men and women desire the perpetuity of past glory, and they are also keenly favored by aristocracies whose members are of the umpteenth generation. The last static steady state universe, strange to say, was conceived at the University of Chicago in 1918 and elaborated in the 1920s by the astronomer William MacMillan (1861–1948).

The foremost problem that besets any static universe in a steady state is the eternal brightness of the stars. The radiation pouring out from stars accumulates in space, and the night sky cannot therefore be dark. It was this old problem, the puzzling darkness of the night sky, that prompted MacMillan to think of the idea that atoms are "generated in the depth of space through the agency of radiant energy." He proposed the theory that stars are formed in the usual way out of interstellar gas; they then evolve and over a long period of time slowly radiate away their entire mass. Out in the depths of space, in some unknown way, starlight is slowly reconstituted back into atoms of matter. The interstellar gas, continually replenished with newborn atoms, condenses to form new stars that in turn melt away into radiation, thus maintaining a perpetual steady state. In one fell swoop, MacMillan was thus able to conserve energy and explain the darkness of the night sky within the framework of an immaculate universe.

MacMillan's "perpetual motion" universe was adopted by the physicist Robert Millikan (1868–1953), who is famous for having in 1905 measured the charge of an electron by studying the motion of electrically charged water droplets (later he used oil droplets). Millikan believed that cosmic rays, discovered in 1911 by Victor Hess, were the "birth cry" of newly created matter in the depths of space and were proof that "the Creator is still on the job."

We know now that a star radiates away only a small fraction of its mass in a luminous lifetime and cannot therefore be regenerated in the way proposed by MacMillan. A star has also a fixed baryon number, and no matter how much it radiates – at the expense of its nuclear and gravitational energy – the number of baryons (in this case nucleons) must remain almost constant, and the star cannot transform entirely into radiation. (A black hole is different and is quite mysterious in this respect.) We also know that matter cannot be recycled in the ingenious fashion proposed by MacMillan (an idea that seems also to have occurred to Otto Nernst). When radiation is very energetic it is capable of creating particles and antiparticles, not just particles by themselves, and if the particles and antiparticles are not separated they will annihilate each other and be converted back into radiation. Starlight in space is far too weak to create particles and antiparticles. Although the total energy in MacMillan's static universe is conserved, the total entropy is not conserved, and therefore the universe cannot be in a steady state. In a static and steady state universe of constant entropy everything is in thermal equilibrium, and there is no arrow of time, no way of distinguishing the past from the future.

The last of the static steady state universes enjoyed a certain amount of popularity until the 1930s, and then, confronted with the mounting evidence in favor of an expanding universe, it quietly faded away.

DE SITTER UNIVERSE

AN EXPANDING VACUOUS STEADY STATE UNIVERSE. The de Sitter universe, proposed in 1917 – the same year as the Einstein static universe – was at first thought to have diminished the status of Einstein's cosmological theory. It consists of flat space and is slightly absurd in the sense that it contains no matter. An empty universe of Euclidean space should not exhibit unusual properties, and yet it does, when the Λ force is not zero. This is seen by putting the curvature K and the density ρ both equal to zero in the Friedmann–Lemaître equations (see the Reflections at the end of this chapter). We then find that space expands and that the Hubble term is $H = (\Lambda/3)^{1/2}$ and is always constant. The repulsive effect of the Λ force causes space to expand with a constant acceleration of $q = -1$, and the scaling factor increases as shown in Figure 15.4.

The de Sitter universe is in a steady state and nothing ever changes; the Hubble and deceleration terms are constant, and there is no contained matter that is diluted by expansion. The *steady state universe* label is commonly used when referring to the model proposed by Bondi, Gold, and Hoyle that expands in exactly the same way as the de Sitter universe, but contains matter that is replenished by continuous creation. Let us not forget, however, that other universes, such as the de Sitter and MacMillan universes, are also in a steady state.

The Einstein universe, having matter but no motion, and the de Sitter universe, having

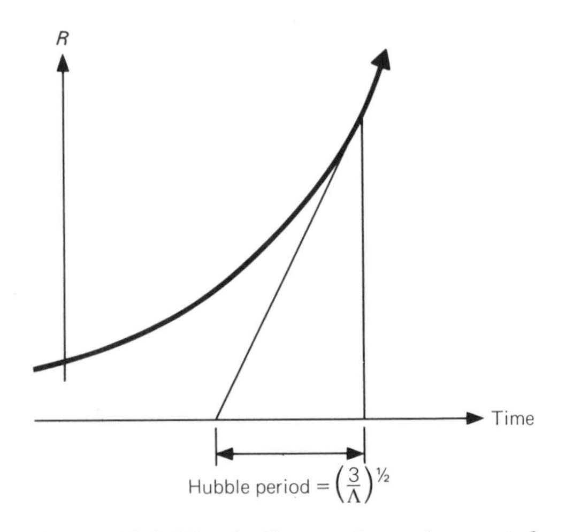

Figure 15.4. The de Sitter universe has an infinite past and an infinite future, and it accelerates at a constant rate of $q = -1$.

motion but no matter, were at first the leading cosmological models; then, in 1927, Lemaître rediscovered the equations originally formulated by Friedmann, and cosmology entered a new era.

FRIEDMANN UNIVERSES

ALEXANDER FRIEDMANN. In the *Friedmann universes,* or the Friedmann models of the Universe, the Λ force is ignored and assumed to be zero. There are three basic types of Friedmann universe, corresponding to the values 0, 1, and -1 of the curvature constant k, as discussed in the previous chapter.

Alexander Friedmann (1888–1925), born into a musical family, was a brilliant scholar with wide interests who became professor of mathematics at the University of Leningrad. According to the *Dictionary of Scientific Biography* he died of typhoid fever, and according to George Gamow, in *My World Line*, he died of pneumonia as a result of a chill caught while flying a meteorological balloon. (Gamow at that time was a student of Friedmann's, and his recollections therefore cannot be lightly set aside.) We are told by Gamow that Friedmann had spotted an error in Einstein's 1917 paper on cosmology, an error that had led Einstein to the conclusion that the universe is necessarily static when the Λ force is introduced. According to Gamow, Friedmann wrote to Einstein about his own more general conclusions but did not receive a reply. Through a colleague who was visiting Berlin, Friedmann succeeded in obtaining from Einstein a "grumpy letter" agreeing with his conclusions. This led to the publication by Friedmann of "On the curvature of space" in 1922 in the German journal *Zeitschrift für Physik.* A second paper by Friedmann, "On the possibility of a world with constant negative curvature," was published in the same journal in 1924. Both papers were remarkable and revolutionary. Furthermore, both were timely because of the discovery of extragalactic redshifts. Although they were published in a leading scientific journal they were ignored, and why they were ignored is a complete mystery. The sad fact is that Friedmann's work had apparently not the slightest impact on cosmology. In honor of Friedmann's pioneer work we refer to uniform expanding universes of zero cosmological constant as *Friedmann universes.*

FRIEDMANN UNIVERSES. The simplest of the Friedmann universes is the Einstein–de Sitter model: It has flat space of $k = 0$, a Hubble term $H = 2/3t$, where t is the age of the universe; and a constant deceleration term $q = \frac{1}{2}$ (see Figure 15.5). The scaling factor varies with time in the manner

$$\frac{R}{R_0} = \left(\frac{t}{t_0}\right)^{2/3} \qquad (15.3)$$

and t_0 is the present age. At any instant the age is $t = 2/3H$ or two-thirds the Hubble period (equal to $1/H$), and for a Hubble term of 20 kilometers a second per million light years (giving a Hubble period of 15 billion years) the present age is 10 billion years.

From the Friedmann equations we have (as shown in the Reflections)

$$K = H^2(2q - 1) \qquad (15.4)$$

and the Einstein–de Sitter universe of $q = \frac{1}{2}$

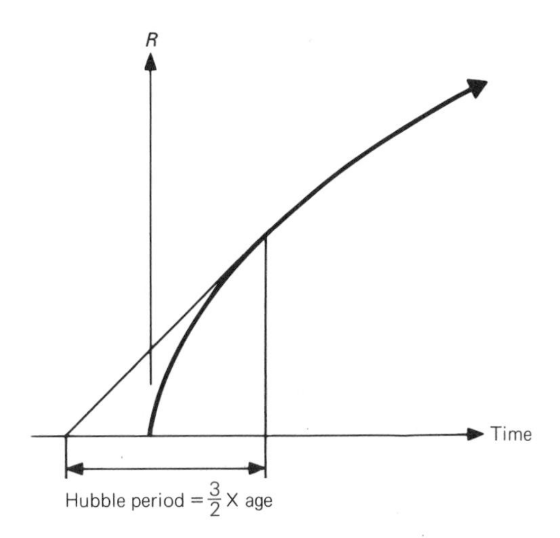

Figure 15.5. The Einstein–de Sitter universe ($q = \frac{1}{2}$).

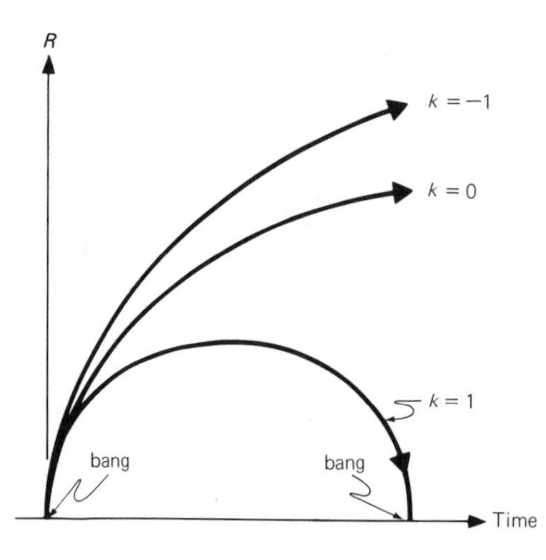

Figure 15.6. Friedmann universes all begin with big bangs.

divides the closed universes of positive curvature K from the open universes of negative curvature K (see Figures 15.6 and 15.7). Hence

q greater than ½: space is spherical and closed ($k = 1$)
q equal to ½: space is flat and open ($k = 0$)
q less than ½: space is hyperbolic and open ($k = -1$)

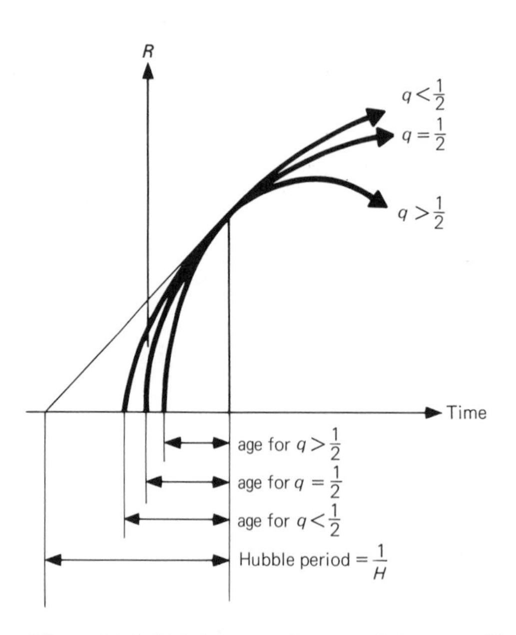

Figure 15.7. Friedmann universes of the same H but different q. The universes of higher q have shorter ages.

From Figure 15.7 it is seen that the greater the deceleration, the shorter the age of the universe, and therefore

q greater than ½: age is less than ⅔ the Hubble period
q equal to ½: age is equal to ⅔ the Hubble period
q less than ½: age is greater than ⅔ the Hubble period

A closed universe has an age shorter than 10 billion years, and an open universe has an age equal to or longer than 10 billion years. These values are approximate only, as we are not completely certain of the value of the Hubble term.

The average density of matter ρ, from the Friedmann equations, is given by

$$4\pi G\rho = 3qH^2 \quad (15.5)$$

and we notice that in the Friedmann universes the deceleration term is always positive. We now define a critical density ρ_{crit}, which is the density of the Einstein–de Sitter universe of $q = ½$:

$$8\pi G\rho_{crit} = 3H^2 \quad (15.6)$$

and from this relation we obviously have

q greater than ½: density is greater than ρ_{crit}
q equal to ½: density is equal to ρ_{crit}
q less than ½: density is less than ρ_{crit}

The critical density is about 10^{-29} grams per cubic centimeter and in round numbers is equivalent to 10 hydrogen atoms per cubic meter.

CLOSED FRIEDMANN UNIVERSE. All Friedmann universes decelerate while expanding, and when the deceleration term is greater than ½ they eventually cease to expand and commence to collapse. Expanding–collapsing closed universes are of particular interest. The curve in Figure 15.8, which shows how R varies with time, is a cycloid; it is a curve similar to that traced out by a point on the rim of a wheel rolling on a flat surface. In these models we can think of the scaling factor as the radius of the universe, and $2\pi R$ is the circumnavigation time for light.

At maximum inflation the radius attains

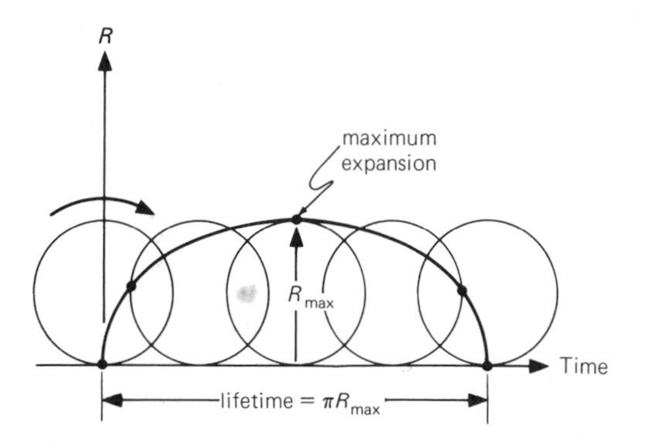

Figure 15.8. A closed Friedmann universe that begins and ends with big bangs ($k = 1$). The curve shown is a cycloid similar to that generated by a point on the rim of a rolling wheel.

a value of

$$R_{max} = R_0\left(\frac{2q}{2q-1}\right) \qquad (15.7)$$

where R_0 is the present radius and q is the present value of the deceleration term. Thus, if $q = 1$, then R_{max} is twice R_0 and the universe is now halfway to its maximum size. At maximum radius the density has a minimum value ρ_{min} given by the relation

$$R_{max} = \left(\frac{3}{8\pi G\rho_{min}}\right)^{1/2} \qquad (15.8)$$

Although the universe at this instant is stationary, and stands poised at maximum inflation, it is not in the least like the Einstein static universe. The mass M of the universe is easily found to be given by

$$R_{max} = \frac{2GM}{3\pi c^2} \qquad (15.9)$$

and the maximum size is therefore one-third the radius of the Einstein universe and is $1/3\pi$ the radius of a black hole of the same mass.

The total lifetime of a closed Friedmann universe, from bang to bang, is πR_{max}, or half the circumnavigation time at maximum inflation. We find, by juggling with the equations, that

$$\text{lifetime} = \frac{2\pi q}{(2q-1)^{3/2}} \text{ Hubble periods} \qquad (15.10)$$

and when $q = 1$, the lifetime is 2π times the Hubble period, or about 100 billion years. The half-age, or time to reach maximum size, is 50 billion years and will be reached in about 40 billion years time. If we exist in such a universe, then the universe is still relatively young and has lasted only 10 percent of its total lifetime. Its constituents, however, are already middle-aged, and by the time the universe begins to collapse the galaxies will have grown old and become dark.

OSCILLATING UNIVERSE. Universes that begin and end in big bangs are referred to as *oscillating universes* (see Figure 15.9). It is sometimes argued that because the universe has already expanded out of one big bang, it should also arise phoenixlike out of the next big bang. It must then evolve and again terminate in a big bang, and so on, repeatedly, bouncing from bang to bang. According to this picture, periods of expansion and contraction repeatedly succeed one another; the universe has already oscillated an infinite number of times and will continue in the future to oscillate indefinitely. A closed universe, which we thought could exist only for a finite duration, has acquired an infinite lifetime by means of continual reincarnation.

Yet each period of oscillation cannot be exactly the same as the previous period, as was shown by Richard Tolman in 1934. Stars and other sources pour out radiation, and the number of photons in space slowly increases during any period. The entropy in one cycle is slightly greater than the entropy in the previous cycle. Despite the devastat-

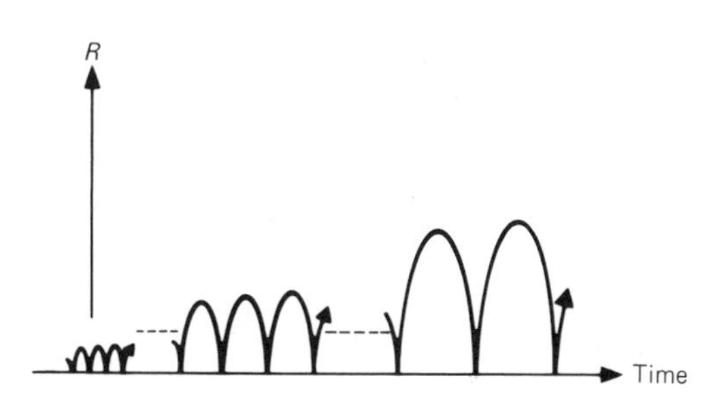

Figure 15.9. The oscillating universe. Each cycle is slightly larger than the preceding cycle owing to the growth of entropy.

ing nature of a big bang and the fact that it obliterates all detail, we can nonetheless conclude that each bang is a little hotter than the previous bang because entropy increases inexorably. The amount of thermal background radiation in the universe therefore increases from cycle to cycle. Richard Tolman in 1932 showed that because of this slow growth in background radiation the universe expands to slightly greater size in each succeeding cycle. The universe steadily gets hotter, R_{max} slowly increases, and the period of each cycle also increases. The cosmic radiation, now at a temperature of only 3 degrees, will get hotter in future cycles until a time is approached when galaxies cannot be born and stars will cease to exist. Thereafter the increase in radiation will be exceedingly slow, and the oscillating universe will have almost constant amplitude and period. Life as we know it will not exist in these starless bright-sky cycles of the future. Back in the past the bangs were cooler and most hydrogen was converted into helium in each bang. Stars, if they formed, were short-lived, and the time available in each cycle was too short for biological evolution. If you ask why we are occupying the universe during the present cycle, the answer is that in an oscillating universe there are a finite number of cycles that can be occupied by life, and we therefore live in one of those cycles.

Although this is a fascinating subject, we must treat with reservation the whole idea of a universe preserving its identity through a sequence of big bangs. We have no knowledge of what happens in a big bang at its state of highest density, when the universe reverts to primordial chaos. Perhaps, in the ultimate and unimaginable chaos of a big bang, there lurks a cosmogenic genie who conjures and launches multitudes of universes, each equipped with its own unique laws and fundamental constants. What comes out of a big bang may have no relation whatever to what went in.

IS THE UNIVERSE OPEN OR CLOSED? Broadly speaking, there are three ways in which we can try to determine whether a Friedmann universe is open or closed. First, we may try to find the value of the deceleration term by observing how things vary with distance – for example, how the apparent brightness of galaxies diminishes with increasing distance. Galaxies have a wide range of intrinsic luminosities and the problem in this case is to select a class of galaxies of known luminosity that can serve as standard candles. The brightest galaxies in the great clusters are usually chosen. To determine the deceleration term we must study these galaxies and observe how their apparent brightness varies over large distances – billions of light years – and this entails looking a long way back into the past. Some questions then arise: How do galaxies evolve? How does their brightness change over long periods of time? Do they get brighter or fainter with age? One might expect them to get fainter, but Jeremiah Ostriker and Scott Tremaine of Princeton University have suggested that the bright galaxies in the centers of great clusters are swallowing smaller galaxies and are hence growing larger and brighter with age. Evolutionary changes in the bright galaxies used as standard candles are not fully understood, and it seems that these changes may be sufficient to be a cause of serious concern. The old method of trying to determine the deceleration term by luminosity-redshift measurements now seems impractical.

Second, we may attempt to find the age of the oldest known systems. Presumably, these systems have ages only slightly shorter than the age of the universe. The population II stars in globular clusters are among the oldest in the Galaxy and have ages of at least 8 billion years and at most somewhere in the region of 16 billion years. The universe is therefore at least 8 billion years old; hence the deceleration term cannot be much larger than $\frac{1}{2}$, and is probably less. On the whole, age determinations tend to indicate that our universe is open, if it is of the Friedmann type.

Third, we may try to determine the average density of matter in the universe and compare it with the critical density of the Einstein–de Sitter universe. The observed density of the universe, owing to the galaxies, is at most about 1 hydrogen atom per cubic meter, and this is at least 10 times less than the critical density. The evidence in this case again suggests that the universe is open. The deuterium abundance, discussed in Chapter 18, also suggests that the universe is open.

Most of the present-day evidence indicates rather strongly that our universe is open, if it is of the Friedmann type, and that it therefore is of infinite extent in space and will expand for an infinite future. There is a traditional dislike in cosmology for open universes, and several attempts have been made to adjust and reinterpret the observational data. A smaller Hubble term would mean that the universe expands more slowly and is of greater age. Over the decades we have seen a steady decline in the value of the Hubble term, as a result of improvements in observations, and it is tempting to think that perhaps we still have not reached the correct value of H. If, for example, the Hubble term were eventually reduced to 10 kilometers a second per million light years, the Hubble period will be 30 billion years, which is more than sufficient for a closed universe of age less than 20 billion years. Such an extreme reduction seems unlikely; it would in any case fail to reduce the critical density to the value of the observed density. This leads to the "missing-mass" problem: Where is all the matter needed to close the universe? We cannot put it inside galaxies because their mass is already known from the motions of stars. The missing-mass problem applies also to the clusters of galaxies, particularly the great clusters such as Virgo and Coma, which appear to lack sufficient mass to make them gravitationally bound. The apparent masses of some clusters are only 10 percent of what is required to hold them together and prevent the galaxies from flying away. It is possible that the missing mass lies on the outskirts of galaxies in the form of low luminosity stars or exotic objects such as black holes; it is possible that the missing mass in the great clusters lies in the gas dispersed between the galaxies; and it is possible that neutrinos do not have zero mass, and the cosmic neutrinos have a density greater than the critical density.

Vigorous debates on whether the universe is open or closed still continue and as yet are inconclusive. Most participants in this exciting controversy have assumed in the past few years that the Λ force is zero and the universe is of the Friedmann type.

FRIEDMANN–LEMAÎTRE UNIVERSES

THE LEMAÎTRE UNIVERSE. Georges Lemaître (1894–1966) was ordained as a priest in 1922, and in 1927, the year in which he obtained his Ph.D at the Massachusetts Institute of Technology, he published his major work on the expansion of the universe. In the midst of discussions on the meaning and merit of the Einstein and de Sitter universes, his work went unnoticed until Eddington drew attention to it three years later and arranged for it to be translated into English. Lemaître was the first to advocate an initial high-density state, which he called the "primeval atom," and he is therefore said to be "the father of the big bang." He rediscovered the cosmological equations that had been developed earlier by Friedmann, and we shall refer to them in their general form, which includes the cosmological constant, as the *Friedmann–Lemaître equations.*

Lemaître singled out from the many possible solutions of the Friedmann–Lemaître equations a closed universe containing a repulsive Λ force (in this case the cosmological constant is positive; see Figure 15.10). This universe has the same basic ingredients as the Einstein universe, with the important difference that Λ is slightly greater than the value chosen by Einstein, and therefore the Lemaître universe cannot be static. It begins as a big bang and has two stages of expansion. In the first stage the expansion decelerates because gravity is stronger than the repulsion of the Λ force; it

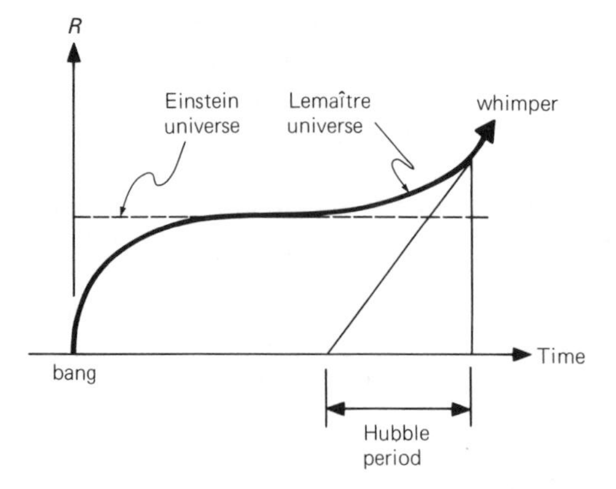

Figure 15.10. The Lemaître hesitation universe.

then approaches slowly the radius of the Einstein universe. At about this time the repulsion becomes greater than gravity and a second stage of expansion begins. The universe now expands away from the Einstein radius, at first slowly and then at an increasing rate. The Lemaître universe therefore begins as a big bang, develops eventually into a whimper, and on the way hesitates while passing through the size of the Einstein universe. It combines neatly the properties of the Einstein and de Sitter universes: It is closed, like the Einstein universe; it has cosmic repulsion, as do both universes; and under the urge of this repulsion it later inflates into a whimper, like the de Sitter universe.

Lemaître's "hesitation universe" had something to offer everybody and was popular because its age in the second stage is greater than the Hubble period. By adjustment of the cosmological constant, so that it exceeds only slightly the Einstein value, the period of hesitation can be prolonged considerably. Until the 1950s it was believed that the Hubble period was 2 billion years, and consequently a universe of prolonged age owing to delayed expansion was rather attractive.

Lemaître thought that galaxies probably formed during the coasting period or hesitation era. Late in the 1960s the Lemaître universe was revived in order to explain why quasars appeared to have redshifts that concentrated near the value of $z = 2$. The idea was that quasars were born in the hesitation era and had a maximum redshift of 2 because the radius of the universe during the hesitation era was one-third its present value. Because light might also circumnavigate more than once during a long hesitation era, the distant quasars and radio sources could produce ghost images. Numerous quasars have since been found, and there is now less pronounced clustering of redshifts at $z = 2$; several quasars of redshifts greater than 2 have also been discovered, and the evidence for a hesitation era has become less impressive.

The trouble with a long hesitation era is that it suffers from the instability that besets the Einstein universe. The hesitation era has been visualized as an age ruled by Titans, in which galaxies are born and quasars reign supreme; an age in which the universe is seized and shaken by violent events. The hesitation era cannot therefore last for long, and it is questionable whether it can prolong the age of the universe by a significant amount.

THE EDDINGTON UNIVERSE. Lemaître was attracted by big bangs, perhaps for religious reasons, whereas Eddington disliked them and found them esthetically displeasing. The two men worshipped in different temples of cosmology, and to this day there exist two cults: the "bangers" and the "antibangers." Instead of adopting an abrupt beginning, Eddington in 1930 championed a universe in

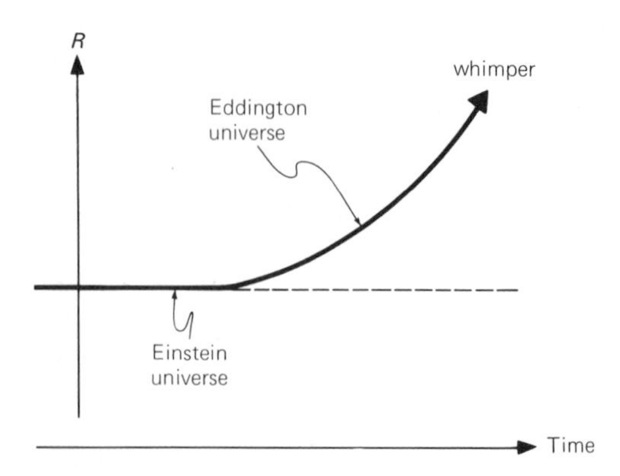

Figure 15.11. The Eddington antibang universe.

which "evolution is allowed an infinite time to get started," which is "necessary if the universe is to have a natural beginning" (see Figure 15.11). His preferred universe exists initially for an infinite period of time as a static Einstein world; then, as a result of an accidental disturbance, it awakes from a long sleep and begins to expand. It exists first in the Einstein static state and later switches to the de Sitter state, in which repulsion dominates over gravity; in this way, it brings together the two worlds that previously had preoccupied cosmologists.

Eddington discovered the instability of the Einstein universe and it is therefore strange that he favored a universe existing initially for an indefinitely long period in the unstable static Einstein state. No galaxy formation could have occurred and no life existed in that prolonged and precariously balanced world that awoke only 10 billion years ago. Eddington was forced to postulate an infinite past to exorcise the specter of a catastrophic beginning. He was the first but not the last of modern cosmologists to be terrified by the nightmare of cosmic birth and death. His universe exists in a slumber state; it awakes at the twilight of dawn, ages gracefully, and ends in a whimper. But however one twists and turns there is no escape from the implacable law of cosmogenesis: Creation cannot be set aside as an event within time that occurred in the infinite past, for the universe contains time, and time whether finite or infinite is created with the universe.

The Lemaître and Eddington models have a common virtue: They contain the cosmological constant that serves as a natural yardstick by which everything is measured. How do particles know what size to be? How do the fundamental constants know what values to have? According to Eddington, the cosmological constant determines the structure of particles and the scale of the universe.

CLASSIFICATION OF UNIVERSES

There are various ways of classifying universes; here we consider the geometric, kinematic, and dynamic systems of classification.

GEOMETRIC. The geometric system is based on the simple geometrical distinction between closed and open: Closed universes have a curvature constant $k = 1$ and open universes have either $k = 0$ or $k = -1$. This primary distinction between finite and infinite space as a method of classification has been popularized in recent years, and was first stressed vividly in 1931 by Ernest Barnes, bishop of Birmingham in England, in a discussion entitled "Evolution of the universe" at a British Association meeting. He said, "It is fairly certain that our space is finite, though unbounded. Infinite space is simply a scandal to human thought." There was no doubt which of the two he favored, and he declared that only in a finite universe could we hope to understand the range of God's activity.

KINEMATIC. The second system is kinematic; that is, it is descriptive and without regard to dynamic considerations. In this system of classification we could, for example, arrange the universes according to whether they are static, expanding, or collapsing. Previously, however, we have used the *bang* and *whimper* nomenclature, and it seems preferable therefore to make an extension by including the possibility of a static or quasi-static state. By combining the bang, static, and whimper states in various ways we obtain fourteen classes:

1	bang–bang	Friedmann ($k = 1$) model
2	bang–static	
3	bang–whimper	Friedmann ($k = 0$, -1) models
4	bang–static–bang	
5	bang–static–whimper	Lemaître model
6	static	Einstein model
7	static–bang	
8	static–whimper	Eddington model
9	whimper	de Sitter model
10	whimper–bang	
11	whimper–static	
12	whimper–whimper	
13	whimper–static–bang	
14	whimper-static–whimper	

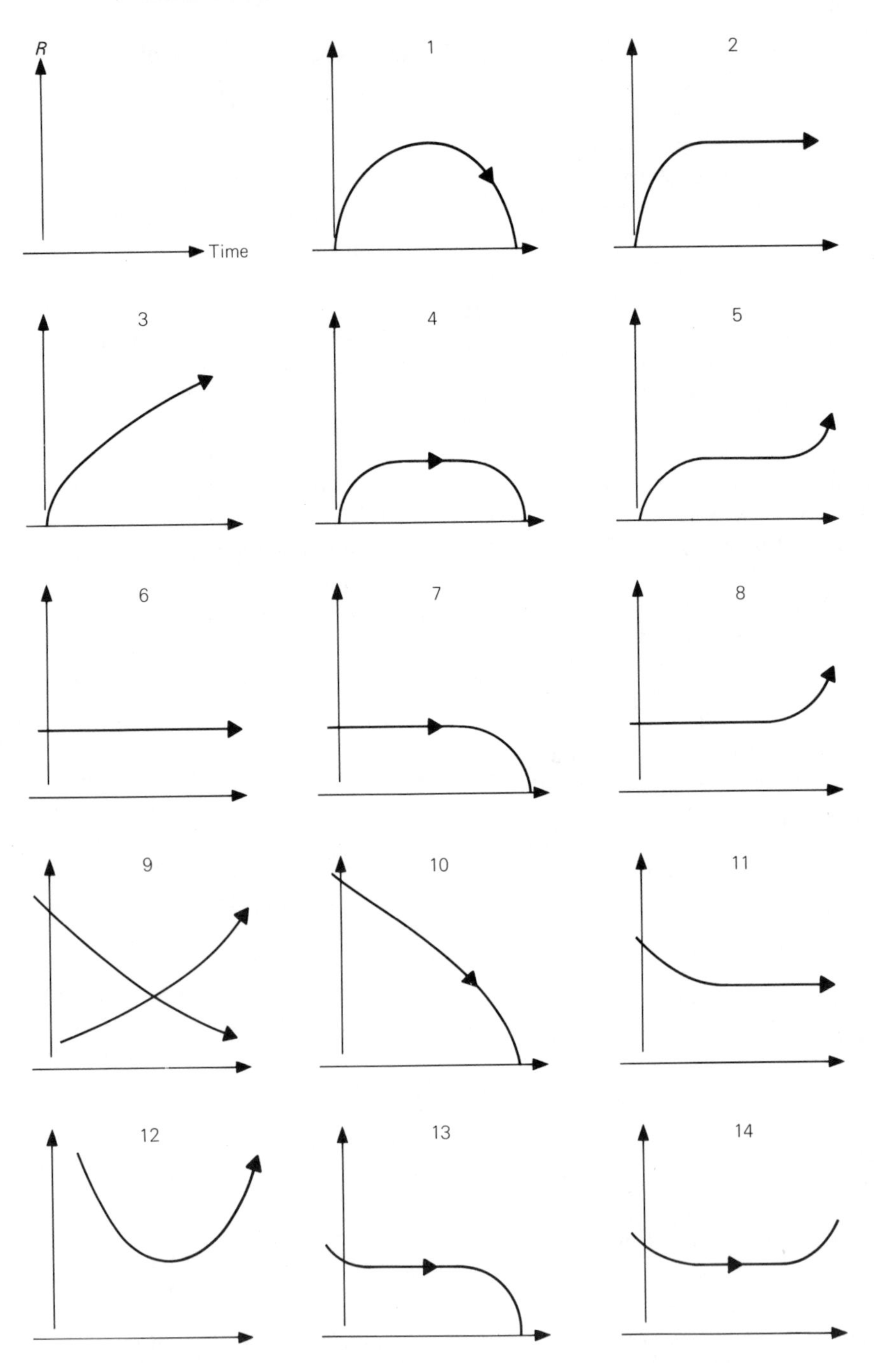

Figure 15.12. A gallery of universes classified according to their bang–static–whimper states.

The manner in which the scaling factor varies with time in these different classes is shown in Figure 15.12. The universe is now expanding, and the acceptable nine classes are hence 1, 2, 3, 4, 5, 8, 9, 12, and 14; the remaining five classes, 6, 7, 10, 11, and 13, are ruled out because they have no expansion periods.

DYNAMIC. The third system is dynamic and founded on the Friedmann–Lemaître equations. There are three values of the curvature constant k; for each of these the cosmological constant can have two significant specific values:

A Λ equals zero
B Λ equals the Einstein value Λ_E

and three significant ranges of value:

C Λ less than zero

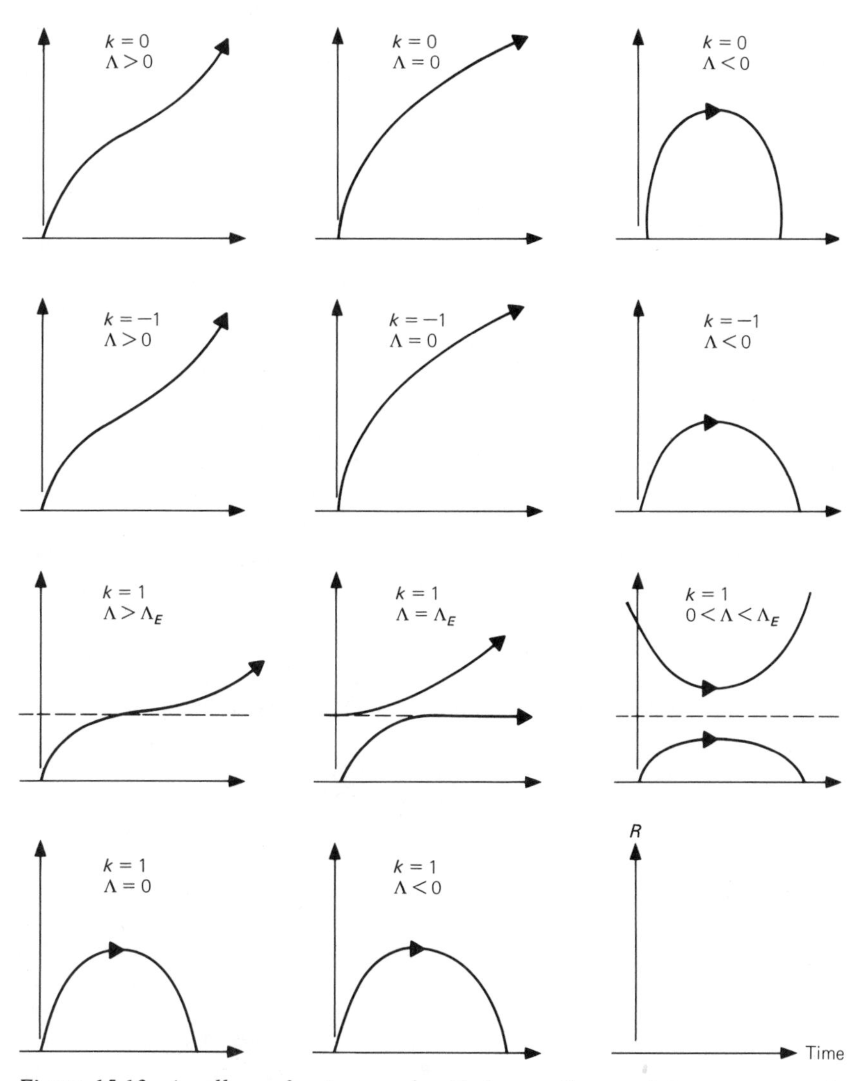

Figure 15.13. A gallery of universes classified according to their curvature and cosmological constants.

D Λ greater than zero but less than the Einstein value Λ_E

E Λ greater than the Einstein value Λ_E

Fifteen possible classes are thus given. The classes *B*, *D*, and *E*, for $k = 0$ and $k = -1$, are dynamically similar, and we are left with eleven distinct classes, as shown in Figure 15.13. In the classes *B*, *D*, and *E* the Λ force is repulsive and opposes gravity, and in *C* the Λ force is attractive and augments gravity. We notice that not all kinematic descriptions are possible according to the dynamic classification scheme.

THE LAMBDA FORCE

The cosmological constant Λ was introduced by Einstein to secure a static universe, and he at first believed it an essential ingredient of cosmology. After the discovery of expansion he swung to the opposite opinion and regarded its introduction into cosmology as an unfortunate mistake; it had misled him, and without it he might have predicted the expansion of the universe.

The Λ force has had a chequered career; it was rejected by its inventor, yet retained by other cosmologists. "I would as soon think of reverting to Newtonian theory as of dropping the cosmical constant," said Eddington in *The Expanding Universe;* he was convinced that without it the constants of nature would lack fixed values. It was popular also because it solved the problem of the age of the universe when the Hubble period was thought to be only 2 billion years. Some theoreticians have said that Λ is an essential part of general relativity and that, without it, the theory would be incomplete. When astronomers later discovered that the Hubble period is nearer 10 billion years, the Λ force became less popular. It came to be widely regarded as a fanciful addition that entails unwarranted complications. It has the peculiar and unappealing property, it is said, of affecting everything but being itself affected by nothing. This, incidentally, can be said of any constant of nature.

The cosmological constant greatly enlarges the variety of possible universes, making the variety so large and bewildering that many people regard this wide choice as a reason for rejecting the cosmological constant outright. A theory of the universe, if it is basic, ought not to bewilder us with numerous options, but should indicate a unique model or a narrow range of choices, as offered by the Friedmann models. For this reason the cosmological constant tended to vanish from cosmology, and for many years observers concentrated on measuring the Hubble and deceleration terms, spurred on by the modest hope of being able to determine whether the universe is an open or closed Friedmann model. Great emphasis was therefore placed on deciding whether *q* is larger or smaller than the crucial value of ½. More recently, the cosmological constant has staged a return, and its popularity is now on the upswing. This is because the observational information has become more complex and the problem of reconciling observations with theory has reawakened interest in the mysterious Λ force.

Reasons for retaining or rejecting Λ are still of a philosophical flavor, and a person with no strong opinion on the subject finds it difficult to take sides in debates on the reality of the Λ force. Failing further theoretical developments we must wait until observations are more decisive. Decades have passed, and many more years may lie ahead before observers will know the answer. If science cannot decide for us, then we are back to philosophy, and the most appealing philosophy advocates a provisional retention of Λ as a continual reminder of the complexity of the universe.

The universe contains matter and perhaps very little antimatter. Is it possible that this cosmic favoritism is due to the Λ force, representing a hidden richness in the properties of spacetime that in some way affects the structure of particles? A universe that contains everything is surely at least as complex as elementary particles – therefore why should we seek utmost simplification by discarding the Λ force? It is not inconceivable that general relativity breaks down on the cosmic scale. Alternatively, we have perhaps failed to identify what "matter" means on the right side of the Einstein equation. Here we recall Einstein's own words: "Not for a moment, of course, did I doubt that this formalism was merely a makeshift to give the general principle of relativity a preliminary closed form." The vacuum state seethes with countless virtual particles patiently waiting to seize on shreds of energy and emerge into the real world, and for all we know this universal sea of virtual particles enriches the properties of spacetime and endows it with a Λ force. The retention of this fifth force of nature is a gesture that admits we do not know everything about the universe.

REFLECTIONS

1 *"In giving this survey of cosmologies we are convinced that the underlying theory forms an integral part of the theory of relativity, and that although the choice of a particular model may for the present be influenced by the predilection of the individual, we can hope that the future will reveal additional evidence to test its validity and to lead us to a satisfying solution." These words by Howard Robertson ("Relativistic cosmology," 1933) reflect the general views of most cosmologists, who would also agree with Dennis Sciama (*Modern Cosmology, *1971) who says: "A rigid theory has not yet been discovered. For instance, general relativity, which is the best theory of space, time and gravitation that has so far been proposed, is, as we shall see, consistent with an infinite number of different possibilities, or models, for the history of the Universe. Needless to say, not more than one of these models can be correct, so that the theory permits possibilities that are not realized in Nature. In other words, it is too wide. We can put this in another way. In the absence of a theory anything can happen. If we introduced a weak theory too many things can still happen. A strong enough theory has not yet been discovered."*

2 *Discuss the Einstein static universe. Will stars, pouring out radiation into space, cause the Einstein universe to expand or collapse?*

* *Consider an Einstein universe in which gravity is repulsive (change G to −G) and the Λ force is attractive (change Λ to −Λ). Is it open or closed? Could life exist in such a universe?*
* *"Whatever pushed the universe over the brink, why did it topple it on the expansion side of the abyss rather than on the other?" (Jagjit Singh,* Great Ideas and Theories of Modern Cosmology, *1970). If there are many Eddington universes, and we exist in one, would it matter if half the initial Einstein universes collapsed?*
* *Suppose that in the past at a redshift of 9 the universe was in an Einstein static state. What was its radius and density?*
* *Why is it important to determine the Hubble and deceleration terms as accurately as possible?*
* *What is the difference among static, steady state, and static steady state universes?*
* *Are you a banger or an antibanger?*
* *Discuss the Lemaître and Eddington universes.*
* *Do you agree with the provocative Bishop Barnes about the possibility of understanding God in a finite universe?*
* *What are the advantages and disadvantages of the Friedmann models?*
* *Can you think of other ways to classify universes (for example, universes that contain matter, antimatter, or matter plus antimatter)?*
* *What kind of universe would you prefer to live in? Is there anything to stop you?*

3 *"Much later, when I was discussing cosmological problems with Einstein, he remarked that the introduction of the cosmological term was the biggest blunder he ever made in his life. But this 'blunder,' rejected by Einstein, is still sometimes used by cosmologists even today, and the cosmological constant denoted by the Greek letter*

Λ rears its ugly head again and again and again" (George Gamow, My World Line, *1970). The Λ force is referred to by various names, such as the* cosmological constant, cosmological term, cosmical constant, *or* cosmical term.

* *On the subject of galaxy formation, William McCrea, in a paper entitled "Cosmology Today" (1970) writes: "Lemaître was, I think, the first to seek to relate this problem to the general evolution of the universe as studied by relativistic cosmology. He considered that the time spent near the Einstein state in the Lemaître model was the time when galaxies and clusters of galaxies were formed out of gas-clouds. For in this phase there is a near balance between gravitational attraction and cosmical repulsion (represented by the Λ-term) that provides the sort of instability that may lead to condensations."*

* *"But here I shall provisionally retain Λ as an unknown parameter, both to show its influence on the models and to allow for the possibility that this theoretically allowable term may more legitimately arise in some future, more comprehensive field theory" (Howard Robertson, American Association for the Advancement of Science meeting held in 1954).*

* *"The tentative conclusion is reached that, if general relativity is to be treated as a self-contained theory, then the 'cosmical terms' that contain the cosmical constant should be omitted. But if general relativity is only part of what is needed to construct a theoretical model of physical reality, then the cosmical terms ought to be retained as affording additional freedom in linking up with other parts of physical theory" (William McCrea, "The cosmical constant," 1971).*

* *"A theory has only the alternatives of being right or wrong. A model has the third possibility: it may be right, but irrelevant" (Manfred Eigen, in* The Physicist's Conception of Nature, *ed. Jagdish Mehra, 1973).*

4 *Cyclic universes are not uncommon in Indo-Aryan mythology. Each cycle of the Hindu universe is a* kalpa, *or day of Brahma, that lasts 4320 million years. Vishnu, who controls the universe, has a life of a hundred "years," each of which contains 360 days of Brahma. After 36,000 cycles, lasting roughly 150 trillion ordinary years, the world comes to an end and only the Absolute Spirit survives. After an indefinite period of time a new world and a new Vishnu emerge and the cyclic scheme is repeated. It is interesting to note that a day of Brahma is not far short of a Hubble period in modern cosmology. Each day of Brahma is divided into 1000 great eons, known as the* maha yugas, *each of which is subdivided into 4 eons, or* yugas, *and a yugas lasts for roughly a million years.*

5 *We exist at a time when the stars are still shining. In 1930, Arthur Eddington said that astronomers must "count themselves as extraordinarily fortunate that they are just in time to observe this interesting but evanescent feature of the sky." In* Measure of the Universe, *J. D. North says that Eddington appears "to rely on a Principle of the Improbability of Good Fortune." Discuss these remarks critically.*

6 *In the previous chapter we presented the Friedmann–Lemaître equations for an expanding universe that is homogeneous and isotropic. The equations are*

$$\ddot{R} = -\frac{4\pi G\rho R}{3} + \frac{\Lambda R}{3}$$

$$\dot{R}^2 = \frac{8\pi G\rho R^2}{3} + \frac{\Lambda R^2}{3} - k$$

and with the exception of MacMillan's steady state universe they apply to all models of the Universe discussed in the present chapter. In these models the density ρ varies with time as $1/R^3$. We also have

$$\text{curvature:}\quad K = k/R^2$$
$$\text{Hubble term:}\quad H = \dot{R}/R$$
$$\text{deceleration term:}\quad q = -\ddot{R}/RH^2$$

On substituting these terms in the preceding equations, we find

$$K = 4\pi G\rho - H^2(q+1) \qquad (15.11)$$

$$\Lambda = 4\pi G\rho - 3H^2 q \qquad (15.12)$$

and these are the Freidmann–Lemaître

equations in terms of the observable quantities K, H, q, and ρ.

∗ *The* Einstein universe *has no expansion or contraction and the Hubble term H is therefore zero. The Friedmann–Lemaître equations (15.11) and (15.12) now simplify to*

$$K = 4\pi G\rho \qquad (15.13)$$

$$\Lambda = 4\pi G\rho \qquad (15.14)$$

Hence $K = \Lambda$, *and the universe has positive curvature equal to the cosmological constant and consists of spherical space. The scaling factor is now the radius of the universe:*

$$R = (4\pi G\rho)^{-1/2} \qquad (15.15)$$

as shown in equation (15.1).

∗ *The* de Sitter universe *has a density equal to zero and space is flat with* $K = 0$; *from the Friedmann–Lemaître equations (15.11) and (15.12) we find*

$$q = -1 \qquad (15.16)$$

$$\Lambda = 3H^2 \qquad (15.17)$$

This is an accelerating universe with a constant Hubble term of $H = \sqrt{(\Lambda/3)}$.

In the Friedmann universes, *as we call them, the cosmological constant is zero and therefore*

$$K = H^2(2q - 1) \qquad (15.18)$$

$$4\pi G\rho = 3H^2 q \qquad (15.19)$$

We notice that when space is flat, and therefore K is zero, the deceleration term is $q = \frac{1}{2}$. *This is the* Einstein–de Sitter universe *of*

$$8\pi G\rho = 3H^2 \qquad (15.20)$$

Because the density ρ is proportional to $1/R^3$, *we find that the age of the universe is* $t = 2/3H$, *and the scaling factor varies with time in the manner*

$$\frac{R}{R_0} = \left(\frac{t}{t_0}\right)^{2/3} \qquad (15.21)$$

where t_0 *is the present age and* R_0 *is the present value of the scaling factor. We now have*

$$\frac{\rho}{\rho_0} = \left(\frac{t_0}{t}\right)^2 \qquad (15.22)$$

and at time $t = 0$ *we see that the density is infinitely great.*

In Friedmann universes of positive curvature, q is greater than $\frac{1}{2}$, *and in universes of negative curvature q is less than* $\frac{1}{2}$. *We notice that in all universes having the same value for the Hubble term, the density increases with q; closed universes therefore have higher density than open universes. We can regard the Einstein–de Sitter universe as having a* critical density:

$$\rho_{crit} = 3H^2/8\pi G \qquad (15.23)$$

We can express the density of all other Friedmann universes as

$$\rho = 2q\rho_{crit} \qquad (15.24)$$

and hence

$$K = H^2\left(\frac{\rho}{\rho_{crit}} - 1\right) \qquad (15.25)$$

We see that when the density is greater than the critical value the universe is closed, and when it is less the universe is open.

7 *We are the music-makers*
And we are the dreamers of dreams,
Wandering by lone sea-breakers,
And sitting by desolate streams;
World-losers and world-forsakers,
On whom the pale moon gleams:
Yet we are the movers and shakers
Of the world forever, it seems.
—Arthur O'Shaughnessy (1844–81), *Ode*

FURTHER READING

Bondi, H. *Cosmology*. Cambridge University Press, Cambridge, 1960.

Eliade, M. *The Myth of the Eternal Return, or Cosmos and History*. Princeton University Press, Princeton, N.J., 1971.

Gamow, G. *My World Line: An Informal Autobiography*. Viking Press, New York, 1970.

Gott, J. R., Gunn, J. E., Schramm, D. N., and Tinsley, B. M. "Will the universe expand forever?" *Scientific American*, March 1976.

Gunn, J. E. "Will the universe expand forever?" *Mercury*, November–December 1975.

Harrison, E. R. "Universe, origin and evolution of." *Encyclopaedia Britannica*, 1974.

Jaki, S. L. *Science and Creation: From Eternal Cycles to an Oscillating Universe.* Science History Publications, New York, 1974.

McCrea, W. H. "Cosmology: a brief review." *Quarterly Journal of the Royal Astronomical Society 4,* 185 (1963). Presedential address to the Royal Astronomical Society.

McCrea, W. H. "Cosmology today." *American Scientist,* September–October 1970.

Margon, B. "The missing mass." *Mercury,* January–February 1975.

North, J. D. *The Measure of the Universe: A History of Modern Cosmology.* Oxford University Press, Clarendon Press, Oxford, 1965.

Robertson, H. P. "Cosmology." *Encyclopaedia Britannica,* 1957.

Schlegel, R. "Steady-state theory at Chicago." *American Journal of Physics 26,* 601 (December 1958). Contains a discussion, with references, of MacMillan's steady state theory.

Sciama, D. W. "The renaissance of observational cosmology." *Revista Nuovo Cimento numero special 1,* 371 (1969).

Tinsley, B. M. "From big bang to eternity." *Natural History,* August–September 1974.

Tinsley, B. M. "The cosmological constant and cosmological change." *Physics Today,* June 1977.

Trimble, V. "Cosmology: man's place in the universe." *American Scientist,* January–February, 1977.

Whitrow, G. J. *The Structure and Evolution of the Universe: An Introduction to Cosmology.* Hutchinson, London, 1959.

SOURCES

Eddington, A. S. "On the instability of Einstein's spherical world." *Monthly Notices of the Royal Astronomical Society 90,* 668 (1930).

Eddington, A. S. *The Expanding Universe.* Cambridge University Press, Cambridge, 1933. Reprint. University of Michigan Press, Ann Arbor Paperback, Ann Arbor, 1958.

Field, G. B. "Intergalactic matter." *Annual Review of Astronomy and Astrophysics 10,* 227 (1972).

Gamow, G. *My World Line: An Informal Autobiography.* Viking Press, New York, 1970.

Harrison, E. R. "Classification of uniform cosmological models." *Monthly Notices of the Royal Astronomical Society 137,* 69 (1967).

Lemaître, G. "A homogeneous universe of constant mass and increasing radius accounting for the radial velocity of extra-galactic nebulae." *Monthly Notices of the Royal Astronomical Society 91,* 483 (1931).

Longair, M. S. "Observational cosmology." *Reports on Progress in Physics 34,* no. 12 (1971).

McCrea, W. H. "Cosmology today." *American Scientist,* September–October 1970.

McCrea, W. H. "The cosmical constant." *Quarterly Journal of the Royal Astronomical Society 12,* 140 (1971).

Mehra, J., ed. *The Physicist's Conception of Nature.* D. Reidel, Dordrecht, Netherlands, 1973.

North, J. D. *The Measure of the Universe: A History of Modern Cosmology.* Oxford University Press, Clarendon Press, Oxford, 1965.

Rindler, W. "Relativistic cosmology." *Physics Today,* November 1967.

Robertson, H. P. "Relativistic cosmology." *Reviews of Modern Physics 5,* 62 (1933). A classic paper with an extensive bibliography.

Sciama, D. W. *Modern Cosmology:* Cambridge University Press, Cambridge, 1971.

Shapiro, S. L. "The density of matter in the form of galaxies." *Astronomical Journal 76,* 291 (1971).

Singh, J. *Great Ideas and Theories of Modern Cosmology.* Constable, London, 1961. Reprint. Dover Publications, New York, 1970.

Stabell, R., and Refsdal, S. "Classification of general relativistic world models." *Monthly Notices of the Royal Astronomical Society 132,* 379 (1966).

Tammann, G. A. "The Hubble constant and the deceleration parameter," in *Confrontation of Cosmological Theories with Observational Data,* ed. M. S. Longair. D. Reidel, Dordrecht, Netherlands, 1974.

Tolman, R. C. "Models of the physical universe." *Science 75,* 367 (1932).

Tolman, R. C. *Relativity Thermodynamics and Cosmology.* Oxford University Press, Clarendon Press, Oxford, 1934.

Weinberg, S. *Gravitation and Cosmology: Principles and Applications of the General Theory of Relativity.* John Wiley, New York, 1972.

16 THEORIES OF THE UNIVERSE

He thought he saw an Elephant,
That practised on a fife:
He looked again, and found it was
A letter from his wife.
"At length I realize," he said
"The bitterness of Life!"
— Lewis Carroll (1832–98), *Sylvie and Bruno*

UNIVERSES IN COMPRESSION

PRESSURE INCREASES DECELERATION. The pressure deep inside a star pushes in an outward direction against the pull of gravity. In this case pressure exerts a force because it is greater at one place than another and has a *gradient*. This is the kind of pressure difference, or gradient, that pushes water along pipes and is used in steam engines and numerous other devices. But in a uniform universe where all places are alike, and the pressure is the same everywhere, there are no pressure gradients, and no forces are exerted.

Because pressure in the universe produces no net force, at least on the cosmic scale, how then does it affect the expansion of the universe? We must realize that pressure is a form of energy, and energy – or its mass equivalent – is a source of gravity. When a sealed vessel containing gas is heated, the gas molecules move more rapidly, and pressure increases. This is another way of saying that energy in the container has increased. The contained gas has consequently slightly greater mass and weighs slightly more than when cooler. Similarly, a universe containing pressure has increased gravity; when translated into the language of general relativity, this means that spacetime has a more active dynamic curvature.

In cosmology we gauge the importance of pressure by its equivalent energy. The speed of sound in a gas is given approximately by

$$\text{sound speed} = \left(\frac{p}{\rho}\right)^{1/2}$$

where p stands for pressure and ρ stands for density. From this it follows that

$$\frac{\text{sound speed}}{\text{light speed}} = \left(\frac{p}{\rho c^2}\right)^{1/2}$$

where ρc^2 is the total energy in a unit of volume. The importance of pressure, as compared with energy density, depends on the ratio of the speeds of sound and light. When the sound speed is relatively small the pressure is also relatively small. In a gas the sound speed is about equal to the average speed of the constituent particles, which usually move much more slowly than light, and in most instances pressure is considerably less than the total energy density and can therefore be ignored in cosmology. But pressure is not always negligible; in the early universe, for example, the temperature was very high and all particles moved rapidly, and pressure as a consequence was considerable.

Universes that have no matter, but contain radiation only, have the maximum possible pressure for a given energy density. When the radiation is uniform – and the flux of radiation is the same in all directions – the pressure is equal to one-third the energy

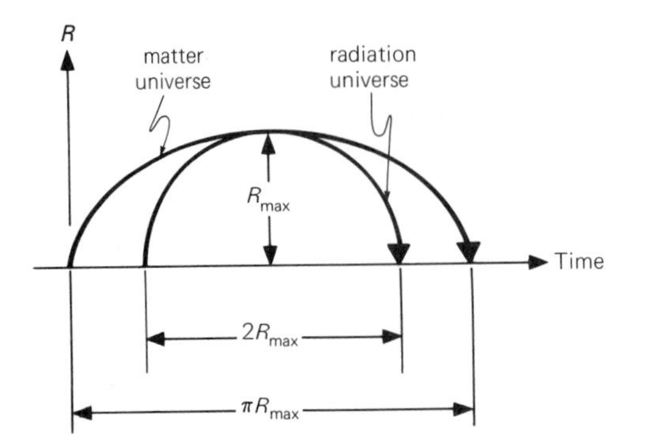

Figure 16.1. Two closed universes, one containing radiation only and the other containing matter only; both expand to the same maximum size R_{max}. The radiation universe has the shorter lifetime.

density. Richard Tolman, an American cosmologist, in the early 1930s investigated universes that contain radiation only and found that these *radiation-dominated* universes behave much like *matter-dominated* universes, with one important difference: A radiation universe of the same density as a matter universe has greater deceleration. This is because its large pressure is a source of gravity and its rate of expansion slows down more rapidly than that of a matter universe. We have seen that a closed Friedmann universe containing matter and no pressure expands to a maximum size $R_{\max}$ and has a lifetime of $\pi R_{\max}$. A closed Friedmann universe containing only radiation also expands and collapses in the same way and has a finite lifetime (see Figure 16.1). Let us suppose that it has the same maximum size $R_{\max}$ as the matter universe. The radiation universe has a lifetime of $2R_{\max}$ – shorter than that of the matter universe owing to the greater deceleration of the radiation universe.

Pressure has an effect opposite to what one might expect. Common sense suggests that a collapsing universe containing pressure should collapse more slowly than a collapsing universe without pressure. Yet pressure causes faster collapse. This is because the universe is uniform and there are no pressure gradients; further, unlike a boiler, the universe has no walls against which the pressure can push, and the only effect of pressure is an increase in the gravitational forces that control the universe. Uniform radiation produces a gravitational force that is twice as strong as that produced by matter of the same mass density and zero pressure.

The Einstein–de Sitter universe, having flat space and no cosmological constant, is convenient for studying the effect of pressure. When it contains matter of zero pressure the scaling factor R is proportional to $t^{2/3}$, where t is the age measured from the big bang. We have previously found that

$$\text{age} = \frac{2}{3} \times \text{Hubble period}$$

where the Hubble period, obtained from the Hubble term, is at present about 15 billion years. The deceleration term q is constant and equal to ½. In an Einstein–de Sitter universe containing radiation only, the scaling factor R is proportional to $t^{1/2}$; hence

$$\text{age} = \frac{1}{2} \times \text{Hubble period}$$

and if the Hubble period is 15 billion years the age is 7.5 billion years. The deceleration term has increased and is equal to 1.

Now consider a more realistic universe, such as our own, which contains both matter and radiation. The principal contribution to the radiation in our universe is the 3-degree cosmic radiation; starlight and radio waves are of only minor importance. At present the density of matter greatly exceeds that of radiation and our universe is matter dominated. In Chapter 13 we saw how the density of matter varies as $1/R^3$, whereas the density of radiation varies as $1/R^4$. As we proceed back into the past, R gets smaller, and the density of radiation rises more rapidly than the density of matter. Sufficiently far back there was a time when radiation was more dense than matter and the universe was radiation dominated. Universes containing cosmic radiation, however little, are at some time in their early stages radiation dominated.

UNIVERSES IN TENSION

THE STRANGE WORLDS OF NEGATIVE PRESSURE. We consider next the effect of negative pressure in the universe. All of us are familiar with positive pressure, as in stars and steam engines, and the notion of negative pressure is at first startling. William McCrea, the British cosmologist, argued in 1951 that negative pressure in the universe, equivalent to a state of cosmic tension, cannot be ruled out on the basis of ordinary experience. A cosmic tension, everywhere the same, does not participate directly in determining the behavior of galaxies, stars, and steam engines. We are aware of the existence of ordinary pressures when they vary from place to place and have gradients, as in stars and the Earth's atmosphere, and when they are acting on walls, as in boilers. But when pressure is everywhere the same, unconfined and without gradients, it produces no perceptible effect except in the dynamic behavior of the universe. The same may be said of a negative pressure; it may exist, but we cannot detect it except in the way it affects the dynamics of the universe.

In the discussion of the Einstein equation we said that the left side of the equation represents dynamic spacetime and the right side represents "matter." Whereas the left side is crystal clear, the right side is murky: Nobody knows quite what "matter" means in this context, and on various occasions strange things have been placed on the right side of the Einstein equation. Convention demands that we keep the right side as simple and tidy as possible and place there only such things as density and positive pressure that are familiar in the everyday world. A uniform cosmic stress of the nature of a negative pressure, having no effect on the structure of planets and stars, is unfamiliar and is usually excluded. But McCrea argued that the universe is perhaps governed by forces not directly manifested in the laboratory, and we therefore cannot rely on common sense to tell us what should be on the right side of the Einstein equation when used in cosmology. A negative cosmic stress might exist that could alter entirely all previous models of the Universe.

Let us consider the effect of negative pressure in an expanding universe. A universe in a state of tension must release energy as it expands. A piece of elastic when stretched gets warm, and this is because the work done while the elastic is stretched releases energy. A similar thing happens in a universe of negative pressure; as it expands, energy is released, and this energy might take the form of newly created matter. In a universe having a very large tension, equal to the energy density (i.e., $p = -\rho c^2$), the energy released by expansion is sufficient to keep the density of matter constant. This is McCrea's brilliant explanation of continuous creation of matter in the steady state universe: The energy released by expansion appears as newly created matter, and the average density of matter stays constant. In 1951, McCrea wrote: "This discussion appears to show that *the single admission that the zero of absolute stress may be set elsewhere than is currently assumed* on somewhat arbitary grounds *permits all of Hoyle's results to be derived within the system of general relativity theory.* Also, this derivation gives the results an intelligible physical coherence."

If we are willing to sacrifice our intuitional belief that pressure must always be positive, and entertain the possibility of cosmic tension, we are confronted with a bewildering variety of new universes (see Figure 16.2). The following are some examples. When tension is equal to one-third of the energy density ($p = -\frac{1}{3}\rho c^2$), gravity becomes ineffective and the universe is controlled by other things, such as the cosmological constant Λ. When tension is less than the energy density but greater than one-third, there are universes that oscillate in size without big bangs; their oscillations are slowly damped, and they become stable Einstein static universes. When the tension is greater than the energy density we encounter those incredible universes in which density increases with expansion, which begin with nothing and expand to

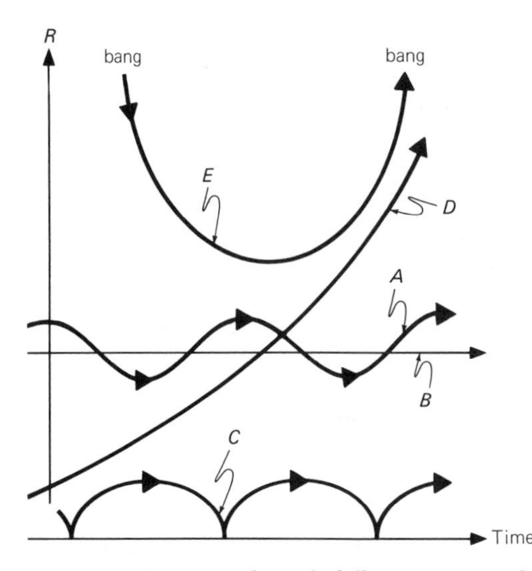

Figure 16.2. Examples of different types of universe in tension. The pressure is $p = (\gamma - 1)\rho c^2$, where ρ is the mass density. A is a closed, oscillating universe of $\gamma > 0$ but $< \frac{2}{3}$. B is a closed, static, stable universe of $\gamma > 0$ but $< \frac{2}{3}$. C is a closed, constant density, oscillating universe of $\gamma = 0$. D is a flat, constant density, steady state universe of $\gamma = 0$. E is a universe of $\gamma < 0$ that expands to become a big bang.

become big bangs, and those in which density decreases with contraction, which collapse to nothing. In 1967, I wrote: "A relevant question is whether at the present stage a proliferation of models in any way advances cosmology. Present observational data are inadequate to make a clear decision even when the [pressure is zero] and the cosmological constant is discarded as an irrelevant complication. With the increased array of models presented in the present scheme the problem of determining an appropriate model from the available data becomes hopeless. But such a situation is not without virtue; it may direct attention to the important matter of bringing more physics into cosmology for the purpose of spanning the gap between the actual universe and our excessively idealized representations."

WORLDS IN CONVULSION

HOMOGENEITY RIDDLE. Homogeneity – meaning that all places are alike – is one of the most remarkable features of the universe. Because the observed universe is isotropic we feel compelled by the location principle to conclude that the universe is on the average the same everywhere. Widely separated regions, billions of light years apart, are in similar states and are dynamically synchronized. We would feel more comfortable with this amazing state of homogeneity if only we could explain why it exists.

When explaining something it is our custom to look back to earlier conditions, simply because causes always precede their effects. Can homogeneity be explained by looking back to a time when the universe was inhomogeneous, when perhaps there were mechanisms at work that created homogeneity out of inhomogeneity, much like the processes that create a calm sea after a storm?

The creation of a calm universe out of a cosmic storm encounters a formidable difficulty, as we now show. To explain homogeneity we must demonstrate how widely separated regions are able to influence one another and produce an average state. The observable universe stretches away to a limit approximately equal to the Hubble length. We see things now within a distance of 15 billion light years, and light or any other kind of signal from things at greater distances has not reached us because the universe is not yet old enough. Hence regions of the universe cannot influence one another if they are separated more widely than a distance that is about equal to the Hubble length. The Hubble length increases with the age of the universe. As we look back in time, seeking the cause of homogeneity, the Hubble length shrinks, and regions now free to interact are found in the past to be isolated. When the universe is 1 year old, only regions close together and separated by distances less than 1 light year are able to exchange signals and thereby in some way or other to attain a state of identity.

If regions now free to interact were once isolated, how then can we hope to explain homogeneity by seeking its cause in the past? This is the homogeneity riddle: We are apparently denied any explanation of homogeneity that appeals to causes acting in the past. There are two main schools of thought concerning homogeneity: the "antichaos school" and the "chaos school."

ANTICHAOS. The antichaos school believes that the universe begins in a homogeneous state. This seems at first a simple-minded way of evading the issue, but actually it conceals a deep thought. We explain the state of individual things by seeking causes active in the past. But the universe is a unity, embracing space and time, and such a procedure is questionable and might even be inappropriate when trying to explain the design of the universe. The universe is not created in time any more than it is created in space – it contains time and space – and therefore we should not seek the cause of its design at a point in time any more than we would seek for it at a point in space. Homogeneity is perhaps fundamental and indispensable, and without it there might not be a universe. Furthermore, without it we might not exist, even if the universe could. This latter possibility has been considered by C. B. Collins and Stephen Hawking, who wrote in 1973, "The fact that we have observed the universe to be isotropic is only a consequence of our existence." In an inhomogeneous universe, they argue, it is probable that galaxies would be nonexistent because all initial condensations would be torn apart before developing into galaxies, and hence stars and living creatures would also be nonexistent. Inhomogeneous universes, if there are any, may not contain life, and that is why our universe is homogeneous.

CHAOS. The chaos school follows the mythological tradition and holds that the initial state of the universe is formless and chaotic. According to this view, in the beginning when "heaven above and earth below had not been formed" there was indescribable chaos. By the operation of natural causes there emerges out of chaos a state of uniformity. We do not know how the miracle of homogeneity is conjured out of chaos, and on this subject modern cosmology is little more enlightened than ancient cosmology. The possibility that uniformity emerges from an initial state of chaos has been stressed by Charles Misner of Maryland University.

The aim of the "chaoticists" is to explain not only homogeneity but also the 3-degree cosmic radiation and the origin of galaxies. The homogenizing mechanism, whatever it may be, releases energy that heats the big bang, and the 3-degree cosmic radiation is a vestigial remnant of this heating process. Also, from the initial inhomogeneity irregularities survive, consisting perhaps of small density fluctuations, which later develop into galaxies.

Homogeneous but anisotropic universes have been studied to see whether the anisotropy can decay and produce an isotropic state like that of our own universe. These "mixmaster universes" (so-named by Charles Misner) thrash backward and forward in giant convulsions; they expand in one direction while oscillating rapidly in the other two directions (imagine a cylinder pulsating in radius while it is stretched), and repeatedly, era after era, each lasting longer than the previous, the directions of expansion and oscillation are interchanged. The question is how much the contained matter and radiation will damp the convulsions by dissipative mechanisms. It seemed at first that neutrinos in the early universe could abate the convulsions and create a state of peaceful isotropy, but later investigations of a more general kind have shown that it is not possible to attain a high degree of isotropy from a preceding state of extreme anisotropy.

If the universe is initially chaotic the homogenizing process must satisfy certain conditions. The 3-degree cosmic radiation is 99.9 percent isotropic, and so there already is a high degree of isotropy, and hence homogeneity, when the universe is only a million years old (the radiation has traveled unimpeded from that early epoch). Most helium is produced when the universe is a few hundred seconds old, and its current abundance sets stringent limits on the anisotropy at the time of its formation. Much of the inhomogeneity must therefore have vanished by the time the universe is only a minute old. Yet the homogenizing mechanism must not be so efficient that all irregularity is obliterated, for irregularities of a suitable nature must survive and develop into galaxies and perhaps even clusters of galaxies. The mechanism must also explain why the cosmic radiation has a temperature of 3 degrees. In addition, it would be nice if it could explain the specific entropy of the universe, but this might be asking for too much, for it requires an understanding of why matter is more favored than antimatter in the design of the universe.

An attractive possibility is that the initial chaos is dissipated extremely rapidly by prolific particle creation, perhaps by the creation of quantum black holes, when the universe is only 10^{-43} seconds old (see Chapter 18). The chaos consists of a foamlike welter of spacetime, of intense gravitational waves and field fluctuations, which create particles just as very strong electric fields create electrons and positrons. Lurking always in the background are virtual particles that by their readiness to spring into reality can rob chaos of its frenetic energy. In this way the universe is perhaps homogenized by the creation of a dense sea of energetic real particles. The maximum energy density that spacetime can contain corresponds to 10^{100} grams per cubic meter, and presumably maximum chaos has this density, which can never be exceeded. Maximum chaos ensures uniform energy density and when dissipated leaves a uniform density of particles. This may be the answer to the homogeneity riddle: The universe is initially everywhere in a state of maximum chaos, and the various regions of the universe do not have to interact with each other in order to attain homogeneity.

While on this subject we should mention a few of the inhomogeneous universes that have been proposed at various times. The hierarchical universes have already been mentioned. There are in addition those, convulsive on the cosmic scale and on scales larger than the Hubble length, that contain unsynchronized regions, or mini-universes, that expand and contract. Remote regions beyond the reach of observation, it has been suggested, are not in dynamic synchronism with our part of the universe, and some are expanding while others are contracting. Collapsed regions become giant black holes isolated behind their event horizons, and hierarchies of black holes may exist nested inside one another. The isotropy of the 3-degree cosmic radiation imposes a high degree of uniformity on the observed part of our universe, and nowadays all such speculations, unfettered by observational constraints, are banished beyond the limits of the observable universe. If the universe is infinite in space, then at distances of trillions of light years, and even trillions of trillions of light years, there may exist regions not synchronized and in identical states; but we are not in a position to know.

KINEMATIC RELATIVITY

MILNE'S STOIC UNIVERSE AND HIS SEARCH FOR THE EXPLANATION OF GRAVITY. Edward Milne of Oxford University, a famous astrophysicist and cosmologist, turned a penetrating eye on general relativity and was not impressed. In 1948, two years before he died, he wrote in *Kinematic Relativity:* "Motion imposed in consequence of a geometry differing from the geometry commonly used in physics was a credible notion. Gravitation as a warping of space was a credible notion, though it gave not the least hint as to the nature or origin of gravitation; why the presence of matter should affect 'space' was left unexplained."

Milne constructed his own theory of the universe, known as *kinematic relativity,* in which gravity is not included as an initial assumption. With a supposedly small number of axioms, such as the cosmological principle and the rules of special relativity, he sought to create a picture of the universe that explained gravity and other laws of nature. He believed, in company with ancient mythologists, that the purpose of cosmology is to explain why things are the way they are, and not just to provide elaborate alternative descriptions of how things work. When measured against what he hoped to attain, his efforts were unsuccessful; but if we measure by what he actually accomplished, there is no doubt that his methods and insight made great impact on cosmology.

Milne's picture of an expanding universe, reduced to its simplest elements, is much easier to understand than general relativity. His universe consists of a spherical cloud of particles that expands within flat space, and the space is infinite and otherwise empty. It is the old Stoic universe, updated and made to obey the rules of special relativity. It has a center and an edge. It begins expanding at a point in space, and all particles are shot out in every direction with velocities ranging from zero to close to that of light; the surface of the cloud, or cosmic edge, expands in space at the velocity of light. About each particle the distribution and recession of all other particles is isotropic. Owing to the relativity effect most particles are crowded close to the edge of the cloud, as shown in Figure 16.3. The finite and bounded Milne universe contains "an infinity of particles in the field of view of any observer, merging towards the limit of visibility into a continuous background."

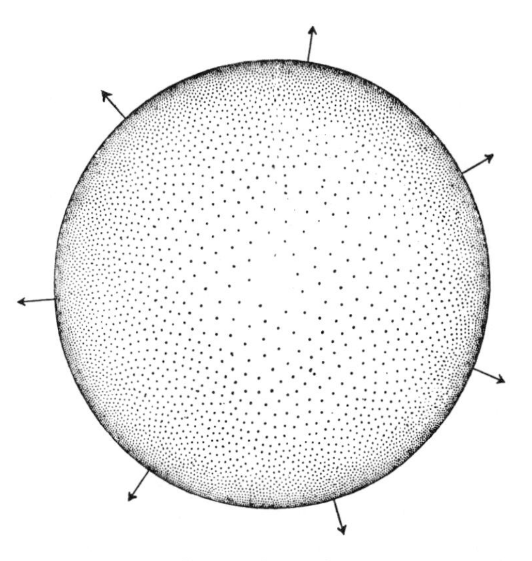

Figure 16.3. Milne's spherical universe expanding outward in space.

Within this descriptive framework of motions (hence the name *kinematic relativity)*, in which all particles move freely, unaffected by forces of any kind, Milne tried to show that each particle manifests a behavior that simulates the effect of gravity. In other words, he explained gravity by starting with a cosmic framework that did not assume the existence of gravity. His arguments were elaborate, and few found them convincing. The physical significance of his cosmological theory is still obscure, and Milne's dislike of general relativity does not change the fact that the Einstein equation at least is comprehensible and appears to reflect faithfully many aspects of the physical world.

Milne identified each particle with a galaxy. Because there are an "infinity of particles," his universe has an infinite mass in a finite cosmic volume. An observer on any particle perceives only a finite density because of the effects of special relativity. It is interesting that Milne's finite and bounded universe of infinite mass can be transformed mathematically, by changing the intervals of space and time, into an infinite and unbounded universe. This universe consists of expanding space, homogeneous and isotropic, and of negative curvature, which is uniformly populated with galaxies. The compact assembly of an infinity of galaxies is distributed over an infinite space. This transformation (which Milne himself pointed out but did not favor, because he could not believe in curved space) has the advantage of making the galaxies stationary in expanding space and the big bang no longer a point in space. If we make spacetime physically real, as general relativity does, space and time are contained within the universe, and this is surely better than the old Stoic idea of a universe contained within space and time.

Having transformed Milne's picture into a universe of dynamic and curved space, we are now able to look at it more closely from the point of view of general relativity. We take the Friedmann–Lemaître equations for zero pressure and set the cosmological constant Λ equal to zero; also, because gravity on the cosmic scale was not required by Milne, we make the gravitational constant G equal to zero. From the equations we find that the curvature $K = -H^2$, where $K = k/R^2$ and k is the curvature constant; and therefore k must be negative, equal to -1, so that space has negative curvature and is infinite in extent. We now have $\dot{R} = 1$, and hence $R = t$, where t is the age of the universe from the time of the big bang. This universe expands at constant rate ($H = 1/t$, $q = 0$), and the Hubble period is always equal to the age of the universe. Thus, when Milne's universe is brought into the fold of general relativity, it has no center or edge, it has infinite space of negative curvature, and it expands at a constant rate with zero deceleration.

CONTINUOUS CREATION

CREATION THEORIES. The steady state expanding universe was jointly proposed in 1948 by Herman Bondi and Thomas Gold. The large-scale aspects of such a universe are independent of the observer's location in space and time and obey the perfect cosmological principle. An expanding universe in a steady state has infinite age, and in 1948 this was an attractive feature of the model because of the time–scale difficulty that beset many evolutionary universes (the Hubble period was thought to be less than the age of the Solar System). An expanding universe in a steady state is possible only if there is a continuous creation everywhere of new matter that maintains a constant density. "Hence," wrote Bondi and Gold, "there must be continuous creation of matter in space at a rate which is, however, far too low for direct observation." The new matter is created not from radiation, as in the MacMillan steady state universe, nor out of anything that preexists, but out of nothing and from nowhere.

The idea of spontaneous creation is exceedingly old and no doubt was rife in the prehistoric Age of Magic. It has been a recurrent theme throughout the history of science, and even in the twentieth century we find many examples. James Jeans in *Astronomy and Cosmogony* surmised that the "centres of the nebulae are of the nature of 'singular points' at which matter is poured into our universe from some other, and entirely extraneous, spatial dimension, so that, to a denizen of our universe, they appear as points at which matter is being continually created." Pascual Jordan of Germany developed in 1939 a scalar–tensor theory that modified general relativity so that matter is not conserved but created. He said, "The conjecture suggests itself that the cosmic creation of matter does not take place as a diffuse creation of protons, but by the sudden appearance of whole *drops* of matter." Jordan's "drops" are stars created in a dense embryonic form. At about this time a group of Japanese mathematicians at Hiroshima also developed a continuous creation theory for a de Sitter–type universe in which galaxies are spontaneously created.

The steady state versus big bang controversy lasted until the late 1960s, and it was frequently claimed that continuous creation assumes no more than is assumed in the instant creation of a big bang, and furthermore has greater esthetic appeal. We have shown previously (Chapter 5) that creation of matter within space and time cannot be regarded as equivalent to the creation of a universe containing space and time. The steady state universe is created in the same way as a big bang universe and creation "little by little" in a steady state universe is therefore not the same as the "instant" creation of a big bang universe.

STEADY STATE EXPANDING UNIVERSE. Bondi and Gold described the main properties of an expanding universe in a steady state. Because nothing changes in the cosmic scene, the curvature K, Hubble term H, and deceleration term q must all remain constant. The curvature K is equal to k/R^2, and because the scaling factor R increases with expansion, the curvature can remain constant only when k is zero, and space is therefore flat and infinite in extent. Because the Hubble term is also constant it follows that $\dot{R}$ is proportional to R, and the scaling factor increases exponentially as in the de Sitter universe; this ensures that the deceleration term has the fixed value of -1.

To maintain the contents of the universe in a steady state, matter must be created at a rate of about 1 hydrogen atom per cubic meter every 5 billion years, equivalent to 1 galaxy per year within the observable universe. It would be hopeless to try to detect this slow rate of creation in the laboratory.

The steady state universe regenerates itself in one-third of a Hubble period; because the expansion time is 15 billion years, the regeneration time is 5 billion years. The average age of everything that endures after creation, such as nucleons and galaxies, is also one-third of a Hubble period. Some galaxies are young and only recently formed; others are exceedingly old; and the average age of all galaxies is 5 billion years. Our Galaxy, which is about 10 billion years old, is therefore twice as old as the average galaxy in the steady state universe. In a big bang universe practically all galaxies have the same age, about 10 billion years. A person believing in the steady state universe might therefore feel concerned that our Galaxy is 10 billion years old, twice as old as the average, and yet coincidentally the right age for a big bang universe.

STEADY STATE THEORIES. Following a "discussion with Mr. T. Gold," Fred Hoyle showed how the theory of general relativity could be modified to allow for a continuous creation of matter. In the year in which Bondi and Gold put forward their steady state theory, Hoyle used the scalar–tensor theory, shortly to be discussed, and found that the constant density of the universe and the Hubble term are related by the equation

$$8\pi G\rho = 3H^3$$

Hoyle's creation theory does not indicate the form in which matter is created. The theory breaks the law of conservation of matter, which is implicit in general relativity, by means of a mathematical device. Many people are uneasy when they see matter created in this way and feel that mathematics is not physics until backed by observations and confirmed by experiments. Steady state advocates have sometimes said that continuous creation means that the universe is necessarily in a steady state, but this is not true: When creation and expansion are not exactly synchronized, it is possible to have expanding universes in which the density either increases or decreases. The steady state theory as it stands does not explain why matter is created at a rate that maintains a steady state, and there is no "genetic code" to guarantee that the universe will replicate itself faithfully every 5 billion years, and will remember what things were like trillions of years ago.

William McCrea advanced the alternative idea that continuous creation is the result of a negative cosmic pressure. Such a cosmic tension, when equal to the energy density, maintains a state of constant density. McCrea's theory does not explain why the tension happens to be the required amount, and like Hoyle's theory it fails to explain why matter is created, and not antimatter.

According to the original work of Bondi and Gold, creation is uniform everywhere, and newly created matter eventually condenses and forms new galaxies. William McCrea in 1964 drew attention to the possibility that the creation process is nonuniform in space and is more active in those regions where matter is already concentrated. "All matter," wrote McCrea in 1964, "is the potential promoter of the creation of matter. All matter is normally in galaxies, and so the creation of fresh matter normally simply promotes the growth of galaxies. But occasionally a fragment of matter becomes detached from its galaxy. Any such fragment is a potential promoter of fresh creation; if it is successful as such it is the embryo of a new galaxy." This idea of galaxies promoting creation fitted into a scenario of galaxies periodically exploding and ejecting fragments.

END OF THE STEADY STATE THEORY. Momentous discoveries have struck down the steady state theory. The 3-degree cosmic radiation indicates strongly that a big bang once existed, and quasars indicate strongly that the universe is evolving. Desperate attempts have been made by Hoyle and Jayant Narlikar to salvage remnants of the theory; they have invoked large-scale inhomogeneities and variations in the laws and constants of nature to show that the universe is eternal and the big bang does not exist. The original elegant simplicity of the steady state theory, attractive to so many people, has been lost, and the theory is now mainly of historical interest. In 1967, Dennis Sciama wrote: "I must add that for me the loss of the steady-state theory has been a cause of great sadness. The steady-state theory has a sweep and beauty that for some unaccountable reason the architect of the universe appears to have overlooked. The universe is in fact a botched job, but I suppose we shall have to make the best of it."

SCALAR–TENSOR THEORY

SCALARS AND TENSORS. Anything continuous that varies from place to place, and has only a single value at each point, is a scalar field. The temperature of our atmosphere and the Newtonian gravitational potential are examples of scalar fields. All scalars are known as zero-order tensors.

When air or water or any other fluid moves, it has at each point in space 3 components of velocity corresponding to the 3 dimensions of space. Velocity is a vector – it has magnitude and direction at each point – and a vector field has at each point 3 values. Maxwell's equations of the electromagnetic field are vector field equations. More generally, in relativity a vector field has 4 components at each point of spacetime corresponding to the 4 dimensions of spacetime. Vectors are known as first-order tensors.

A fluid in motion is often quite complex and its behavior, incorporating expansion, rotation, and shear, is more adequately described by means of second-order tensors. A second-order tensor in relativity has at each point 16 components, and many basic equations in physics, such as the Einstein equation, use second-order tensors. The metric tensor, which contains the metric coefficients used for determining the geometry of spacetime, is a second-order tensor. Owing to the symmetries of spacetime – for example, that the distance from A to B is the same as that from B to A, and that the distance around a circle is the same measured clockwise or counterclockwise – the metric tensor has only 10 distinctly different components at each point.

The number of components in a tensor at each point of spacetime is 4^n, where 4 denotes the number of dimensions of spacetime and the symbol n indicates the order of the tensor. A scalar is zero order with $n = 0$ and has 1 component; a vector is first order with $n = 1$ and has 4 components; and so on. Occasionally, tensors higher than second order are used, and the gravitational field of general relativity is the Riemann curvature tensor of fourth order that has 256 components at each point of spacetime. Owing to various symmetries of spacetime, most of which we take for granted as common sense, many of the components of the Riemann curvature are equal to each other, and there are only 20 distinctly different values at each point of spacetime.

THE SCALAR–TENSOR THEORY OF GRAVITY. The scalar–tensor theory of gravity was advanced in 1939 by Pascual Jordan of Germany. In this theory the idea is to take the Riemannian spacetime of general relativity and lay in it a simple scalar field that varies from place to place. Gravity retains its character of dynamic curvature of spacetime, but is now modified by the introduction of the scalar field.

The scalar field is introduced in a

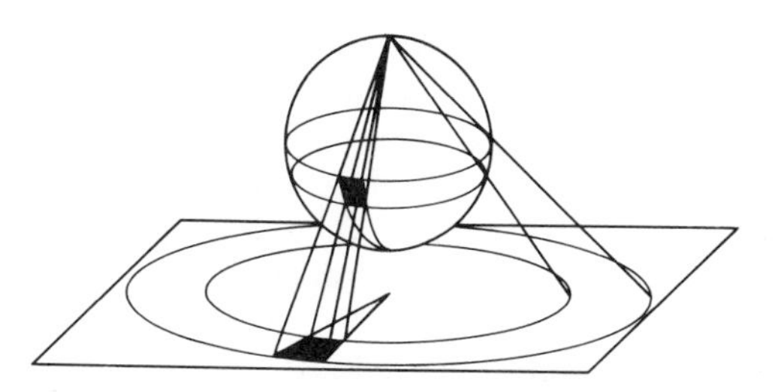

Figure 16.4. A conformal transformation changes space but preserves angles of intersection, as in this stereographic projection that maps a spherical surface onto a plane.

remarkable way by using what is known as a *conformal transformation* (see Figure 16.4). This transformation is made by multiplying the spacetime interval by the scalar. Space and time intervals are together stretched or contracted by an amount depending on the value of the scalar. This kind of transformation is called *conformal* because angles are unaltered, and it affects space and time intervals in the same way and the speed of light remains unchanged. If the scalar is everywhere the same, and constant in time, then spacetime is uniformly changed by a fixed amount and we can regard the process as merely a change of our conventional units of measurement. If everything in the universe is doubled in size, with the exception of a meter stick, all we have to do is relabel the stick as half a meter, and nothing has changed. Calling a centimeter a meter does not change the physical world. But when the scalar field varies from place to place in space and time, it then controls the relative size and duration of things, and the new spacetime interval obtained by such a transformation is quite different from the old. The transformation has now changed the physical properties of the universe in a dramatic fashion.

A conformal transformation changes the magnitude of the basic units that define intervals of space and time (see Figure 16.5). Thus the classical electron radius and the jiffy (a measure of time) are both increased or decreased by the same amount. For this reason a conformal transformation

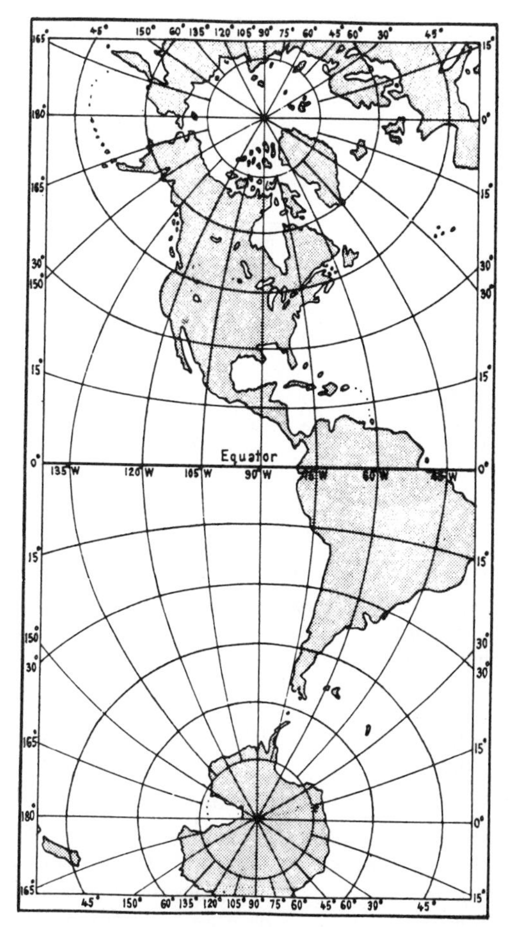

Figure 16.5. This Mercator map is a conformal transformation of the Western Hemisphere.

is sometimes called a *units transformation*. The main purpose of introducing such a transformation by means of a scalar field is to break the rigid constraints of general relativity and widen its physical possibilities. Let us take an ordinary universe in which atoms are of constant mass, everywhere the same, and in which the gravitational constant G has a fixed universal value. We now transform it into a new universe in which the basic units of measurement vary from place to place. In this new universe we do not know that some places have been stretched and other places have been contracted. As we travel around with a meter stick and a clock, each centimeter will

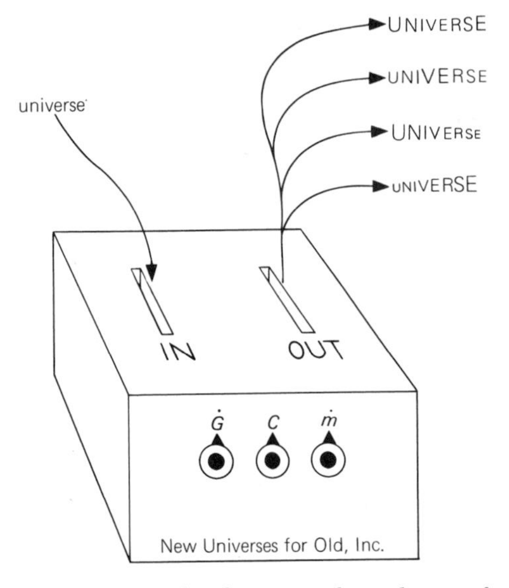

Figure 16.6. The dream machine that conformally transforms one universe into a physically different universe. Only angles and velocities are preserved by the transformation.

still be a fingerbreadth and each heartbeat will still be slightly less than a second. What we notice, however, is that the gravitational constant has different values at different places, that electrons and protons have masses that vary from place to place, and that matter is being created or destroyed.

The *scalar–tensor theory*, as it is called, which alters intervals of spacetime, is like a black box with two openings labeled IN and OUT. We insert a universe into the IN hole, and from the OUT hole comes a new and physically different universe (see Figure 16.6). On the outside of the black box – or "cosmic dream machine" – are a number of controls. Usually in cosmology we are not interested in making transformations that vary things from place to place, because we want to preserve homogeneity. The transformations commonly made are those that cause things to vary only in time. Hence we are interested in only three controls, marked $\dot{G}$, C, and $\dot{m}$, and by their adjustment it is possible to determine the kind of universe that emerges from the dream machine. With only these three knobs to adjust, the scalar field allows three things to happen that normally are forbidden in physics: Adjustment of the $\dot{G}$ knob controls the way the gravitational constant will vary in time; adjustment of the C knob controls the rate at which matter is created or destroyed; and adjustment of the $\dot{m}$ knob controls the way that particle masses will vary in time. By judicious adjustments these effects can be isolated or brought together in various combinations. Our dream machine, more fabulous than anything conceived in the annals of science fiction, is thus able to manufacture multitudes of different universes from any given universe.

JORDAN'S THEORY. Jordan's interest in the scalar–tensor theory was aroused by Paul Dirac's large-number hypothesis, discussed in the next chapter. To maintain the coincidence of the N_1 and N_2 cosmic numbers, Dirac had proposed that the gravitational constant G decreases in value as the universe expands. Jordan developed the scalar–tensor theory in order to show that G could be made to vary in the desired way. At first, the control knobs of the dream machine were not properly adjusted, and universes were produced in which everything varied in an alarming manner. Not only did G vary, as postulated by Dirac, but in addition matter was created and particle masses varied. Despite their heated debates, cosmologists are moderately tolerant of rival inventions, provided that credulity is not taxed excessively. A variation in G, or a creation of matter, or a variation in subatomic masses is each by itself of tolerable interest; but patience is exhausted when two or all three are allowed to happen simultaneously. After all, the essence of science is to seek and cling to laws of conservation and not allow them to come and go like will-o'-the-wisps. Although Jordan pioneered the scalar–tensor theory, his work in cosmology has not made a lasting impact.

THE HOYLE–NARLIKAR THEORY. Hoyle used the scalar–tensor theory to create matter in an expanding steady state universe. The creation rate C was adjusted nicely so that density remained constant. A slip of the hand, however, could quite easily make the creation rate either too fast or too slow, in which case a steady state situation would be lost.

The steady state theory has been overthrown by discoveries that its originators had not anticipated. Hoyle has earned notoriety as its most active supporter and with Narlikar has sought unceasingly for modifications of the steady state theory that will bring it into conformity with the new discoveries. The latest idea is that particles are not created; instead, owing to a universal interaction, their masses change with time. The $\dot{G}$ knob and the C knob are set to zero and the $\dot{m}$ knob is adroitly adjusted. As a result, "the usual mysteries concerning the so-called origin of the universe begin now to dissolve," wrote Hoyle in 1975. The universe is assumed to be static, and atoms, human beings, and stars shrink in size slowly because of the growth in the mass of subatomic particles. The expanding universe with atoms of constant mass has been transformed into a static universe of shrinking atoms. The big bang, or "creation of the universe," which Hoyle dislikes, is banished and becomes a moment when all masses happen to be close to zero. According to this picture the 3-degree cosmic radiation is not a product of the big bang, but actually starlight from an earlier phase of the universe that has been thermalized by scattering off atoms of enormous size at the epoch of zero mass. This "shrinking atom" theory, considered and rejected by Eddington, retains the infinite age of the steady state universe but abandons the idea of continuous creation.

THE BRANS–DICKE THEORY. Robert Dicke, whose skillful control of the dream machine is most impressive, has always left the C knob set at zero. With Carl Brans he has used the scalar–tensor theory as a basis for investigating Mach's principle.

Let m_{grav} be the gravitational mass of a particle determined by its response to gravity, and let m_{inert} be the inertial mass of the particle determined by its response to accelerated motion. In Newtonian and general relativity theories these two masses are made equal and written as m. We recall that Ernst Mach believed that the inertial mass is the result of a particle "feeling" the presence of all other particles in the universe. Distant particles, beyond the Hubble length L, are unobservable and therefore do not contribute to the determination of local inertial mass. Let M be the gravitational mass of the observable universe, and GMm_{grav}/L the gravitational energy of a single particle; it is conjectured, in accord with the spirit of Mach's principle, that the gravitational energy of a particle is equal to its intrinsic energy $m_{\text{inert}} \times c^2$:

$$\frac{GMm_{\text{grav}}}{L} = m_{\text{inert}}c^2$$

The interaction between the particle and the observable universe is assumed to be sufficiently strong to make m_{grav} equal to m_{inert}, and therefore

$$GM = Lc^2$$

Alternatively, we can say that with a value of m_{inert} determined by the interaction, the gravitational constant G is defined so that $m_{\text{inert}} = m_{\text{grav}}$, and both are expressed by the same symbol m. The equation $GM = Lc^2$ is now said to be a Machian law, or a bootstrap relation, which encompasses Mach's principle. It is a purely conjectural law, and many scientists have endeavored to establish for it a more secure theoretical foundation. The problem is to find a way in which the value of G is determined by the universe. Because the universe is expanding, this continual change, it is said, should react back on the value of G so that it also continually changes. Mach's principle, within the framework of an expanding universe, thus suggests that G cannot be constant in time. Brans and Dicke used the scalar–tensor theory because it allows G to vary with expansion, and this coupling of G-variation with expansion, they argued, is in accord with Mach's principle and is the justification for using the scalar–tensor theory.

In Chapter 17 we shall see that

$GM/Lc^2 = 1$ is characteristic of many universes controlled by gravity. One might argue that Mach's principle, in some sense, is already a feature of these universes. Advocates of Mach's principle regard this as suggestive but not sufficient, and a demonstration of the validity of Mach's philosophy requires that G be shown to vary with expansion owing to its dependent nature. But nobody is quite sure what Mach's principle actually means and everybody has a different interpretation. Any variation in the value of G, no matter how small, can be regarded as evidence of a Machian effect. The scalar–tensor theory allows G-variation to be either large or small, and if we adjust the $\dot{G}$ knob so that G-variation is small enough never to be in conflict with observation, we can say that the universe obeys Mach's principle. The Dirac universe (see Chapter 17) demands a variation in G so large that it is in contradiction with observation; the Brans–Dicke universe is more flexible and can have a G-variation as small as we please. Observations of orbital motions within the Solar System show that G-variation, if it exists, is very small; theoretical studies of the helium produced in the early universe also indicate that G-variation must be extremely small. As a consequence, the Brans–Dicke universe has become almost indistinguishable from a universe in which G is constant.

Scientists are reluctant to sacrifice their cherished laws of conservation and become willing only when the evidence is overwhelming; that the evidence for G-variation is now not very impressive has struck a blow at the scalar–tensor theory.

THE NEW THEORIES OF GRAVITY. With the scalar–tensor theory we are able to create from any one universe a large number of physically different universes. One of the most intriguing aspects of the theory is that any scalar–tensor universe can be transformed into any other scalar–tensor universe. A Brans–Dicke type of universe in which G varies can be transformed into a Hoyle–Narlikar type of universe in which m varies, and either type can be transformed into a continuous creation type of universe in which G and m are both constant. All are interchangeable by appropriate transformations.

It has been said that each scalar–tensor type of universe is the result of a new theory of gravity. This is a controversial subject. Such statements imply that each variant of the theory has the status of a new theory of gravity that rivals general relativity. But any one of the many scalar–tensor universes can be transformed into a universe in which G is constant, m is constant, and there is no creation of matter, and we then have what looks like a universe in concordance with general relativity. What happens is that the scalar field does not vanish but is transferred to the right side of the Einstein equation and tucked away in the "matter." Because general relativity never defines the exact nature of matter, it may be said that matter containing a scalar field is still "matter." At one time Einstein derived his equation by a "variational principle" in which matter is assumed to depend on an unspecified number of different scalar fields. It therefore seems that the multitude of scalar–tensor universes do not necessarily represent new theories of gravity but are specialized products of the Einstein master equation.

We may take the view that each scalar–tensor interpretation is equivalent to a new theory of gravity on equal footing with Einstein's equation. In that case we must then appeal to observation and experiments to decide which is the correct gravitational theory. Or we may take the view that all are implicit in the Einstein equation, and observations and experiments must determine what "matter" means in the Einstein equation and what is the appropriate transformation that generates the physical world. In this case the Einstein equation has an infinite number of physical interpretations. The virtue of the scalar–tensor theory is that it impels us to make careful observations and experiments in the hope that one day we shall know how to interpret the Einstein equation.

REFLECTIONS

1 *Why does pressure increase the deceleration term?*

* *Why does an expanding universe in tension release energy? Discuss McCrea's steady state theory.*
* *What is the homogeneity riddle? Discuss the chaotic and antichaotic viewpoints.*
* *What do you think of kinematic relativity and Milne's approach to cosmology?*
* *Discuss continuous creation. Imagine two steady state universes, one contracting and the other expanding, connected to each other by numerous spacetime bridges. As matter disappears from the contracting universe it reappears in the expanding universe.*
* *Discuss tensors. Why is speed a scalar and velocity a vector? What is a conformal transformation?*
* *Discuss the scalar–tensor theory.*
* *If you had a scalar–tensor dream machine and could create other universes, would you hesitate to use it because of the anthropic principle?*
* *Imagine a large number of universes, some of which evolve while others are in a steady state. Consider now the few that are self-aware and ask, What is the chance of being in an evolving universe? Each steady state universe is of infinite duration and therefore contains infinitely more intelligent beings than an evolving universe that is habitable for only a finite period of time. Hence, if steady state universes exist, the probability of occupying one is unity, and the probability of occupying an evolving universe is zero. But our universe is not in a steady state, and it follows that steady state universes of any kind probably do not exist.*

2 *Consider a box of volume V, having perfectly reflecting walls and containing radiation of mass density ρ. The mass of the radiation in the box is $M = \rho V$. We now weigh the box and find that its mass, because of the enclosed radiation, has increased not by M but by an amount $2M$. For example, 1 gram of radiation in a box increases the mass of the box by 2 grams. This unexpected increase in mass occurs because the radiation exerts pressure on the walls of the box and the walls contain stresses. These stresses in the walls are a form of energy that is equal to $3pV$, where p is the pressure of the radiation. The pressure is equal to $\frac{1}{3}\rho c^2$, and the energy in the walls is therefore $\rho c^2 V$ and has a mass equivalent of $M = \rho V$. The mass of the box is therefore increased by the mass M of the radiation and the mass M of the stresses in the walls, giving a total increase of $2M$. In the universe there are no walls; nonetheless, the radiation still behaves as if it had a gravitational and inertial mass twice what is normally expected. Instead of using ρ, we must use $\rho + 3p/c^2$, as in the first of the Friedmann–Lemaître equations shown further on in these Reflections. This peculiarity of general relativity explains why in a collapsing star, when all particles are squeezed to high energy, increasing the pressure hastens the collapse of the star.*

3 *Edward Milne (1896–1950) stressed the importance of the cosmological principle (which is the name he used for spatial homogeneity), and also the importance of operational methods of distance measurement similar to those used in radar. Among his many other ideas is that of two time scales: an atomic time scale in which time is measured by atoms and a dynamic time scale in which time is measured by the motions of planets and other bodies. The dynamic intervals of time are the logarithm of the atomic intervals of time; in dynamic time the gravitational constant G remains fixed, whereas in atomic time G increases and is not of fixed value. The idea of two time scales is intriguing but has not been taken seriously by physicists. Both Eddington and Milne asked searching questions, often of the "why" kind that probe deeply, and it is perhaps true to say that little of science can survive such devastating inquisitions. The roots of knowledge are buried in mystery, and when we dig too deep, too soon, the tree of knowledge falls down.*

Milne in his last book, Modern Cosmology and the Christian Idea of God, *published posthumously in 1952, wrote: "There is a remarkable difference between physics and philosophy. On the one hand, physicists agree with one another in general at any one time, yet the physical theories of any one decade differ profoundly from those of each succeeding decade – at any rate in the twentieth century. On the other hand, philosophers disagree with one another at any one time, yet the grand problems of philosophy remain the same from age to age. . . . The man of science should be essentially a rebel, a prophet rather than a priest, one who should not be ashamed of finding himself in opposition to the hierarchy. . . . The hard-baked or hard-boiled scientist usually holds that science and religion, whilst on nodding terms, have no immediate bearing on one another. On the contrary, one cannot study cosmology without having a religious attitude to the universe. Cosmology assumes the rationality of the universe, but can give no reason for it short of a creator of the laws of nature being a rational creator. God himself is limited by reason in the divine act of creation. God cannot do the impossible."*

4 *"The application of the laws of terrestrial physics to cosmology is examined critically. It is found that terrestrial physics can be used unambiguously only in a stationary homogeneous universe. . . . As the physical laws cannot be assumed to be independent of the structure of the universe, and conversely the structure of the universe depends upon the physical laws, it follows that there may be a stable position. We shall pursue the possibility that the universe is in such a stable, self-perpetuating state, without making any assumptions regarding the particular features which lead to this stability. We regard the reasons for pursuing this possibility as very compelling, for it is only in such a universe that there is any basis for the assumption that the laws of physics are constant; and without such an assumption our knowledge, derived virtually at one instant of time, must be quite inadequate for an interpretation of the universe and the dependence of its laws on its structure, and hence inadequate for any extrapolation into the future or the past" (Herman Bondi and Thomas Gold, "The steady-state theory of the expanding universe," 1948).*

* *"By introducing continuous creation of matter into the field equations of general relativity a stationary universe showing expansion properties is obtained without recourse to a cosmical constant. . . . The following work is concerned with this aspect of the matter and arose from a discussion with Mr. T. Gold who remarked that through continuous creation of matter it might be possible to obtain an expanding universe in which the proper density of matter remained constant. This possibility seemed attractive, especially when taken in conjunction with aesthetic objections to the creation of the universe in the remote past. For it is against the spirit of scientific enquiry to regard observable effects as arising from 'causes unknown to science,' and this in principle is what creation-in-the-past implies" (Fred Hoyle, "A new model for the expanding universe," 1948).*

* *"It is the purpose of a scientific hypothesis to stick out its neck, that is to be vulnerable. It is because the perfect cosmological principle is so extremely vulnerable that I regard it as a useful principle. It is something that could in practice be 'shot down' by experiment and observation far more easily than the ordinary cosmological principle, and I think you will agree with that" (Herman Bondi, in* Rival Theories of Cosmology, *a 1959 BBC discussion).*

* *"Therefore, when looking at the distant parts of the universe we see them as they were a long time ago. Any meaningful comparison of these distant parts with the ones nearby presupposes that the laws of physics are the same in the two cases. Since the universe, by definition, includes the study of all observable phenomena, we may expect the laws of physics also to be somehow determined by the universe as a whole. If the state of the universe was once very much different from what it is now, what guarantee do we have that the laws of*

physics were the same in the past as they are now?" (Jayant Narlikar, "Steady state defended," 1974). Narlikar repeats the argument originally made by Bondi and Gold that only in a universe eternally the same can the laws of nature remain invariant. The argument contains, however, a serious flaw. It is assumed that "the universe as a whole" extends in space but not in time. This is the human view of the universe that sees it changing with time, but a cosmic view of the universe of spacetime is unchanging. The unity of the universe, which perhaps determines the laws of nature, embraces spacetime and is not limited in either space or time.

5 *In Chapter 15 we considered the Friedmann–Lemaître equations for universes of zero pressure. These equations were derived by Newtonian arguments. According to Newtonian theory pressure has no gravitational effect, and we must therefore turn to general relativity for guidance when pressure is important. If the pressure p is isotropic, and everywhere the same, the Friedmann–Lemaître equations are*

$$\ddot{R} = -\frac{4\pi}{3}G\left(\rho + 3\frac{p}{c^2}\right)R + \frac{1}{3}\Lambda R \qquad (16.1)$$

$$\dot{R}^2 = \frac{8\pi}{3}G\rho R^2 + \frac{1}{3}\Lambda R^2 - k \qquad (16.2)$$

and all symbols are the same as before. These two equations can be combined to give

$$\frac{d}{dt}(\rho c^2 R^3) + p\frac{dR^3}{dt} = 0 \qquad (16.3)$$

which is the first law of thermodynamics for a homogeneous universe, as discussed in Chapter 13.

Let us suppose that when the pressure is large it is related to the density by the equation

$$p = (\gamma - 1)\rho c^2 \qquad (16.4)$$

where γ is a constant. In a universe containing matter of zero pressure, $\gamma = 1$, and in a universe containing radiation only, $\gamma = 4/3$. As in earlier discussions, let K be the curvature, H the Hubble term, and q the deceleration term. The Friedmann–Lemaître equations now take the form

$$(3\gamma - 2)\,K = \gamma\Lambda + H^2\,(2q - 3\gamma + 2) \qquad (16.5)$$

$$\Lambda = (3\gamma - 2)4\pi G\rho - 3qH^2 \qquad (16.6)$$

and the density ρ varies as $R^{-3\gamma}$.

When γ equals 1, we obtain the previous results for a zero-pressure universe:

$$K = \Lambda + H^2\,(2q - 1) \qquad (16.7)$$

$$\Lambda = 4\pi G\rho - 3qH^2 \qquad (16.8)$$

and when γ is 4/3, as in a radiation-dominated universe, we find

$$K = \frac{2}{3}\Lambda + H^2\,(q - 1) \qquad (16.9)$$

$$\Lambda = 8\pi G\rho - 3qH^2 \qquad (16.10)$$

A static Einstein universe has H = 0, and hence

$$K = \frac{1}{R^2} = 4\pi G\rho\gamma \qquad (16.11)$$

Of two Einstein universes with the same density, one containing matter ($\gamma = 1$) and the other radiation ($\gamma = 4/3$), the radiation universe has the smaller radius of curvature R.

Friedmann universes of $\Lambda = 0$ are given by

$$(3\gamma - 2)K = H^2(2q - 3\gamma + 2) \qquad (16.12)$$

$$(3\gamma - 2)4\pi G\rho = 3qh^2 \qquad (16.13)$$

and the zero-pressure universes of $\gamma = 1$ have already been discussed. In the radiation universes of $\gamma = 4/3$, we find

$$K = H^2\,(q - 1) \qquad (16.14)$$

$$8\pi G\rho = 3qH^2 \qquad (16.15)$$

These universes are closed (K is positive) when q is greater than 1, and open (K is negative) when q is less than 1. The Einstein–de Sitter radiation universe of K = 0 has a deceleration term of 1, and because the density ρ is proportional to $1/R^4$, we find that R varies as $t^{1/2}$.

6 *A small energy change dE, corresponding to a small volume change dV, is*

$$dE = -pdV$$

and when pressure p is negative, we see that energy increases with expansion. As before, we write $p = (\gamma - 1)\rho c^2$*, and a negative pressure – or cosmic tension – corresponds to* γ *having a value less than 1.*

We have seen that the density ρ *varies as* $R^{-3\gamma}$*, and hence when* $\gamma = 0$*, the density is constant. Universes of* $p = -\rho c^2$ *therefore expand and contract and have constant density; when they expand, matter is continually created, and when they contract, matter is continually destroyed. For these constant density universes the Friedmann–Lemaître equations become*

$$K = -H^2(q + 1)$$

$$\Lambda = -8\pi G\rho - 3qH^2$$

A flat ($K = 0$)*, static* ($H = 0$)*, and stable universe, such as the Newtonian universe, is possible when* $\Lambda = -8\pi G\rho$*. The steady state universe has constant H and q, and this is possible only when* $K = 0$*, and hence* $q = -1$*. If we assume that the cosmological constant* Λ *is zero (although nothing in the steady state theory demands that it should have this value), then we obtain Hoyle's result of* $8\pi G\rho = 3H^2$*, which was obtained from the scalar–tensor theory.*

FURTHER READING

Bondi, H. *Cosmology.* Cambridge University Press, Cambridge, 1960.

Kaufmann, W. "The Hoyle–Narlikar cosmology." *Mercury,* May–June 1976.

SOURCES

Bondi, H., Bonnor, W. B., Lyttleton, R. A., and Whitrow, G. J. *Rival Theories of Cosmology: A Symposium and Discussion of Modern Theories of the Universe.* Oxford University Press, London, 1960.

Bondi, H., and Gold, T. "The steady-state theory of the expanding universe." *Monthly Notices of the Royal Astronomical Society 108,* 252 (1948).

Brans, C., and Dicke, R. H. "Mach's principle and a relativistic theory of gravitation." *Physical Review 124,* 925 (1961).

Collins, C. B., and Hawking, S. W. "Why is the universe isotropic?" *Astrophysical Journal 143,* 317 (1973).

Dicke, R. H. *The Theoretical Significance of Experimental Relativity.* Gordon and Breach, New York, 1964.

Dicke, R. H. "Gravitational theory and observation." *Physics Today,* January 1967.

Dicke, R. H. *Gravitation and the Universe.* American Philosophical Society, Philadelphia, 1970.

Dingle, H. "On science and modern cosmology." *Monthly Notices of the Royal Astronomical Society 113,* 393 (1953). A presidential address attacking the steady state theory.

Harrison, E. R. "Classification of uniform cosmological models." *Monthly Notices of the Royal Astronomical Society 137,* 69 (1967).

Hoyle, F. "A new model for the expanding universe." *Monthly Notices of the Royal Astronomical Society 108,* 372 (1948).

Hoyle, F. "On the origin of the microwave background." *Astrophysical Journal 196,* 661 (1975).

Jeans, J. H. *Astronomy and Cosmogony.* Cambridge University Press, Cambridge, 1929. See p. 360.

Jones, B. J. T., and Peebles, P. J. E. "Chaos in cosmology." *Comments on Astrophysics and Space Physics 4,* 121 (1972).

Jordan, P. "Formation of the stars and development of the universe." *Nature 164,* 637 (October 15, 1949). A readable discussion giving references to Jordan's earlier work which began in 1939.

McCrea, W. H. "Relativity theory and the creation of matter." *Proceedings of the Royal Society 206,* 562 (1951).

McCrea, W. H. "Continual creation." *Monthly Notices of the Royal Astronomical Society 128,* 335 (1964).

Milne, E. A. *Relativity, Gravitation, and World-Structure.* Oxford University Press, Clarendon Press, Oxford, 1935.

Milne, E. A. *Kinematic Relativity.* Oxford University Press, Clarendon Press, Oxford, 1948.

Milne, E. A. *Modern Cosmology and the Christian Idea of God.* Oxford University Press, Clarendon Press, Oxford, 1952.

Narlikar, J. V. "Steady state defended," in *Cosmology Now,* ed. J. Laurie. British Broadcasting Corp., London, 1974.

Sciama, D. W. "Cosmology before and after quasars and the cosmic black-body radiation." *Scientific American,* September 1967.

17 THE COSMIC NUMBERS

Do you believe then that the sciences would ever have arisen and become great if there had not beforehand been magicians, alchemists, astrologers, and wizards who thirsted and hungered after secret and forbidden powers?
— Nietzsche (1844–1900), *Joyful Science*

JUGGLING WITH THE CONSTANTS OF NATURE

NATURAL NUMBERS. We measure distances in units such as centimeters and light years, intervals of time in units such as seconds and years, and masses in units such as grams and kilograms. There is nothing sacred about these units, which have been determined by our history, environment, and physiology. If we communicate with beings in another planetary system and inform them that something has a size of so many meters, an age of so many seconds, and a mass of so many kilograms, they will not understand because their units of measurement are undoubtedly different. But they will understand if we say the size is so many times that of a hydrogen atom, the age is so many times that of a certain atomic period, and the mass is so many times that of a hydrogen atom, simply because their atoms are the same as ours. The universe provides us all with the same set of natural units of measurement.

The only objects that appear to be exactly the same everywhere are the atoms and their constituent particles. A natural unit of mass is the nucleon mass, equal approximately to that of the hydrogen atom. A natural unit of length is the size of a subatomic particle, such as a nucleon, and this unit is known as the *fermi,* in honor of Enrico Fermi, and is equal to 10^{-13} centimeters. A natural unit of time is the period required by light to travel a distance of 1 fermi, and Richard Tolman suggested the name *jiffy* for this unit, which is equal to 10^{-23} seconds. Roughly speaking, a human being has a mass of 10^{29} nucleons, a height of 10^{15} fermis, and a lifetime of 10^{32} jiffies. With such natural units from the subatomic world we are able to appreciate the lavish scale on which the universe is constructed. Thus a planet such as the Earth has a mass of 10^{52} nucleons, a star such as the Sun has a mass of 10^{57} nucleons, and a galaxy such as our own has a mass of 10^{68} nucleons.

When the observable universe is measured in natural units we find that it has approximately a mass of 10^{80} nucleons, a size of 10^{40} fermis, and an age of 10^{40} jiffies. These *cosmic numbers* reveal harmony in the scale of the universe, and the more we investigate the cosmic numbers, the more impressive becomes the apparent harmony.

Cosmic numerology is as old as the Babylonians and the Pythagoreans; in many ways it is akin to astrology and has aspects, such as number worship, that rule it out as a scientific subject. The cosmic numbers, however, which deal with the scale and structure of the universe, form a more respectable subject that has received much attention in science. One of the most remarkable contributions to the subject of cosmic numbers was made more than 2000 years ago by Archimedes, who was perhaps the greatest scientist of the ancient world. Archimedes invented a system of numbers, called the "naming of numbers," and in a work entitled *The Sand Reckoner* he used this system to calculate the total number of grains of sand the universe could contain. In effect, he estimated the volume of the universe as it was then known, using a grain of sand as his natural unit of volume. He found that the universe had a volume of 10^{63} grains of sand. What is astonishing to us is that this mass of sand, as estimated by Archimedes, is fortuitously equal to our present estimates of the mass of the observable universe (see the Reflections section of this chapter).

CONSTANTS OF NATURE. Generally, each constant of nature indicates the involvement of a branch of physics:

G: gravity
c: relativity
h: quantum mechanics
m_n, m_e, e: subatomic particles

The gravitational constant G appears in calculations whenever gravity is involved. The speed of light c is always associated with relativity and the propagation of light. Planck's constant h is the cornerstone of quantum mechanics and is associated with the wavelike nature of particles and the corpuscular properties of radiation. Associated with the subatomic particles are the mass m_n of the nucleon, the mass m_e of the electron, and the electric charge e that is positive for a proton and negative for an electron. Nucleons (i.e., protons and neutrons) have a mass 1836 times the mass of the electron, and in each gram of matter there are almost a trillion trillion nucleons. One coulomb of electric charge contains nearly 10 million trillion elementary e charges, and this is the number of electrons that flow each second through the filament of a 100-watt electric light bulb.

The constants of nature can be arranged to form natural numbers (often referred to as *dimensionless numbers*) that are independent of our units of measurement. To start with, we give two examples. The ratio of the nucleon and electron masses

$$\frac{m_n}{m_e} = 1836$$

is obviously a natural number that is quite independent of whether we measure mass in grams or kilograms. The second example is the Sommerfeld *fine structure constant* denoted by α:

$$\alpha = \frac{2\pi e^2}{hc} = \frac{1}{137}$$

The fine structure constant appears whenever radiation interacts with particles, and the combination of c, h, and e indicates a wavelike (h) interaction between particles (e) and light (c). The value 1/137 is a natural number, used by terrestrial and extraterrestrial scientists alike, and is independent of the units of measurement. Planck's constant h frequently occurs in the form $h/2\pi$; for convenience we write $\hbar = h/2\pi$, and the fine structure constant is hence $\alpha = e^2/\hbar c$.

A characteristic size of atoms is the radius a_0 of the hydrogen atom:

$$a_0 = \frac{\hbar^2}{m_e e^2} = 0.5 \times 10^{-8} \text{ centimeters}$$

known as the Bohr orbit radius. The absence of G and c indicates that gravity and relativity are not of primary importance in the structure of atoms. A characteristic wavelike size of an electron, traveling close to the speed of light, is the *electron Compton length,*

$$\lambda_e = \frac{\hbar}{m_e c}$$

We notice that $\lambda_e = \alpha a_0$, and the electron Compton length is 137 times smaller than the hydrogen atom. The wavelength of an electron moving at speed v is $h/m_e v$, and because this equals the circumference of the hydrogen atom, an electron in a hydrogen atom has a speed $v = \alpha c$, and is therefore equal to the speed of light divided by 137. The *classical electron radius* is the size of an electron as calculated prior to the introduction of quantum mechanics. It is obtained by assuming that all the energy $m_e c^2$ of the electron is in the form of electrical energy equal to e^2/a, thus giving a radius a expressed by

$$a = \frac{e^2}{m_e c^2} = 3 \times 10^{-13} \text{ centimeters}$$

which is equal to 3 fermis. Electrons are wavelike and do not behave in a commonsense way like billiard balls; the classical electron radius is therefore usually not a proper measurement of the size of electrons. It is significant, nonetheless, that the classical electron radius is characteristic of the size of a nucleon (and of an electron sometimes, when it manifests itself in a corpuscular fashion), and we shall use it as a natural unit of length that denotes the intrinsic size of subatomic particles. We note that $a = \alpha\lambda_e = \alpha^2 a_0$, and hence the proton that is the nucleus of the hydrogen atom is α^2 times smaller than the atom.

So far we have found two natural numbers (1836 and 1/137) that involve relativity, quantum mechanics, and the properties of subatomic particles. We now construct a third number that involves the gravitational constant G. The electrical and gravitational forces between a proton and an electron are both attractive and are proportional to the inverse square of their separating distance r. The electrical force between a proton and an electron is e^2/r^2; the gravitational force is $Gm_n m_e/r^2$; and the ratio of these two forces gives the large natural number

$$\frac{e^2}{Gm_n m_e} = 0.2 \times 10^{40}$$

Electrical forces between neighboring particles are always vastly stronger than their gravitational forces. On the atomic scale electrical forces dominate, but on larger scales the positive and negative charges neutralize each other and the electrical forces therefore tend to become weak. Gravity on the other hand cannot be neutralized, and the larger the number of particles in a system the stronger becomes their combined gravitational force. Although gravity is weak and its effect is negligible among a few particles, in large systems of many particles it becomes relatively strong and sometimes of dominant importance. This is why the universe is governed by gravity and not by electrical forces.

If a nucleon were a nonrotating black hole it would have a Schwarzschild radius of $2Gm_n/c^2$. The coefficient 2 is not important and we shall say that $a_g = Gm_n/c^2$ is the gravitational length of a nucleon. This means that, if the nucleon had a radius a_g, then gravity would be of primary importance in determining its size. The actual radius is a, and we see that

$$\frac{a}{a_g} = \frac{e^2}{Gm_n m_e} = 0.2 \times 10^{40}$$

The nucleon, although small by ordinary standards, is vastly larger than it would be if it were collapsed into a black hole. It would seem therefore that gravity is not important in determining the structure of particles. Sometimes it is more convenient to use the *nucleon Compton length* $\lambda_n = \hbar/m_n c$ as a measure of the size of a nucleon, in which case we find

$$\frac{\lambda_n}{a_g} = \frac{\hbar c}{Gm_n^2} = 1.5 \times 10^{38}$$

This large number serves to illustrate that quantum mechanical and not gravitational forces are of dominant importance in the structure of particles.

CLUSTER HYPOTHESIS. We have obtained two groups of natural numbers from the constants of nature. The first consists of relatively small numbers clustered around unity:

$$\frac{m_e}{m_n}, \frac{e^2}{\hbar c}, \frac{\hbar c}{e^2}, \frac{m_n}{m_e}$$

and these numbers are 1/1836, 1/137, 137, and 1836, respectively. The second, with slight elaboration, consists of relatively large numbers clustered about 10^{40}, which are of the kind

$$\frac{e^2}{Gm_n^2}, \frac{\hbar c}{Gm_n^2}, \frac{e^2}{Gm_n m_e}, \frac{\hbar c}{Gm_n m_e}, \frac{e^2}{Gm_e^2}$$

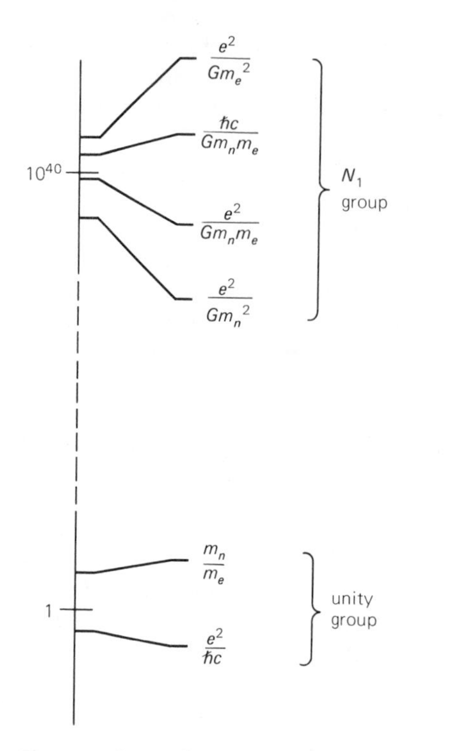

Figure 17.1. The unity and N_1 groups of numbers.

These numbers are 1/1836, 137/1836, 1, 137, and 1836 multiplied by 0.2×10^{40}, respectively. Each group consists of numbers covering a range that is quite small in comparison with the wide separation of the two groups, as shown in Figure 17.1. We shall refer to the first set of numbers as the *unity group* and to the second set of numbers as the *N_1 group.*

The clustering of natural numbers into two groups of relatively narrow spread is sufficiently remarkable for us to postulate a *cluster hypothesis.* This tentative hypothesis states that *all natural numbers compounded from the constants of nature are members of either the unity or the N_1 group.* Other numbers, such as $N_1^{1/2}$ or N_1^2, may be obtained from N_1 and cannot be regarded as basic in the same sense as N_1. We have no theory to support the cluster hypothesis, but presumably the explanation must have something to do with the basic design of the universe.

VARIATION OF THE CONSTANTS OF NATURE. On various occasions scientists have considered the possibility that the constants of nature, either singly or in combination, change with time as the universe evolves. To these scientists it has seemed reasonable that if the constants are related in some unknown Machian way to the properties of the universe as a whole, then they should change their values as the universe evolves. Let us see whether this idea contradicts the cluster hypothesis.

A frequent suggestion is that the gravitational constant G decreases in value as the universe expands. The unity group does not contain G, and therefore it remains unaffected by G-variation. All members of the N_1 group are inversely proportional to G; therefore G-variation merely moves the N_1 group without altering its tight clustering. Hence G-variation does not contradict the cluster hypothesis, and all natural numbers retain their membership in two widely separated groups. Some consequences of G-variation are mentioned later in connection with the Dirac universe.

It has been suggested that the elementary charge e changes its value slowly with time. This idea was once more popular than at present and was proposed as a possible way of explaining the redshifts of distant galaxies. If e was smaller in the past, then atoms were larger (a_0 was bigger) and their emitted wavelengths were longer, making distant sources appear to have recession redshifts. We notice that when e varies, then all numbers containing e^2 will wander outside the unity and N_1 groups, and at some time these groups will become dispersed beyond recognition. Hence e-variation contradicts the cluster hypothesis. Furthermore, if e varies, then the value of the fine structure constant also changes. Many details in the spectra of atoms are sensitive to the value of the fine structure constant α, and by studying the atomic emissions of distant quasars we have learned that α has not changed over long periods of time. The abundance of the isotopes of many terrestrial elements is also sensitive to the value of

α; so we know that it cannot have varied much during the lifetime of the Earth. Furthermore, both the structure of stars and the nuclear energy stars release are sensitive to the value of α, and changes of only a small percentage are sufficient to eliminate the possibility of long-lived luminous stars. All this increases our confidence in the cluster hypothesis, for if α remains constant, the unity and N_1 groups then preserve their tight clustering.

We are left with the possibility that subatomic particle masses change with time. It is now known that the combination $\alpha m_n/m_e$ does not change much, if at all, because the redshifts of distant galaxies are the same when measured in optical and 21-centimeter radiation. The ratio of the nucleon and electron masses is therefore constant because α, as shown by other methods, is itself constant. When m_n/m_e remains constant, the clustering of the unity and N_1 groups is preserved.

The evidence in support of the cluster hypothesis is impressive, and we must therefore draw the tentative conclusion that all natural numbers derived from the constants of nature reside in two groups that are tightly clustered. The only constant of nature that can vary without destroying the clustering is the gravitational constant.

MAGIC NUMBERS. Max Planck in 1913 showed that the constants G, c, and $\hbar$ can be combined to create natural units of length, time, and mass. The *Planck length,* as it is called, is

$$a^* = \left(\frac{G\hbar}{c^3}\right)^{1/2} = 2 \times 10^{-33} \text{ centimeters}$$

and the *Planck mass* is

$$m^* = \left(\frac{\hbar c}{G}\right)^{1/2} = 2 \times 10^{-5} \text{ grams}$$

The Planck unit of time is obtained by dividing the Planck length a^* by the speed of light. These units can be found in the following way. We suppose that a particle exists of mass m^* that has a gravitational length equal to its Compton length. Its gravitational length is $a^* = Gm^*/c^2$, and its Compton length is $a^* = \hbar/m^*c$; by equating the two we find the Planck units. Such a particle is a black hole of quantum mechanical size and is therefore a *quantum black hole.* From the constants G, c, and $\hbar$ we infer that its structure is determined by gravity, relativity, and quantum mechanics. Creation of quantum black holes requires an enormous energy, and so far they have not been observed. They are possibly the basic constituents of spacetime; if spacetime could be examined on the scale of 10^{-33} centimeters it would perhaps consist of a sea of virtual quantum black holes. It is conceivable that in the beginning the universe consists of an extremely dense sea of real quantum black holes (see Chapter 18). On this matter we cannot be certain because we lack a quantum mechanical theory of gravity.

On comparing quantum black holes with nucleons we find

$$\frac{\text{nucleon Compton length}}{\text{Planck length}} = \left(\frac{\hbar c}{Gm_n^2}\right)^{1/2} = N_1^{1/2}$$

$$\frac{\text{Planck mass}}{\text{nucleon mass}} = \left(\frac{\hbar c}{Gm_n^2}\right)^{1/2} = N_1^{1/2}$$

The nucleon is roughly 10^{20} times larger than a quantum black hole, but the mass of a quantum black hole is roughly 10^{20} times greater than that of a nucleon.

The number of nucleons in a star can be found in the following way. A star is held together by gravity and is supported by the pressure resulting from the motion and interaction of its individual particles. Unlike other self-gravitating bodies, such as planets, it consists mainly of hot hydrogen gas, and its constituent particles are protons and electrons that move freely. As these particles rush around they have close encounters and are strongly affected by their electrical forces. A star is therefore dominated by gravity on the large scale and by electrical forces on the small scale. Despite their diversity in size and luminosity, stars have only a relatively small range of masses. Those only 1/10 the mass of the Sun are not hot enough to count as luminous stars, and those more than 10 times the mass of the Sun are rare and burn their hydrogen rapidly. Because the structure of a star is governed by gravitational and electrical forces, a reasonable guess would be that the number of particles in a star depends in some way on the number N_1 that expresses the relative strengths of electrical and gravitational forces. Let N_{star} be the number of nucleons in a typical star; it can be shown (see the Reflections section) that

$$N_{\text{star}} = \left(\frac{\hbar c}{G m_n^2}\right)^{3/2} = 2 \times 10^{57}$$

Thus $N_{\text{star}} = N_1^{3/2}$, and the typical mass of a star is therefore given by

$$M_{\text{star}} = m_n N_{\text{star}} = 4 \times 10^{33} \text{ grams}$$

This, astonishingly enough in view of the roughness of the calculation, is twice the mass of the Sun.

We have derived two numbers $N_1^{1/2}$ and $N_1^{3/2}$, and by interchanging e^2 and $\hbar c$, and m_n and m_e, we can construct sets of numbers that belong to the $N_1^{1/2}$ and $N_1^{3/2}$ groups. For example, a stellar mass has a gravitational length GM_{star}/c^2, and this is approximately the radius it has when it is in the form of a black hole. This radius is $N_1^{3/2}$ times the gravitational length of a nucleon, N_1 times a Planck length, and $N_1^{1/2}$ times the size of a nucleon.

The constants of nature form natural numbers in the unity and N_1 groups, and we have found other groups of physical significance, such as $N_1^{1/2}$ and $N_1^{3/2}$, which derive from the unity and N_1 groups. Breakdown in the cluster hypothesis would destroy the "magic-number" sequence: 1, $N_1^{1/2}$, N_1, $N_1^{3/2}$. We shall see that the magic numbers are even more remarkable than this sequence suggests.

THE COSMIC CONNECTION

A STRIKING COINCIDENCE. The observable universe extends out to a distance of 15 billion light years. This distance is the Hubble length denoted by L, and when it is

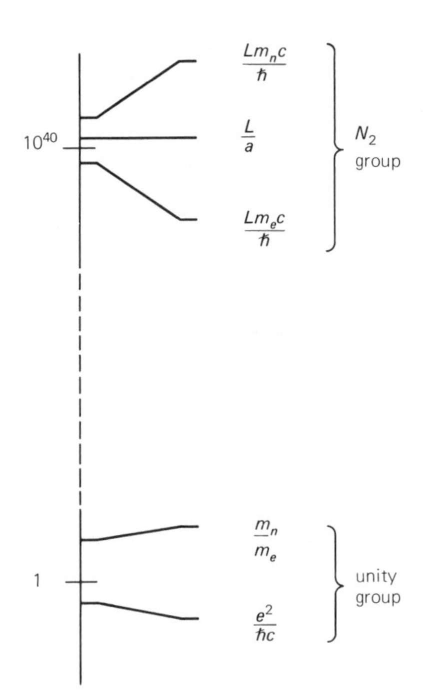

Figure 17.2. The unity and N_2 *groups of numbers.*

expressed in units of the nucleon size, we find

$$\frac{L}{a} = N_2 = 5 \times 10^{40}$$

The size of the observable universe is hence approximately 10^{40} times the size of a nucleon; alternatively, it is 4×10^{38} electron Compton lengths, or 7×10^{41} nucleon Compton lengths. Roughly speaking, the observable universe has a size of 10^{40} fermis, and the Hubble period is therefore 10^{40} jiffies, where a jiffy (10^{-23} seconds) is the approximate time taken by light to travel a distance of 1 fermi. The characteristic size and age of the universe, when measured in natural units, are represented by a group of numbers clustered in the neighborhood of 10^{40}. We shall refer to these numbers as the N_2 group (see Figure 17.2).

The approximate coincidence

$$N_1 = N_2$$

is rather striking. Both groups have extremely large values and their approximate

Table 17.1. *Cosmic numbers*

1	$N^{1/2}$	N	$N^{3/2}$	N^2
$\frac{\hbar^2 H}{G m_n m_e^2} = \frac{N_1}{N_2}$	$\frac{a}{a^*} = N_1^{1/2}$	$\frac{a}{a_g} = N_1$	$\frac{r}{a_g} = N_1^{3/2}$	$\frac{L}{a_g} = N_1 N_2$
	$\frac{r}{a} = N_1^{1/2}$	$\frac{r}{a^*} = N_1$	$\frac{L}{a^*} = N_1^{1/2} N_2$	$\frac{M}{m_n} = N_1 N_2$
	$\frac{L}{r} = \frac{N_2}{N_1^{1/2}}$	$\frac{L}{a} = N_2$	$\frac{M_\odot}{m_n} = N_1^{3/2}$	$\frac{\rho_j}{\rho} = N_2^2$
	$\frac{m^*}{m_n} = N_1^{1/2}$	$\frac{M_\odot}{m^*} = N_1$	$\frac{M}{m^*} = N_1^{1/2} N_2$	$\frac{\rho^*}{\rho_n} = N_1^2$
	$\frac{M}{M_\odot} = \frac{N_2}{N_1^{1/2}}$	$\frac{\rho_n}{\rho} = \frac{N_2^2}{N_1}$		
		$\frac{\rho_j}{\rho_n} = N_1$		
		$\frac{\rho^*}{\rho_j} = N_1$		

equality suggests a hidden fundamental relation between the constants of nature and the structure of the universe. A fortuitous coincidence cannot be ruled out, of course, but seems unlikely with numbers so large.

THE LARGE-NUMBER HYPOTHESIS. In 1937 the famous physicist Paul Dirac ventured into cosmology and postulated the *large-number hypothesis*. This hypothesis, according to Dirac, declares: "*Any two of the very large dimensionless numbers occurring in Nature are connected by a simple mathematical relation, in which the coefficients are of order unity.*" The large numbers referred to by Dirac are the members of the N_1 and N_2 groups. Dirac's speculative hypothesis implies that a physical relation exists between the two groups and both remain permanently equal to each other; it denies that the equality of the two groups is a mere coincidence. The N_1 group comes from the world of subatomic particles, and the N_2 group measures the size of the universe; at present we have no physical theory that connects the two.

The cluster hypothesis states that all natural (i.e., dimensionless) numbers, derived from the constants of nature, fall into the unity and N_1 groups; the large-number hypothesis is the cosmic connection which states that the N_1 and N_2 groups are physically related and permanently equal.

THE COSMIC-NUMBER DILEMMA. The N_2 group of numbers measures the scale of the universe. But the universe is expanding and therefore N_2 increases with time. When the universe was only 1 year old, for instance, N_2 had a value of 10^{30}, whereas N_1, which does not vary, still had a value of 10^{40}. The present-day coincidence of N_1 and N_2 is therefore rather puzzling. Are we to believe that the coincidence is a peculiarity of the present era and could not exist in the remote past and future? In this case we must reject the large-number hypothesis. Or should we contrive in some way to preserve the equality of N_1 and N_2 so that either both are constant or they vary together? This is the cosmic-number dilemma: Are N_1 and N_2 equal by chance or are they permanently equal and related in a way that we do not yet understand? We shall consider how Dirac and others have attempted to resolve the cosmic-number dilemma. First, however, we show that the large-number hypothesis leads to an impressive magic-number sequence.

MAGIC NUMBERS

Because N_1 and N_2 are equal, there exists a sequence 1, $N^{1/2}$, N, $N^{3/2}$, N^2, . . . , where N stands for either N_1 or N_2.

We have deliberately not assigned precise values to either N_1 or N_2 because each stands for a group of numbers clustered about 10^{40}. Let N represent any number somewhere in the range 10^{38} to 10^{42} that comes from either the N_1 or the N_2 group. In other words, $N_1 = N_2 = N$, and the equality sign in this case means approximate equality. We now find that there is a sequence of magic numbers

$$\ldots N^{-2}, N^{-3/2}, N^{-1}, N^{-1/2}, 1, N^{1/2}, N, N^{3/2}, N^2, \ldots$$

The numbers preceding 1 are obtained by simply inverting those that follow, and we need consider only the numbers that are unity and greater. We are given a sequence of numbers of the kind 1, 10^{20}, 10^{40}, 10^{60}, 10^{80}, . . . , all compounded from N_1 and N_2. Some examples of how these numbers may be derived are found later in this chapter, and more information is given in Table 17.1 (see Table 17.2 for identification of unfamiliar symbols). The equality sign = in most cases, in the text and the tables of this chapter, is used to indicate "approximately equal to."

Table 17.2. *List of symbols*

length	$a_g = \dfrac{Gm_n}{c^2}$	gravitational size of nucleon
	$a^* = \left(\dfrac{G\hbar}{c^3}\right)^{1/2}$	Planck length
	$a = \dfrac{e^2}{m_e c^2}$	classical electron radius
	$r = \dfrac{GM_\odot}{c^2}$	gravitational size of star
	$L = \dfrac{c}{H}$	Hubble length
time	$t^* = \dfrac{a^*}{c}$	Planck unit of time
	$j = \dfrac{a}{c}$	jiffy unit of time
mass	m_e	electron mass
	m_n	mass of nucleon
	$m^* = \left(\dfrac{\hbar c}{G}\right)^{1/2}$	Planck mass
	$M_\odot$	stellar mass
	M	mass of observable universe
density	$\rho = \dfrac{M}{L^3}$	density of the universe
	$\rho_n = \dfrac{m_n}{a^3}$	density of nucleon
	$\rho_j = \dfrac{c^2}{Ga^2}$	jiffy density
	$\rho^* = \dfrac{c^5}{G^2\hbar}$	Planck density

UNITY GROUP. The fine structure constant and the ratio of the nucleon and electron masses are numbers that belong to the unity group. In this same group there is now also N_1/N_2, of which a representative sample is

$$\frac{e^4}{GLm_n m_e^2 c^2} = 0.4$$

Other members in the unity group are obtained by interchanging the particle masses m_n and m_e and using also the fine structure constant. John Stewart of the Princeton University Observatory was the first to draw attention – in 1931 – to the possible importance of unity numbers of the N_1/N_2 kind. From the *Stewart numbers* we find

$$\text{characteristic particle mass} = \left(\frac{e^4}{GLc^2}\right)^{1/3}$$

This has often been cited as possible evidence that the masses of subatomic particles are in some way related to the scale of the universe.

An additional unity number of a different kind is found in the following way. According to most theories the expansion of the universe is controlled by gravity, and if numerical coefficients and other numbers such as π are ignored, this is expressed by the approximate relation $G\rho = H^2$, where ρ is the average density of the universe. The Friedmann universes, particularly the Einstein–de Sitter universe, are examples. Because the Hubble length L equals c/H, we have

$$\frac{G\rho L^2}{c^2} = 1$$

The mass of the observable universe is approximately $M = \rho L^3$, which gives

$$\frac{GM}{Lc^2} = 1$$

This says that the mass of the observable universe has a gravitational length equal to the Hubble length. Some scientists, particularly Dennis Sciama of Oxford University and Robert Dicke of Princeton University, have argued that this last equation holds rigorously and is not just an approximate relation. Their argument is based on Mach's principle and states that the gravitational constant G is determined by the distribution of matter in the universe in a way not explained by general relativity. If this interpretation of Mach's principle is correct, then it is plausible, they argue, that G is equal to Lc^2/M at every instant; and therefore G must vary with expansion.

$N^{1/2}$ GROUP. We have already encountered numbers of the $N^{1/2}$ kind. In addition, the radius of a black hole of stellar mass is roughly $r = GM_{\text{star}}/c^2$, and because M_{star} is equal to $m_n N_1^{3/2}$, we easily find

$$\frac{L}{r} = \frac{N_2}{N_1^{1/2}} = N^{1/2}$$

Also, because the mass of the observable universe is $M = Lc^2/G$,

$$\frac{M}{M_{\text{star}}} = \frac{N_2}{N_1^{1/2}} = N^{1/2}$$

Put in words rather than symbols, the observable universe is 10^{20} times the radius of a black hole of stellar mass, and the mass of the observable universe is 10^{20} times that of a star.

N GROUP. The N_1 and N_2 numbers have been discussed, and of their combinations that yield N, we single out the ratio of the density of a nucleon to the density of the universe. If we neglect small numbers such as π, the density of a nucleon is m_n/a^3 and the density of the universe is M/L^3, from which we get

$$\frac{\text{density of nucleon}}{\text{density of universe}} = \frac{N_2^2}{N_1} = N$$

The density of a nucleon is therefore roughly 10^{40} times the density of the universe.

$N^{3/2}$ GROUP. We have seen that the mass of a star is $N^{3/2}$ times the mass of a nucleon. From previous results we also find that the mass of the observable universe, measured in Planck units of mass, is expressed as

$$\frac{\text{mass of universe}}{\text{Planck mass}} = N_1^{1/2}N_2 = N^{3/2}$$

and the size of the observable universe, measured in Planck units of length, is given by

$$\frac{\text{size of universe}}{\text{Planck length}} = N_1^{1/2}N_2 = N^{3/2}$$

N^2 GROUP. The total number of nucleons in the observable universe is the famous *Eddington number* N^2. It is obtained by dividing the mass of the observable universe by the nucleon mass, and we find

$$\text{number of nucleons in the universe} = N_1N_2 = N^2$$

Eddington attached great importance to this number. Because the Eddington universe is closed he was able to argue that N^2, equal roughly to 10^{80}, is the actual number of nucleons in a finite universe. But, in general, this number is no more than a rough estimate of the nucleons in the observable universe of radius equal to the Hubble length. Archimedes' 10^{63} grains of sand contain, by pure chance, 10^{80} nucleons, and

Archimedes' number and Eddington's number are therefore equivalent to each other.

OTHER NUMBERS. These are only a few examples of the various combinations of N_1 and N_2 that have physical significance. It is interesting to note that there are no numbers of the $N_2^{1/2}$, $N_2^{3/2}$ kind, and also there are no obvious combinations that yield $N^{5/2}$ and $N^{7/2}$. There are a total of 21 numbers shown in Table 17.1, of which 11 involve powers of only N_1, 2 involve powers of only N_2, and 8 involve combinations of the powers of N_1 and N_2.

CLUSTER AND LARGE-NUMBER HYPOTHESES. The cluster hypothesis asserts that all natural numbers formed by the constants of nature are tightly clustered in the unity and N_1 groups. This tendency to cluster is then carried over into the physically significant 1, $N_1^{1/2}$, N_1, $N_1^{3/2}$, . . . sequence. The coincidence $N_1 = N_2$ now implies that the cluster hypothesis applies to the unity and N groups, where N stands for either N_1 or N_2, and as a consequence there is now a quite impressive magic-number sequence 1, $N^{1/2}$, N, $N^{3/2}$, N^2, . . .

The large-number hypothesis locks together permanently N_1 and N_2 and states, in effect, that the cluster hypothesis applies to the unity and N groups, and not just to the unity and N_1 groups. Let us suppose that the large-number hypothesis is wrong and N_1 and N_2 are not permanently equal. In the unity group are numbers such as N_1/N_2; in the $N^{1/2}$ group are numbers such as $N_1^{1/2}$ and $N_2/N_1^{1/2}$; in the N group are numbers such as N_1, N_2, and N_2^2/N_1; and so on for the other groups. It is clear that if N_1 and N_2 are not permanently equal, then in the remote past and in the remote future all groups will lose their tight clustering and become smeared out and indistinguishable. If the large-number hypothesis is wrong, the magic-number sequence is merely a quirk of the present era and is not of major cosmic significance. Many people find this difficult to believe and are reluctant to throw away the elegance of the magic-number sequence by considering it a passing whim in the design of the universe. After all, there is very little that we know about the harmony and structure of the universe, and why discard one of our most important clues?

WHY THE COINCIDENCE?

EDDINGTON'S ANSWER. How can we make N_1 and N_2 permanently equal and thereby preserve the magic-number sequence? There are various answers.

Eddington was the first to insist that the proper cosmic yardstick is not the Hubble length L, which continually changes, but the cosmological constant Λ. Hence N_2 is not equal to L/a; rather,

$$N_2 = \frac{c}{a\Lambda^{1/2}}$$

The Eddington and Lemaître universes contain the cosmological constant, and therefore, according to this interpretation, they have constant values of N_2. In both universes N_2 is roughly equal to N_1, so that the large-number hypothesis is satisfied. Eddington's solution of the cosmic-number dilemma is attractive, although not fully persuasive, because of our uncertainty concerning the existence of the Λ force.

DIRAC'S ANSWER. Dirac suggested that N_1 increases with time and in this way remains permanently equal to N_2. The only way N_1 can change is by slow variation of one or more of the constants of nature. The cluster hypothesis eliminates all such variations except G-variation, and Dirac assumed that the gravitational constant G decreases slowly in an expanding universe. We have seen that N_1 is proportional to $1/G$, and N_2 is proportional to L; in the Dirac universe $N_1 = N_2$ because GL is constant.

From the unity group of numbers we have the relation $G\rho L^2 = c^2$, where ρ is the density of the universe. Because GL is constant, ρL is constant in the Dirac universe. But the density ρ varies as $1/R^3$, where R is the scaling factor; and the expansion distance L

varies as $1/H$, where the Hubble term is $H = \dot{R}/R$. This means that $\dot{R}R^2$ is constant, and hence the scaling factor R varies as $t^{1/3}$, where t is the age of the universe. We find

$$\text{age of Dirac universe} = \frac{1}{3} \times \text{Hubble period}$$

and for a Hubble period of 15 billion years this gives an age of 5 billion years. We also find that G is proportional to $1/t$, and the gravitational constant has an infinitely large value at the beginning of expansion.

The Dirac universe is only slightly older than the Solar System. This rather young age of the universe is deceptive, however, because as we go back in time the gravitational constant increases and all self-gravitating systems speed up their evolution. The distance of the Earth from the Sun changes as $1/G$; the length of the year varies as $1/G^2$; and 4 billion years ago the Sun was 5 times nearer and the year was 25 times shorter than at present. The luminosity of stars varies as G^7; two and a half billion years ago the Sun was 128 times more luminous than at present, and even half a billion years ago the oceans on Earth were at boiling point owing to the brightness and closeness of the Sun. In this case there is little doubt that, in the early days of the Solar System, Venus would have lost its atmosphere, so that it would now be a seared planet like Mercury, and the Earth a billion or so years ago would have had an atmosphere like Venus's at present and a surface glowing at dull red heat.

Variation of G cannot affect atomic and molecular processes or the biochemistry of life. Self-gravitating systems evolve rapidly in the Dirac universe. But life evolves no more rapidly than in a universe of constant G. From the fossil record it is known that algae existed on Earth more than 3 billion years ago. But 3 billion years ago in the Dirac universe the Earth's surface was scorched by the intensely bright Sun and life was impossible. This strongly suggests, as first shown by Edward Teller, that G does not vary in the way proposed by Dirac. N_1 has been made permanently equal to N_2 at the price of eliminating the creatures who observe the coincidence $N_1 = N_2$.

Other theories have been advanced in which the gravitational constant varies at a much slower rate than suggested by Dirac. These theories are of the scalar–tensor kind discussed in the previous chapter. A slow variation of G is not in discord with the existence of life on Earth, but from the standpoint of the cluster and large-number hypotheses, a slow variation of G that fails to maintain the equality $N_1 = N_2$ is little better than no variation at all.

THE STEADY STATE ANSWER. In the de Sitter universe, and also the continuous creation steady state universe proposed by Bondi, Gold, and Hoyle, nothing ever changes in the cosmic scene. In particular, the Hubble term is constant and the Hubble length L is always the same. Therefore the relationship $N_2 = L/a$ is unchanging and the equality of $N_1 = N_2$ is permanently secured.

DICKE'S ANSWER. If N_2 changes with time, then we should at least try to explain why the coincidence $N_1 = N_2$ exists at present. Robert Dicke of Princeton University proposed in 1961 an ingenious explanation that appeals to the anthropic principle. In the distant past when the universe was young, the number N_2 was small, but nobody was around to take note of its discordant value. In the distant future when the universe is old and the stars are dead, the number N_2 will be large, and again nobody will be around to notice the difference. Dicke argued that N_2 is approximately equal to N_1 during that period of cosmic history when intelligent life exists.

Life, as far as we understand it, cannot begin until the first generation of stars has evolved and produced elements, such as carbon, oxygen, and nitrogen, that are essential for biological structures. These elements are ejected into space by the dying stars and are then incorporated in other stars and their planetary systems that follow later. Thus life cannot begin until the first generation of stars has died.

Stars have a typical lifetime given by the equation

$$\text{stellar lifetime} = \frac{1}{1000} \times \frac{M_{\text{star}}c^2}{\text{luminosity}}$$

The total energy of a star is its mass multiplied by c^2, of which only 1 percent is released by nuclear reactions that convert hydrogen into helium and heavier elements; only about 10 percent of the hydrogen is converted. This released energy, about one-thousandth of $M_{\text{star}}c^2$, divided by the luminosity, gives the lifetime of the star.

Light streaming away from the surface of a star pushes on particles in its atmosphere. When a star has maximum brightness, the radiation pressure pushing outward at the surface equals the gravitational pull inward. A star of greater brightness would blow itself away. This maximum brightness is known as the *Eddington luminosity* and is equal to $GM_{\text{star}}m_nc/a^2$, where a is the classical electron radius. A normal star has a luminosity typically one-thousandth of the Eddington value:

$$\text{luminosity} = \frac{1}{1000} \times \frac{GM_{\text{star}}m_nc}{a^2}$$

When this result is substituted in the stellar lifetime equation, we find

$$\text{stellar lifetime} = N_1 \text{ jiffies}$$

Life cannot commence until the age of the universe equals the lifetime of the first generation of stars, and life in general therefore begins at some time after

$$\text{age of universe} = \text{stellar lifetime}$$

The present age of the universe is 10 or more times greater, but a factor of 10 is not important and lies well within the spread of the unity and N groups of numbers. Because the age of the universe is N_2 jiffies, we have

$$N_1 = N_2$$

as the condition for life to exist. Life in its intelligent form then notices the remarkable coincidence of these two large numbers. Eventually N_2 will become much larger than N_1, but by then all stars will have died and life will have perished. The cosmic numbers N_1 and N_2 are therefore in agreement while intelligent life exists.

According to Dicke's anthropic argument the universe is a fit place for habitation by life, and is self-aware, while the values of N_1 and N_2 are in agreement. It explains not only why these two cosmic numbers are observed to be equal, but also helps us to understand why they are so large: We can exist only in a universe that is generously endowed with the eons of time sufficient for stellar and biological evolution.

Strictly speaking, Dicke's argument claims too much: It provides a necessary but not a sufficient condition for life. Elements such as carbon and oxygen are necessary for life but are also necessary to make rocks. The approximate agreement of N_1 and N_2 is sufficient for the formation of rocks but is not sufficient for the origin and evolution of life. Universes can exist in which N_1 equals N_2 and yet there might be nobody around to notice the coincidence.

THE COSMIC YARDSTICK. The Hubble length – or the size of the observable universe – is commonly used as a cosmic yardstick (see Figure 17.3). But it has the undesirable and unfortunate property of changing with time. Is it possible therefore that we have failed to identify the correct cosmic yardstick when we determine the value of N_2? We have seen that the cosmological constant provides a fixed measure of the scale of the universe, and perhaps Eddington was right when he said that Λ is the natural standard that determines the sizes and masses of subatomic particles. We might then use the anthropic principle to show that life exists only in those universes in which the value of Λ is such that N_1 and N_2 are equal.

There is yet another way we might approach this whole subject. The universe contains spacetime and its contents are displayed as events and worldlines within spacetime. Why then should we choose as characteristic of the size of the universe the Hubble length that has different values at different places in spacetime? A diameter of a vase does not necessarily tell us the size of

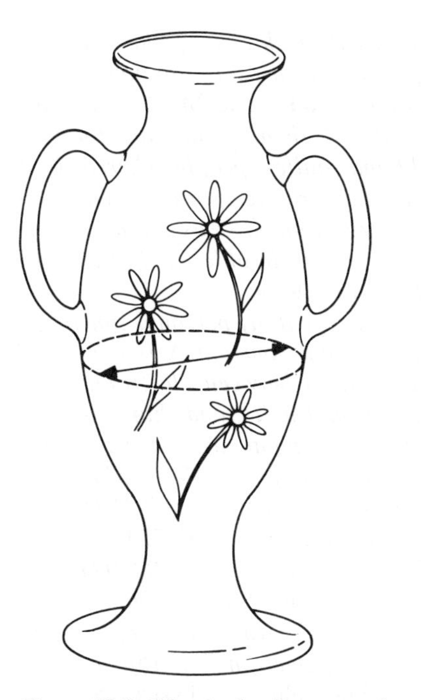

Figure 17.3. What is the diameter of a vase? The question illustrates the difficulty of using the Hubble length as a cosmic yardstick.

a vase. We should perhaps try to rid ourselves of prerelativity modes of thought and take an enlarged view of a universe of global spacetime. We then ask ourselves, What is of cosmic and invariant magnitude in the universe?

Expanding universes of infinite space and infinite time do not possess an invariant geometric scale. Using Eddington's kind of argument we could say that these universes possess nothing that determines the sizes and masses of subatomic particles; hence there is no natural value for N_1, and such universes probably do not contain life.

Expanding universes of finite space and infinite time also do not possess an invariant geometric scale, and perhaps also do not contain well-determined subatomic particles; hence they are without life.

But expanding universes of finite lifetime, and of finite or infinite space, do contain an invariant spacetime scale. It is the lifetime of the universe. A universe that expands and then collapses has an invariant cosmic length equal to its duration. In such a universe there is an intrinsic scale that could determine the sizes and masses of subatomic particles, and life is hence possible.

Expanding universes that are finite in both space and time are doubly attractive because in all directions of spacetime there exists approximately the same maximum cosmic length. The closed Friedmann universe is a particularly simple example; it has a lifetime of πR_{max} and a maximum spatial scale determined by R_{max}, and the natural and only cosmic yardstick is R_{max} (see Figure 17.4). In this case we should therefore write $N_2 = R_{max}/a$; and $N_1 = N_2$ is then a constant relation throughout spacetime.

REFLECTIONS

1 *In his article "Any physics tomorrow?" (1949), George Gamow writes: "If and when all the laws governing physical phenomena are finally discovered, and all the empirical constants occurring in these laws are finally expressed through the four independent basic constants, we will be able to say that physical science has reached its end, that no excitement is left in further explorations, and that all that remains to a physicist is either tedious work on minor details or the self-educational study and adoration of the magnificence of the completed system. At that stage physical science will enter from the epoch of Columbus and Magellan into*

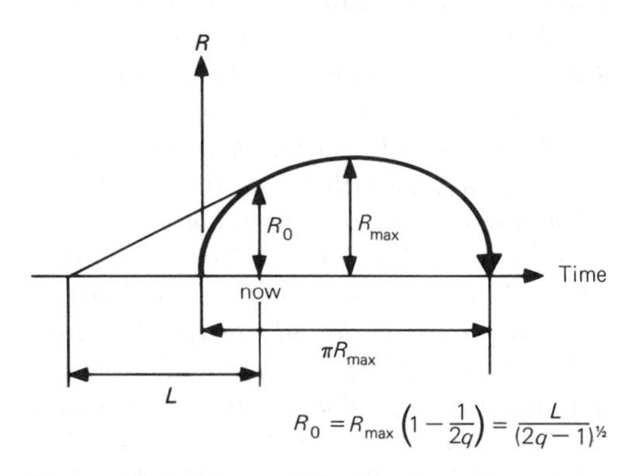

Figure 17.4. The Hubble length L varies with time, but in a closed universe R_{max} is the natural cosmic length because it is constant.

Figure 17.5. Cover of Physics Today, *January 1949. (With permission from American Institute of Physics.)*

the epoch of the National Geographic Magazine." Do you think this is likely ever to happen?

2 *"All the systems of units which have hitherto been employed, including the so-called absolute c.g.s. [centimeter–gram–second] system, owe their origin to the coincidences of accidental circumstances, inasmuch as the choice of the units lying at the base of every system has been made, not according to general points of view which would necessarily retain their importance for all places and all times, but essentially with reference to the special needs of our terrestrial civilization . . . These quantities [Planck units of length, time, and mass] therefore must be found always the same, when measured by the most widely differing intelligences according to the most widely differing methods" (Max Planck,* The Theory of Heat Radiation, *1913, based on lectures at the University of Berlin in 1906–7).*

* *"From the intrinsic evidence of his creation, the Great Architect of the Universe now begins to appear as a pure mathematician" (James Jeans,* The Mysterious Universe, *1930). Mathematicians are thought to be cold and soulless people; whereas God as a creator, an architect, or a designer is a common attribution, God as a mathematician caused a considerable stir in theological circles in the 1930s.*

* *"Finally, I repeat my personal conviction that the cosmical constant Λ is connected with the relation between electromagnetic and gravitational units, and that sooner or later a theory giving an accurate value of Λ will be forthcoming" (Arthur Eddington,* "The expansion of the universe," *1931).*

* *"I believe there are 15,747,724,136,275,-002,577,605,653,961,181,555,468,044,717,-914,527,116,709,366,231,425,076,185,631,-031,296 protons in the universe, and the same number of electrons" (Arthur Eddington,* The Philosophy of Physical Science, *1939).*

* *"It is proposed that all the very large dimensionless numbers which can be constructed from the important natural constants of cosmology and atomic theory are connected by simple mathematical relations involving coefficients of the order of magnitude unity. The main consequences of this assumption are investigated and it is found that a satisfactory theory of cosmology can be built up from it" (Paul A. M. Dirac, "A new basis for cosmology," 1938).*

* *"There are several amusing relationships between the different scales. For example, the size of a planet is the geometric mean of the size of the universe and the size of an atom; the mass of man is the geometric mean of the mass of a planet and the mass of a proton. Such relationships, as well as the basic dependences on α and α_G [where $1/\alpha_G = N_1$] from which they derive, might be regarded as coincidences if one did not appreciate that they can be deduced from known physical theory" (Bernard Carr and Martin Rees, "The anthropic principle and the structure of the physical world," 1979).*

3 *Discuss the cluster and large-number hypotheses.*

* *Derive the results shown in Table 17.1*
* *Suppose that N_2 can have any value from 1 to 10^{100}, and we require N_2 to equal N_1 within a factor of 100. What is the chance that they are equal?*
* *What is the cosmic-number dilemma?*
* *Discuss the anthropic principle and the cosmic numbers. Why is life unlikely when N_1 and N_2 are grossly unequal?*
* *How can we make N_1 and N_2 permanently equal?*
* *Do you think that numbers of the kind $N^{1/4}$ can be significant? Consider that there are $N^{1/4}$ stars in a typical galaxy and also $N^{1/4}$ typical galaxies in the observable universe. The specific entropy, which is a pure and natural number, has also the value $N^{1/4}$.*

4 *"There are some, king Gelon, who think that the number of sand is infinite in multitude; and I mean by the sand not only that which exists about Syracuse and the rest of Sicily but also that which is found in every region whether inhabited or uninhabited. Again there are some who, without regarding it as infinite, yet think that no number has been named which is great enough to exceed its multitude" (Archimedes, the opening words of* The Sand Reckoner*).*

In the third century B.C. the Greeks had a numerical system with which they were able to count easily up to a myriad, where a myriad was 10,000. This system could be extended to a myriad myriads, or a hundred million, but became awkard and cumbersome for larger numbers. Archimedes introduced a new system, called the "naming of numbers," which greatly extended the range of numbers beyond a myriad myriads. In this new system the number expressed as p units of the qth order and the rth period is given by

$$\text{number} = pM^{2[(q-1)+(r-1)M^2]}$$

where $M = 10^4$, and p, q, and r are integers in the range of 1 to M^2. The largest number in Archimedes' system is M^2 units of the M^2 order and M^2 period, and hence

$$\text{largest number} = M^{2M^4} = 10^{8\times10^{16}}$$

which is 10 followed by 80,000 trillion zeros. Because the large numbers in modern cosmology are 10 followed by 40 or 80 zeroes, it is evident that Archimedes' system is adequate for almost all purposes.

Archimedes then discussed the size of the universe: "Now you are aware that 'universe' is the name given by most astronomers to the sphere whose center is the center of the earth and whose radius is equal to the straight line between the center of the sun and the center of the earth. This is the common account, as you have heard from astronomers. But Aristarchus of Samos brought out a book consisting of some hypotheses, in which the premisses lead to the result that the universe is many times greater than that now so called. His hypotheses are that the fixed stars and the sun remain unmoved, that the earth revolves about the sun in the circumference of a circle, the sun lying in the middle of the orbit, and that the sphere of the fixed stars, situated about the same center as the sun, is so great . . ." There is some ambiguity concerning Aristarchus's estimate of the size of the universe, and Archimedes assumed that what he meant was

$$\frac{\text{size of universe}}{\text{size of Solar System}} = \frac{\text{size of Solar System}}{\text{size of Earth}}$$

Distances were measured in stadia (1 stadium is about 200 meters); we find that the Aristarchean heliocentric universe, converted into modern units, had a radius of 1 light year. Archimedes then found that this universe, if filled completely with sand, would contain a number of grains of sand equal to a thousand myriad units of the eighth order and first period:

$$\text{number of grains of sand} = 10^7 \times 10^{8(8-1)} = 10^{63}$$

The modern universe, which is 10^{10} times larger than the Aristarchean universe, could therefore contain 10^{93} grains of sand. But the modern universe has a density 10^{-30} times that of the density of sand, and therefore the Aristarchean universe filled with sand contains the same mass as the modern observable universe.

Archimedes' number of 10^{63} grains of sand is remarkable for the following *reason. He assumed that a poppy seed has a diameter of one-tenth of a fingerbreadth and a volume equal to that of a myriad grains of sand. If we suppose that a fingerbreadth is 1 centimeter, and the density of sand is 3 grams per cubic centimeter, we find that each grain of sand is rather small, with a mass* 2×10^{-7} *grams. On dividing this by the mass of a nucleon we find that each grain of sand contains* 10^{17} *nucleons. Hence, according to Archimedes' calculations, the Aristarchean universe, if filled with sand, would contain* $10^{63+17} = 10^{80}$ *nucleons, which is the Eddington number. Archimedes knew nothing about nucleons, and yet by pure chance his number of* 10^{63} *grains of sand is equivalent to the Eddington number.*

"I conceive that these things, king Gelon, will appear incredible to the great majority of people who have not studied mathematics, but that to those who are conversant therewith and have given thought to the question of the distances and sizes of the earth the sun and moon and the whole universe the proof will carry conviction. And it was for this reason that I thought the subject would be not inappropriate for your consideration" (closing words of The Sand Reckoner*).*

Most of the work of Archimedes (about 287–212 B.C.) has been lost, but from what is known it seems that he was the greatest of all mathematicians and scientists before the time of Newton. There are several stories of how he perished in the sack of Syracuse. According to Plutarch (A.D. 46–127): "For it chanced that he was by himself, working out some problem with the aid of a diagram, and having fixed his thoughts and his eyes upon the matter of his study, he was not aware of the incursion of the Romans, or of the capture of the city. Suddenly a soldier came upon him and ordered him to go with him to Marcellus. This Archimedes refused to do until he had worked out his problem, whereupon the soldier flew into a passion, and drew his sword, and slew him."

5 *The typical energy of a particle in a star is the energy required by a nucleon to escape from the surface to infinity. When R is the radius of a star, of mass M, this energy is* GMm_n/R*. In the deep interior charged particles rush around with this energy and have close encounters with one another. Let d be their nearest distance of approach; their kinetic energy is then equal to* e^2/d*, and hence*

$$\frac{e^2}{d} = \frac{GMm_n}{R}$$

The number of nucleons in the star is $N_{star} = M/m_n$*, and we find*

$$N_{\text{star}} = \frac{R}{d} \times \frac{e^2}{Gm_n^2}$$

Now let l be the mean separating distance between neighboring particles, such that $N_{star} = (R/l)^3$*, and we find*

$$N_{\text{star}} = \left(\frac{l}{d}\frac{e^2}{Gm_n^2}\right)^{3/2} = \left(\frac{l}{d}\right)^{3/2} \times 10^{54}$$

If l and d are almost equal, the particles are squeezed together as in metallic hydrogen and have little freedom to move. A body consisting of 10^{54} *nucleons has a mass one-thousandth that of the Sun and is a large planet such as Jupiter.*

Stars are luminous and hot, not cold like Jupiter, and the radiation in their deep interiors plays an important role in the determination of stellar structure. As a consequence, the wavelength of the radiation in their central regions is approximately equal to the separating distance l between particles. Or, put differently, the number of photons in a star is roughly the same as the number of particles. Each photon at the center of a star has therefore an average energy of $\hbar c/l$*, and this is equal to the energy* e^2/d *of the particles. This gives us the approximate relation* $d = \alpha l$*, which defines a star; neighboring particles deep inside a hot star are thus separated by an average distance that is 137 times their distance of closest approach. The number of nucleons in a star is therefore*

$$N_{\text{star}} = \left(\frac{\hbar c}{Gm_n^2}\right)^{3/2} = N_1^{3/2}$$

and $N_{star} = 10^{57}$. This result was first derived by Pascual Jordan in 1939.

FURTHER READING

Alpher, R. A. "Large numbers, cosmology and Gamow." *American Scientist,* January–February 1973.

Bondi, H. *Cosmology.* Cambridge University Press, Cambridge, 1960. Chapter 7, "Microphysics and cosmology," is brief and to the point.

Harrison, E. R. "The cosmic numbers." *Physics Today,* December 1972.

SOURCES

Alpher, R. A., and Gamow, G. "A possible relation between cosmological quantities and the characteristics of elementary particles." *Proceedings of the National Academy of Sciences 61,* 363 (1968).

Carr, B. J., and Rees, M. J. "The anthropic principle and the structure of the physical world." *Nature 278,* 605 (April 12, 1979).

Carter, B. "Large number coincidences and the anthropic principle in cosmology," in *Confrontation of Cosmological Theories with Observational Data,* ed. M. S. Longair. D. Reidel, Dordrecht, Netherlands, 1974.

Dicke, R. H. "Dirac's cosmology and Mach's principle." *Nature 192,* 440 (November 4, 1961).

Dirac, P. A. M. "The cosmological constants." *Nature 139,* 323 (February 20, 1937).

Dirac, P. A. M. "A new basis for cosmology." *Proceedings of the Royal Society, A165,* 199 (1938).

Dyson, F. J. "The fundamental constants and their time variation," in *Aspects of Quantum Theory,* ed. A. Salam and E. P. Wigner. Cambridge University Press, Cambridge, 1972.

Eddington, A. S. "The expansion of the universe." *Monthly Notices of the Royal Astronomical Society 91,* 412 (1931).

Eddington, A. S. *The Philosophy of Physical Science.* Cambridge University Press, Cambridge, 1939.

Gamow, G. "Any physics tomorrow?" *Physics Today,* January 1949.

Gamow, G. "History of the universe." *Science 158,* 766 (1967).

Heath, T. L. *The Works of Archimedes.* Cambridge University Press, Cambridge, 1897. Reprint. Dover Publications, New York, 1953. Contains *The Sand-Reckoner.*

Jeans, J. *The Mysterious Universe.* Cambridge University Press, Cambridge, 1930. Reprint. Macmillan, London, 1937.

Jordan, P. "Formation of the stars and development of the universe." *Nature 164,* 637 (October 15, 1949).

Planck, M. *The Theory of Heat Radiation.* Dover Publications, New York, 1959.

Stewart, J. Q. "Nebular red shift and universal constants." *Physical Review 38,* 2071 (1931).

Teller, E. "On the change of physical constants." *Physical Review, 73,* 801 (1948).

Weisskopf, V. F. "Of atoms, mountains and stars: a study in qualitative physics." *Science 187,* 605 (1975).

Whittaker, E. *From Euclid to Eddington: A Study of Conceptions of the External World.* Cambridge University Press, Cambridge, 1958. Reprint. Dover, New York. A readable discussion of Eddington's theory of cosmic numbers is given in pt. 5.

THE EARLY UNIVERSE

He passed the flaming bounds of space and time:
The living throne, the sapphire-blaze,
Where angels tremble while they gaze,
He saw; but blasted with excess of light,
Closed his eyes in endless night.
— Milton (1608–74) *Progress of Poesy*

THE PRIMEVAL ATOM

Time will run back, and fetch the age of gold.
— John Milton, "On the morning of Christ's nativity"

The universe expands; we naturally therefore suppose that in the past it was in a more congested state than at present. If we could journey back in time we would notice the universe becoming steadily more and more dense; eventually we would arrive at the high-density state referred to as the big bang. This conclusion seems unavoidable. It might be a mistake, however, to forget entirely the many debates among cosmologists concerning the reality of a big bang. Eddington was firmly opposed to the idea of a universe originating in a dense condition, and many people who were first drawn to science by Eddington's popular books have felt disinclined to set his views aside lightly. The steady state theory undoubtedly attracted those who were united in their dislike of the big bang idea, and there are still some astronomers and cosmologists who believe that a big bang interpretation of the observations is hasty and naive.

Alexander Friedmann led the way to the notion of a big bang. But his work made little impact until some years later, when Abbé Georges Lemaître investigated and championed a dense beginning. Lemaître referred to the cosmic origin as the "primeval atom," whereas others used quaint expressions such as the "cosmic egg" and the "big squeeze." The early universe was imagined by Lemaître to be exceedingly dense, and he advanced the idea that it was like a large radioactive atomic nucleus. This cosmic nucleus, or primeval atom, exploded and formed fragments that later became the galaxies. Toward the end of his life he wrote: "These considerations, besides providing a natural beginning, supply what can be called an inaccessible beginning. I mean a beginning which cannot be reached, even by thought, but which can be approached in some asymptotic manner." Lemaître was a priest, and some cosmologists were inclined to regard his views as a suspicious amalgam of science and religion.

The big bang type of universe held uneasy sway in cosmology until the rise of the steady state universe in the late 1940s. From earlier chapters we recall that big bang universes were plagued with the time-scale difficulty (the Hubble period was observed to be less than the age of the Solar System), which was resolved by Walter Baade in 1952. The steady state idea circumvented this difficulty and consequently attracted considerable interest. In the late 1940s a small group of scientists, the "Gamow school," began to make important contributions to the theory of a big bang type of universe. The principal members of this group were the charismatic George Gamow of George Washington University and the young scientists Ralph Alpher and Robert Herman at the Johns Hopkins Applied Physics Laboratory. The original dense state of the universe was referred to as the "big squeeze" until Fred Hoyle coined the title "big bang." Gamow and his colleagues referred to the dense substance of the early universe as "ylem" (pronounced "i lem"), meaning the original material from which the elements were made, but this term has not been widely adopted.

The Gamow school studied the physics of the early universe and made many important contributions to our present knowledge. Their major breakthrough, in 1948, was the realization that the big bang had a very high temperature. This led to two remarkable conclusions: First, there existed a *radiation era* of the big bang during which the density of radiation greatly exceeded the density of matter; and second, this radiation survives, cooled by expansion, so that the universe is now bathed in the afterglow of the big bang. They estimated that this cosmic radiation, or afterglow of the big bang, had a temperature somewhere between 5 and 50 degrees Kelvin.

Since 1941 it has been known that molecules in interstellar space are excited by radiation of a low intensity equivalent to that of thermal radiation of only a few degrees Kelvin. At one time it was thought that this excitation of interstellar molecules might be due to starlight, but we now know that it is caused by the cosmic radiation. Another clue was provided by radioastronomers, who discovered a persistent noise, or hiss, in their antennas that at first had no obvious explanation. The noise was at last identified as the cosmic radiation in the following way. Arno Penzias and Robert Wilson of the Bell Telephone Laboratories patiently eliminated all possible sources of noise in their radio receiver and horn-shaped antenna at Holmdel, in New Jersey, and finally obtained an irreducible noise having a temperature of 3 degrees. They were naturally puzzled by the cause of this noise of extraterrestrial origin. At that time, P. James Peebles, a young astrophysicist at Princeton University, was repeating the calculations made by Alpher and Herman more than a decade earlier, and was investigating the possibility of helium production in a hot big bang and the possibility of a background of low-temperature cosmic radiation. This work came to the attention of Penzias and Wilson, and in 1965, Robert Dicke and his colleagues at Princeton University were able to identify the radio signal discovered by Penzias and Wilson as the cosmic radiation that has survived from the big bang. In 1978, Penzias and Wilson received the Nobel prize for their momentous discovery.

It is now widely believed that this discovery has established that we live in a big bang universe and has confirmed also the prediction of the Gamow school that the big bang was at a high temperature. The bang–antibang controversy has been set to rest, and there is now little doubt in the minds of most cosmologists that the universe began in a very dense and hot state. In this chapter we outline the "standard model" of the early universe, which takes us back to the time when the universe was only $1/_{10,000}$ of a second old. We are not sure what happened before then, and current ideas are still highly speculative and uncertain. Some of these speculative ideas are mentioned later in the chapter.

THE LAST TEN BILLION YEARS

A SAFARI IN TIME. As a rough guide we use the rule of thumb that the average density of the universe is inversely proportional to the square of its age, as in the Einstein–de Sitter universe (see Figure 18.1). Thus, when the universe was 1 billion years old, or about one-tenth its present age, the average density was 100 times greater than now. For illustrative purposes we assume that the present density is equivalent to 1 hydrogen atom per cubic meter. At an age of 10

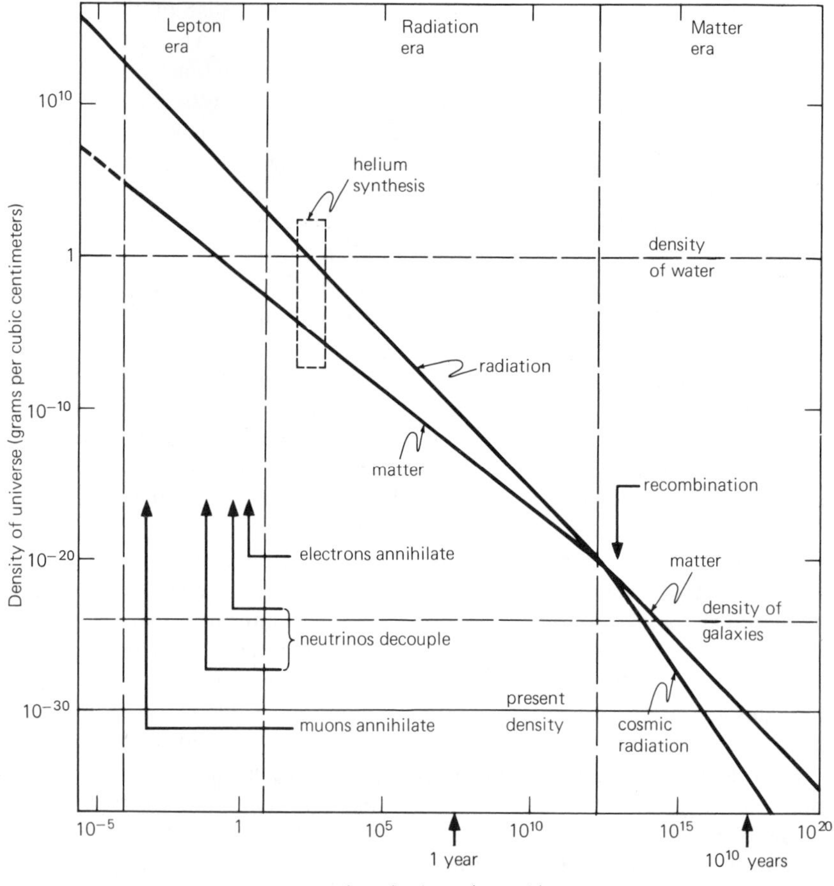

Figure 18.1. Standard model of the early universe showing how density varies with age. The total density ρ is proportional to $1/t^2$, where t is the age of the universe, as in the Einstein–de Sitter universe. Each time the age is increased by a factor of 10, the density is reduced by 100. In the matter-dominated era the density ρ is proportional to T^3, where T is the temperature of the cosmic radiation; hence T is proportional to $1/t^{2/3}$. In the radiation era the density is due almost entirely to radiation and ρ is proportional to T^4; hence the temperature T is proportional to $1/t^{1/2}$. Note that in the matter-dominated era the radiation density decreases with time faster than the matter density and is proportional to $1/t^{8/3}$, and in the radiation era the matter density decreases with time slower than the radiation density and is proportional to $1/t^{3/2}$.

million years, the universe had a density a million times greater, or 1 hydrogen atom per cubic centimeter. A typical average density within galaxies is also 1 hydrogen atom per cubic centimeter, and hence we know that galaxies in their present form did not exist at a time earlier than 10 million years. The galaxies came into being in their present form when the universe was about 1 billion years old.

We shall travel back and attempt to reconstruct the history of the universe. As our time machine clicks off the billions of years we see the galaxies moving closer together. We see the birth and death of countless stars and observe the origin of life on Earth and possibly the origin, and even death, of life in myriads of planetary systems. When the universe is 5 billion years old, and about half its present age, we notice the birth of the Solar System.

We continue on our safari into the past, watching the galaxies as they get younger and closer together. In the days when galaxies are young, long before the Sun is born, lifeforms of many kinds presumably arise and evolve, and it is a matter of absorbing interest to speculate on their fate and the possibility that many still exist in a highly advanced state of intelligence. When the universe is about 1 billion years old we see the galaxies swelling into gigantic orbs of gas. Somewhere between 100 million and 1 billion years is the heroic age of galaxy formation. We pause and watch immense and slowly turning orbs of gas, lit up with constellations of newborn stars in their central regions. Gas descends between these stars and settles into either the rotating disks of spiral galaxies or the quasar-bright nuclei of giant elliptical galaxies.

Earlier still, when the universe is 10 million years old, it is at room temperature of 300 degrees Kelvin. Little is known of this strange prenatal era of the galaxies. Perhaps there is swirling gas illuminated with flickers of light caused by shock waves. It has been suggested that strewn throughout space there perhaps exist black holes of all sizes forged in the big bang. At an age of 1 million years the universe has begun to glow red-hot. Its density is 100 hydrogen atoms per cubic centimeter and the temperature is just over 1000 degrees – or more than 700 degrees Centigrade.

Between 1 million and 100,000 years of age the universe is flooded with brilliant yellow light, at a temperature of 3000 degrees, and we at last stand at the threshold of the big bang. Behind us – in the dark future – lies the matter-dominated universe; before us lies the incandescent radiation era in which the universe is dominated by radiation. From this epoch, at the frontier of the big bang, descends directly the cosmic radiation, cooled by expansion, that now has a redshift of 1000 and a temperature of 3 degrees.

THE FIRST MILLION YEARS

DECOUPLING AND EQUAL DENSITIES. We have journeyed back to a time when the universe is less than 1 million years old, and we are now about to enter the big bang. The transition to the big bang is marked by two events occurring close together in time. The first is that radiation no longer travels freely through space. This is because the gas, consisting almost entirely of hydrogen and helium, is now hot enough to be partially ionized. Some of the hydrogen atoms are dissociated into free protons and electrons, and the free electrons interact with the lightrays, impeding and scattering them in all directions. This is why nowadays we cannot see the big bang as it was at earlier times: The radiation was continually scattered and was unable to travel directly to us. At the end of the big bang the gas has become relatively cool and is only weakly ionized, and radiation is hence able to decouple from matter and travel unimpeded through space. From this *decoupling epoch* (also known as the *recombination epoch* because protons and electrons recombine to form hydrogen atoms) comes directly the cosmic radiation that we now observe.

The other event of importance at the termination of the big bang is the approximate equality of the mass densities of matter and radiation: 1 cubic centimeter of matter weighs as much as 1 cubic centimeter of radiation. The density of matter varies as $(1 + z)^3$, and because its present value is assumed to be 1 hydrogen atom per cubic meter, the density of matter at a redshift $z = 1000$ is therefore 1000 hydrogen atoms per cubic centimeter. The radiation density varies as $(1 + z)^4$, and at $z = 1000$ it is a trillion times greater than now. At the present temperature of 3 degrees the cosmic radiation has a density slightly less than one-thousandth that of matter; hence, at a redshift not much greater than 1000, the radiation and matter densities are equal. It follows that the decoupling epoch and the *epoch of equal densities* both occur close together at the end of the big bang.

RADIATION ERA. In the radiation era the universe is dominated by radiation. The radiation era begins when the expanding universe is about 1 second old and lasts for almost a million years, and throughout that time the radiation is more dense than matter. At the beginning of this era the temperature is 10 billion degrees, the density of radiation is 10^5 grams per cubic centimeter (one-tenth of a ton per thimbleful), and the density of matter is one-tenth of a gram per cubic centimeter (one-tenth that of water). At the end of the radiation era, as we have seen, the temperature has dropped to a few thousand degrees, and radiation and matter have equal densities equivalent to a thousand or so hydrogen atoms per cubic centimeter. The reason the radiation density overtakes and then greatly exceeds the matter density, as we go back in time, is that the radiation density increases as T^4 (where T represents temperature), whereas the matter density increases more slowly, as T^3.

With trepidation we have ventured into the big bang and found ourselves in a silent world of radiant serenity. At first, everywhere is incandescent, and as we plunge deeper and go further back in time, the intensity soars and becomes brighter than the central furnaces of the hottest stars. Although matter is of relatively low density, it is nonetheless important, because the electrons incessantly scatter the rays of light and the radiation behaves like a dense fluid.

ORIGIN OF HELIUM. An event of great importance in the radiation era is the production of primordial helium. About 25 percent of all matter is transformed into helium in a few hundred seconds after the beginning of the radiation era. For 10 billion years the stars have worked industriously at converting their hydrogen into helium and have succeeded in transforming only 2 or 3 percent of all hydrogen in the universe. Yet 10 times as much hydrogen is burned into helium in only 200 or so seconds in the early universe. The universe at this time is like a hydrogen bomb, and the energy released by the fusion of hydrogen is immense. But the energy already existing in the radiation is vastly greater, and the thermonuclear detonation raises the temperature less than 10 degrees and increases the energy by only one hundred-millionth.

The amount of helium produced can be estimated in the following way. At the beginning of the radiation era there exists a dense sea of radiation and a gas of much lower density that consists of protons, neutrons, and electrons. Because of earlier conditions to be discussed shortly, there are about 2 neutrons to every 10 protons. The neutrons and protons continually collide together and are easily transformed into deuterons, which are the nuclei of atoms of heavy hydrogen known as deuterium. The deuterons are also easily dissociated back into neutrons and protons by the energetic radiation. When the universe is a few seconds old the neutrons and protons combine continually to form deuterons, and the deuterons dissociate continually back into neutrons and protons. But after 100 seconds the temperature has dropped to 1 billion degrees and the radiation is then insufficiently energetic to break up the deuterons. All neutrons and protons are then able to combine and form deuterons.

Meanwhile the neutrons have been slowly decaying into protons and electrons. This is because neutrons, when free and not bound in atomic nuclei, are unstable: They decay in about 1000 seconds. By the time neutrons are able to combine permanently with protons, their abundance has decreased, and of every 16 nucleons, 2 are neutrons and 14 are protons. The 2 neutrons combine with 2 protons, and the result is 2 deuterons and 12 protons.

Deuterons are reactive at high temperature and rapidly combine to produce helium. At 1 billion degrees the temperature is insufficient to dissociate the deuterons but is high

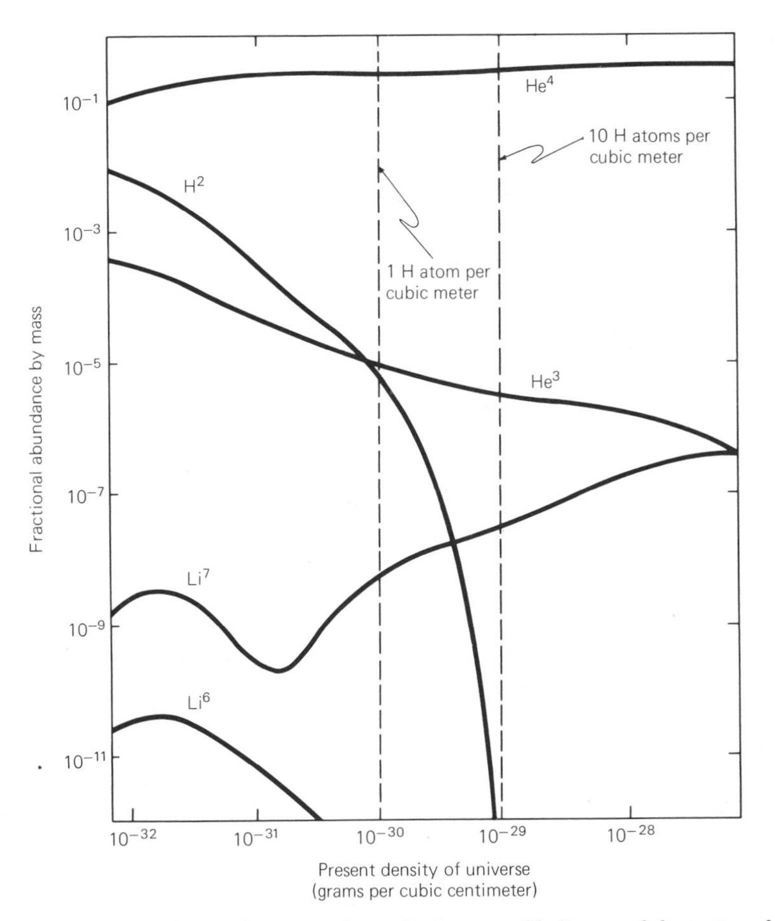

Figure 18.2. The production in the radiation era of helium and deuterium fractions by mass, showing their dependence on the present average nucleon density of the universe. Note that the deuterium abundance changes rapidly with nucleon density. Lithium and helium-3 nuclei are also produced in low abundance. Adapted from David Schramm and Robert Wagoner, "What can deuterium tell us?" Physics Today *(1974).*

enough to cause them to burn quickly into helium nuclei. Two deuterons combine to form 1 helium nucleus, thus yielding 1 helium nucleus and 12 protons. The whole process is over in 200 seconds, and in that time 25 percent of all matter is converted into helium (4 of every 16 nucleons form a helium nucleus) and the remainder consists predominantly of hydrogen. Slight amounts of deuterium, helium-3 (helium nucleus of 2 protons and 1 neutron), and lithium are also produced. The amount of helium created in the early radiation era is not very sensitive to the density of matter, and as our rough calculation has shown, it is mainly controlled by temperature and depends on the initial neutron–proton ratio. Not all deuterons succeed in combining to form helium nuclei, and a small fraction survives and accounts for the deuterium that now exists. Between one-hundredth and one-thousandth of 1 percent of all hydrogen is at present in the form of deuterium. The amount of deuterium that escapes burning into helium in the early universe is very sensitive to the density of matter. When the density of matter is low, there are fewer deuterons to collide with one another and a larger fraction survives; when the density of matter is high, there are more deuterons to collide with one another and a smaller fraction survives.

It will be recalled that the specific entropy is the number of photons to each nucleon. It is a constant quantity and has a value of approximately 1 billion. The number of photons is known from the 3-degree cosmic radiation, but we are uncertain about the number of nucleons because of the possibility of "missing mass" existing nowadays in nonluminous states. The abundance of helium in the universe is not very sensitive to the density of matter and is therefore not a good indicator of the specific entropy. The deuterium abundance, on the other hand, is a very sensitive indicator (see Figure 18.2). Thus, if the specific entropy is low, the density of matter is high, and the deuterium that survives in the early universe is small; if the specific entropy is high, the density of matter is low, and the deuterium that survives is large. By measuring the fraction of hydrogen now in the form of deuterium we can estimate the present average density of matter. The results obtained so far indicate that the density of matter is less than the critical Einstein–de Sitter value. According to the simplest models possible – the Friedmann universes – the deuterium abundance therefore indicates that the universe is open and will expand forever. It is conceivable that some of the deuterium observed has been created in other ways, by supernovas for example, but at present this is an uncertain subject. We cannot see ways in which deuterium might easily be manufactured, other than in the big bang, but we can see ways in which it might easily be destroyed. Stars are quite wasteful of matter, and a considerable amount of it is cycled through stars and returned to the interstellar medium. This cycling of matter, as a result of star formation, stellar winds, supernovas, and other forms of mass ejection, is known as *astration*. Some of the primordial deuterium is destroyed by astration, and consequently more was made in the early universe than is presently observed. The observed high deuterium abundance, despite astration, reinforces the widely held belief that the universe is open. Nevertheless, we can be sure that the last word has not yet been said on the problem of the deuterium abundance.

THE FIRST SECOND

THE LEPTON ERA. Immediately preceding the radiation era exists what is often referred to as the *lepton era*. Leptons are light particles – hence the name – such as electrons, positrons (positive electrons), and neutrinos. The lepton era begins when the universe is $1/10{,}000$ of a second old, when the temperature is 1 trillion degrees and the density is 1000 tons per thimbleful, and lasts until the beginning of the radiation era, when the universe is 1 second old and has a temperature of 10 billion degrees.

In the lepton era the temperature is high enough for electron pair production. Electron pairs (i.e., electrons and positrons) are continually created and annihilated, and there is an incessant interchange of energy among photons, electron pairs, and neutrinos. Everything is in thermal equilibrium and there are approximately equal numbers of photons, electrons (both kinds, positive and negative), and neutrinos. Buried in this dense ferment of seething photons and leptons are the nucleons – the protons and neutrons – and for every nucleon there are roughly a billion photons, a billion electrons, and a billion neutrinos. Each nucleon continually collides with leptons; when it is a neutron it captures a positron and becomes a proton, and when it is a proton it captures an electron and becomes a neutron. At any instant approximately half the nucleons are neutrons and the other half protons.

A neutron is about one-seventh of 1 percent heavier than a proton; thus it is able when it is in a free state to decay into a proton and an electron. Slightly more energy is therefore needed to create a neutron than is needed to create a proton, and because of this small energy difference, each nucleon in the lepton era tends to be a proton slightly longer than a neutron. Hence there are slightly more protons than neutrons. Throughout most of the lepton era the protons outnumber the neutrons only very slightly. At the end of the lepton era, however, the temperature has dropped sufficiently for the difference in the neutron and proton masses to become important. Many electrons now have not enough energy to convert protons into neutrons, whereas the conversion of neutrons into protons by positron bombardment is much easier; as a result, at the beginning of the radiation era there are only 2 neutrons to every 10 protons.

DISAPPEARANCE OF MUONS AND ELECTRONS. So far, only the electrons and their neutrinos have been mentioned; but other leptons, known as muons, also exist and have their own kind of neutrinos. The muon is similar to the electron, except that it is over 200 times more massive and is unstable, decaying in a millionth of a second into an electron. The antiparticle of the electron is the positron, or positive electron, and the antiparticle of the muon is the antimuon, or positive muon. At a temperature approaching 10^{10} degrees there is copious pair production of electrons but no pair production of muons, owing to their much greater mass. Not until the temperature approaches 10^{12} (a trillion) degrees is the thermal energy sufficient to start copious pair production of the muons and their neutrinos.

Right at the beginning of the lepton era, when the universe is $1/10{,}000$ of a second old and the temperature is a trillion or so degrees, the whole universe is flooded with photons, electrons and their neutrinos, muons and their neutrinos, and also other particles, such as the pions that are a relic of an earlier era. As the temperature declines, the negative and positive muons annihilate each other and vanish from the universe. Hitherto, at higher temperature, the muons have been created at the same rate at which they have annihilated each other, but at a temperature less than a trillion degrees their annihilation is more rapid than their creation. The energy released by the disappearance of the muon population is shared among the surviving particles, which are mostly photons and electrons and their neutrinos.

After the disappearance of the muons the temperature continues to drop as the universe expands. At a temperature of 10 billion degrees the electron pairs begin to annihilate faster than they are created, and by the time the temperature has declined to about 5 billion degrees, the electron hordes have vanished. The energy released by their disappearance is given almost entirely to the photons.

DECOUPLING OF NEUTRINOS. An event of interest in the lepton era is the decoupling of the muon neutrinos and the electron neutrinos. All neutrinos are uncharged and are weakly interacting particles that move at the speed of light. There are two kinds, electron neutrinos and muon neutrinos, and each kind has its antiparticles: electron antineutrinos and muon antineutrinos. The neutrinos interact very weakly with matter, and only when their energy is high and matter is dense is it possible for them to participate sufficiently rapidly in the give-and-take of particle reactions in thermal equilibrium.

The early lepton era is densely packed with muons and their neutrinos, and these muon neutrinos interact with the muons and are continually absorbed and emitted. Then, as the universe expands and cools, the muons begin to disappear. Owing to the high temperature, the muons and their neutrinos continue to interact with each other until almost all muons have vanished. After the muons have vanished there is left a dense sea of muon neutrinos that ignores the electrons and everything else. With the decline of the muon dynasty, hordes of muon neutrinos are released to wander forever freely through the universe, interacting with virtually nothing.

After the decoupling of the muon neutrinos the temperature continues to decline, and eventually the electrons begin to disappear. Up to this time the electrons and their neutrinos have interacted intimately with each other. But now, the moment the electrons begin to vanish, their neutrinos quickly decouple because of the lower temperature. The electron neutrinos then join the muon neutrinos, and together they form a background of cosmic neutrinos. They flood the universe and are still with us: About a million trillion cosmic neutrinos pass through each human body every second.

There are four kinds of neutrinos (electron neutrinos and antineutrinos, and muon neutrinos and antineutrinos), and for every 8 photons of the cosmic radiation there are, according to theory, 3 of each kind still with us. Hence, for every 2 photons there are 3 neutrinos, and in every cubic centimeter of the universe there are now 600 neutrinos. Their interaction with matter is so weak that at present we know of no way in which they can be detected. Presumably one day it will be possible to detect them, and this will provide us with a means of probing the first second of the universe.

All four kinds of neutrinos decouple toward the end of the lepton era when the temperature is in the region of 10 billion degrees. They have since traveled freely at the speed of light and have lost energy because of the expansion of the universe. Their redshift, since decoupling, is approximately 10^{10}. The temperature of the neutrinos is now 2 degrees, however, and not the 3 degrees of the photons of the cosmic radiation. The reason for this difference in temperature is as follows. When the neutrinos decouple they continue to cool because of the expansion of the universe. Everything else also cools with the expansion, and the neutrino temperature remains in step with the temperature of the photons and electrons. But not for long: Shortly after the neutrinos decouple most of the electrons vanish; their energy is released and given to the photons; and none of this released energy is shared with the neutrinos. The radiation consisting of photons is therefore heated to a temperature higher than that of the neutrinos, and thereafter the neutrinos have a temperature that is about 70 percent that of the photons.

THE FIRST TEN-THOUSANDTH SECOND

Our journey back through the radiation and lepton eras has brought us to a time when the universe is $^1/_{10,000}$ of a second old. The temperature has progressively risen and is now a few trillion degrees; and the density has soared to a billion tons per thimbleful, or a thousand trillion times that of water. It is a tribute to modern physics that we can, in broad outline, trace the history of the universe back to this early epoch. What we have studied so far is known as the standard model of the early universe. There is no standard and widely accepted theory, however, to guide us in what lies ahead in the first ten-thousandth of a second.

We stand so close to the beginning of time that it seems absurd to want to continue the journey further. Yet everything happens with such rapidity in the very early universe that perhaps more of cosmic history occurs in the first thousandth of a second than has occurred in 10 billion years since (see Figure 18.3). Before us lies a blur of intense action, and we are not sure exactly what happens, simply because we do not yet know enough physics. Inspired by the words of Pascal – "if our view be arrested there let our imagination pass beyond" – we shall consider some speculative ideas concerning the earliest moments of the universe.

THE HADRON ERA. On proceeding further back in time we leave the lepton era and enter what has been called the *hadron era.* Hadrons, as their name implies (from *hadro* in Greek, meaning stout or strong), are heavy particles, and in addition to their weak and electromagnetic interactions, they interact with a force known as the strong interaction. Familiar hadrons are the nucleons – the protons and neutrons. There are also, in the nuclei of atoms, particles called pions that are hadrons with masses 270 times that of electrons. The pions skip to

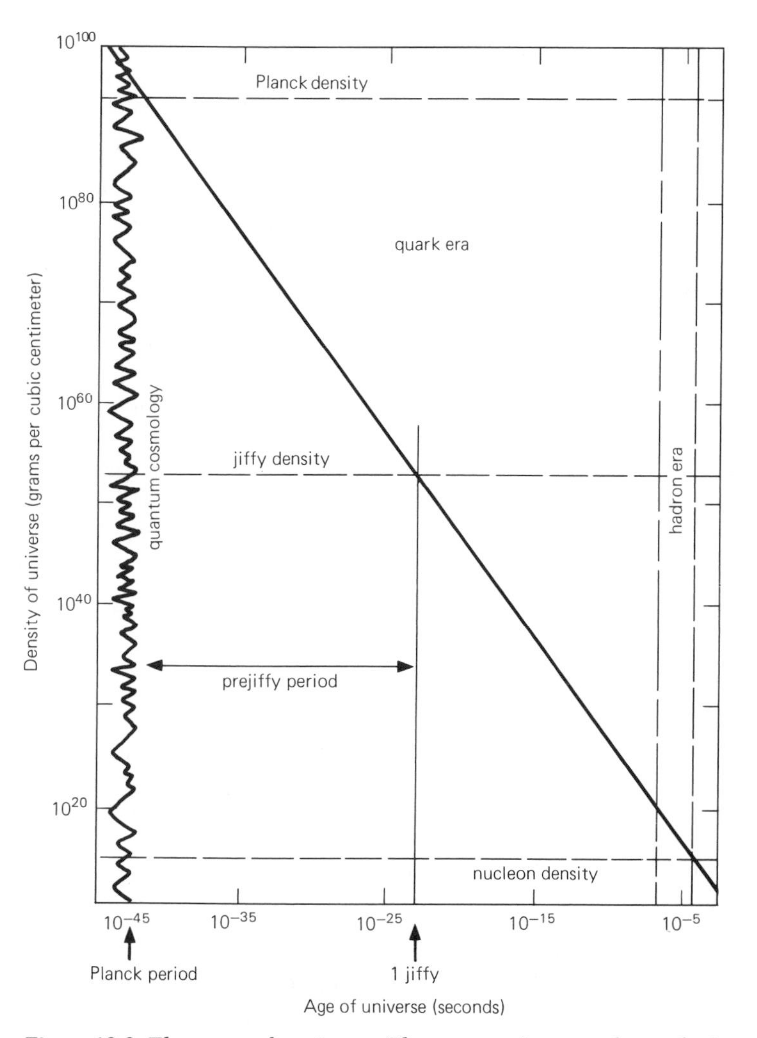

Figure 18.3. The very early universe. There may exist a quark era, during which the universe is densely populated with free quarks. A hadron era of brief duration follows. We cannot trace the history of the universe back to infinite density and temperature because the beginning consists perhaps of a primordial chaos of space and time.

and fro among the nucleons and hold the nucleus together despite the electrical repulsion of the protons. In the hadron era the universe is flooded with hadrons, because the temperature is now high enough for the creation of pions, nucleons and antinucleons, and other hadrons and their antiparticles. Leptons also exist, but the universe is now dominated by the presence of a dense sea of hadrons.

Let us pause for a moment in the hadron era, when the universe is approaching an age of $^{1}/_{10,000}$ of a second. There is everywhere a dense conglomeration of photons, leptons, pions, and also protons and neutrons and their antiparticles, the antiprotons and antineutrons. These latter particles – the nucleons and antinucleons – continually annihilate each other, and the energy released is continually used to create fresh nucleons and antinucleons. The universe consists of warring matter and antimatter.

But as the temperature drops, annihilation overtakes creation, and at the end of the hadron era all matter and antimatter have practically vanished. The pions are still abundant, owing to their lighter mass, but they soon suffer the same fate; they vanish in the early stages of the lepton era. This raises an interesting question in cosmology: Why does matter still exist today? In other words, why was it not all annihilated long ago, at the end of the hadron era? The most simple answer is that the universe originally contained slightly more matter than antimatter, and the slight excess of matter has survived annihilation and now constitutes the present material universe. But why the universe should favor matter slightly more than antimatter we do not know. It is possible to calculate the excess of matter over antimatter in the hadron era, and we find, as the following argument shows, that the excess of matter is about 1 part in a billion.

In Chapter 13 it was said that most of the entropy of the universe resides in the 3-degree cosmic radiation. We must amend this remark and state that the entropy is mostly in the cosmic radiation and the cosmic neutrinos. For our purpose we can say that the entropy of the universe is equal to the number of cosmic photons and neutrinos, of which there are 1000 per cubic centimeter. As we go back in time the total entropy remains constant, which means – and this may not be obvious to many readers – that the total number of particles of all kinds remains also approximately constant. The particles that flood the universe may change their nature, from hadrons in the hadron era, to leptons in the lepton era, to photons in the radiation era; but their total number is approximately conserved. If there are N particles in the hadron era then there are still N particles in the universe today. The N particles of the congested hadron era have become, in the greatly expanded universe, n nucleons and $N - n$ photons and neutrinos. (For convenience we neglect other particles, such as electrons.) At present there are approximately 1 billion cosmic photons and neutrinos to each nucleon, or $N/n = 10^9$. Because n is so small compared with N, we can say that N nucleons and antinucleons in the hadron era have become the N photons and neutrinos that now exist. The neglect of other particles in the late stages of the hadron era, such as the pions and leptons, introduces in the discussion only relatively small errors.

To see what happens, let N^+ be the number of nucleons and N^- be the number of antinucleons in the hadron era. Of course, if particle numbers are infinite, as in an open universe of infinite space, then we must think of a finite comoving volume of space that contains the N^+ and N^- particles. The total number of nucleons of both kinds is given by

$$N = N^+ + N^-$$

The excess of nucleons over antinucleons can be written as

$$\Delta N = N^+ - N^-$$

and this is the conserved *baryon number* of the comoving region. In a universe that has no preference for either matter or antimatter the baryon number is zero and $N^+ = N^-$. The fractional difference of matter and antimatter in the hadron era is expressed by

$$\frac{\Delta N}{N} = \frac{N^+ - N^-}{N^+ + N^-}$$

and is the ratio of the baryon number and the total number of baryons.

The original N hadrons have vanished, with the exception of the small baryon number ΔN, and have now become N photons and neutrinos. The baryons that survive are the present n nucleons ($\Delta N = n$); and because $n/N = 10^{-9}$, we have $\Delta N/N = 10^{-9}$. Hence the fractional difference of matter and antimatter in the hadron era is the present ratio of the number of nucleons and the number of photons and neutrinos, and is about one-billionth.

According to this argument the universe has changed in a dramatic fashion. Once, long ago during its earliest moments, the universe consisted of densely intermingled matter and antimatter of almost equal amounts, and their slight difference, about 1 part in a billion, was of little immediate consequence. But this small difference has survived and now makes up the galaxies, stars, planets, and living creatures. Without this difference the universe would consist of little more than photons and neutrinos, and we would not be here. All the immense energy of the hadron era, released by the annihilation of matter and antimatter, has passed into the cosmic photons and neutrinos that have been cooled by expansion. The 3-degree cosmic radiation and the ghostly neutrino background, which at present appear to be so unimportant, represent the unimaginable energy that existed in the very early universe; whereas the matter that we prize so highly is the result of a freakishly small difference that was once of no apparent cosmological importance.

THE QUARK ERA. Most particles observed in nature fall into two classes: the leptons and the hadrons. The leptons are relatively few in kind and have no apparent internal structure, whereas the hadrons are of bewilderingly numerous kinds and have complex internal structure. At the beginning of this century the only known subatomic particle was the electron; since then, decade by decade, slowly at first, new particles have been discovered. To cope with a motley crowd of more and more particles, new methods of classification and new laws of conservation have been introduced in attempts to maintain a state of reasonable order. High-energy accelerators have revealed hundreds of different particles, mostly short-lived hadrons, that indicate a surprising complexity in the subatomic realm. The situation is not unlike that in the nineteenth century when chemists and physicists sought to find order in the atomic realm. Science advances, so it seems, by decomposing intricate systems into an activity of simpler systems. The world is decomposed into molecules that break apart into atoms, which in turn break apart into subatomic particles; and now we are confronted with subatomic complexity that possibly can be understood as an activity of even smaller particles that are the primary constituents of hadrons.

The quark hypothesis, proposed independently by Murray Gell-Mann and George Zweig of the California Insititute of Technology, is now widely accepted. The idea is that hadrons are constructed from more elementary particles called quarks. In the initial version of the theory there were three different quarks, each having its own antiquark (thus making a total of six). Mesons, a class of hadrons of integer spin and zero baryon number, which includes the pions, are constructed by pairing a quark of one kind with an antiquark of another kind. Baryons, a class of hadrons of half-integer spin, which includes the nucleons, are constructed by arranging quarks and antiquarks into groups of three. Quarks are electrically charged, but their charge is fractional, either $2/3$ or $-1/3$ the elementary charge of the proton. The idea of arranging a few fundamental quarks into groups is attractive and so far has been successful in explaining the various properties of the many hadrons.

The quarks are distinguished by the names *up, down,* and *strange,* or the symbols u, d, and s, and their antiquarks are denoted by $\bar{u}$, $\bar{d}$, and $\bar{s}$. Developments have led to the introduction of a new quark having the property of *charm.* The four basic characteristics, up, down, strange, and charm, are called flavors. Each flavor has three distinct colors, however, and hence there are 12 kinds of quarks, or 24 when their antiquarks are included. Quarks, like leptons, have no internal structure so far as we know, and symmetry arguments suggest that the number of quark flavors should equal the number of leptons.

A new lepton, 4000 times the mass of the electron, has been discovered – the tau lepton, with its own distinctive neutrino. Hence the total number of different leptons is now six. Evidence also suggests the existence of two additional quark flavors, referred to as *top* and *bottom* (or *truth* and *beauty*); the result is six quark flavors to match the six leptons.

No quark in an isolated state has been discovered. It is thought that the force acting between quarks is of an unusual kind that does not decrease in strength as the distance between quarks increases. The quark force is therefore unlike all other forces in nature, which decrease in strength as the separating distance between particles increases. Suppose that, by brute force, we try to separate two quarks; the energy required to pull them apart is more than sufficient to create new quark and antiquark pairs, and instead of isolating the two original quarks, we have created new quarks. It is like trying to isolate the ends of a piece of string: When the string is pulled hard enough, it snaps into two pieces, creating two new ends. In this analogy the piece of string is a particle and its ends are the quarks. Or it is like trying to separate the north and south poles of a magnet: By breaking the magnet into two pieces we create two magnets, each with its opposite poles; in this case the magnet represents the particle and its poles represent the quarks.

New developments in the theory of particles affect our understanding of the early universe considerably. Let us see what happens in the hadron era according to the quark theory. As we approach the hadron era, traveling back in time, we encounter first a dense sea of pions at a temperature of a few trillion degrees and a density of about a billion tons per cubic centimeter. Then, in the hadron era, at a slightly earlier time of higher temperature and density, the universe is deluged with heavier hadrons, such as the nucleons. All hadrons now touch and even overlap their immediate neighbors, and the confined quarks of each hadron are very close to the confined quarks of neighboring hadrons. In such a state it is possible that the hadrons dissolve and their confined quarks break forth and form a dense sea of free quarks. The boundaries of individual hadrons have in effect melted away and the universe has become a single huge multiquark particle, similar in some ways to the primeval atom envisaged by Lemaître. All the structureless particles – photons, leptons, and quarks – now abound everywhere in a state of thermal equilibrium and are continually annihilated and recreated.

Vast numbers of quarks exist in the very early universe. If it were possible for them to exist in an isolated state, then some should have survived, and it is estimated they would now be as abundant as gold atoms. That none have been found confirms that quarks cannot exist in an isolated state.

We have entered the hadron era and found that it is short-lived, giving way quickly to a quark era that consists of structureless particles of various kinds. In this new era all particles are crushed close together; they densely overlap each other and cannot combine together to form structured particles such as hadrons. Structures of some kind may exist of which we are unaware. It is unlikely, however, that structured particles will exist in the universe when it is younger than 1 jiffy. A jiffy, you may remember, is 10^{-23} or 10 trillion-trillionths of a second. The density at the jiffy epoch, when the universe is 1 jiffy old, is N times the density in the hadron era, or N times the density of a nucleon, where N is the cosmic number 10^{40}. Because the nucleon density is about 10^{15} grams per cubic centimeter, it follows that the *jiffy density* is 10^{55} times the density of water. At this high density a star would occupy a volume equal to that of an ordinary hydrogen atom. The *jiffy temperature* is $N^{1/4}$ times the temperature of the hadron era – about 10^{22} degrees.

We recall that a jiffy is the time taken by light to travel a distance equal to the size of a subatomic particle such as a nucleon. When the universe is only a jiffy old, the Hubble period is 1 jiffy, and the radius of the observable universe is 1 light jiffy, or 10^{-13} centimeters. We note that at the jiffy epoch the observable universe is the size of a nucleon. If we consider any point in the universe at that time, then at a distance of 1 light jiffy from that point everything recedes at the speed of light, and at greater distances the recession is faster than the speed of light. This strange situation becomes even stranger when we consider the universe at an age less than 1 jiffy. The observable universe is then less than the size of a nucleon. Under these conditions it is a safe guess that structural relations between quarks, such as ordinary hadrons, cannot exist. After the jiffy epoch it is conceivable that the quark era contains some structural relations between the quarks, but before the jiffy epoch such structures cannot exist. The jiffy epoch is therefore a *hadron barrier:* Ordinary hadrons, such as pions and nucleons, cannot exist in the universe when it is younger than 1 jiffy, for each is larger than the observable universe. If quarks are not the ultimate constituents of matter, but themselves have internal structure, then we would have to consider the possibility of a quark barrier – provided, of course, that we have the temerity to study the universe at such exceedingly early epochs.

QUANTUM COSMOLOGY. Our discussion has entered a speculative vein and it would be difficult to resist facing up to one of the most challenging of all questions: How far can we travel back in time? Is it possible to journey back to "zero time" when the density and temperature are both of infinite value? What lies before us in our journey, when the universe is younger than 1 jiffy, is veiled from view by our overwhelming ignorance (see Figure 18.4). But undeterred, like seafarers of old who voyaged across unknown and hazardous seas, we shall push on to the frontier of time.

When the universe is 1 ten-million-trillion-trillion-trillionth (or 10^{-43}) of a second old, our journey ultimately comes to a halt before a totally impenetrable wall, the Planck barrier. The universe is now $N^{-1/2}$ of a jiffy old, or N^{-1} of the age of the hadron era. It has a density of 10^{94} grams per cubic centimeter, which is N times the jiffy density or N^2 times the density of the hadron era. The present-day observable universe contains N^2 nucleons, and the entire mass of our observable universe therefore occupies a volume at the Planck epoch equal to that of a single ordinary nucleon. The Planck

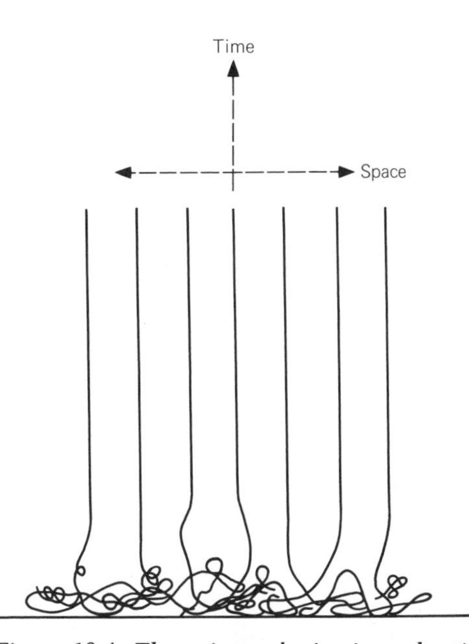

Figure 18.4. The universe begins in a chaotic state. Imagine, as in this illustration, a large number of strings hanging vertically. At floor level the strings are coiled, tangled, and knotted together in a dense layer. In this analogy each string is a worldline along which intervals of time are measured. High above the floor there is an orderly arrangement, as in ordinary spacetime, where events occur sequentially and affect each other in a deterministic manner; but close to the floor the worldlines are jumbled into closed and open loops and there is no common sequential property that we can identify with time. Everything has become irrational and indeterministic.

temperature is 10^{32} degrees, or $N^{1/2}$ times the temperature of the hadron era.

Before us lies the mysterious realm of quantum cosmology, about which we know almost nothing. The quantum fluctuations of spacetime, of a scale equal to the Planck length, are now of cosmic magnitude, and space and time are scrambled together discontinuously and nonsequentially in a chaotic fashion. John Wheeler of Princeton University, who has investigated this subject, visualizes spacetime under these conditions as a chaotic foam. The energy density is so immense that the virtual particles of spacetime, the quantized black holes of Planck mass, have all become real. Spacetime has become a foam of quantized black holes, and space and time no longer exist in the sense that we normally understand. There is no "now" and "then" and no "here" and "there," for everywhere is torn into discontinuities. We cannot go further because an orderly historical sequence of events, such as hitherto has unfolded during our journey backward in time, no longer exists. Here in the domain of quantum cosmology lie perhaps the secrets that foretell the design and architecture of the universe.

TO THE END OF TIME

I saw eternity the other night
Like a great ring of pure and endless light.
All calm, as it was bright;
And round beneath it, Time in hours, days,
years,
Driven by the spheres
Like a vast shadow moved; in which the world
And all her train were hurled.
— Henry Vaughan (1622–95), "The world"

INTO THE FUTURE. Let us now explore the future, venturing further than any time traveler of science fiction has ever dared. As our time machine again clicks off the billions of years we observe the galaxies drifting further and further away. In about 5 billion years the Sun swells into a red giant and engulfs the inner planets. The Earth by now has lost its atmosphere and oceans in the intense red glare, and life has either perished or fled to other places.

The stars begin to fade like guttering candles and are snuffed out one by one. Out in the extreme depths of space the great celestial cities, the galaxies, cluttered with the memorabilia of ages, are gradually dying. Tens of billions of years pass in the growing darkness. Occasional flickers of light pierce the fall of cosmic night, and spurts of activity delay the sentence of a universe condemned to become a galactic graveyard.

We consider two possibilities: first that the universe dies not with a whimper but a bang; and second that the universe expands eternally, and lifeless galaxies voyage in frozen darkness to the end of time. The picture presented by this second possibility has changed in recent years: Apparently dead galaxies do not last forever. Nevertheless, apart from the initial brilliance of the big bang, and a subsequent galactic era lasting a few tens of billions of years, the universe is an everlasting whimper of dark despair. The whimper universe is surely the most appalling state of damnation ever conceived by the human mind.

RETURN OF THE BIG BANG. We suppose that at some time in the future, say 40 or 50 billion years, the universe ceases to expand and begins to collapse. Dead and dying galaxies halt their headlong flight and begin to approach each other once again. The cosmic radiation that has cooled to little more than 1 degree now begins to get warmer. At first only the nearest galaxies are blueshifted, while all others, seen as they were in the past, are still redshifted. As time passes, more and more galaxies appear blueshifted.

Slowly the galaxies approach each other; and first the superclusters, then the great clusters, and finally all clusters overlap each other and their boundaries melt away. When the cosmic radiation has risen to 100 degrees, the galaxies begin to overlap each other and their boundaries also melt away. The universe now consists of a more or less uniform distribution of stars and clusters of stars. But the stars are old and most of them are dark dwarfs, neutron stars, and black holes of all sizes. The collapse continues, and the stars begin to move in their random directions faster and faster, like particles in a gas that is slowly compressed. Rushing stars occasionally collide and erupt in brilliant explosions of light. But most stars, strange to say, do not collide with each other but accelerate to higher and higher speeds, and as they tear headlong through the interstellar gas they are slowly ablated and worn away.

Amidst this uproar the universe becomes incandescent. The radiation era has at last returned in the throes of unimaginable devastation. After a lifespan of 100 or so billion years the big bang is back again, and the universe reverts ultimately to primordial chaos. What follows then we do not know. It may be meaningless to inquire, since time has lost its sequential orderliness and everything has dissolved into a turmoil of discontinuity. Perhaps, in the ultimate chaos, there lurks the cosmogenic genie who conjures universes into being and from the primordial ferment spins new worlds of every possible kind.

WHIMPER OF DEATH. We turn now to the possibility of a whimper universe. In trillions of years the galaxies will be lifeless and as dead as doornails, and the question is whether they will remain in this state for eternity. Given sufficient time – and in eternity there is always more than sufficient – all systems must lapse into their lowest energy states and entropy must rise remorselessly to its maximum value. In eternity there are no bottlenecks and obstructions. Hydrogen gas at room temperature, for example, will transform into helium in a time of 10^{20} years. A hundred million trillion years sounds a long time, but it is nothing, less than the proverbial blink of an eye, compared with eternity.

Our Galaxy and the great galaxy in Andromeda are in orbit about each other and are slowly spiraling together because of the gravitational radiation they emit as a result of their motion. In 10^{30} years, if nothing else happens, the two galaxies will come together and fuse into one galaxy. Clusters of galaxies, because their members orbit about each other, also emit gravitational radiation, and in about 10^{35} years they will shrink into supergalaxies. There is an additional and even more important reason why clusters of all kinds are slowly shrinking. Fast-moving galaxies tend occasionally to escape, and because this evaporation process "cools" the cluster it shrinks continually in size. The surviving galaxies in the cluster contract eventually into supergalaxies.

Stars in binary systems also spiral together, owing to the emission of gravitational radiation, and in about 10^{30} years will amalgamate. Stellar clusters and galaxies will slowly shrink, owing to gravitational radiation and the evaporation of their fastest members. In a time less than 10^{40} years most stars in galaxies will have been swallowed into supermassive black holes. Eventually, all galaxies and supergalaxies will have become superholes surrounded by vast halos of evaporated dark stars.

It is unlikely that the nucleon is a permanently stable particle. On a time scale of 10^{10} years, or N jiffies (N equals approximately 10^{40}), baryon conservation works well and is a reliable conservation law. But on a time scale of 10^{30} or more years, equal roughly to $N^{3/2}$ jiffies, the protons perhaps decay into other particles, such as photons and electron pairs. The electron pairs, comprising electrons and positrons, then annihilate and become photons. Stars therefore never cool to zero temperature but have a temperature of about a millionth of a degree, owing to the slow annihilation of their protons. In 10^{30} years stars and all other bodies not in the form of black holes will have dissolved into radiation. The cosmic radiation that now has a temperature of 3 degrees will have been cooled by expansion to 1 thousand-trillionth (10^{-15}) of a degree.

We are left with a universe of black holes bathed in exceedingly weak radiation. Stephen Hawking has shown that black holes have a temperature and are therefore slowly radiating away their mass. A black hole of solar mass radiates away its entire mass in a time of 10^{66} years, and a superhole of mass equal to that of a giant galaxy radiates away its entire mass in 10^{100} years. Ultimately, when the cosmic radiation has cooled to a temperature of 1 trillion-trillion-trillion-trillion-trillionth (10^{-60}) of a degree, the universe will consist only of photons, neutrinos, and gravitational waves, and all else will have vanished. There will exist no matter and antimatter, and the universe will at last be baryon symmetric and in a state of maximum entropy – for eternity. *Sic transit gloria mundi!*

REFLECTIONS

1 *"If the world had begun with a single quantum, the notion of space and time would altogether fail to have any meaning at the beginning; they would only begin to have sensible meaning when the original quantum had been divided into a sufficient number of quanta. If this suggestion is correct, the beginning of the world happened a little before the beginning of space and time" (Georges Lemaître, "The beginning of the world from the point of view of quantum theory," 1931). The single quantum referred to is Lemaître's primeval atom. The suggestion that time and space have no meaning in the beginning is interesting, but the remark that the world began a little before the beginning of time is surely also without meaning. In* The Primeval Atom *(1951), Lemaître wrote: "The atom world broke up into fragments, each fragment into still smaller pieces. Assuming, for the sake of simplicity, that this fragmentation occurred in equal pieces, we find that two hundred and sixty successive fragmentations were needed in order to reach the present pulverization of matter into our poor little atoms which are almost too small to be broken further. The evolution of the world can be compared to a display of fireworks that has just ended: some few red wisps, ashes, and smoke. Standing on a cooled cinder, we see the slow fading of the suns, and we try to recall the vanished brilliance of the origin of the worlds."*

* *"Many people would argue that it makes no physical sense to talk about half an hour which took place ten billion years ago. To answer that criticism, let us consider a site, somewhere in Nevada where an atomic bomb was set off several years ago. The site is still 'hot' with long-lived fission products, and it took only about one microsecond for the nuclear explosion to produce all the fission products" (Paul Vogel, Amherst College student, 1978).*

2 *In 1948, Gamow and his colleagues Alpher and Herman were attracted by the idea that all elements are synthesized in the big bang. "We conclude first of all that the relative abundance of various atomic species (which are found to be essentially the same all over the observed region of the universe) must represent the most ancient archaeological document pertaining to the history of the universe. These abundances must have been established during the earliest stages of expansion when the temperature of the primordial matter was still sufficiently high to permit nuclear transformations to run through the entire range of chemical elements" (George Gamow, "The evolution of the universe," 1948). In the same year Alpher and Herman wrote, "The temperature in the universe at the present time is found to be about 5° K," which was remarkably close to the value discovered by Penzias and Wilson in 1965. In their calculations they assumed that matter in the big bang is initially in the form of neutrons; that element synthesis begins at a temperature of 1 billion degrees, low enough to avoid the dissociation of deuterons; and that approximately 50 percent of all matter is converted into helium.*

Gamow and his colleagues were able to explain the origin of helium – a remarkable achievement – but their hope of explaining the origin of all elements was shattered by an insuperable difficulty. Elements heavier than helium, such as carbon, nitrogen, and oxygen, cannot be produced abundantly because of the absence of stable elements of atomic weights 5 and 8. "The trouble lies in the fact that the nucleus of mass 5, which would be the next stepping stone, is not available. Due to some peculiar interplay of nuclear forces, neither a single proton nor a single neutron can be rigidly attached to the helium nucleus, so that the next stable nucleus is that of mass 6 (the lighter isotope of lithium), which contains two extra particles. On the other hand, under the assumed physical conditions, the probability that two particles will be captured simultaneously by a helium nucleus is negligibly small, and the building-up process seems to be stopped short at that

Table 18.1. *Leptons and lepton numbers*

Particles		Antiparticles	
e^- electron	+1	e^+ positron	−1
ν electron neutrino	+1	$\bar{\nu}$ electron antineutrino	−1
μ^- muon	+1	μ^+ antimuon	−1
ν_μ muon neutrino	+1	$\bar{\nu}_\mu$ muon antineutrino	−1

Table 18.2. *Hadrons*

Baryons		Mesons	
proton	p	pions	π^+, π^o, π^-
neutron	n	kaons	K^+, K^-
lamda	Λ^o	eta meson	η^o
sigmas	$\Sigma^+, \Sigma^o, \Sigma^-$		
xi	Ξ^o, Ξ^-		
omega	Ω^-		

point" (George Gamow, The Creation of the Universe, *1952). In the big bang the temperature and density continually drop, and by the time helium has been synthesized the physical conditions are no longer suitable for the production of carbon. In stellar evolution, on the other hand, the temperature and density progressively rise, and eventually the physical conditions necessary for the transformation of helium into carbon, and into heavier elements, are attained. This is why we now believe that elements heavier than helium are the products of stellar evolution.*

3 *Protons (p) and neutrons (n) combine to form deuterons (d):*

$$p + n \rightarrow d$$

and in this way deuterium is created in the early radiation era. At first, the deuterons are dissociated back into protons and neutrons by energetic radiation:

$$d \rightarrow p + n$$

but when the temperature has dropped to a billion degrees, the deuterons are not easily dissociated and they are free to combine and form helium nuclei. Two deuterons combine to form 1 helium nucleus:

$$d + d \rightarrow \text{He}$$

If we have 14 protons to every 2 neutrons we finally obtain 12 protons and 1 helium nucleus:

14 protons + 2 neutrons

→ 12 protons + 2 deuterons

→ 12 protons + 1 helium nucleus

Of the 16 original nucleons, 4 are in the helium nucleus, and hence we have 25 percent helium and 75 percent hydrogen.

4 *The leptons – consisting of electrons, muons, their neutrinos, and their antiparticles – and the* lepton numbers *are shown in Table 18.1. The total electron lepton number is always conserved, as in the reaction $p + e^- = n + \nu$, where the electron lepton number of each side is +1. The total muon lepton number is also conserved, as in the reaction $p + \mu^- = n + \nu_\mu$, where the muon lepton number of each is +1.*

The lepton interactions interconvert neutrons and protons:

$$n + e^+ \rightarrow p + \bar{\nu}, \qquad n + \nu \rightarrow p + e^-$$

$$p + e^- \rightarrow n + \nu, \qquad p + \bar{\nu} \rightarrow n + e^+$$

Note the conservation of the electric charge and the electron lepton number. Similar reactions exist for the muons and the muon neutrinos. When the electrons begin to vanish at the end of the lepton era, and the neutrinos decouple, the neutrons and protons are frozen in the ratio of 2 neutrons to every 10 protons. The neutrons then decay according to $n \rightarrow p + e^- + \bar{\nu}$, at a rate of about 10 percent every 100 seconds, and at the time of helium production there are approximately 2 neutrons to every 14 protons.

Some important hadrons are shown in Table 18.2. It is seen that hadrons fall into two classes: baryons and mesons. The baryons have antiparticles; each baryon has a baryon number *of +1, and its antiparticle has a baryon number of −1. The baryon number is conserved in all interactions.*

Table 18.3. *Quarks*

Flavor	Charge	Baryon number
u up	2/3	1/3
d down	−1/3	1/3
c charm	2/3	1/3
s strange	−1/3	1/3
t truth	2/3	1/3
b beauty	−1/3	1/3

Note: A quark is the cry of a seagull. A more appropriate word for such odd elementary particles might have been *quirk*, meaning oddity.

Table 18.4. *Some hadrons and their quark compositions*

	Hadron	Quark composition
baryons	*p* proton	*uud*
	n neutron	*udd*
	Λ° lamda	*uds*
mesons	π^{+} positive pion	$u\bar{d}$
	π° neutral pion	$u\bar{u} + d\bar{d}$
	π^{-} negative pion	$d\bar{u}$
	K^{+} positive kaon	$u\bar{s}$
	K^{-} negative kaon	$s\bar{u}$

Thus, when a proton and an antiproton are created, their combined baryon number is zero. Baryons have half-integer spin: The spin angular momentum of nucleons, for example, is ½h. Mesons have zero baryon number and possess integer spin, such as 0, h, and so on. To create particles of mass m abundantly in a hot gas, the existing particles – such as photons – must have thermal energies of at least mc^2. The thermal energy is kT, where k is the Boltzmann constant, and hence nucleon pairs are created when the temperature T has reached the value $2m_nc^2/k$, which is about 10^{13} degrees.

The flavors, fractional electric charges, and baryon numbers of quarks are displayed in Table 18.3. The situation is more complicated than shown, because each flavor has three colors, and each has its antiquark. Some examples of how hadrons are constructed from quarks are shown in Table 18.4. Each quark has a baryon number of 1/3 and each antiquark ($\bar{u}$, $\bar{d}$, etc.) has a baryon number of −1/3. Note that the quark composition gives a baryon number of +1 for each baryon and zero for each meson.

* *Can we be sure that quarks and leptons are the ultimate constituents of matter? There are numerous quarks of various flavors and colors and each is apparently a complicated entity of several properties. Perhaps they are not elementary but are composite structures of particles that are even more fundamental. Is it possible that particles are like Chinese boxes, endlessly enclosing particles of a more fundamental nature? "We do not know how much further we shall have to probe into subatomic phenomena before we reach an end to novelties, if indeed that will ever happen. Nor do we know if we as individuals, and as a species, are capable of scientific investigations to the point where this happens. These are questions for future humans to answer" (Gerald Feinberg,* What Is the World Made Of?, *1977).*

5 *"Surely something is wanted in our conception of the universe. We know positive and negative electricity, north and south magnetism, and why not some extra terrestrial matter related to terrestrial matter as the source is to the sink, gravitating towards its own kind, but driven away from the substances of which the solar system is composed. Worlds may have formed from this stuff, with elements and compounds possessing identical properties with our own, indistinguishable in fact from them until they are brought into each other's vicinity. If there is negative electricity, why not negative gold, as yellow and valuable as our own, with the same boiling point and identical spectral lines; different only in so far that if brought down to us it would rise up into space with . . . acceleration?" (Arthur Schuster, "Potential matter: a holiday dream," 1898). Schuster was the*

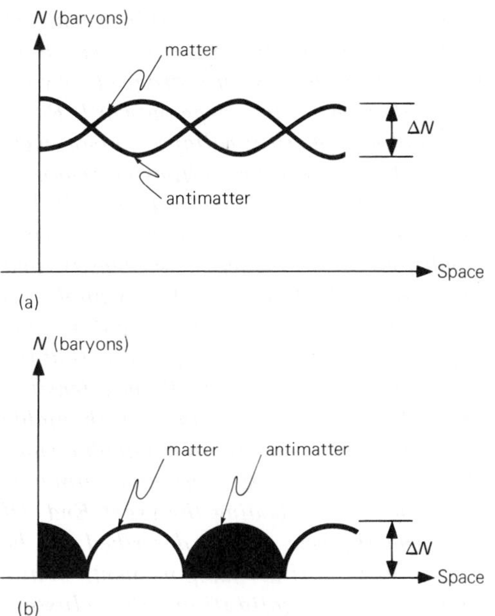

Figure 18.5. In (a) it is supposed that the densities of matter and antimatter vary periodically in the very early universe and that, on the average, there are equal quantities of matter and antimatter. After annihilation has occurred, only the differences survive, as shown in (b), and there are isolated regions of matter and antimatter.

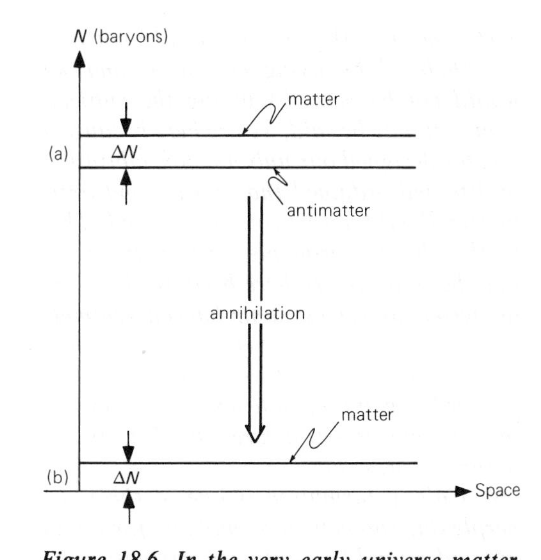

Figure 18.6. In the very early universe matter and antimatter have almost equal densities, as shown in (a), with a difference of about 1 part in a billion. After matter and antimatter have annihilated each other only the small difference remains, as shown in (b).

first to speculate on the possibility of antimatter. We now know that antimatter does not have negative gravity and is attracted to matter in the same way as ordinary matter.

* *The idea that the universe is baryon symmetric (which means that it contains matter and antimatter in equal amounts) is attractive. But, to avoid complete annihilation in the hadron era, it is necessary for some separation to occur so that isolated regions of matter and antimatter can survive. Several separation mechanisms have been suggested, but so far none can be regarded as entirely satisfactory. The simplest proposal is shown in Figure 18.5 (which should be compared with Figure 18.6), in which the densities of matter and antimatter in the hadron era vary in space relative to each other in a periodic manner, with on the average equal amounts of each kind. After annihilation has ceased in the hadron era, there are left isolated pockets of matter and antimatter that later form into galaxies and antigalaxies. This proposal as it stands is not entirely satisfactory because it fails to explain why separation occurs in the first place. Recently developed grand unified theories provide a new explanation of why matter is more favored than antimatter. The hyperweak force, which is dominant in the prejiffy era and consists of the strong, electromagnetic, and weak forces all rolled into one, does not distinguish between quarks and leptons, and does not respect the conservation of baryon number. Because of expansion, and the effect of particle asymmetries, slightly more quarks than antiquarks are created. The hyperweak force is still feebly active and this accounts for the possible slow decay of matter.*

* *If the universe is baryon asymmetric, then presumably we shall one day understand why matter is favored over antimatter. Meanwhile we can fall back on either the theistic or the anthropic principle as a makeshift explanation. If the 1-part-in-a-billion difference in the amounts of matter and antimatter had not existed in the very*

early universe, the universe would not now be inhabited by living creatures, and we would not be here discussing the subject. Hence, it may be said, we are here because a creator designed our universe for habitation by life and ordained that matter and antimatter should not be equally favored. This is the theistic principle. Alternatively, it may be said, we are here because there are numerous universes, some baryon symmetric and others baryon asymmetric, and only those universes that favor matter or antimatter by an appropriate amount can evolve into a state of self-awareness. This is the anthropic principle.

6 *Galaxy formation is one of the most perplexing subjects in cosmology, for we do not understand how galaxies originate in an expanding universe. Presumably, before galaxies are born, there exist precursor irregularities of some kind – the initial conditions – that lead to the formation of galaxies when the universe is about 1 billion years old. But we are not sure of the nature of the initial conditions: They may consist of density variations; or specific entropy variations; or velocity variations, as in turbulence; or matter–antimatter variations; or something else. The irregularities that lead to the formation of galaxies might be vestigial from an earlier and more chaotic state, or they might have grown as the result of an instability in the universe.*

7 *When a star in an advanced state of evolution falls inward, it terminates its collapse as a white dwarf, a neutron star, or a black hole. A black hole is the ultimate state of collapse; to an external observer it is frozen in a perpetual state of collapse, whereas to an internal observer who falls with the star the collapse continues and ends in a singularity of maximum density. Such singularities, predicted by the laws of nature, are themselves beyond the reach of the known laws of nature and hence are not understood.*

Singularity theorems have been developed by Roger Penrose, Stephen Hawking, and George Ellis. Penrose was the first to show that a singularity is inevitable whenever a region is enclosed within a trapped surface. Inside a trapped surface (see Chapter 9) all lightrays are dragged inward faster than they can escape outward. Rotation and nonspherical collapse cannot avert a singularity when the collapsing region is surrounded by a trapped surface. It has also been shown that when certain general conditions are satisfied – if density and pressure are both positive, for example – a collapsing universe ends in a singular state. It is possible, although not definite, that the singular state occurs at the Planck density.

8 *"Then, indeed, amid unfathomable abysses, will be glaring unimaginable suns. But all this will be merely a climactic magnificence foreboding the great End. Of this End the new genesis described can be but a very partial postponement. While undergoing consolidation, the clusters themselves, with a speed prodigiously accumulative, have been rushing towards their own general centre – and now, with a million-fold electric velocity, commensurate only with their material grandeur and their spiritual passion for oneness, the majestic remnants of the tribe of Stars flash, at length, into a common embrace. The inevitable catastrophe is at hand . . . Are we not, indeed, more than justified in entertaining a belief – let us say, rather, in indulging a hope – that the processes we have here ventured to contemplate will be renewed forever, and forever, and forever; a novel Universe swelling into existence, and then subsiding into nothingness, at every throb of the Heart Divine" (Edgar Allan Poe [1809–49],* Eureka*). This is a remarkable essay that anticipates the expansion, collapse, and possible oscillation of the universe. It will be recalled that Poe also perceived the solution of the dark-sky paradox.*

9 *Spacelike cosmic edges were discussed in Chapter 5, at which stage it was shown that in modern cosmology we no longer believe in the possibility of such edges. Timelike cosmic edges, however, are another matter – one that poses some rather tricky problems. In a big bang universe time* *has a beginning, and such a universe has therefore a timelike edge. If it recollapses back to a big bang it has two timelike edges. These edges are not clifflike, or wall-like, but are similar to the gradual fading of the Aristotelian edge. The physical world of an orderly sequence of events merges slowly into a disorderly world of spacetime chaos. We do not understand timelike cosmic edges simply because we do not understand spacetime singularities. At least we can say that time is contained within the physical universe, and that it is nonsense to ask what preceded the beginning of time. Perhaps this is why we have no physical theory of the singularity state – it takes us beyond physics into metaphysics.*

10 *Why do we travel in time but not in space to reach the big bang? If the universe is open, and therefore of infinite extent, was the big bang also of infinite size?*

* *Why was the observational discovery of the cosmic radiation by Penzias and Wilson so important?*
* *How did Gamow and his colleagues Alpher and Herman estimate the temperature of the cosmic radiation, and why were they unable to be more precise?*
* *What is the present theory about the origin of the elements heavier than helium?*
* *Why are cosmic neutrinos of lower temperature than the photons of the cosmic radiation?*
* *If the universe were baryon symmetric, and matter and antimatter were completely annihilated, the 3-degree cosmic radiation would now be warmer by about 1 part in a billion. But in our universe matter originally exceeded antimatter by one-billionth; if this matter that has survived were now annihilated and converted into thermal radiation, the temperature would be 20 degrees. Explain this temperature difference.*
* *Discuss the matter–antimatter difference in the very early universe from the point of view of the anthropic principle.*
* *Show that the* jiffy density *is N times the density of a nucleon, and the* Planck density *is N times the jiffy density. Using the Eddington number, show that all matter in our observable universe once occupied a volume equal to one nucleon at the Planck density. The Eddington number fails to take account of the cosmic photons and neutrinos, and the total number of particles in the observable universe is nearer $N^{9/4}$. What volume did these particles occupy at the Planck epoch? (Answer: The volume had a radius 1000 times that of a nucleon. This is roughly the size of a closed universe at the beginning.)*
* *Why is it that we cannot trace the history of the universe back to infinite density?*
* *What will happen to the universe in the future?*

FURTHER READING

Alpher, R. A. "Large numbers, cosmology and Gamow." *American Scientist,* January–February 1973.

Alpher, R. A., and Herman, R. "Reflections on 'big bang' cosmology," in *Cosmology, Fusion and Other Matters: George Gamow Memorial Volume,* ed. F. Reines. Colorado Associated University Press, Boulder, 1972.

Burbidge, G. "Was there really a big bang?" *Nature 233,* 36 (September 3, 1971).

Davies, P. C. W. *The Runaway Universe.* J. M. Dent, London, 1978.

Davies, P. C. W. *The Forces of Nature.* Cambridge University Press, Cambridge, 1979.

Gamow, G. *The Creation of the Universe.* Viking Press, New York, 1952.

Harrison, E. R. "The early universe." *Physics Today,* June 1968.

Islam J. N. "The ultimate fate of the universe." *Sky and Telescope,* January 1979.

Lemaître, G. *The Primeval Atom.* Van Nostrand, New York, 1951.

Partridge, R. B. "The primeval fireball today." *American Scientist 57,* Spring 1969.

Pasachoff, J., and Fowler, W. "Deuterium in the universe." *Scientific American,* May 1974.

Salam, A. "Quarks and leptons come out to play." *New Scientist,* December 16, 1976.

Schramm, D. N., and Wagoner, R. V. "What can deuterium tell us?" *Physics Today,* December 1974.

Silk, J. "The evolution of the universe: a layperson's guide to the big bang." *Griffith Observer,* July 1975.

Silk, J. *The Big Bang.* W. H. Freeman, San Francisco, 1979.

Wagoner, R. V. "Cosmological element production." *Science 155,* 1369 (1967).

Weinberg, S. *The First Three Minutes: A Modern View of the Origin of the Universe.* Basic Books, New York, 1977.

Whittaker, E. T. *The Beginning and End of the World.* Oxford University Press, Clarendon Press, Oxford, 1942.

SOURCES

Alpher, R. A., and Herman, R. "Evolution of the universe." *Nature 162,* 774 (November 13, 1948).

Alpher, R. A., and Herman, R. "Big bang cosmology and the cosmic black-body radiation." *Proceedings of the American Philosophical Society 119,* 325 (October 1975).

Collins, C., and Hawking, S. W. "Why is the universe isotropic?" *Astrophysical Journal 180,* 317 (1973).

Dicke, R. H., Peebles, P. J. E., Roll, P. G., and Wilkinson, D. T. "Cosmic black-body radiation." *Astrophysical Journal 142,* 414 (1965).

Drell, S. D. "When is a particle?" *American Journal of Physics 46,* 597 (January 1976).

Eddington, A. S. "The end of the world: from the standpoint of mathematical physics." *Nature,* Supplement, p. 447, March 24, 1931.

Ellis, G. F. R. "Singularities in general relativity." *Comments on Astrophysics and Space Physics 8,* 1 (1978).

Feinberg, G. *What Is the World Made Of? Atoms, Leptons, Quarks, and Other Tantalizing Particles.* Doubleday, Anchor Press, New York, 1977.

Fowler, W. A., and Stephens, W. E. "Origin of the elements: resource letter." *American Journal of Physics 36,* 1 (April 1968).

Gamow, G. "The evolution of the universe." *Nature 162,* 680 (October 30, 1948).

Gamow, G. *The Creation of the Universe.* Viking Press, New York, 1952.

Harrison, E. R. "Standard model of the early universe." *Annual Review of Astronomy and Astrophysics 11,* 155 (1973).

Hawking, S. W., and Sciama, D. W. "Collapsing stars and expanding universes." *Comments on Astrophysics and Space Physics 1,* 1 (1969).

Islam, J. N. "Possible ultimate fate of the universe." *Quarterly Journal of the Royal Astronomical Society 18,* 3 (1977).

Lemaître, G. "The beginning of the world from the point of view of quantum theory." *Nature 127,* 706 (May 9, 1931).

Lemaître, G. *The Primeval Atom.* Van Nostrand, New York, 1951.

Nambu, Y. "The confinement of quarks." *Scientific American,* November 1976.

Peebles, P. J. E. *Physical Cosmology.* Princeton University Press, Princeton, N.J., 1971.

Penrose, P. "Singularities and time-asymmetry," in *Einstein Centenary Volume,* ed. S. W. Hawking and W. Israel. Cambridge University Press, Cambridge, 1979.

Penzias, A. A., and Wilson, R. W. "A measurement of excess antenna temperature at 4080 Mc/s." *Astrophysical Journal 142,* 419 (1965).

Rees, M. J. "The collapse of the universe: an eschatological study." *Observatory 89,* 193 (1969).

Schuster, A. "Potential matter: a holiday dream." *Nature 58,* 367 (August 18, 1898).

Schwitters, R. F. "Fundamental particles with charm." *Scientific American,* October 1977.

Stecker, F. W. "Baryon symmetric big bang cosmology." *Nature 273,* 493 (June 15, 1978).

Steigman, G. "Observational tests of antimatter cosmologies." *Annual Review of Astronomy and Astrophysics 14,* 339 (1976).

Trimble, V. "The origin and abundance of the chemical elements." *Reviews of Modern Physics 47,* 877 (1975).

Weinberg, S. "Unified theories of elementary particle interactions." *Scientific American,* July 1974.

19 HORIZONS IN THE UNIVERSE

I am a part of all that I have met;
Yet all experience is an arch wherethro'
Gleams that untravelled world, whose margin fades
For ever and for ever when I move.
— Tennyson (1809–92), *Ulysses*

WHAT ARE COSMIC HORIZONS?

HORIZONS. We look out in space and back in time and do not see the galaxies stretching away endlessly to an infinite distance in the infinite past. We look out to a finite distance and see only those things that are within the observable universe. Like the sea-watching folk in Robert Frost's poem, we "cannot look out far" and "cannot look in deep."

The observable universe is usually only a portion of the whole universe. We are at the center of our observable universe, and its distant boundary is a horizon beyond which lie objects that cannot be observed. Observers in other galaxies are located at the centers of their observable universes, which are bounded by horizons. A person on a ship at sea observes the sea stretching away to the horizon, where the sky meets the sea, and is at the center of an "observable sea." People on other ships at sea are at the centers of their "observable seas," which are bounded by horizons of a similar nature. Despite this analogy we shall find that cosmic horizons are not quite so simple as the horizon of the sea.

PARTICLE AND EVENT HORIZONS. The subject of cosmic horizons was rather confusing until Wolfgang Rindler cleared up the muddle in 1956. He showed that when we discuss the observable and the unobservable we must realize that there are two kinds of things observed: worldlines and events. Worldlines represent objects such as particles and galaxies that endure over long periods of time; they occupy at any instant a place in space and have extension in time. An event is a brief happening, such as the flash of a firefly or the explosion of a supernova that occupies a place in space and a place in time. Worldlines are in effect strings of events. The events of interest are those that emit rays of light, and the worldlines of interest are those that represent luminous bodies such as galaxies.

To determine what is observable we must specify the nature of the things to be observed. If they are particles or galaxies that endure and therefore have worldlines, we obtain one kind of result; if they are events that occur briefly, we obtain another kind of result. For instance, if one is asked, "Have you met Mr. X?" the answer could be different to that given in response to the question, "Did you see Mr. X at his wedding?" The first question asks if a worldline has been observed, and the second asks if an event has been observed.

There are two types of horizon, a *worldline horizon* and an *event horizon,* and both are important. Rindler referred to the worldline horizon as the *particle horizon,* and as this latter term is now widely adopted we shall also use it. It must be understood, however, that the word *particle* in this case means worldline and represents anything that endures. We shall first define the particle and event horizons and then make clear their meaning with the aid of a static universe. Our discussion throughout this chapter is concerned only with horizons in universes that are uniform (i.e., isotropic and homogeneous).

A particle horizon is a spherical surface, with the observer inside at the center, which at a certain instant divides the whole of space into two regions. The region inside contains all galaxies that are observed, and the region outside contains all galaxies that are unobserved. The particle horizon is thus a frontier in space and encloses the observable universe. The horizon at sea is of this type; it is a frontier that divides all ships at sea into two groups: those that are observed and the remainder that are unobserved.

The event horizon divides all events into two groups: those observable at some time or other and those that are never observed. An observer sees events displayed on a backward lightcone. The event horizon is therefore not a surface in space but a backward lightcone that separates events that can be observed from events that are never observed. The event horizon is not quite so obvious as the particle horizon, with its sea-horizon analogy, but this need not cause concern: Subsequent sections will clarify the whole matter.

HORIZONS IN A STATIC UNIVERSE. The two types of horizon are most easily demonstrated in an infinite and static universe. Let us forget for the moment that the universe is expanding and suppose that we live in a static Newtonian universe containing uniformly distributed stars. For the sake of simplicity we also suppose that the stars have been luminous for 10 billion years. We can assume either that the universe was created with its luminous stars or that the stars became luminous simultaneously in a preexisting dark universe. The situation is as shown in Figure 19.1. The universe consists

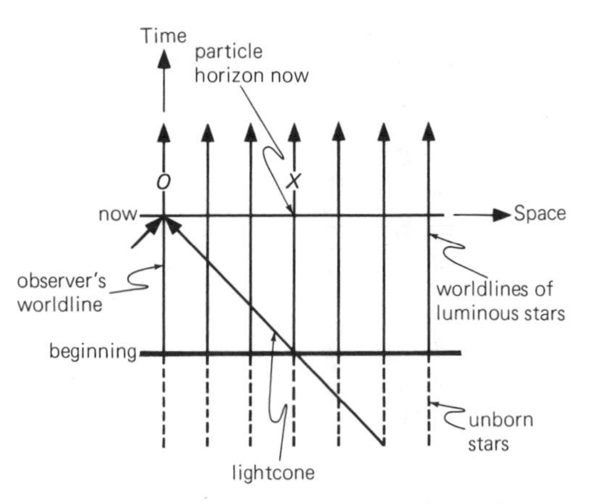

Figure 19.1. A static universe that has a beginning and consists of uniformly distributed stars. We are the observer O who looks out "now" and sees the worldlines of luminous stars intersecting our lightcone. At the particle horizon is worldline X. Stars with worldlines beyond X cannot be seen.

of worldlines of luminous stars that commence at the "beginning."

The worldline labeled O is our star – or Solar System – from which we observe the universe. From O, at the instant "now," we look out in space and back in time and observe other stars in our backward lightcone. We observe the stars because their worldlines intersect our lightcone, and we see each at a particular instant in its lifetime. All stars have been shining for 10 billion years and it is therefore possible to look out and see the stars stretching away to a distance of 10 billion light years. Stars at greater distances cannot be seen because we look back to a time before they existed.

A particle horizon divides all luminous sources into those observed and those unobserved. In the Newtonian universe the particle horizon is therefore at the distance indicated by worldline X, and in the present example it lies at a distance of 10 billion light years. Stars at distances less than 10 billion light years are observed and are hence within the observable universe, and stars at greater distances are unobserved and are outside the observable universe.

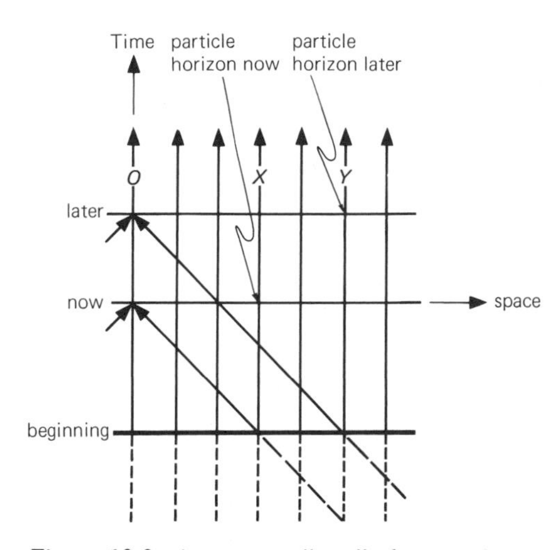

Figure 19.2. At moment "now" observer O sees no further than worldline X. Subsequently, at the moment labeled "later," the observer sees beyond X to worldline Y. The particle horizon therefore recedes in a static universe, and the observable universe, bounded by the particle horizon, is expanding.

Let us wait a period of time – a billion years, say – and repeat our observations. We are then at the instant "later," shown in Figure 19.2, when stars have been shining for 11 billion years, and we see them stretching away to a distance of 11 billion light years. The particle horizon has receded to a distance of 11 billion years. It is evident that the particle horizon sweeps outward at the velocity of light, and although the universe is static, the observable universe is expanding. In most universes, static and nonstatic, the particle horizon sweeps outward and overtakes the stars and galaxies at the velocity of light.

We turn now to the event horizon and ask whether in the static universe there are events that can never be observed by an observer on a worldline such as *O*. If such events exist, then we should be able to divide the universe into two parts: one that contains all the events observable from *O* and another that contains the remaining events, which are unobservable from *O*. The surface separating the two parts will then be the event horizon for observer *O*.

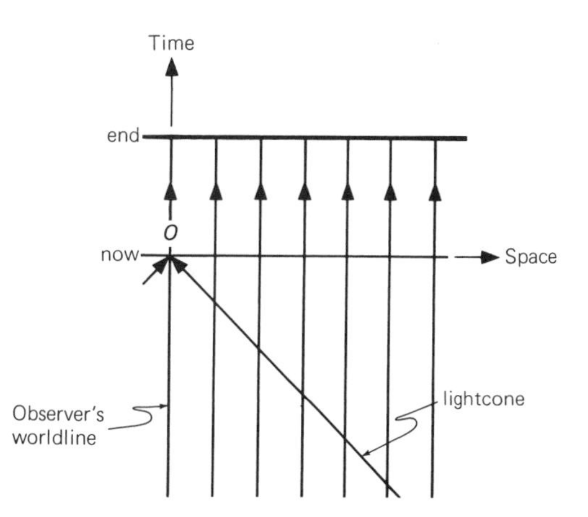

Figure 19.3. A static universe that has an ending. The time labeled "end" is the observer's last possible moment of observation.

If stars shine forever in the future we realize immediately that there is no event horizon. *O*'s lightcone in this case advances up *O*'s worldline, and any point in spacetime that is an event will eventually lie on the lightcone and be observed. Hence, in a static universe in which stars are luminous for an infinite future, there is no event horizon, and every event at some time or other is observed by any observer.

We consider now the case where the Newtonian universe of luminous stars comes to an "end." Either the universe is terminated by its creator, or the stars and observers cease to exist and the universe becomes dark. All worldlines as a result come to an end, as shown in Figure 19.3. It is immediately obvious that in such a universe there is an event horizon, and it is *O*'s lightcone at the last possible moment. Inside the event horizon are the events that can be observed, and outside are the events that are never observed, as shown in Figure 19.4. The lightcone cannot advance further into the future, and all events outside this ultimate lightcone are never seen.

Despite its simplicity, the Newtonian universe serves to illustrate reasonably clearly the nature of cosmic horizons. With it we have learned that beyond the particle

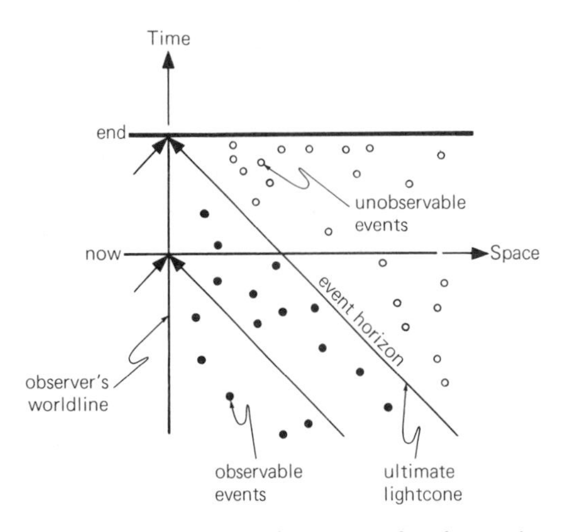

Figure 19.4. The event horizon is the observer's lightcone when the universe ends. Inside this ultimate lightcone are the events that can be observed, and outside are the events that are never observed.

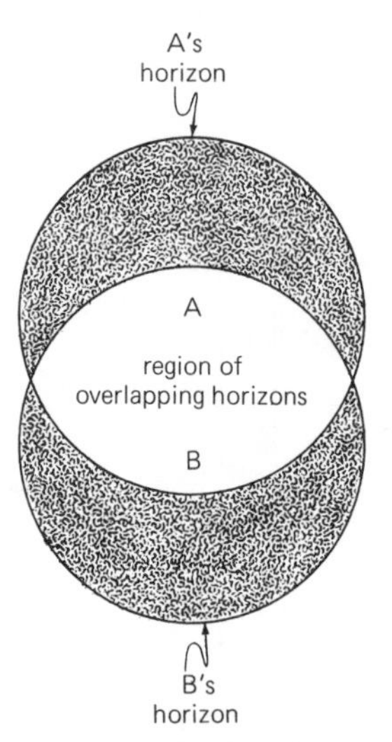

Figure 19.5. Albert (A) and Bertha (B) have overlapping horizons, but each is able to see things that the other cannot. Can they, by communicating with each other, enlarge their individual horizons into a joint horizon? If they can, then their horizons are not true cosmic horizons.

horizon are worldlines (particles, stars, galaxies) that cannot now be seen at any stage of their existence, and beyond the event horizon are events (happenings of short duration) that can never be seen at any stage of the observer's existence.

THE HORIZON RIDDLE

While to deny the existence of an unseen kingdom is bad, to pretend that we know more about it than its bare existence is no better.
— Samuel Butler (1835–1902), *Erewhon*

Consider two observers, A (for Albert) and B (for Bertha), who are widely separated from each other in the universe. Each has a horizon of some kind such that A cannot see objects beyond his horizon and B cannot see objects beyond her horizon. Although they see each other, each sees objects that the other cannot, as illustrated in Figure 19.5. We now ask: Can B communicate to A information that will extend A's knowledge of objects beyond his horizon? If so, then surely a third observer C may communicate to B information that B in turn relays to A. A sequence of observers B, C, D, E, . . . may in this way extend A's knowledge of the universe to indefinite limits. According to this argument A has no horizon! This is the horizon riddle that puzzles many students.

The riddle, it must be admitted, stems from our experience with horizons on the surface of the Earth. Thus if A and B are on ships at sea, within sight of each other, they each observe the sea stretching away to the horizon. A sees objects that B cannot see, and similarly B sees objects that A cannot see; they can communicate with flags and thereby keep each other aware of objects not directly visible. By communication, A and B pool their observations and succeed in extending their horizons. In this way a pre-twentieth-century admiral had a horizon that embraced his entire fleet.

When we speak of objects that are seen or not seen we have in mind those things that endure and are therefore represented by worldlines. Hence the horizon riddle applies to the particle horizon of the universe. Let us consider the particle horizon in the static Newtonian universe and show that the riddle has a simple answer. All stars were born, we suppose, 10 billion years ago, and the particle horizon is hence at a distance of 10 billion light years. Observers A and B see each other and have similar horizons that overlap. Suppose that B is at a distance of 6 billion light years from A. Observer B sends out information that travels at the speed of light, and this information takes 6 billion years to reach A. Therefore, A receives from B information that was sent 6 billion years ago when the luminous universe was 4 billion years old. B's particle horizon in the past, when the information was sent, was at a distance of 4 billion light years. Hence B's horizon at that time does not extend beyond A's horizon. Actually, B's horizon in the past, when information was sent, just touches A's horizon at the time that information is received. With this argument, and the aid of Figure 19.6, it is apparent that neither B nor any other observer can extend A's particle horizon. The particle horizon is thus a real horizon, and information cannot be gained from other observers concerning what lies beyond this horizon.

Cosmic horizons are information barriers. The horizon riddle is usually easily resolved; when it cannot be resolved, as in

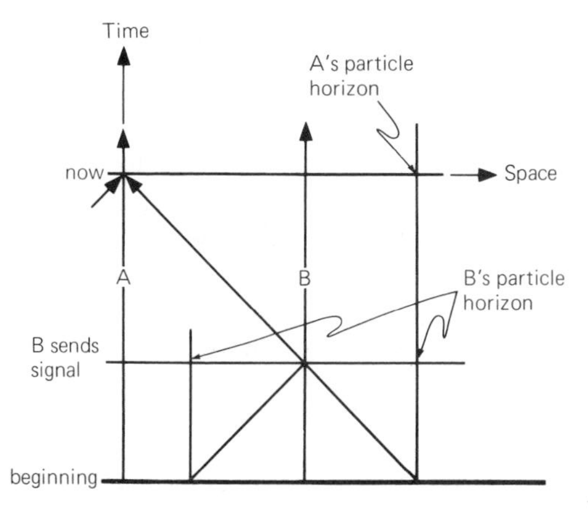

Figure 19.6. Proof that Bertha cannot help Albert to see beyond his horizon (and similarly, that Albert cannot help Bertha to see beyond her horizon). The horizons in this case are particle horizons: B communicates information to A and sends it along A's lightcone; but at the time when B sends the information, she cannot see beyond A's horizon.

the case of ships at sea, then we are not talking about true horizons that are information barriers.

"A SLOW SORT OF COUNTRY"

HUBBLE SPHERE. Static universes are not realistic, and horizons in static universes, though illustrative, are not very important. In a preambling fashion we consider here what happens in an expanding universe. A more precise treatment follows in subsequent sections.

In earlier chapters we have encountered briefly the possibility of horizons in expanding universes and have occasionally referred to the "observable universe." According to the velocity–distance law of the universe the recession velocities of galaxies increase with distance. If the distance is doubled the recession velocity is doubled. The recession velocity eventually equals the velocity of light, and this happens at the Hubble length denoted by $L = c/H$, where H is the Hubble term. The Hubble length is not known precisely; we have assumed that it has a value of 15 billion light years. Galaxies beyond the Hubble length are receding faster than the velocity of light. The velocity–distance law of the universe applies everywhere in space at an instant in time; it tells us how fast galaxies are receding at the present moment, and we must remember that their recession velocities were different at the time they emitted the light that we now receive.

We are at the center of the Hubble sphere, a sphere that has a radius equal to the Hubble length, and has therefore a radius of 15 billion light years. Inside the Hubble sphere are all the galaxies that at present recede more slowly than the velocity of light, and outside the sphere are all the galaxies that at present recede faster than the velocity of light. In earlier chapters we have rather crudely identified the Hubble sphere with the observable universe. But the observable universe is bounded by the particle horizon, and this, as we shall see, is not at the edge of the Hubble sphere. Indeed, if the Hubble sphere and the observable universe were of the same size, the observable universe would be infinitely large in a static universe. But static universes, with their infinitely large Hubble spheres, have particle horizons at finite distance, and therefore in general the Hubble sphere is not the same as the observable universe.

COUNTRY OF THE RED QUEEN. Consider a galaxy outside the Hubble sphere that emits a ray of light in our direction. We are the observer, as shown in Figure 19.7, at the center of the Hubble sphere. The lightray travels through space at the velocity of light, but the space through which it moves is receding from us faster than light. Although the lightray hurries toward us it is actually receding. As Eddington said: "Light is like a runner on an expanding track with the winning-post receding faster than he can run." The edge of the Hubble sphere is thus a kind of horizon. All lightrays emitted in our direction within the Hubble sphere approach us, whereas all lightrays emitted outside the Hubble sphere recede from us. The lightrays emitted in our direction at the edge of the Hubble sphere stand still; they hurry toward us at the same velocity as that at which expanding space carries them away. The edge of the Hubble sphere is the country of the Red Queen: "Now, here, you see, it takes all the running *you* can do, to keep in the same place," said the Red Queen to Alice.

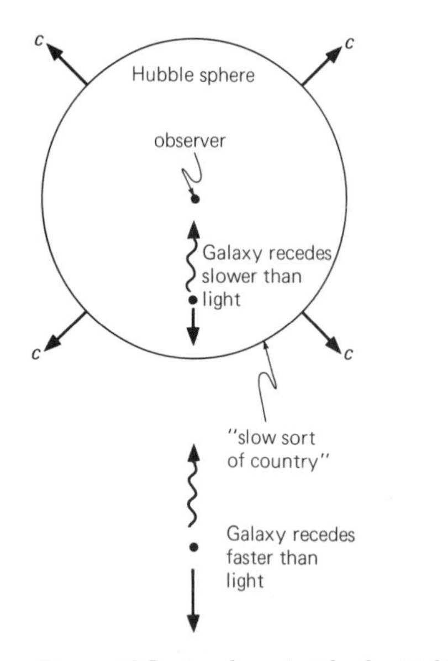

Figure 19.7. A galaxy inside the Hubble sphere recedes from us slower than light velocity, and the light emitted in our direction by the galaxy is able to approach us. A galaxy outside the Hubble sphere recedes from us faster than light velocity, and the light emitted in our direction by the galaxy actually recedes, although it is hurrying toward us. The edge of the Hubble sphere is the country of the Red Queen.

This picture of a horizon in an expanding universe is simple to understand but unfortunately is misleading. We have viewed everything in space at the same instant – in a sort of snapshot picture – and have not allowed for the important fact that the observer sees things as they were in the past and cannot see them as they are now at great distances. It is true that galaxies outside the Hubble sphere recede faster than light can travel, and the lightrays they send in our direction are also receding, but this does not mean that all galaxies beyond the Hubble sphere are necessarily always hidden from view.

The rate of expansion of most universes is not constant. In a decelerating universe in which the Hubble term decreases with time, the Hubble length steadily increases, and the Hubble sphere consequently expands. Lightrays outside the Hubble sphere, moving in our direction, may therefore eventually be overtaken and lie inside the Hubble sphere. These rays will at last begin to approach us, and later we shall receive them. On an expanding track the runner sees the winning-post receding, but the runner must keep running and not despair because the expansion of the track may be slowing down, and the winning-post in that case can ultimately be reached. This argument enables us to realize that the Hubble sphere is not necessarily the observable universe.

DISTANCES. Before discussing horizons in more detail we should make clear what we mean by distance. Cosmologists use various kinds of distance indicators, such as the luminosity distance and the distance by apparent size, but we shall consider only "pure" distances of the kind displayed in spacetime diagrams.

We look out in any direction in the expanding universe and see galaxies of increasing redshift. It is sometimes said we see galaxies of increasing redshift stretching away to greater and greater distances. Do we? The answer is yes if we mean their present distances as used in the velocity–distance law. But the answer is emphatically no if we mean their distances at the time they emitted the light that is now seen. We cannot see galaxies at their present distances; we see them only at the distances they had in the past when they emitted the light now seen, and these distances do not continually increase with redshift in a big bang universe.

We therefore have in mind two kinds of distance when we speak of a distant galaxy: its present distance and its distance at the time of emission of the light we now see. The first will be called the *reception distance* (the distance at the time of reception of light) and the second the *emission distance* (the distance at the time of emission of light). These two distances have the simple relation

$$1 + z = \frac{\text{reception distance}}{\text{emission distance}}$$

where z is the observed redshift. The reception distance is always greater than the emission distance in an expanding universe.

We have asked whether distance increases with redshift. To answer this question we must look at the spacetime diagram, and the conclusion we shall reach is anticipated by the following remarks. The reception distances of galaxies in an expanding universe always increase with redshift, and the larger their redshifts, the larger are their reception distances. The emission distances at first also increase with redshift, and then a remarkable thing happens: The emission distances stop increasing with redshift and begin to decrease! Faint galaxies of large redshifts are actually closer to us at the time of emission than are brighter galaxies of smaller redshifts. This strange situation, unlike any encountered by Alice in her adventures, happens in all expanding universes for which the deceleration term q is greater than -1. It does not happen in the de Sitter and steady state universes for which H is constant and q is equal to -1.

A spacetime diagram convenient for our purpose is shown in Figure 19.8, in which worldlines diverge in all directions from a big bang. Space is represented by any spherically curved surface that is perpendicular to the worldlines, and time is measured radially along the worldlines. Each worldline is a galaxy fixed in space, and as time advances, the galaxies recede from each other because of the expansion of space. An expanding balloon is a helpful analogy; galaxies are denoted by points marked on the surface of the balloon, and as the balloon is inflated, the "galaxies" recede from each other. The two-dimensional surface of the balloon represents our three-dimensional space, and the radial direction represents time and should not be confused with the third dimension of space.

One of the worldlines – it does not matter which – is chosen as the observer's and

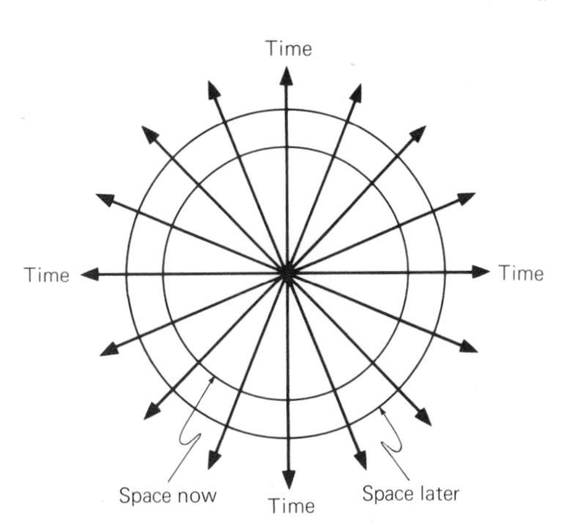

Figure 19.8. A big bang with worldlines diverging in all directions. Do not let this diagram mislead you into thinking that the big bang occurs at a point in space: Time is measured along the worldlines, and space is represented by any spherically curved surface perpendicular to the worldlines.

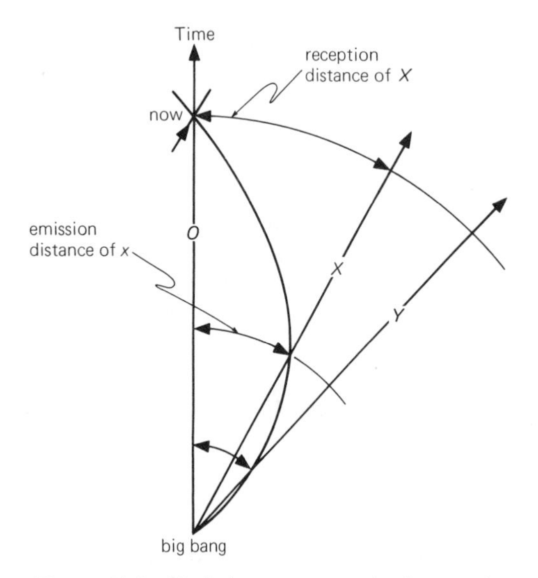

Figure 19.9. O's lightcone curves back into the big bang, and for this reason we are able to observe the cosmic radiation that has traveled freely since the big bang. The diagram shows the reception and emission distances of galaxy X. Although galaxy Y has a greater reception distance, its emission distance is smaller than that of X. Thus Y, which is now further away than X, was closer to us than X at the time of the emission of the light that we now see.

labeled *O*, as in Figure 19.9. The observer's lightcone at a certain instant (call it "now") stretches back and intersects other worldlines, such as those of *X* and *Y*. Because of the expansion of space the lightcone does not stretch out straight, as in a static universe, but instead is curved and reaches back into the big bang. The observer is thus able to look back into the big bang, and the lightrays received from the big bang, which travel forward along the lightcone, comprise the cosmic radiation. On the lightcone are events that emit light, and the closer the events are to the big bang, the larger is their observed redshift. Hence redshift increases steadily as we proceed along the lightcone toward the big bang.

Figure 19.9 shows the emission and reception distances of galaxy *X*; the emission distance is the distance in space when *X* emits the lightrays that *O* now receives, and the reception distance is the distance in space of *X* at the time *O* receives the lightrays. The reception distance of the second galaxy *Y* is greater than that of *X*, and *Y* at present (at the moment "now") is further away than *X*. *Y*'s worldline intersects the lightcone at a point closer to the big bang, and its redshift, as seen by *O*, is greater than that of *X*. From this it is clear that redshift increases with reception distance, and galaxies of large redshifts are at present further away than galaxies of small redshifts.

As we proceed back along the lightcone the emission distance at first increases, then finally reaches a maximum value, and thereafter begins to decrease. *Y*'s emission distance in Figure 19.9 is less than that of *X*, although *Y* is now further away and has the greater redshift. In the Einstein–de Sitter universe the maximum emission distance is 8/27 of the present Hubble length, almost 5 billion light years, and occurs at a redshift of 5/4.

Galaxies at maximum emission distance have a recession velocity equal to the velocity of light. They were at the edge of the observer's Hubble sphere at the time of emission. Galaxies now observed with redshifts less than that corresponding to the maximum emission distance had recession velocities less than the velocity of light at the time of emission (and were inside the Hubble sphere), and those with larger redshifts had recession velocities greater than the velocity of light at the time of emission (and were outside the Hubble sphere). Notice in Figure 19.9 that when *Y* emits light toward *O* the lightcone is diverging away from *O*'s worldline; the lightrays leaving *Y* at first move away from *O* and then, after reaching maximum emission distance, converge on *O*.

PARTICLE HORIZONS

RECEDING PARTICLE HORIZON. We recall that beyond the particle horizon are galaxies that cannot be observed. In more technical language we say that beyond the particle horizon are worldlines that do not intersect the observer's lightcone. Worldlines that do

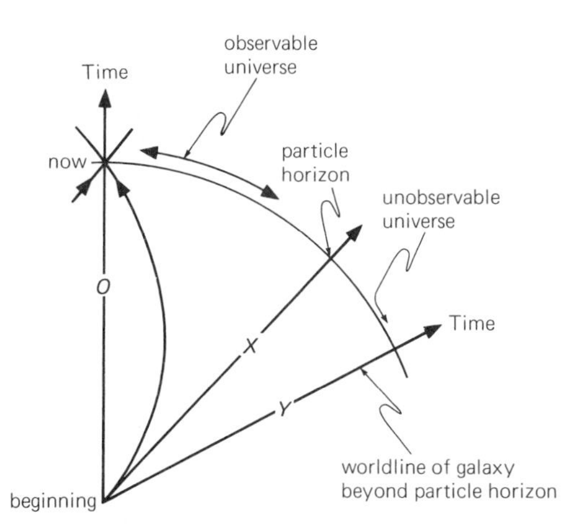

Figure 19.10. At the instant "now" the particle horizon is at worldline X. In a big bang universe all galaxies at the particle horizon have infinite redshift.

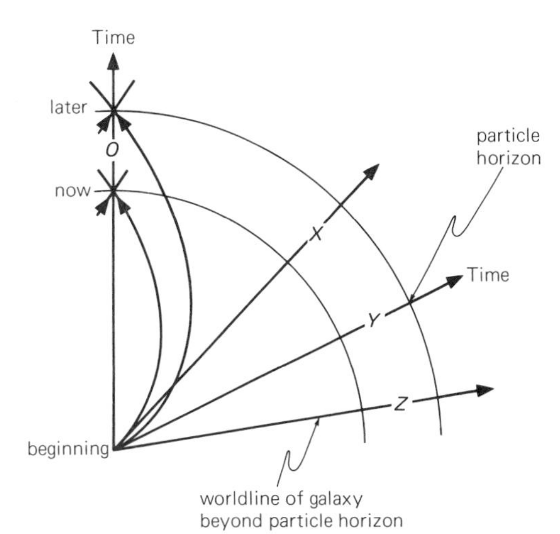

Figure 19.11. At the instant labeled "later" the particle horizon has receded to worldline Y. Note that the distance of a particle horizon is a reception distance; it is the distance measured in space at the time of observation.

intersect the observer's lightcone represent galaxies within the observable universe and are visible at some state in their evolution, whereas worldlines that do not intersect the observer's lightcone represent galaxies outside the observable universe and at the time of observation are not visible at any stage in their evolution.

The type of spacetime diagram shown in Figures 19.8 and 19.9 is useful for illustrating the nature of the particle horizon in an expanding universe. In such a diagram, as in Figure 19.10, we notice immediately that space "now" is divided into two regions: The first, surrounding the observer, contains all worldlines that intersect the observer's lightcone; the second, further away from the observer, contains all worldlines that do not intersect the lightcone. The first region is the observable universe, the second region is the unobservable universe, and the two are separated by the particle horizon. The distance of the particle horizon is measured in the observer's present space (at time "now") and is the reception distance of galaxies of infinite redshift.

We consider next a later instant, labeled "later" in Figure 19.11, when the universe has expanded further. It is clear that the observer's lightcone now intersects more worldlines and the particle horizon has receded to a greater distance. Because the particle horizon has receded, the observable universe has expanded. We notice that the observable universe expands faster than the actual universe, and as the particle horizon recedes, it sweeps out faster than the receding galaxies. In fact, the particle horizon sweeps past the galaxies at the velocity of light. This is a general rule: the observable universe overtakes the galaxies at the velocity of light.

In the Einstein–de Sitter universe, with which it is simple and convenient to work, the particle horizon is always at twice the Hubble length. The observable universe, in other words, has a radius twice that of the Hubble sphere. The recession velocity of the galaxies at the edge of the observable universe, from the velocity–distance law, is hence twice the velocity of light, and the edge of the observable universe – which is the particle horizon – is moving away at three times the velocity of light. The situation is similar, but not exactly the same, in the other two Friedmann universes at the present stage of cosmic evolution.

UNIVERSES WITHOUT PARTICLE HORIZONS. Some universes, for example, the Milne, de

Sitter, and steady state universes, do not have particle horizons. All worldlines intersect an observer's lightcone and all galaxies in the universe are visible. To show diagrammatically why such universes are possible, we use a different type of spacetime. This spacetime, as shown in Figure 19.12, consists of comoving rather than ordinary space. All comoving objects have constant comoving distances, and in this new spacetime all worldlines are parallel. An observer's lightcone again does not reach out straight, as in a static universe; in this case it opens up and spreads out, as shown. In many universes, such as the Friedmann kind, the lightcone reaches back to the beginning of time at a finite comoving distance, and a particle horizon exists at the worldline X. In some other universes, however, such as the Milne universe (in which R increases at a constant rate), the lightcone reaches the beginning of time at an infinite comoving distance, and consequently there is no particle horizon. The observable universe fills the entire actual universe and all galaxies are visible. The de Sitter and steady state universes are of this kind, but are more complicated, and will be considered when we discuss event horizons.

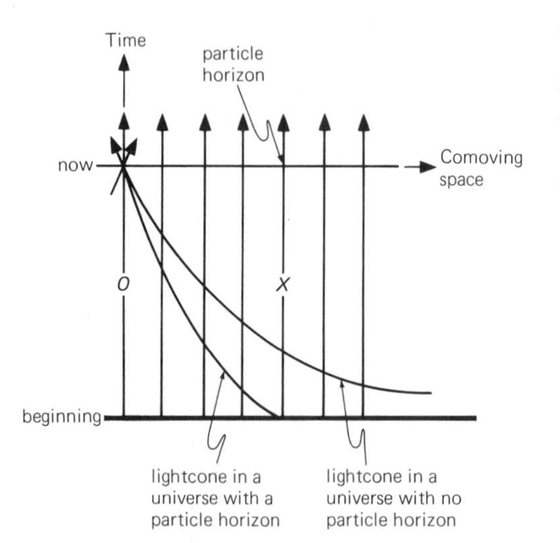

Figure 19.12. A spacetime diagram of comoving space (in which all worldlines are parallel) and ordinary time. Some universes have particle horizons, and in these the lightcone stretches back to the beginning at a finite distance indicated by worldline X. In universes that do not have particle horizons, the lightcone stretches out to infinity and intersects all worldlines.

SPACETIME DIAGRAMS WITH STRAIGHT LIGHTCONES. The time has come to introduce the reader to a useful trick when discussing horizons of any kind. We have just used a spacetime diagram (Figure 19.12) in which comoving space takes the place of ordinary space. This type of diagram has the advantage that all worldlines are parallel to one another, but retains the disadvantage that the lightcone spreads out in a way so peculiar that often it is difficult to realize what is happening. Why not – as the next logical step – alter time so that the lightcone is straight, as in a static universe? We then have a spacetime diagram of comoving space and "altered time" that looks like an ordinary spacetime diagram of a static universe. And we know how to handle horizons in a static universe.

The spacetime interval between two events close together is given by

$$(\text{spacetime interval})^2 = (\text{time interval})^2 - (\text{space interval})^2$$

in which space intervals are measured in light travel time. All lightrays follow paths – technically known as null geodesics – for which the spacetime intervals are zero. Hence, for a lightray and anything else that travels at the speed of light,

$$(\text{time interval})^2 = (\text{space interval})^2 = (R \times \text{comoving space interval})^2$$

If we now change the intervals of time, such that

$$\text{altered time interval} = \frac{\text{time interval}}{R}$$

light rays will follow paths determined by

$$(\text{altered time interval})^2 = (\text{comoving space interval})^2$$

and hence

$$\text{altered time interval} = \pm \text{comoving space interval}$$

(the plus sign is for the forward lightcone and the minus sign for the backward lightcone). This last relation between intervals of

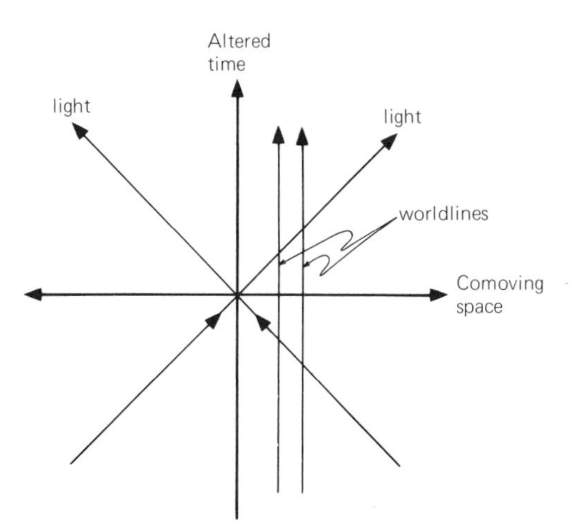

Figure 19.13. A spacetime diagram that consists of comoving space and "altered time." If we straighten out the lightcone by altering the intervals of time, the spacetime diagram of a nonstatic universe looks like that of a static universe. This allows us to study horizons in nonstatic universes just as in static universes.

altered time and comoving space is for a spacetime, like that shown in Figure 19.13, in which the lightcones are straight, as in a static universe. The advantage of this type of diagram is that it allows us to treat horizons the way we do in a static universe. There are four possibilities that must be considered.

When a universe has a beginning and an ending in altered time, the spacetime

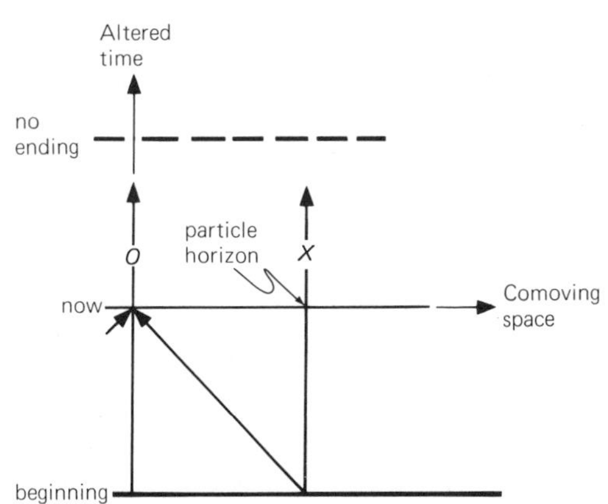

Figure 19.15. A spacetime diagram of comoving space and altered time that has a beginning but no ending. The particle horizon is at the worldline X, and there is no event horizon.

diagram is of the type shown in Figure 19.14. The closed Friedmann universe that begins and ends with a bang belongs to this class.

When a universe has a beginning but no ending in altered time, the spacetime diagram has a lower boundary but no upper boundary, as shown in Figure 19.15. The Einstein–de Sitter universe and the Friedmann universe of negative curvature that begin with bangs and expand forever are examples in this class.

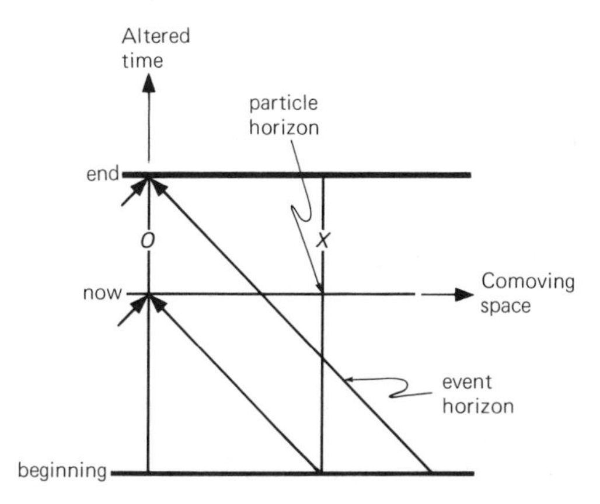

Figure 19.14. A spacetime diagram of comoving space and altered time that has a beginning and an ending. The worldline X is at the particle horizon. Note also that there is an event horizon.

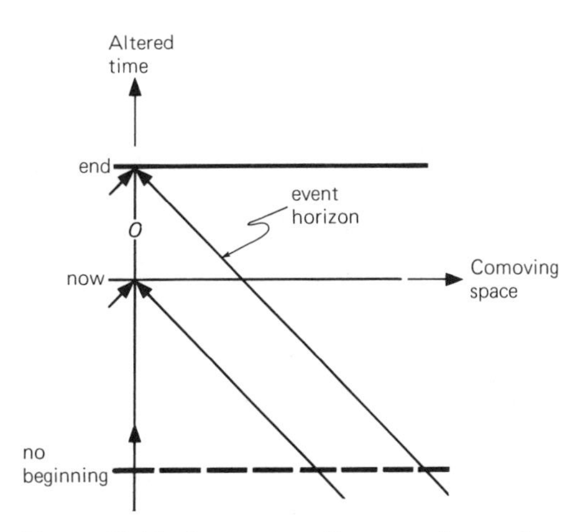

Figure 19.16. A spacetime diagram of comoving space and altered time that has an ending but no beginning. There is an event horizon but no particle horizon.

When a universe has an ending but no beginning in altered time, the spacetime diagram has an upper boundary and no lower boundary, as shown in Figure 19.16. The de Sitter and steady state universes are examples in this class.

Finally, there are some universes, such as the Einstein static universe and the Milne universe, that have neither beginnings nor endings in altered time, as shown in Figure 19.17.

It is now easy to see which universes have particle horizons. The necessary condition for a particle horizon is that altered time have a beginning, as in Figures 19.14 and 19.15. The lightcone in this case stretches back; terminates at the lower boundary, where the universe begins; and fails to intersect all worldlines. Many universes, including the Friedmann universes, have particle horizons. When, however, altered time has no beginning, and there is no lower boundary, as in Figures 19.16 and 19.17, the lightcone stretches back without limit and intersects all worldlines in the universe. In this case there is no particle horizon.

By straightening out the lightcone (which is always possible when we know how the scaling factor changes with time), we make

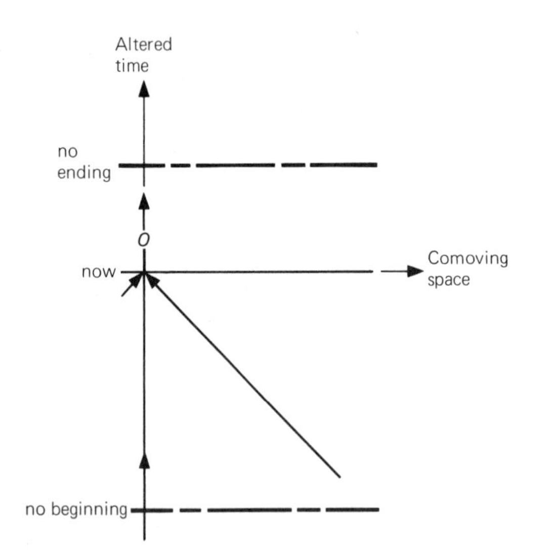

Figure 19.17. A spacetime diagram of comoving space and altered time that has no beginning and no ending. There are no particle and event horizons.

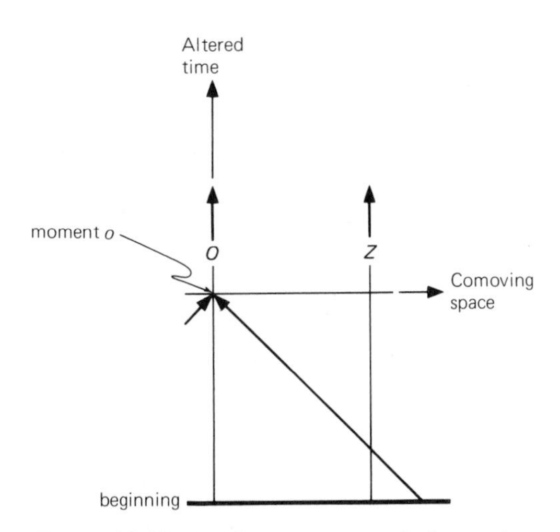

Figure 19.18. As the moment o of observation advances up the observer's worldline O, the particle horizon recedes. Once a worldline such as Z is within the particle horizon, it remains inside forever. This means that a galaxy inside the observable universe will remain always inside and therefore always observable while it exists.

it immediately obvious whether a particle horizon exists: There is a particle horizon whenever altered time has a beginning.

As the observer's lightcone advances into the future, and therefore advances also in altered time, the particle horizon recedes and more and more galaxies become visible. Once a worldline is within the particle horizon – and hence within the observable universe – it always remains within the particle horizon. As Figure 19.18 shows, galaxies can never move out of the observable universe; those now observed will remain always observable.

EVENT HORIZONS

THE ULTIMATE LIGHTCONE. Beyond the event horizon lie events that an observer is never able to see. With our new spacetime diagrams, consisting of comoving space and altered time, event horizons are easy to understand. Consider an observer's worldline O and an event labeled a that is somewhere in spacetime, as in Figure 19.19. O's lightcone advances into the future, and when it reaches the moment labeled "later,"

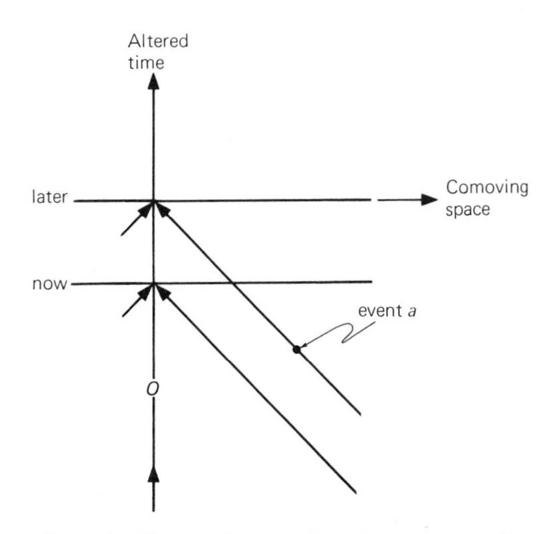

Figure 19.19. An observer O and an event a. O's lightcone advances up worldline O and at the instant labeled "later" event a is seen.

the observer sees the event a. It follows that in the course of time all events, no matter where they are in spacetime, are disclosed to the observer. If this were always the situation, then no event horizon could exist. In cosmology, however, there are exceptions, and they occur when altered time has an ending. In Figure 19.20 we see what happens. O's lightcone cannot advance

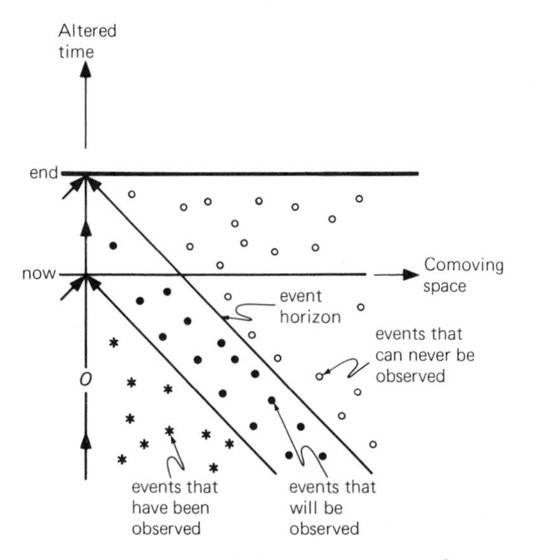

Figure 19.20. O's lightcone cannot advance beyond the end of altered time. Hence there are events that can never be seen by observer O, and the ultimate lightcone is the event horizon.

beyond the upper limit, and hence there are events that can never lie on O's lightcone and will never be observed. The necessary condition for an event horizon is that altered time have an ending; the event horizon is nothing more than the observer's ultimate lightcone at the end of altered time. All those events inside the ultimate lightcone, or event horizon, are at some time observed, and all those events outside are never observed.

BLUESHIFTS AND REDSHIFTS AT EVENT HORIZONS. Universes that collapse into big bangs, such as the closed Friedmann universe, always have event horizons. Consider an observer, of worldline O, who is in a collapsing universe. At some instant o the observer receives signals from an event a, as shown in Figure 19.21. The redshift of the received lightrays is given by

$$z = \frac{R_0}{R} - 1$$

where R_0 is the value of the scaling factor at the time of reception and R is the value at

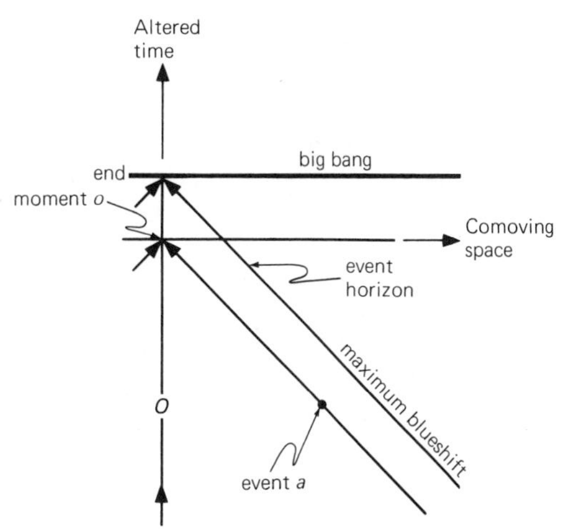

Figure 19.21. All collapsing universes that terminate in big bangs have an end in altered (and ordinary) time, and therefore possess event horizons. At moment o, observer O sees event a and sees it blueshifted. As the moment of observation o approaches the end, events are seen with increasing blueshift, and at the last possible moment, all events on the lightcone are seen to happen infinitely rapidly. In this case the event horizon has maximum blueshift.

the time of emission. Because the universe is collapsing, R_0 is less than R, and the redshift is negative. A negative redshift means that light is blueshifted. As O's last moment approaches, and the point o moves toward the end, the value R_0 of the scaling factor shrinks, and all events close to the event horizon have large blueshifts. At the event horizon itself the blueshift is maximum (the redshift has the minimum value of -1), and all events are seen to happen infinitely rapidly.

There are universes of the whimper kind – they expand forever and do not terminate in big bangs – that also have endings in altered time. The de Sitter and steady state universes are of this kind and therefore have event horizons. By repeating the argument made previously we find the events close to the event horizon have large redshifts. This is because the value R_0 of the scaling factor at the time of observation is always greater than the value R at the time of emission, and as the point o approaches the end of altered time, the ratio R_0/R becomes large and approaches infinity.

An alternative way of looking at the de Sitter and steady state universes is to consider a spacetime diagram that consists of comoving space and ordinary time, as in Figures 19.12 and 19.22. The observer O at the moment o sees event a. As the moment of observation o advances into the unlimited future, the lightcone moves upward and approaches the event horizon more and more slowly. Close to the event horizon observed events happen very slowly, and at the event horizon they are frozen into an immobility of infinite redshift.

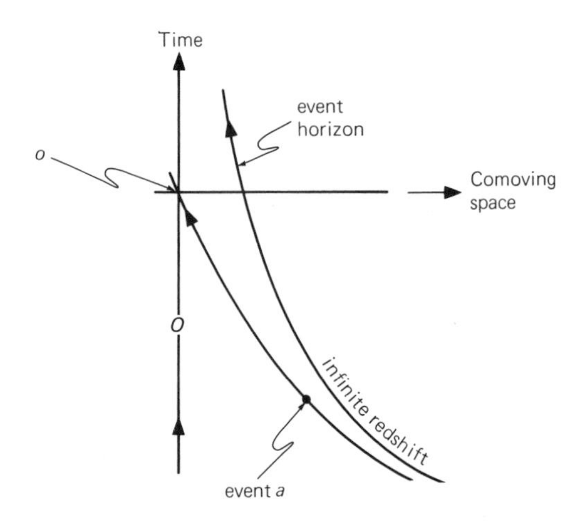

Figure 19.22. The de Sitter universe in a spacetime diagram of comoving space and ordinary time. The event horizon is as shown, and as moment o of observation advances into the infinite future, the observer's lightcone approaches the event horizon more and more closely. Events close to the event horizon are observed to have large redshifts because of expansion, and the event horizon has infinite redshift.

DE SITTER AND STEADY STATE UNIVERSES. The de Sitter universe is at first puzzling, and it must be admitted that our comments so far have not greatly clarified what happens. The Hubble term H is constant and the Hubble sphere has a radius that is always the same. At the edge of the Hubble sphere the recession velocity equals the velocity of light, and calculations show that in the de Sitter universe the redshift at the Hubble length is infinite. The Hubble sphere is therefore the observable universe. But why is the edge of the Hubble sphere an event horizon and not a particle horizon? Consider also this problem: As the universe expands the observed galaxies are carried further and further away from the observer and become progressively more redshifted; what happens to these galaxies – do they finally cross the edge of the Hubble sphere and disappear from sight?

In the de Sitter universe the edge of the Hubble sphere is a true horizon. Yet it is not a particle horizon because at any instant all galaxies in the universe are visible to an observer. Any galaxy now beyond the Hubble sphere, no matter how far away, has part of its worldline inside the Hubble sphere and is therefore observable. The galaxies recede and move out of the Hubble sphere, and yet, oddly enough, the observer never sees them crossing the Hubble edge.

The spacetime diagram that consists of comoving space and altered time has shown us that the de Sitter universe has an event horizon but no particle horizon. To under-

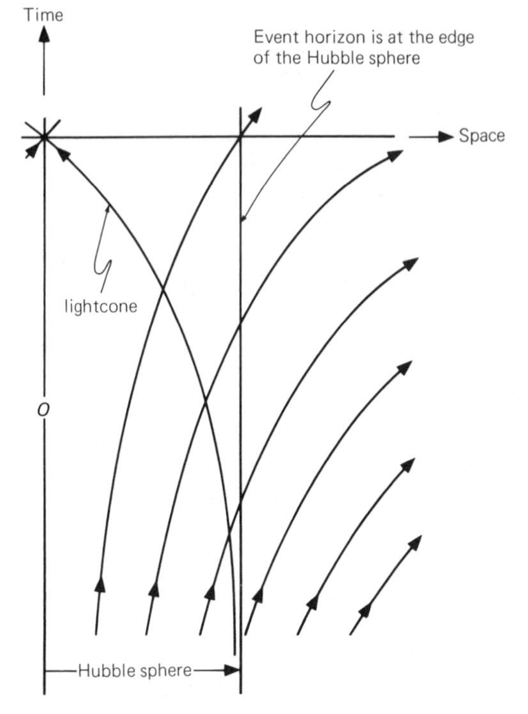

Figure 19.23. The de Sitter universe displayed in ordinary space and ordinary time. The Hubble sphere has constant radius about the observer's worldline O. All worldlines diverge away from O, and every worldline at some time in the past intersects the edge of the Hubble sphere. The observer's lightcone curves back, as shown, and approaches asymptotically the edge of the Hubble sphere. The observer sees all worldlines, and hence there is no particle horizon. Because events outside the Hubble sphere are never observed, the edge of the Hubble sphere is the event horizon.

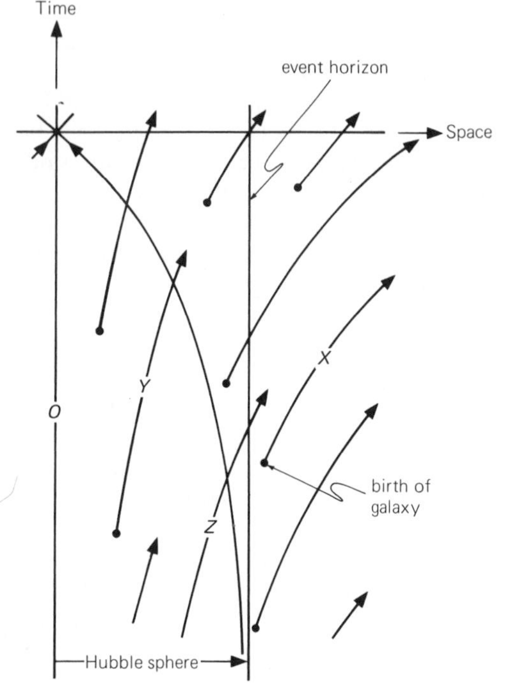

Figure 19.24. The steady state expanding universe is different from the de Sitter universe only in that galaxies are continually created so as to maintain a constant density. If we are to consider only worldlines of luminous galaxies, a worldline begins when a galaxy is born and terminates when it dies. Most galaxies, such as X, are born outside the Hubble sphere and are never seen; some galaxies, such as Y, are born inside the Hubble sphere and may die before they reach the Hubble edge; and other galaxies, such as Z, cross the Hubble edge while luminous. As in the de Sitter universe, the number of galaxies of infinite redshift at the event horizon is infinitely great. There are hence an infinite number of galaxies crowded at the edge of the Hubble sphere, and the observable universe contains an infinite number of galaxies.

stand the nature of the horizon it is more convenient to revert to a diagram of ordinary space and ordinary time, as in Figure 19.23. The Hubble sphere is of constant radius, as shown, and all worldlines of galaxies diverge away from the observer's worldline *O*. The observer's lightcone is curved and approaches asymptotically the edge of the Hubble sphere. The lightcone consequently intersects the worldlines of all galaxies in the universe, and yet never extends beyond the Hubble sphere. Because all worldlines are intersected by the observer's lightcone, and hence all galaxies are observed, there is no particle horizon. The edge of the Hubble sphere is an event horizon because it is the observer's ultimate lightcone, and all events outside the Hubble sphere are never observed.

Figure 19.23 makes it apparent that the observer sees the galaxies approaching the edge of the Hubble sphere, with increasing redshift, but never sees them crossing the edge. At the edge are crowded an infinite number of galaxies of infinite redshift.

Galaxies, of course, do not shine forever, and their worldlines as luminous sources are therefore of finite length. This is something for the reader to think about, and suitable amendments, where necessary, can easily be made in what has been previously said. Worldlines of finite length are of particular interest in the steady state universe.

The steady state universe expands in the same way as the de Sitter universe and has an event horizon at the edge of the Hubble sphere. Galaxies, however, are continually created everywhere in order to maintain a constant density, and hence it is not true to say that all galaxies are observable in the steady state universe. Most galaxies are created outside the observer's Hubble sphere and are never seen at any time, as indicated by worldline X in Figure 19.24. Also, a galaxy born inside the Hubble sphere may die and become nonluminous before it has reached the Hubble edge, as indicated by worldline Y. Even so, the number of luminous galaxies that cross the Hubble edge, and have worldlines such as Z, is still infinite. At the event horizon, where the redshift is infinite, and where lightrays that try to reach us stand still, there is an infinite number of galaxies.

REFLECTIONS

1 *"A horizon is here defined as a frontier between things observable and things unobservable. Two quite different types of horizon exist which are here termed event-horizon and particle-horizon" (Wolfgang Rindler, "Visual horizons in world-models," 1956).*

* *"God not only plays dice. He also sometimes throws the dice where they cannot be seen" (Stephen Hawking, 1975).*

2 *Do we distinguish between events (happenings) and worldlines (objects) in ordinary language?*

* *Explain particle and event horizons. What would happen if light traveled infinitely fast?*

* *In a static universe of stars that have been shining for a past eternity there is no particle horizon; if the stars shine for a future eternity there is no event horizon. Explain.*

* *Discuss the horizon riddle.*

* *Although cosmic horizons form one of the most fascinating topics in cosmology, they are usually not discussed in elementary texts. On occasion they are referred to briefly, and then it is implied that the edge of the Hubble sphere is a horizon. Discuss the difference between the Hubble sphere and the observable universe. When are they the same?*

* *In which direction do galaxies cross the particle horizon?*

* *As time passes, the observable universe contains more and more galaxies; is this true also in a collapsing universe?*

* *At death each person has a horizon. What kind of horizon?*

* *Consider the dark night sky paradox in the Newtonian universe. What kind of horizon prevents an observer from seeing a sufficient number of luminous stars to cover the sky?*

* *Once we understand horizons in a static universe, why is it then relatively easy to understand them in nonstatic universes?*

* *Give examples of event horizons at which (a) the redshift z is infinite, (b) $z = 0$, (c) $z = -1$.*

* *Discuss the maximum distance that can be seen in an expanding big bang universe. Could we argue that this maximum distance is a special type of event horizon?*

* *How is it possible in an Einstein–de Sitter universe for a source of redshift $z = 2$ to be at an emission distance less than that of a source of $z = 1$?*

* *Can you think of cosmic horizons that might exist because of the observer's forward lightcone? Such horizons determine the observer's ability to influence events and particles elsewhere in the universe. Use Figures 19.14–17 with forward lightcones.*

3 *Consider all big bang universes in which the scaling factor R is proportional to t^n, where t is the age of a universe and n is a*

constant number less than 1. In such universes

$$\text{particle horizon distance} = \frac{Ln}{1 - n} \quad (19.1)$$

and this is the radius of the observable universe. In this expression L is the Hubble length and is the radius of the Hubble sphere. It is seen that when n is greater than $\frac{1}{2}$*, the observable universe is larger than the Hubble sphere. According to the velocity–distance law, the recession of galaxies at the particle horizon is now given by*

$$\text{recession velocity at particle horizon} = \frac{cn}{1 - n} \quad (19.2)$$

where c is the velocity of light. The recession of the particle horizon itself, however, is given by the relation

$$\text{velocity of particle horizon} = \frac{c}{1 - n} \quad (19.3)$$

and this is the velocity of expansion of the observable universe. Note this relation: recession velocity at particle horizon + velocity of light = velocity of particle horizon. The particle horizon sweeps out past the galaxies at the velocity of light. In the Einstein–de Sitter universe of $n = \frac{2}{3}$*, the particle horizon is at a distance 2L (30 billion light years), the recession velocity of galaxies at the particle horizon is 2c (twice the velocity of light), and the horizon itself moves away at 3c (thrice the velocity of light). Discuss what happens in the Dirac universe of* $n = \frac{1}{3}$.

When n is equal to or greater than unity, there is no particle horizon, and the observer sees all luminous objects in the universe. Thus, in Milne's universe of n = 1, there is no particle horizon. Milne disliked universes with horizons and regarded the absence of a horizon in his universe as a distinct advantage.

The maximum distance of the lightcone from an observer's worldline is expressed by

$$\text{maximum emission distance} = Ln^{1/(1-n)} \quad (19.4)$$

and the redshift of sources at this distance is given by

$$\text{redshift at maximum emission distance} = n^{-n/(1-n)} - 1 \quad (19.5)$$

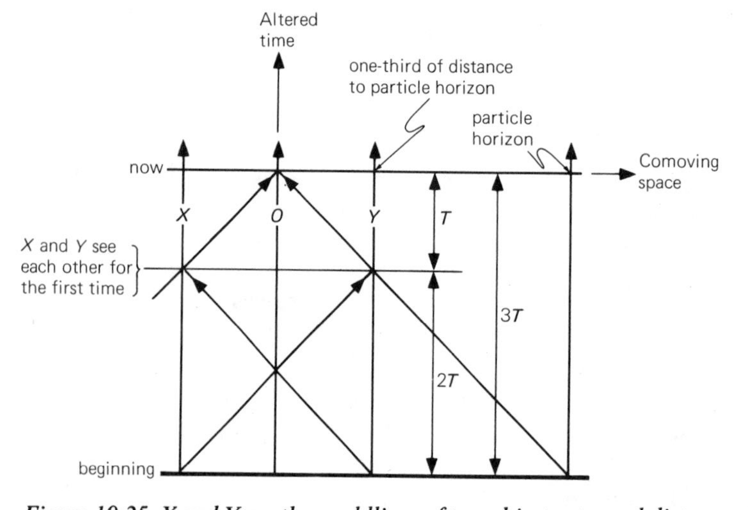

Figure 19.25. X and Y are the worldlines of two objects at equal distances in opposite directions from observer O. The time required for light to travel from X to O, and from Y to O, is denoted by T. At the moment "now," when O observes X and Y, both X and Y have just begun to see each other for the first time if the universe has an age of 3T. In this case the distance of X and Y from O is one-third the particle horizon distance. When the universe is younger than 3T, O is able to see X and Y but X and Y cannot see each other.

Thus, when n = ½, the maximum emission distance is ¼L, and sources at this distance have redshift z = 1. When n = ⅔, as in the Einstein–de Sitter universe, the maximum emission distance is 8/27 times L, and the corresponding redshift is 5/4. The recession velocity of sources at maximum emission distance, at the time of emission, is always equal to the velocity of light; these sources are therefore at the edge of the observer's Hubble sphere at the time they emit the light that is now seen.

4 *Two objects are at equal distances in opposite directions from us, as shown by worldlines X and Y in Figure 19.25. We are able to see these objects, but can they see each other? Let T be the time (in units of altered time) that light takes to travel to us from X and Y. The time that light takes to travel from X to Y and from Y to X is obviously 2T. When the universe is older than 3T we observe X and Y after they observe each other for the first time; when the universe is younger than 3T we observe X and Y before they observe each other. There is thus a maximum distance beyond which the observed objects X and Y do not know that each other exists. By examining the diagram it can be seen that this maximum distance is one-third the distance of the particle horizon (this is a reception distance and applies at the present moment). Hence the answer to our question is that objects in opposite directions, further away than one-third the particle horizon distance, cannot see each other. In the Einstein–de Sitter universe this distance is 10 billion light years, corresponding to a redshift of 1.25.*

This raises the interesting problem of why the universe is isotropic and homogeneous. Regions in opposite directions of sufficiently large redshift are isolated from each other; they have not influenced each other and are unaware of the other's existence; and yet in an isotropic universe they are in identical states. This is the puzzling feature of isotropy and homogeneity: How can regions be alike when they do not know that each other exists?

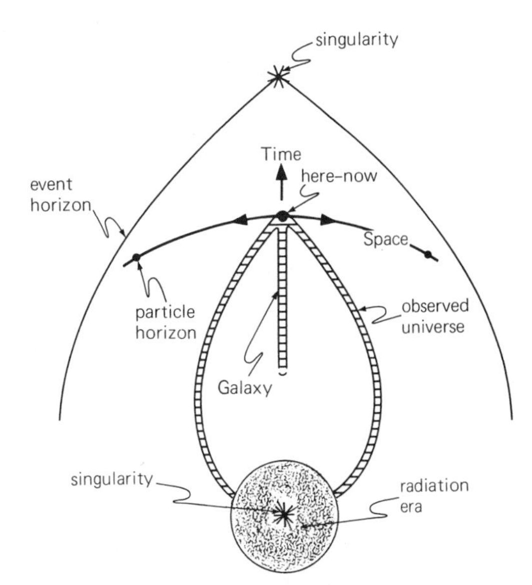

Figure 19.26. The observed universe consists of only those events that lie on the observer's backward lightcone. A small region about the observer's worldline contains events not on the lightcone whose existence can be inferred from the immediate environment. The history of the Galaxy, the Solar System, the Earth, and the human race is confined to this region. All the rest of spacetime contains events that at present are unobserved. If there is an event horizon, then beyond this horizon are events that can never be observed.

"We are unable to obtain a model of the universe without some specifically cosmological assumptions which are completely unverifiable" (George Ellis, "Cosmology and verifiability," 1975). The problem is that we observe isotropy, which we cannot explain, and we assume homogeneity, which we cannot verify. We observe only those events that lie on our backward lightcone, as in Figure 19.26, and the rest of spacetime – except for a small region about our worldline – is unobserved.

SOURCES

Centrella, J. C. "Visual horizons in cosmological models: a study." Senior honors thesis, Department of Physics and Astronomy, University of Massachusetts, 1975. Most of

this chapter is based on Joan Centrella's honors thesis.

Ellis, G. F. R. "Cosmology and verifiability." *Quarterly Journal of the Royal Astronomical Society 16,* 245 (1975).

Hawking, S. W. See M. MacCallam. The breakdown of physics." *Nature 257,* 362 (October 2, 1975).

Rindler W, "Visual horizons in world-models." *Monthly Notices of the Royal Astronomical Society 116,* 662 (1956).

LIFE IN THE UNIVERSE

Life, like a dome of many-coloured glass,
Stains the white radiance of eternity.
— Shelley (1792–1822), "Adonais"

ORIGIN OF LIFE

SPECIAL CREATION. The belief that life originated on Earth as a supernatural event is known as the theory of *special creation.* There are numerous versions of this theory, but according to most myths the nonliving world was created first and the creation of life followed. Mythology distinguishes between the nonliving and the living, and creation in the beginning is therefore usually presented as a twofold act. Catastrophe theories of the eighteenth and nineteenth centuries elaborated on these myths and claimed that there had been many acts of creation in the past; after each catastrophe had destroyed the terrestrial environment, newly created life arose in a more highly evolved state than before. Hence evolution occurred not naturally but supernaturally in the mind of the Creator.

Life even in its rudest forms was thought to be composed of substances different from those of nonliving objects. To this day we retain the distinction between inorganic and organic chemistry, although this is now a matter of convenience only, and organic chemistry deals with the numerous compounds that contain carbon. It came as a shock when the chemist Friedrich Wöhler in 1828 first made urea – a simple organic substance – from inorganic chemicals. Subsequent discoveries showed that chemicals are interchangeable between inorganic and organic things and thereby unify the nonliving and living worlds at the atomic and molecular levels. Living things are not made of sacred substances that require special creation in a nonliving material world. This has placed severe constraints, to say the least, on the special creation theory.

SPONTANEOUS CREATION. From the oldest myths on the origin of life comes the theory of *spontaneous creation.* For ages it was believed that living things were created sporadically in a magical manner. "Even now multitudes of animals are formed out of the earth with the aid of showers and the Sun's genial warmth," said Lucretius, the Epicurean poet. In the earlier civilizations spontaneous creation was regarded as part of the cyclic fecundity of nature – a quickening of activity – in a world animated by vestigial magical elements.

This theory, elaborated by Aristotle, was later widely accepted even by men such as Descartes and Newton. Worms in the earth, eels in mud, maggots in apples, flies around waste matter, and many other lower links in the chain of being were thought to be spontaneously created. It was viewed as a residual creative urge, surviving from the special creation in the beginning, and it was believed that witches and wizards had the

power to influence this creative urge in its demonic form.

William Harvey, who discovered in 1628 the circulation of the blood, speculated that all life forms, seemingly spontaneously created, might actually be born from seeds and eggs too small to be seen with the naked eye. The Italian physician Francesco Redi tested this hypothesis in 1668 and found, using meat isolated in flasks or covered with gauze, that maggots are born from eggs laid by flies. A new difficulty soon arose: Van Leeuwenhoek of Holland, with carefully made microscopes, discovered a teeming world of small organisms. It seemed to many people that this new world of microscopic life was maintained by spontaneous generation and was perhaps the missing link between the living and nonliving worlds.

The deathblow was delivered by Louis Pasteur in 1884, who demonstrated beyond doubt that microorganisms are not spontaneously created. By using filtered air, and by isolating nutrient fluids from direct contact with the atmosphere, he showed that the generation of microorganic life could be avoided. When exposed to the atmosphere the nutrient fluids became cloudy and swarmed with microscopic life. The conclusion was clear: These lower forms of life are conveyed hither and thither by the atmosphere and breed wherever there is a nutrient medium.

PANSPERMIA THEORY. Special and spontaneous creation seemed natural and even necessary in a world where life, according to the Mosaic chronology, had existed for only thousands of years. The Newtonian universe offered a counterview. The Atomist theory was revived and the Newtonian universe opened up the dizzy prospect of life existing throughout infinite space over indefinitely long periods of time. It was soon suggested that life in a primitive form is propagated from planet to planet, from star to star, flourishing wherever it encounters a hospitable environment. This line of thought culminated with the panspermia theory proposed early this century by the Swedish chemist Svante Arrhenius. He suggested that spores and other minute organisms in a dormant state are capable of traveling through interstellar space, propelled by the pressure of starlight. These organisms originate on planets and then are wafted through the heavens, like interstellar pollen, and initiate life wherever possible. Hence we on Earth have kinship with the great races of the skies.

The diaspora of life, as visualized by Arrhenius, is strictly a transport theory and does not explain the origin of life. Life flourishes on planets such as Earth because they are visited by primitive messengers, and life is not created afresh each time it ascends the chain of being. The problem of creation is therefore left unresolved.

It has been said that minute organisms, however hardy, are incapable of surviving the hazards of interstellar travel, and that strong stellar winds, fierce ultraviolet radiation, and penetrating cosmic rays will inevitably destroy them. This argument is not altogether convincing. Stars are born from condensations of gas and dust in cool and dense clouds that are rich in organic and inorganic molecules. In these clouds it might be possible for complex molecules and even cells, attached perhaps to the surfaces of dust grains, to survive and be transported over long periods of time. Almost certainly we have not heard the last word on the panspermia theory.

BIOCHEMICAL ORIGIN OF LIFE. The physical manifestation of life undoubtedly consists of molecules that obey the laws of physics. These molecules are assembled into elaborate structures that apparently also obey the laws of physics. Hence we are free to investigate in a scientific manner how such structures might have originated in the physical world. This is the biochemical theory of the origin of life. With an abundance of prebiotic (before life) organic chemicals on the early Earth, and a variety of favorable conditions lasting over long periods of time,

we cannot reject the possibility that self-replicating molecular systems formed naturally. Once such systems are created there is no apparent reason why they should not evolve into cells and more complex organisms by means of natural selection. Charles Darwin wrote in 1871 in a letter, "But if (and oh! what a big if!) we could conceive in some warm little pond, with all sorts of ammonia and phosphoric salts, light, heat, electricity, &c present, that a protein compound was chemically formed ready to undergo still more complex changes." The idea of a biochemical origin of life was developed by John Haldane, Alexander Oparin, and many others, and is now widely accepted. Haldane, in "Origin of life" (1954), wrote, "Critics will say that a self-reproducing machine is still a machine, and that there is an absolute gulf between any possible activity of such a machine and the most elementary feeling or desire, let alone human consciousness. Of such critics I ask, 'Do you think that your idea or perception of a stone is like a stone?' "

THE EARLY EARTH. Life originated on Earth almost 4 billion years ago. The terrestrial environment was probably rich in organic chemicals – a veritable organic Eden – and conditions were favorable for the biochemical origin of life. We do not know how the first steps were taken; this is still a speculative subject that nonetheless is guided by a growing body of knowledge. We can only guess what happened, and all reconstructions stretch the imagination to the limit.

The Earth trembled ceaselessly with earthquakes and the surface was cratered incessantly with the infall of meteors of all sizes. The primeval atmosphere of hydrogen, nitrogen, ammonia, methane, and other gases, with only a trace of oxygen, was wracked by great storms; was fed by fiery volcanic plumes; and, a flicker with continual lightning, reflected the ruddy glare of widespread lava flows. Volcanic ash and torrential rain fell on the smoldering surface and steaming seas. Strong ultraviolet radiation, electrical discharges, radioactivity, shockwaves, and all the razzmatazz of an exuberant Earth conspired to establish an immense biochemical industry that produced myriads of organic compounds in mountainous quantities. These compounds, including amino acids and nucleotides, concentrated in warm and shallow seas and converted them into organic soups.

Because the Earth's crust was thin, mountains were not very high; perhaps there was less water then (it had not yet all come from the interior), and great oceans were a thing of the future. But shallow seas were everywhere, scattered in their thousands, covering much of the surface. Each was a pool of "primeval broth," insulted in every conceivable way: boiled, cooled, diluted, shaken, decanted, and occasionally thrown into the sky by some violent event. Every moment trillions of biochemical experiments were performed under all possible circumstances. In the seas, in rock fissures, and on the surfaces of dust particles, molecules of exotic form had their hour and were then incinerated, buried under ash, and enfolded within the Earth. Irresistibly the wizardry of biochemistry advanced, becoming more ingenious with the passage of millions of years; nucleotides joined into chainlike molecules of weird codings and affected the assembly of amino acids into proteins of freakish design.

At some stage – and why not? – a molecule became self-replicating and was therefore able to multiply rapidly. In this way an entire sea was dominated by a species of replicating molecule. Perhaps each sea found its own solution to the challenge of replication; and volcanic eruptions, inundations, and strong winds enabled the seas to exchange genetic codings and compete with one another.

Conceivably, more than once the whole biotic enterprise was destroyed. The cratered face of a visiting planetesimal, hundreds of kilometers in diameter, loomed in the sky and moments later a titanic explosion devastated the entire terrestrial surface. After millions of years, when volcanic fury had created a new atmosphere and new seas, it would all begin again.

The bombardment era, during which the

planets and moons swept up the debris left over from the formation of the Solar System, ended about 4 billion years ago. During the next half billion years, or less, in a way not yet understood, cells originated in their simplest form. Perhaps an increasing dilution of the seas and a decreasing abundance of organic chemicals favored the survival of replicating structures that had learned to retain their own small environment of enriched organic substances. The invention of the membrane, which enclosed the enriched environment, transferred the individuality of life from the seas to the cells, and life as we now recognize it had begun.

EVOLUTION OF LIFE

The building blocks of organisms are relatively simple molecules known as amino acids and nucleotides that are made from atoms of hydrogen, carbon, nitrogen, oxygen, and other elements in lesser amounts. The amino acids join together to form the long chainlike molecules of the numerous proteins; the nucleotides join together to form the long chainlike molecules of the nucleic acids DNA and RNA. The DNA molecules have the shape of a double helix and contain the genetic coding of the entire organism.

The word *cell* was introduced in 1665 by Robert Hooke, and the cell theory – that all organisms are constructed from cells – was advanced in the nineteenth century. Unicellular (single-celled) organisms vary greatly in size, from small bacteria, measuring only hundreds of times the diameter of the hydrogen atom, to large ostrich eggs. Multicellular (many-celled) organisms contain numerous interacting cells of various functions; in the human body there are about 10^{14} cells, each containing on the average 10^{14} atoms.

There are two basic types of cell: the prokaryotes, which were the first to evolve and are relatively simple; and the eukaryotes, which evolved later and are more complex. A prokaryote does not have a nucleus, and its DNA is dispersed in the interior of the cell; a eukaryote, on the other hand, has its DNA confined within a small region of the cell known as the nucleus. Of the five kingdoms of life, the Monera (such as bacteria and blue-green algae) consist of prokaryotes; and the kingdoms of Protista (such as diatoms, amoeba, seaweed, and slime molds), Fungi (such as toadstools and mushrooms), Plantae (such as mosses and trees), and Animalia (such as worms, insects, and mammals) consist of organisms containing eukaryotes.

There are two basic types of organism: the autotrophs and the heterotrophs. Autotrophs, such as plants, are self-nourished: Their food supply consists mainly of inorganic chemicals, and their energy comes usually from sunlight. Heterotrophs, such as animals, are other-nourished: Their food supply consists of organic substances. Autotrophs obtain their energy from sunlight by photosynthesis, in which carbon dioxide and water are converted into sugars:

$$\text{carbon dioxide} + \text{water} + \text{sunlight} \rightarrow \text{sugars} + \text{oxygen}$$

and energy is stored in the sugars. Respiration by heterotrophs is the opposite process:

$$\text{sugars} + \text{oxygen} \rightarrow \text{carbon dioxide} + \text{water} + \text{energy}$$

Photosynthetic autotrophs create the sugars, discharge oxygen into the atmosphere, and are consumed as food by heterotrophs that breathe in the oxygen. There are many variations of this simplified picture.

Symbiosis is the living together of dissimilar organisms for mutual benefit (parasitism is when the association is not to mutual advantage). The dependence of flowering plants on insects, which feed on and pollinate flowers, is an example of communal symbiosis. In host–guest symbiosis the host can be a heterotroph and the guest an autotroph, as in the case of lichens, where the host is a fungus (providing moisture and minerals) and the guest is an alga (providing food by photosynthesis). Some of the component structures (such as mitochondria and chloroplasts) of eukaryote cells have their own genetic material, and it has been suggested that the eukaryotes evolved long ago by incorporating prokaryotes in a host–guest relationship.

How cells originated and evolved is not known, and the following remarks are speculative (see Table 20.1). The first cells were perhaps primitive heterotrophs of bacterial type that depended on an abundant supply of organic molecules. Simple autotrophs, similar to photosynthetic blue-green algae, came later as the supply of organic chemicals declined. According to the fossil record these latter cells first appeared about 3.5 billion years ago. Perhaps the hard-pressed heterotrophs that had evolved in a world of plenty survived by host–guest symbiosis, which provided many autotrophs protection in return for their special functions. Some cells attached themselves to the heterotrophs as threadlike flagella, and the combined organism became mobile and hence able to search for food; others became residents within the heterotrophs. In this way colonies of prokaryotes might have combined to produce more sophisticated cells that had a better chance of surviving. How sexual and asexual cell division came about is still a mystery, and the elaboration and perfection of these and other processes may have occupied most of the Proterozoic era that ended less than a billion years ago.

Toward the end of the Proterozoic era the highest forms of life were the single-celled eukaryotes. Suddenly, it seems, with the protection of the newly created ozone layer in the upper atmosphere, which filtered out the ultraviolet radiation, the eukaryotes evolved into a profusion of multicellular organisms. A great variety of invertebrates (creatures without backbones) appeared in a hundred million years or so, and an explosion of vertebrates and plants followed. The oxygen in the atmosphere continued to increase throughout the Paleozoic and Mesozoic eras, attaining its present level at about the beginning of the Cenozoic era. The drifting land masses, driven by motions in the Earth's mantle, formed and reformed, creating a changing pattern of continents and oceans. On the continents were great forests, inland seas, marshes, and rivers, which swarmed with fish, insects, amphibia, and reptiles. In the Triassic period came the dinosaurs. They were exceedingly diverse: Some were small and fleet-footed, some were able to fly, some were scaled and armored, and many were colossal in size. The dinosaurs continued to flourish through the Jurassic and Cretaceous periods as the most vigorous and intelligent creatures of those times. By the end of the Cretaceous period they had almost vanished, leaving the birds as their descendants.

With the decline of the dinosaurs came the grasses and flowering plants, and the world was transformed by bright colors and the foliage of deciduous trees. The mammals began to flourish in the early Cenozoic era, and about 20 million years ago, in the forests and on the savannas, the first manlike creatures appeared. More recently the climate cooled, to the point where a repetition of ice ages occurred. And in the last 10,000 years, since the most recent retreat of the northern glaciers, civilizations have arisen. Equipped with a million-year-old body and a hundred-thousand-year-old brain, mankind now bestrides the Earth and is master of the biosphere.

Jean Lamarck, a French naturalist, emphasized the importance of biological evolution. In 1809 he revived the old idea that skills and other aptitudes acquired by individual creatures are inherited by their descendants. In other words, evolution is *self-directed* by individual effort. This theory, known as *Lamarckism,* has left a lasting impression, and many people still accept it as true. But we know now that Lamarckism is incorrect and does not explain the way that life evolves. The skills acquired by the efforts of one generation are not inherited genetically by the next generation; at most, all that is inherited is the capacity to acquire skills.

Natural selection, a new theory of evolution, was independently proposed in 1858 by Charles Darwin and Alfred Wallace. Because Darwin presented a year later a detailed argument in his book *The Origin of Species,* and because he had pondered on the theory for many years, he has received

Table 20.1. *Geological time scales*

Time (millions of years before present)	Era	Period	Epoch	Events
	Cenozoic (*ceno* = recent)	Quaternary	Recent Pleistocene (*pleisto* = most)	Homo sapiens
5		Tertiary	Pliocene (*plio* = more)	
10			Miocene (*mio* = less)	hominids primates
			Oligocene (*oligo* = few)	grass, flowering plants
			Eocene (*eos* = dawn)	mammals
50			Paleocene	dinosaurs disappear
100	Mesozoic (*meso* = middle)	Cretaceous		birds
		Jurassic		dinosaurs
		Triassic		reptiles
	Paleozoic (*paleo* = ancient)	Permian		fish, amphibians, insects
500		Carboniferous to Cambrian	Devonian Silurian Ordovician	vertebrates, forests invertebrates metazoans ozone layer
1000	Precambrian era or Proterozoic era (*protero* or *proto* = earliest)			oldest photosynthetic plants oldest fossil cells oldest rocks origin of Earth
5000	Pre-Solar era			origin of Galaxy

most of the credit for the discovery. Darwin and Wallace were widely traveled naturalists and were intrigued by the diversity of animal and plant species. Both hit on the idea of natural selection after reading *An Essay on Population,* written in 1798 by Thomas Malthus.

The theory of natural selection brought together two streams of thought. The first is that all individuals of a species are never exactly alike but differ by small and seemingly haphazard variations. The second is that the growth of populations is checked continually by environmental constraints. From this confluence of thought came the new theory: Those differences that aid individuals to survive and reproduce are shared increasingly among the members of a species, and those that do not are progressively eliminated.

In any species there are individual variations that are advantageous and others that are disadvantageous, and by elimination of the latter a species evolves and becomes better adapted. The giraffe's long neck is not the result of striving to reach greater heights, as supposed by Lamarck, but is the outcome of natural selection that has nothing to do with the personal desires of giraffes. These animals have lived in competition with other species that browsed off lower vegetation, and successive small increases in height over many generations have been advantageous to the survival of giraffes. Natural selection operates on interbreeding populations, and though it may produce many strange twists and turns, it is as inexorable as any other law of nature.

How fortunate, we exclaim, that natural selection preserves only advantageous variations! Let us not forget, however, that *advantageous* is defined by what natural selection preserves. The *survival of the fittest* is a phrase that must be used with care, for many peculiar creatures have evolved and then become extinct, such as the Irish Elk, with its immense antlers that were more a burden in the end than an advantage. If we use *survival of the favored* instead, it must be understood that the favoring is done by the environment – a witless environment that nonetheless is extraordinarily complex: It encompasses the world and contains other members of the same species arranged in pecking order, and numerous other species; and the intricate whole is a biosphere in a continual state of change. Natural selection is not teleological (directed by final goals) and does not guarantee progress of a desired kind. Each step in evolution is directed by what survives and reproduces most extensively at that instant under the reigning circumstances; accordingly, many variations and species are eliminated that later might have proved superior to those that actually survived.

Darwin's theory did not explain the origin of the individual variations in a species. It is now known that mutations are the ultimate cause of these variations. Mutations, or changes in the genetic code that often are of disastrous consequences, are the natural changes that must inevitably occur in systems of many interacting atoms. At one time it was thought that all variations would be smeared out by interbreeding and would blend together to produce similar individuals. But Gregor Mendel, an Austrian monk, showed in 1865 that variations do not blend and disappear with interbreeding. Mendel planted the seeds of tall and short pea plants and, after crossbreeding their peas, noticed that the resulting plants were not of intermediate height. He found that tall and short plants were produced in numbers that have a fixed ratio. Mendel's discovery passed unnoticed at the time and did not attract attention until 1900. Inheritance, with its multitudinous variations, is determined by genes or elements of the genetic code, and individuals are the result of their genetic makeup responding to the tumultuous welter of the environment.

INTELLIGENT LIFE

Why human beings developed large brains about a hundred thousand years ago or more is still an unresolved puzzle. They lived under primitive social conditions and seemingly had no urgent need for large brains

equal to those of modern humans. Early humans made and used tools, and it is often said that this explains the origin of their great brains. Apes also use tools, however, and they do not have such large brains. The chipping of stones to make tools and weapons is only one manifestation of intelligence; it is doubtful whether this by itself can account for the development of the human brain. The circumstances that prompted human brain growth must have been unusual and of such a kind that intelligence was naturally selected because of its survival value. We can imagine primates who were vulnerable because of their naked skins and absence of sharp claws and long teeth. Because they were defenseless, their survival depended on cooperative action that entailed the development of speech. The breakthrough to large brains probably came when our ancestors learned to speak and invented language. Communication with a large vocabulary of symbolic sounds, and the reconstruction of the external world in terms of symbolic images, became an effective way of surviving.

In food-hunting and -gathering societies the young were taught the language and were initiated into the laws and myths, and the old were cherished as guardians of the cultural heritage. The society survived because of the vigor of its youth and the knowledge of its elders. In the tens of thousands of years of the unrecorded past there were surely great singers, great artists, great thinkers, and dynamic and visionary leaders, who perhaps surpassed all those known in the short span of recorded history. It would be a great mistake to suppose that brains have been highly active only in the last 10,000 years.

To grasp the situation let us suppose that the evolution of life is compressed into the time span of a single day. Every 24 seconds of the day is equivalent to a million years of biological evolution. For 18 hours unicellular organisms evolve and then, at 6:00 in the evening, multicellular creatures first appear. The mammals arrive at 11:00 and manlike creatures emerge about 1 minute before 12:00. Human intelligence is of relatively recent origin and blossomed in the last 2 to 3 seconds. In this compressed picture of evolution we cannot help but wonder what human intelligence will be like only 1 second after 12:00. Will it have advanced to a state beyond the reach of our imagination, or will it vanish like a match flaring briefly in the long night? There is a chance that society will lapse back into a precivilized age of barbaric magic. If the human race should revert to earlier ways of living and thinking, its subsequent evolution might follow a different direction into a less intelligent form of life.

Science explains the world around us by decomposing it into an activity of smaller and smaller parts. In this way the world is reduced to molecules, atoms, subatomic particles, and various interaction forces. When this method of explanation is applied to living organisms it is known as reductionism. We explain organisms by decomposing them into numerous cells, each of which is a system of molecules; the molecules in turn are decomposed into atoms and their subatomic particles. The recognizable properties of life and mind are apparently lost in the process, yet remain implicit, in a potential sense, in the nature of particles and their fields of force. Many people oppose this reductionist philosophy and argue that life and mind cannot be explained as a mere dance of atoms and waves.

Vitalism is the theory that life is essentially a nonphysical force that permeates and animates the physical world. This vital ingredient is the psyche according to Aristotle, or the élan vital according to the Irish-French philosopher Henri Bergson. In Bergson's theory of creative evolution the world of material things is orderly and deterministic, whereas the élan vital, the breath of life, is creative and free. "The vitalist principle," he wrote, "may indeed not explain much, but it is at least a sort of label affixed to our ignorance, so as to remind us of this occasionally, while mechanism invites us to ignore that ignorance." Pierre Teilhard de Chardin, a Jesuit paleontologist who became a vitalist, proposed that mind is interwoven in the physical universe and evolves to an "omega" state where individuality becomes submerged in a cosmic social mind.

Vitalism introduces nonphysical agents into the physical universe and therefore violates the containment principle of Chapter 5. Many biologists are opposed to vitalism and regard it as an attempt to enliven the physical universe with magic forces – a sort of restoration of the Age of Magic. Moreover, one might ask how it is possible for a vital force – a kind of efflorescence of material things – to explain the mind that is conceiving the vital force. This kind of question applies also to the extreme reductionist philosophy that makes the sweeping claim that life and mind consist only of an activity of physical things. These materialists, like the vitalists, must answer the containment riddle: Where in the physical world is the mind conceiving the physical world? Materialists and vitalists confuse mental imagery with reality and believe that they themselves are fully portrayed in their own mental images.

LIFE BEYOND THE EARTH

EXOBIOLOGY. Since the Age of Mythology, people have believed in the existence of life beyond the Earth. Gods and goddesses lived in the skies long before the medieval universe adopted them, and modern science fiction is often little more than a resurrection of mythology within a framework of pseudoscience. From a cosmological viewpoint it is improbable, we must admit, that Earth is the only place in the universe where life exists. It would be preposterous to suppose that our Galaxy, out of multitudes of galaxies, is the only one in the universe where life has arisen. Finding ourselves in a vast universe we feel impelled to believe that we are not alone, and therefore the ancient myths were in this respect probably correct.

Speculation about life in other galaxies is interesting but at present far beyond all means of verification. Speculation about life elsewhere in our Galaxy, on which cosmology has less to say, is a possibility that is not, however, beyond all means of verification. This has awakened interest, and in recent years attention has focused on the possibility of communicating with intelligent life elsewhere in the Galaxy. From the marriage of astronomy and biology has come the new subject of exobiology.

Biologists are acutely aware of the hazards of evolution, and to most of them it is not in the least obvious that multicellular organisms must arise and eventually become intelligent. Their strong reservations concerning life elsewhere have been overshadowed by the euphoric optimism of many astronomers. The latter have taken their cue from the Atomists who long ago visualized a universe of endless worlds teeming with

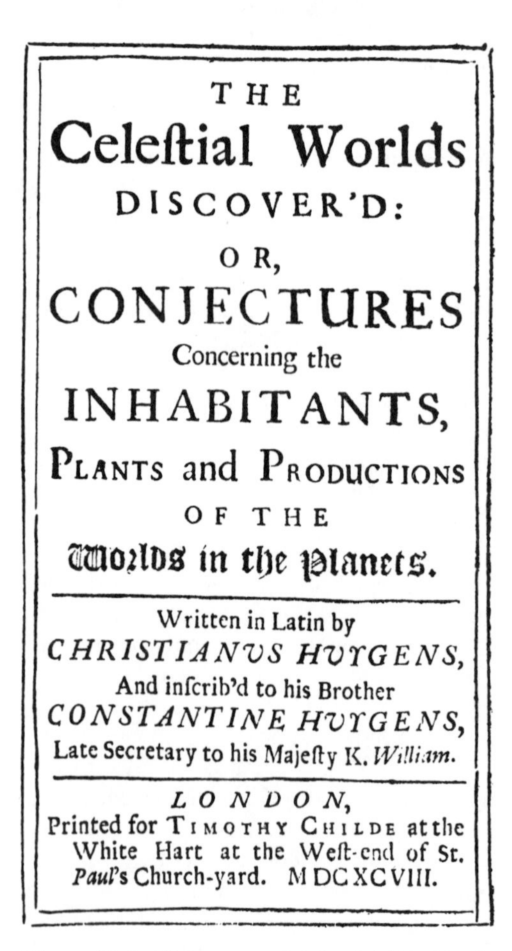
THE
Celeftial Worlds
DISCOVER'D:
OR,
CONJECTURES
Concerning the
INHABITANTS,
PLANTS and PRODUCTIONS
OF THE
Worlds in the Planets.

Written in Latin by
CHRISTIANUS HUYGENS,
And infcrib'd to his Brother
CONSTANTINE HUYGENS,
Late Secretary to his Majefty K. William.

LONDON,
Printed for TIMOTHY CHILDE at the White Hart at the Weft-end of St. Paul's Church-yard. MDCXCVIII.

Figure 20.1. Title page of Celestial Worlds, *by Christiaan Huygens, 1698.*

abundant life (see Figure 20.1). In the words of Christiaan Huygens, who wrote in 1698: "Why [should] not every one of these stars or suns have as great a retinue as our sun, of planets, with their moons, to wait upon them?" These planets, he said, "must have their plants and animals, nay and their rational creatures too, and these as great admirers and diligent observers of the heavens as ourselves." From this astronomical outlook we see our Galaxy lavishly strewn with sunlike stars, formed much as the Sun was, and therefore probably encircled with planets, many of which are earthlike. Intelligent beings live on Earth, and therefore surely intelligent beings must also live in other planetary systems? A technological civilization is here on Earth; why not then other such civilizations elsewhere, and accordingly, why not try to communicate with them? The debate concerning the existence of extraterrestrial technological civilizations – or ETCs – within the Galaxy has thrown up several interesting viewpoints, some of which are mentioned in the following sections.

ETCs. Let N stand for the number of stars in the Galaxy. The number of extraterrestrial technological civilizations that have arisen in the lifetime of the Galaxy can be expressed by the equation

$$\text{number of ETCs} = N \times A \times B$$

where A is a fraction determined by astronomical considerations and B is a fraction determined by biological considerations. The fraction A can be written:

$$A = p_1 \times p_2 \times p_3$$

Here p_1 is the fraction of all stars in the Galaxy similar to the Sun, stars that are not too blue, not too red, and not members of close binary systems; a reasonable estimate is that p_1 has a value $1/10$. The second component, p_2, is the fraction of these sunlike stars with an earthlike planet; it is often supposed that p_2 has a value of unity, but we shall be conservative and assign to it a value $1/10$. The third component, p_3, is the fraction of such planets occupying a habitable zone, neither too close (like Venus) nor too far (like Mars) from the parent star; for this we shall also assume a value $1/10$. In the Galaxy there are approximately 100 billion stars ($N = 10^{11}$), and therefore

$$\text{number of ETCs} = 10^8 \times B$$

This rough estimate gives 100 million planets in the Galaxy where conditions are similar to those on Earth and where life could have originated.

Our difficulties begin when we attempt to estimate a value for the biological fraction B, which can be written:

$$B = p_4 \times p_5 \times p_6 \times p_7$$

In this expression, p_4 is the probability that life originates in a unicellular form; p_5 is the probability that it evolves into complex multicellular organisms, such as mammals; p_6 is the probability that life develops intelligence comparable with that of humans; and p_7 is the probability that intelligent life discovers science and develops an advanced technological civilization.

In optimistic studies it is often assumed that $p_4 = 1/10$, $p_5 = p_6 = p_7 = 1$, and hence $B = 1/10$. The number of extraterrestrial technological civilizations in this case is 10 million. These civilizations have existed at different times in the history of the Galaxy, and the number existing at any instant, including the present, is given by

$$\text{number of ETCs at any moment} = \text{number of ETCs} \times \frac{t}{T}$$

where t is the average lifetime of such a civilization, and T is the age of the Galaxy, which is approximately 10 billion years. Let us continue in this optimistic vein and suppose that technological civilizations last on the average for 1 million years. The number existing at any moment, including the present, is therefore 1000. A simple calculation then shows that they are separated from each other by an average distance of 1000 light years. Technological civilizations lasting for a million years are therefore able to communicate with one another.

A pessimistic and perhaps more realistic view of the value of B is as follows. We again assume for the chance that life originates on an earthlike planet the value $p_4 = 1/10$. There is no guarantee that life will evolve over billions of years into multicellular organisms, and in recognition of the hazards and traps involved we shall guess that $p_5 = 1/10$. There is nothing that compels us to conclude that advanced intelligence is inevitable, simply because numerous species have survived over long periods of time without it. The environment must in some way administer rude shocks in the right way at the right time to the right species, so that natural selection will favor the development of large brains. The probability of this happening is extremely small, but in order not to overdo the pessimism we shall assume that $p_6 = 1/10$. Finally, we must ask, what is the chance p_7 that intelligent life develops science and its handmaiden of advanced technology? Science was not discovered by the cultures of Africa, America, China, India, Japan, or most other places, and its discovery was by no means a simple and straightforward set of events. Science made its first faltering steps in the Heroic Age of Greece as a consequence of the efforts of a few individuals, who rejected the magic of gods and whose efforts were ridiculed by the majority of their compatriots. Science was revived in Europe in the face of organized hostility, and in only a few places was tolerance sufficient to allow a relatively few individuals to rediscover the scientific method and to lay down the foundations of the Newtonian universe. Science arose because of accidental and improbable circumstances that existed in Greece and later in Europe. The probability p_7 of intelligent life creating a scientific universe is perhaps no greater than $1/10$. With $p_4 = p_5 = p_6 = p_7 = 1/10$, we have $B = 10^{-4}$, and the number of technological civilizations that have arisen in the Galaxy is therefore 10,000. For the Galaxy to contain only one ETC at any time, each must last for 1 million years. A pessimistic appraisal of the lifetime of an advanced technological civilization is a few centuries, say a thousand years. This indicates that we, an advanced technological civilization, are alone in the Galaxy, and after our demise the next will occur somewhere in about 1 million years.

GALACTIC COLONIZATION. Estimates of the total number of technological civilizations that have existed in the Galaxy range from 10 million to 10,000, and the lower number is more likely to be accurate. The number of civilizations without advanced technology is undoubtedly greater, and because they endure for longer periods of time, it is possible that many exist at the present moment. We have stressed the importance of technological civilizations in this discussion, not just because of the prospect of communication, but also because the attainment of an advanced technology such as our own marks a critical point in biological evolution. The development of science opens up vistas and avenues of advancement that are denied nontechnological civilizations.

On the other hand, the development of science creates new hazards, and the chances are that most technological civilizations are short-lived. When there are many people with the power to devastate a planetary environment, by misadventure and by the employment of nuclear weapons, civilizations cannot be expected to last longer than a few centuries. Perhaps only 1 in every 10 survives the first thousand years, and the others either self-destroy or revert back to a more primitive state. Of the 10,000 technological civilizations that have arisen in the Galaxy, it is possible therefore that only 1,000 have survived.

We must consider what happens to those technological civilizations, about 1,000 that survive and do not self-destroy in their first thousand years.

A thousand years is sufficient to develop interstellar space travel. Fusion power and other technologies, still beyond our reach, will enable a technological civilization to construct large space vehicles that can travel at one-thousandth the speed of light. A journey of 10 light years distance, from one planetary system to another, will last 10,000 years. This would not be unthinkable in a large space vehicle having its own biosphere and containing a social unit of millions of people. A halt at their destination might last no more than 10,000 years before embarkation on the next interstellar journey. In this way, step by step, in a mounting wave of space colonies, life will diffuse outward from the home planet at a rate of 10 light years every 20,000 years. Given this rate of diffusive migration the entire Galaxy will be colonized in 50 million years – a time less than 1 percent of the age of the Galaxy. Despite many hazards and setbacks on different fronts, the growth and magnitude of such an enterprise will ensure that it survives and continues to spread. This not implausible picture leads to the conclusion that the Galaxy is colonized by highly intelligent life that originated from about 1000 initial technological civilizations.

Confronted with this intriguing scenario, we feel impelled to ask why we are not aware of the existence of other intelligent life in the Galaxy. Surely it should rally to our aid and welfare and show us how to solve our most pressing problems?

THE BIOGALACTIC LAW. We are the outcome of natural selection, and to the operation of this law can be attributed the fitness of the human body and brain. Presumably this is true also of the life on other planets that attains a state of advanced intelligence. When a civilization gains control of its planetary environment, the evolutionary game changes, however, and new rules then determine what is fit and unfit. Natural selection now operates on a planet-wide scale according to a "biogalactic law" that will be referred to as *galactic selection.* This speculative law of galactic selection states simply: *Intelligent life forms that are destructively aggressive do not colonize the Galaxy.* This law operates conceivably in two modes; the first is unconscious and automatic, and the second is conscious and deliberate.

Natural selection and galactic selection are both weeding-out processes. When a species becomes intelligent, and develops a technological civilization, the environment that has hitherto directed natural selection falls under the control of the civilization. All previous checks and balances are overridden, creating an unstable situation in which irrational, irresponsible, and aggressively destructive behavior leads to disastrous consequences. At this stage galactic selection takes over. Aggressively destructive life forms, irrationally and irresponsibly prone to warfare, are unlikely to survive for very long after they have assumed control over their biospheres. They will not survive long enough to attain command of interstellar travel. Locked in their biospheres they are like virulent organisms in planetary test tubes, and they either destroy themselves or revert back to primitive conditions before they are able to embark on interstellar voyages of conquest. This means that any alien creatures who have already colonized the Galaxy are probably peaceful, not aggressive, and that they do not build oppressive galactic empires of subject races.

Galactic selection as described is natural and automatic. It is a fail-safe sort of law ensuring in effect that destructive aggression is self-terminating. Let us suppose that occasionally an aggressive technological civilization, owing to unusual circumstances, is able to break free from its planetary test tube. Galactic selection might then operate in its conscious and deliberate mode. The creatures who have already colonized the Galaxy, reckoning up the woeful cost of an interstellar race of vandals, may find that they have only one recourse: the deliberate termination of the aggressive civilization. Probably they will not themselves watch and wait for such a premature breakaway, but will place automated devices in the neighborhood of solar systems where life is awakening into an intelligent state. These monitoring machines will read the signs of technological advancement, and if there is excessive aggression, will then await the automatic self-termination. When on rare occasions technological advancement and aggression continue unarrested, and there are signs of preparation for long voyages of interstellar travel, the machines will then follow their programmed instructions and proceed to effect termination. How this is done is a matter of more than academic interest to the human race in the next few centuries.

Mythology has accustomed us to the notion that the gods are involved in the affairs of Earth and are concerned with the welfare of mankind. The probable existence of advanced intelligent life elsewhere in the Galaxy is therefore puzzling, and we find it difficult to understand why it ignores us. Such indifference offends our self-esteem and fails to conform to ancient anthropocentric philosophy. The explanation may actually be quite simple. Galactic selection, operating in its conscious mode, places an embargo on all forms of direct contact with civilizations that are still confined to their planets. Such civilizations must not be encouraged or aided to quit their planets prematurely. They must prove their fitness to mingle with alien creatures, and there is no better way of demonstrating unfitness than self-destruction.

Astronomy and biology, while stressing different viewpoints, lead to the conclusion that advanced technological civilizations may have colonized the Galaxy. Because of a galactic selection process these alien intelligent creatures are probably angelic and not demonic. The future beckons us with unlimited promise. Everything in the universe is on the side of mankind except perhaps mankind itself.

REFLECTIONS

1 *"I have no doubt that in reality the future will be vastly more surprising than anything I can imagine. Now my own suspicion is that the universe is not only queerer than we suppose, but queerer than we can suppose" (John Haldane,* Possible Worlds and Other Papers, *1927).*

* *What is life? – this question concerns us all and is one of the most important in cosmology. We divide the world into living and nonliving things but still have no widely accepted meaning of the word* life. *Complex organisms are composed of cells, which are composed of molecules, which are composed of atoms; and it is not clear at what level of complexity life first emerges. The cell is a miracle of the physical world and required billions of years to evolve; dare we exclude it, assert that it is nonliving, and claim that life is manifest only in complex multicellular organisms?*

Living organisms feed, grow, move, reproduce, and behave in response to their environment. Many things admittedly nonliving exhibit similar properties. An automobile moves and consumes food; a crystal grows; a candle flame needs nourishment, reacts to its environment, and self-reproduces with sometimes alarming consequences. Manmade automatons are extremely intricate, and computers now play chess with each other. With so many nonliving things mimicking the characteristics usually ascribed to organisms, it is difficult to pinpoint exactly what defines life. Are we to believe that self-reproduction and evolution are the hallmarks of life? According to biochemistry, self-reproduction is possible in highly organized chemical systems, and according to biology, evolution operates automatically by means of natural selection. The physical world, it seems, has an astonishing power for creating organized complexity, and there is nothing of a physical nature that sets life apart from the rest of the physical world.

When we search within ourselves for the meaning of life, and are not limited to observing external events, we find that life is essentially psychic and consists of thoughts, emotions, and all that contributes to a state of self-awareness. But we are not sure what psychic *means any more than what* life *means. Many people believe the word* psychic *denotes a nonphysical realm that interfaces with and mirrors the physical world; others hold that it will be explained when more is understood about the physical world.*

* *"I have been saying that modern science*

broke down the barriers that separated the heavens and the earth, and thus it united and unified the universe. And this is true. But, as I have said, too, it did this by substituting for our world of quality and sense perception, the world in which we live, and love, and die, another world – a world of quantity, of reified geometry, a world in which, though there is a place for everything, there is no place for man. Thus the world of science – the real world – became estranged and utterly divorced from the world of life, which science has been unable to explain – not even to explain away by calling it 'subjective'. . . This is the tragedy of the modern mind which 'solved the riddle of the universe,' but only to replace it by another riddle: the riddle of itself" (Alexander Koyré, Newtonian Studies, *1965).*

2 *On two occasions at least it was thought that the vital force of life had been discovered. Luigi Galvani, an Italian anatomist, discovered accidentally in 1786 that the amputated hind legs of a frog would kick convulsively when in contact with a source of electricity. The legs also twitched when in contact with two different metals. Galvani thought that he had found the vital force of life and referred to it as "animal electricity." Nowadays we say that a person is* galvanized *into action. Robert Brown, a Scottish botanist, discovered in 1827 that minute particles of pollen, suspended in water, have continual irregular motion. He thought that this ceaseless jittery behavior was due to a vital force. The Brownian motion of small particles is due to the random kicks of atoms and provides visible evidence of the atomic nature of matter.*

3 *"The atmospheres of celestial bodies as well as whirling cosmic nebulae can be regarded as the timeless sanctuary of animate forms, the eternal plantations of organic germs" (Justus von Liebig,* Letters on Chemistry, *1861).*

* *"Who would deny that such bodies, floating everywhere in the universal space, do not leave behind them the germs of life wherever the planetary conditions are already suitable to promote creation?" (Herman von Helmholtz,* Formation of Planetary Systems, *1884). Helmholtz thought that microscopic organisms had been brought to Earth by meteorites.*

* *"In this manner life may have been transplanted for eternal ages from solar system to solar system and from planet to planet of the same system. But as among the billions of grains of pollen which the wind carries away from a large tree – a fir tree, for instance – only one may on an average give birth to a new tree, thus of the billions, or perhaps trillions, of germs which the radiation pressure drives out into space, only one may really bring life to a foreign planet on which life had not yet arisen, and become the originator of living things on that planet . . . Finally, we perceive that according to this version of the theory of panspermia, all organic beings in the whole universe should be related to one another" (Svante Arrhenius,* Worlds in the Making, *1908).*

4 *Jean Lamarck (1744–1829) was the first to use the word* biology, *and is best known for his work distinguishing between vertebrates and invertebrates and for a theory of evolution referred to as Lamarckism. Lamarck believed that evolution is goal-oriented and self-directed.*

* *"I say the power of the population is indefinitely greater than the power of the earth to produce subsistence for man . . . the population when unchecked increases in geometrical ratio, subsistence only increases in an arithmetical ratio" (Thomas Malthus [1766–1834],* An Essay on Population, *1798).*

* *"If variations useful to any organic being ever do occur, assuredly individuals thus characterized will have the best chance of being preserved in the struggle for life; and from the strong principle of inheritance, these will tend to produce offspring similarly characterized. This principle of preservation, I have called, for the sake of brevity, Natural Selection" (Charles Darwin [1809–82],* The Origin of Species*).*

* *"There is grandeur in this view of life, with its several powers, having been origi-*

nally breathed into a few forms or into one; and that, while this planet has gone cycling on according to the fixed laws of gravity, from so simple a beginning endless forms most beautiful and most wonderful have been, and are being, evolved" (final sentence of Darwin's The Origin of Species*).*

* *"These checks – war, disease, famine and the like – must, it occurred to me, act on the animals as well as man. Then I thought of the enormously rapid multiplication of animals, causing these checks to be much more effective in them than in the case of man; and while pondering vaguely on this fact there suddenly flashed upon me the idea of the survival of the fittest – that the individuals removed by these checks must be on the whole inferior to those that survived" (Alfred Russell Wallace [1823–1913]).*

5 *"How, then, was an organ developed so far beyond the needs of its possessor? Natural selection could only have endowed the savage with a brain a little superior to that of an ape, whereas he actually possesses one but little inferior to that of the average member of our learned societies" (Alfred Wallace).*

* *Language "is not merely a reproducing instrument for voicing ideas but rather is itself the shaper of ideas, the program and guide for the individual's mental activity, for his analysis of impressions, for his synthesis of his mental stock in trade... We dissect nature along lines laid down by our native languages" (Benjamin Lee Whorf,* Language, Thought, and Reality, *1956). Whorf, who worked in an insurance office in Hartford, Connecticut, found many striking differences between the European and Hopi languages. He wrote: "We cut up and organize the spread and flow of events as we do, largely because, through our mother tongue, we are parties to an agreement to do so, not because nature itself is segmented in exactly that way for all to see. Languages differ not only in how they build their sentences but also in how they break down nature to secure the elements to put in those sentences." Edward Sapir of Yale University had earlier written: "Human beings do not live in the objective world alone, nor alone in the world of social activity as ordinarily understood, but are very much at the mercy of the particular language which has become the medium of expression for their society... The worlds in which different societies live are distinct worlds, not merely the same world with different labels attached." Language is thus the framework that contains and shapes our thoughts. Human languages and the size of the human brain (in relation to body weight) are the two things that distinguish us from all other animals on this planet. It is possible that both evolved together and that each is the consequence of the other.*

6 *Why is science not automatically guaranteed when life becomes intelligent? See* "Wrong Number?" *(1979), by Robert Wesson, who says: "The odds that another creature like* Homo sapiens, *with individual and social drives and capacities, would attain an electronic civilization may be compared to those of winning a lottery... In sum, the likelihood of an intelligent creature attaining an electronic or higher civilization may be much less than is often assumed, and the number of such civilizations to be expected in our galaxy is therefore correspondingly reduced." We have assumed in the text that the probability is as high as 1/10 because nontechnological civilizations can exist for tens, perhaps hundreds, of thousands of years in which science might accidentally begin and proceed to flourish.*

7 *"Perhaps we should first attempt reciprocal communication with nonhuman organisms here on earth – say with a vegetable or a scarab beetle, or a termite queen–mother who represents the highest natural societal organization known on this planet. Foolish suggestions, yes, but they suggest the difficulty and probable impossibility of interplanetary communication" (Harlow Shapley,* View from a Distant Star, *1963).*

* *"If intelligence of these creatures were sufficiently superior to ours, they would choose to have little if any contact with us"*

(Brookings Institution, 'The Implications of a Discovery of Extraterrestrial Life," *1961).*

* *"The probability of success is difficult to estimate; but if we never search, the chance of success is zero" (Giuseppe Cocconi and Philip Morrison, "Searching for interstellar communications," 1959).*
* *"Of course, I do not believe in flying saucers; however, I realize that they may exist, whether I believe in them or not" (Peter Sturrock, Lecture at the Aspen Center of Physics, 1977).*

8 *What are your views on the origin of life on Earth?*

* *Does survival of the fittest mean "might is right?"*
* *Is aggression an essential characteristic of intelligence? Many people believe that the answer is yes and think that adversary relations are the essence of a vigorous society.*
* *In the early decades of the twentieth century there occurred an explosion of new theories in physics, and in more recent decades we have witnessed an explosion of technological devices. Many people cannot distinguish between science and technology and vilify the former for the ills of the latter, which they have themselves created. Do you agree with them?*
* *Consider the following gruesome situation. A devilish assassin travels back in time and eliminates at birth 100 of the greatest artists in recorded history. As a result the world of today becomes a duller place. But nothing else changes. The assassin then travels back in time and eliminates at birth 100 of the greatest writers and poets. The world of today is again made a duller place. Finally the assassin travels back and eliminates at birth 100 of the greatest scientists in recorded history. When the assassin returns to the world of today he discovers that this last excursion has wiped out at least 90 percent of the world's present population. The remainder, living in the Age of Mythology, are members of slave- or serf-powered societies, periodically terrified by epidemic diseases, and ruled by despots who claim either to be gods or to have special relations with gods. (This is a matter for debate and I have deliberately adopted a provocative point of view.) An alternative and more optimistic viewpoint is expressed in* The Promise of the Coming Dark Age *(1976), by L. S. Stavrianos, who argues that a return to barbarism offers the promise of a new renaissance.*
* *Elliptical galaxies contain large numbers of population II stars, and star formation generally occurred long ago. Does this mean that such galaxies do not contain life, or is it possible that life in these systems is entirely different?*
* *Should we seek to make contact with intelligent life outside the Solar System?*
* *Will space travel solve the population problem and allow people to have as many children as they wish?*
* *Estimate the number of planets on which (a) life has originated, (b) intelligent life has emerged.*
* *Why is intelligence not automatically guaranteed by evolution? See* This View of Life *(1963), by George Gaylord Simpson, who argues that the probability of "humanoid" creatures is exceedingly small. In the text it is assumed that this probability is as large as 1/10, and this is not an impossible number if we consider the billions of years (not just the millions of years of the recent past) in which planetary life has a chance of becoming intelligent.*
* *Discuss galactic selection and compare it with natural selection.*
* *Imagine that you are a wise and compassionate being in the Galaxy. What would you do if an intelligent but destructively aggressive life form evolved on a planet to the point where it was in a state of imminent self-destruction? What would you do if that life form survived self-destruction and was bent on the conquest of other worlds?*

9 *Our revels now are ended. These our actors,*
As I foretold you, were all spirits and
Are melted into air, into thin air;
And, like the baseless fabric of this vision,

The cloud-capped towers, the gorgeous palaces,
The solemn temple, the great globe itself,
Yea, all which it inherit, shall dissolve
And, like this insubstantial pageant faded,
Leave not a rack behind. We are such stuff
As dreams are made on, and our little life
Is rounded with a sleep.
— The Tempest, *Shakespeare's last play*

FURTHER READING

Beadle, G., and Beadle, M. *The Language of Life: An Introduction to the Science of Genetics.* Doubleday, Garden City, N.Y., 1960.

Blakemore, C. *Mechanics of the Mind.* Cambridge University Press, Cambridge, 1977.

Bracewell, R. N. *The Galactic Club: Intelligent Life in Outer Space.* W. H. Freeman, San Francisco, 1974.

Crick, F. *Of Molecules and Men.* University of Washington Press, Seattle, 1966.

Huang, S. S. "Life in space and humanity on the Earth." *American Scientist,* June 1965.

Napier, J. *The Roots of Mankind.* Harper and Row, New York, 1973.

O'Neill, G. K. "The colonization of space." *Physics Today,* September 1974.

Orgel, L. E. *The Origins of Life: Molecules and Natural Selection.* John Wiley, New York, 1973.

Ponnamperuma, C. *The Origins of Life.* E. P. Dutton, New York, 1972.

Sagan, C., and Drake, F. "The search for extraterrestrial intelligence." *Scientific American,* May 1975.

Schrödinger, E. *What Is Life?* Cambridge University Press, Cambridge, 1946.

Schrödinger, E. *Mind and Matter.* Cambridge University Press, Cambridge, 1958.

Shapley, H. *Of Stars and Men: The Human Response to an Expanding Universe.* Beacon Press, Boston, 1958.

Shklovsi, I. S., and Sagan, C. *Intelligent Life in the Universe.* Holden-Day, New York, 1966.

Sneath, P. H. A. *Planets and Life.* Thames and Hudson, London, 1970.

Sullivan, W. *We Are Not Alone.* McGraw-Hill, New York, 1966.

SOURCES

Arrhenius, S. *Worlds in the Making.* Harper and Brothers, New York, 1908.

Brookings Institution. *The Implications of a Discovery of Extraterrestrial Life.* Prepared for NASA, March 24, 1961, 87th Congress, U.S. Government Printing Office, Washington, D.C.

Cameron, A. G. W., ed. *Interstellar Communication: A Collection of Reprints and Original Contributions.* W. A. Benjamin, New York, 1963.

Cocconi, G., and Morrison, P. "Search for interstellar communications." *Nature 184,* 844, (September 19, 1959).

Darwin, C. *The Origin of Species by Means of Natural Selection; or, the Preservation of Favoured Races in the Struggle for Life.* 1859. Reprint. Penguin Books, Harmondsworth, Middlesex, 1968.

Frankfort, H., Frankfort, H. A., Wilson, J. A., and Jacobsen, T. *Before Philosophy.* Penguin Books, London, 1949. First published as *The Intellectual Adventure of Ancient Man.* University of Chicago Press, Chicago, 1946.

Gillie, O. *The Living Cell.* Funk and Wagnalls, New York, 1971.

Gould, S. J. *Ever since Darwin: Reflections in Natural History.* W. W. Norton, New York, 1977.

Haldane, J. B. S. *Possible Worlds and Other Papers.* Chatto and Windus, London, 1927.

Haldane, J. B. S. "Origin of life," in *New Biology,* no.16. Penguin, London, 1954.

Heilbroner, R. *An Inquiry into the Human Prospect.* W. W. Norton, New York, 1975.

Hoerner, S. von. "Population explosion and interstellar expansion." *Journal of the British Interplanetary Society 28,* 691 (1975).

Koyré, A. *Newtonian Studies.* Chapman and Hall, London, 1965.

Kuiper, T. B. H., and Morris, M. "Searching for extraterrestrial civilizations." *Science 196,* 616 (1977).

Mayr, E. "Darwin and natural selection." *American Scientist,* May–June 1977.

Murray, B., Gulkis, S., and Edelson, R. E. "Extraterrestrial intelligence: an observational approach." *Science 199,* 485 (1978).

Oparin, A. I. *The Origin of Life.* 2nd ed. Dover Publications, New York, 1953.

Oró, J., Miller, S. L., Ponnamperuma, C., and Young, R. S., *Cosmochemical Evolution and the Origins of Life.* D. Reidel, Dordrecht, Netherlands, 1974.

Papagiannis, M. "Are we all alone, or could they be in the asteroid belt?" *Quarterly Journal of the Royal Astronomical Society 19,* 277 (1978).

Ponnamperuma, C., and Cameron, A. G. W. *Interstellar Communication: Scientific Perspectives.* Houghton Mifflin, Boston, 1974.

Shapley, H. *View from a Distant Star: Man's Future in the Universe.* Basic Books, New York, 1963.

Simpson, G. G. *This View of Life: The World of an Evolutionist.* Harcourt, Brace and World, New York, 1963.

Stavrianos, L. S. *The Promise of the Coming Dark Age.* W. H. Freeman, San Francisco, 1976.

Teilhard de Chardin, P. *The Phenomenon of Man.* Harper and Row, New York, 1961.

Wesson, R. G. "Wrong number? A skeptic argues against the likelihood of advanced extraterrestrial civilizations." *Natural History,* March 1979.

Whorf, B. L. *Language, Thought, and Reality: Selected Writings.* Ed. J. B. Carroll. M.I.T. Press, Cambridge, Mass., 1956.

APPENDIX: CONSTANTS OF NATURE

velocity of light	$c = 3.00 \times 10^{10}$ cm sec^{-1}
gravitational constant	$G = 6.67 \times 10^{-8}$ dyn cm^2 g^{-2}
Planck constant	$h = 6.63 \times 10^{-27}$ erg sec
electron charge	$e = 1.60 \times 10^{-19}$ coulombs
	$= 4.80 \times 10^{-10}$ e.s.u.
mass of electron	$m_e = 9.11 \times 10^{-28}$ g
mass of hydrogen atom	$m_p = 1.67 \times 10^{-24}$ g
Boltzmann constant	$k = 1.38 \times 10^{-16}$ erg deg^{-1}
1 angstrom	$A = 10^{-8}$ cm
1 electron volt	$eV = 1.60 \times 10^{-12}$ erg

INDEX